Inorganic Aspects
of Biological and
Organic Chemistry

Inorganic Aspects of Biological and Organic Chemistry

Robert P. Hanzlik

Department of Medicinal Chemistry
School of Pharmacy
University of Kansas
Lawrence, Kansas

ACADEMIC PRESS   NEW YORK   SAN FRANCISCO   LONDON   1976

A Subsidiary of Harcourt Brace Jovanovich, Publishers

ACADEMIC PRESS, INC.
111 Fifth Avenue, New York, New York 10003

United Kingdom Edition published by
ACADEMIC PRESS, INC. (LONDON) LTD.
24/28 Oval Road, London NW1

Library of Congress Cataloging in Publication Data

Hanzlik, Robert P
        Inorganic aspects of biological and organic chem-
istry.

        Includes index.
        1.   Chemistry, Inorganic.   2.   Chemistry, Physical
organic.   3.   Biological chemistry.   I.   Title.
QD31.2.H37   546   75-32028
ISBN 0−12−324050−6

To Lois

Contents

IV   Atomic Structure and Structure–Activity Correlations

V   Bonding in Ligands and Metal Complexes

VI   Ligand Exchange Reactions and Factors in Complex Stability

VII   Redox Potentials and Processes

VIII   Influencing Equilibria with Metal Ions: Synthesis via Chelation

IX   Catalysis by Metal Ions, Metal Complexes, and Metalloenzymes

X   Oxygen and Nitrogen

XI Organometallic Complexes: Structure, Bonding, and Reaction Mechanisms

XII Reactions of Ligands in Organometallic Complexes

Preface

During the last several decades chemistry has expanded enormously as a knowledge base for the more detailed areas of natural science. The science of molecules and molecular interactions is now seen as a key to understanding not only our nonliving world but also the nature of life processes and, perhaps eventually, life itself. Along with this broad new outlook we see the distinctions between the classic areas of natural science softening and diffusing around new areas—the "interdisciplines." This mood of reorganization in natural science has also touched the internal structure of chemistry. Some previously distinct specialties have partially merged giving rise to such interdisciplines as bio-organic chemistry, organometallic chemistry, and bioinorganic chemistry. A very interesting part of this progression has been the transition from trying to understand the chemistry of life processes to trying to *mimic* them as in prebiotic and biogenetically patterned synthesis, enzyme model systems, and biomimetic chemistry.

As areas of chemical research, interdisciplines have proved very fertile, spawning not only new theoretical knowledge but practical technology as well, and there is no reason to expect that this will not continue for some time. In the field of chemical education, however, the interdisciplinary approach has not been quite so visible. This is partly due to the inertia of the system. There is a long tradition of strict distinction between inorganic and organic chemistry, with perhaps somewhat less extreme distinction between organic and biochemistry since both are based on carbon compounds. Contrasting, if not confronting this tradition, we now have organometallic chemistry, with its many potential and actual applications in the laboratory as well as in industry, and bioinorganic chemistry, sorting out the roles of metals in such biological and biochemical settings as nerve function, enzyme action, and disease processes.

This book is an outgrowth of several years of teaching an introductory course on the "interdisciplinary aspects of inorganic chemistry" to seniors and graduate students. It was actually written with two groups of people in mind: teachers and practitioners of organic, biological, and inorganic chemistry.

In the area of teaching and the chemistry curriculum, a case could be made for introducing inorganic chemistry immediately after second year organic. This would serve to emphasize the many *similarities* between carbon-based and non-carbon-based compounds in terms of structure and mechanism. By building on and reinforcing concepts just learned, rather than establishing inorganic chemistry as separate from or antithetical to organic chemistry, the students' understanding of and facility with basic chemical theory would be strengthened; a sense of the *universality* of the fundamental principles of structure, bonding, and reactivity would develop. The early introduction of organometallic and bioinorganic chemistry might help to convince the student that a broad-minded approach, based on a solid and unified theoretical background, is both a successful and satisfying way to approach complex problems through chemistry.

For the practicing chemist working in the gray areas between inorganic chemistry and organic or biological chemistry this book represents an attempt to provide a rapid and pertinent introduction to the common ground of the three areas. Here again it is hoped that some value is derived from the approach of developing fundamental principles of inorganic structure, bonding, and reactivity *vis-à-vis* the applications of the same principles and concepts in organic chemistry. Of course, in addition to the basic principles, the practicing chemist is also concerned with the "current state of the art." While this book does not pretend to review each area in such depth, an effort has been made to indicate the limits of our current understanding. This is backed up with references to recent in-depth reviews of specialized areas, as well as references to current research papers which offer important new information and relate it to previous work in the area of concern.

Having expressed my views on the interdisciplinary aspects of chemistry and the organization of this manuscript, I must also express a debt of gratitude to several of my former teachers whom I feel were, and still are, among the *avant-garde* of interdisciplinary chemical teaching and research: Professors Eugene E. van Tamelen, James P. Collman, and Jack Lewis. A special debt of gratitude is due another man from whom I, like others who knew him, have learned a great deal: the late Professor Edward E. Smissman. To us he personified an interdisciplinary approach to chemical–biological problems, inspiring students and colleagues alike with his approach to science and his unfailing warmth and friendship. Finally, I wish to express my deepest appreciation to my wife Lois for her patience, encouragement, and immeasurable assistance during the preparation of this manuscript.

ROBERT P. HANZLIK

Periodic Table

(See pages xvi and xvii.)

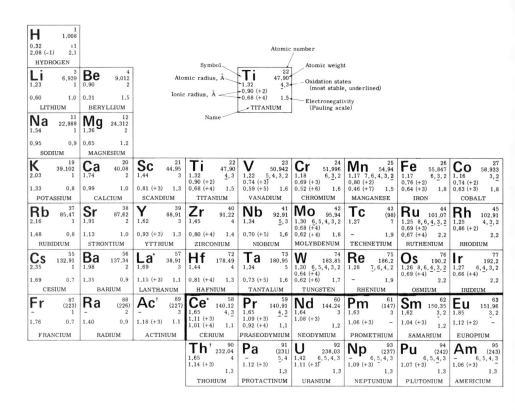

He	2
	4.003
HELIUM	

B	5	C	6	N	7	O	8	F	9	Ne	10
	10.811		12.01115		14.0067		15.9994		18.998		20.183
0.82	3	0.77	±4,2	0.75	±3,5,4,2	0.73	-2	0.72	-1		
		2.60 (-4)		1.71 (-3)							
0.20 (+3)	2.0	0.15 (+4)	2.5	0.11 (+5)	3.0	1.40 (-2)	3.5	1.36 (-1)	4.0		
BORON		CARBON		NITROGEN		OXYGEN		FLUORINE		NEON	

Al	13	Si	14	P	15	S	16	Cl	17	Ar	18
	28.981		28.086		30.9738		32.064		35.453		39.948
1.18	3	1.11	4	1.06	±3,5	1.02	+2,4,6	0.99	±1,3,5,7		
				2.12 (-3)		1.84 (-2)					
0.50 (+3)	1.5	0.41 (+4)	1.8	0.34 (+5)	2.1	0.29 (+6)	2.5	1.81 (-1)	3.0		
ALUMINUM		SILICON		PHOSPHORUS		SULFUR		CHLORINE		ARGON	

Ni	28	Cu	29	Zn	30	Ga	31	Ge	32	As	33	Se	34	Br	35	Kr	36
	58.71		63.54		65.37		69.72		72.59		74.922		78.96		79.909		83.80
1.15	3,2	1.17	2,1	1.25	2	1.26	3	1.22	4	1.20	±3,5	1.16	-2,4,6	1.14	±1,5		
0.72 (+2)		0.96 (+1)								2.22 (-3)		1.98 (-2)					
0.62 (+3)	1.8	0.69 (+2)	1.9	0.74 (+2)	1.6	0.62 (+3)	1.6	0.53 (+4)	1.8	0.47 (+5)	2.0	0.42 (+6)	2.4	1.95 (-1)	2.8		
NICKEL		COPPER		ZINC		GALLIUM		GERMANIUM		ARSENIC		SELENIUM		BROMINE		KRYPTON	

Pd	46	Ag	47	Cd	48	In	49	Sn	50	Sb	51	Te	52	I	53	Xe	54
	106.4		107.87		112.40		114.82		118.69		121.75		127.60		126.90		131.30
1.28	4,2	1.34	3,2,1	1.48	2	1.44	3	1.41	4,2	1.40	±3,5	1.36	-2,4,6	1.33	±1,5,7		
								1.12 (+2)		2.45 (-3)		2.21 (-2)					
0.86 (+2)	2.2	1.26 (+1)	1.9	0.97 (+2)	1.7	0.81 (+3)	1.7	0.71 (+4)	1.8	0.62 (+5)	1.9	0.56 (+6)	2.1	2.16 (-1)	2.5		
PALLADIUM		SILVER		CADMIUM		INDIUM		TIN		ANTIMONY		TELLURIUM		IODINE		XENON	

Pt	78	Au	79	Hg	80	Tl	81	Pb	82	Bi	83	Po	84	At	85	Rn	86
	195.09		196.96		200.59		204.37		207.19		208.98		(210)		(210)		(222)
1.30	4,2	1.34	3,1	1.49	2,1	1.48	3,1	1.47	4,2	1.46	3,5	1.46	2,4	-	±1,3,5,7		
						1.40 (+1)		1.20 (+2)		1.20 (+3)							
0.96 (+2)	2.2	1.37 (+1)	2.4	1.10 (+2)	1.9	0.95 (+3)	1.8	0.84 (+4)	1.8	0.79 (+5)	1.9		2.0		2.2		
PLATINUM		GOLD		MERCURY		THALLIUM		LEAD		BISMUTH		POLONIUM		ASTATINE		RADON	

Gd	64	Tb	65	Dy	66	Ho	67	Er	68	Tm	69	Yb	70	Lu	71
	157.25		158.92		162.50		164.93		167.26		168.93		173.04		174.97
1.61	3	1.59	4,3	1.59	3	1.58	3	1.57	3	1.56	3,2	-	3,2	1.56	3
1.02 (+3)	1.1	1.00 (+3)	1.2	0.99 (+3)	-	0.97 (+3)	1.2	0.96 (+3)	1.2	0.95 (+3)	1.2	0.94 (+3)	1.1	0.93	1.2
GADOLINIUM		TERBIUM		DYSPROSIUM		HOLMIUM		ERBIUM		THULIUM		YTTERBIUM		LUTETIUM	

Cm	96	Bk	97	Cf	98	Es	99	Fm	100	Md	101	No	102	Lw	103
	(247)		(247)		(249)		(254)		(253)		(256)		(254)		(257)
	3		4,3		3										
	-														
CURIUM		BERKELIUM		CALIFORNIUM		EINSTEINIUM		FERMIUM		MENDELEVIUM		NOBELIUM		LAWRENCIUM	

I
Introduction

Metals in Biological Chemistry

In the realm of biology there are several general ways in which metal ions either facilitate or control the essential biochemical processes of living cells. Perhaps the simplest role is one in which the positively charged metal ions attract negatively charged portions of other molecules such as proteins. In this way metals act as a type of cement, holding together two or more different molecules, or two or more parts of the same large molecule. These interactions may bring subunits together to form a large multimolecular array, as in bone or muscle fibers, or they may affect the three-dimensional shape of a single large molecule. Many persons break out in a rash from jewelry or watchbands which contain nickel. Nickel corrodes slightly because of fatty acids present in perspiration. The nickel ions formed bind tightly to protein molecules in dermal and epidermal cells, changing the conformation of some of these proteins. If the conformational change is large enough the new protein derivatives are recognized as "foreign" by the body's immune system, and the result is an inflammatory reaction or rash.

In other situations metal ions may be firmly bound in their molecular environment and yet participate actively in dynamic processes. Metalloenzymes are a good example of this. Here the metal is usually held tightly by several amino acid side-chain groups from the polypeptide backbone of the protein, but is not completely surrounded; thus, other smaller molecules can still approach it. When they do they may undergo a chemical change under the combined influence of the metal ion and perhaps other parts of the protein as well. Metalloenzymes will not function if their metal is removed, and metal

1

replacement usually restores enzymatic activity. It is interesting, however, that while cells from one particular organism may carry out a reaction using a metalloenzyme, cells from a different organism may carry out the same net biochemical transformation using an enzyme which is devoid of metals. The explanation is that the two reactions have different mechanisms although they give identical products.

Another way that rigidly bound metal ions can function in dynamic biological processes hinges on their ability to participate in redox reactions. Transition metals usually can occur in several oxidation states, and their ability to alternately donate or accept an electron makes them very suitable as redox catalysts. Most living organisms are aerobic; that is, they derive energy by oxidizing the molecules they ingest as food. However, unlike combustion, biological oxidation is very indirect. It involves the removal of electrons and protons from the substrates, and the transport of the electrons by means of redox reactions through a chain or series of large molecules, until ultimately they reach an oxygen molecule and reduce it to water. Very often the members of the electron transport chain are proteins that contain iron which interconverts between $Fe(II)$ and $Fe(III)$ as electrons move along the chain. Copper, cobalt, vanadium, manganese, and molybdenum are also important in biological redox processes. Electron transport also occurs in anaerobic organisms, but rather than oxygen the ultimate electron acceptor is often sulfate, nitrate, or even molecular nitrogen.

Not all metal ions are firmly bound in biological systems. One of the most important functions of sodium and potassium involves the rapid movement of these ions from one part of a system to another. This kind of ion flow creates nonuniform ion distributions and this in turn causes the development of electrical potential gradients which can be used to do work. Much of the metabolic energy spent by nervous tissue goes into the production of electrical potential gradients used for the propagation of nervous impulses. In fact, just thinking about it will move a lot of sodium and potassium ions around the cells of your brain. The electron transport connected with respiration is quite distinct from the cation transport connected with nerve and muscle action. The former system traps chemical energy in a storage form, adenosine triphosphate (ATP), while the latter utilizes the energy stored in ATP to perform useful work such as thought or muscle contraction. Yet metal ions are at the very heart of both systems.

Before cataloging some of the important roles of metals in biology there are two more important generalizations which can be made. The first concerns the intracellular vs. extracellular location of the metals (Fig. 1-1). The dividing line is the cell membrane, present around all cells. This membrane is a dynamic, metabolically active structure, and is responsible for the homeostasis of the cell; that is, it is selectively permeable to chemical species in both directions so

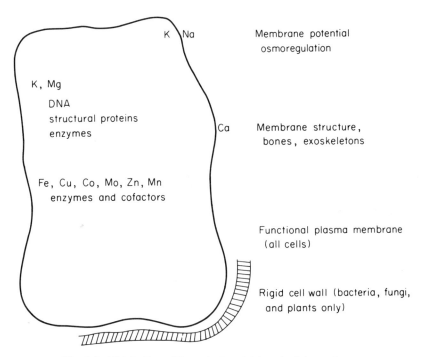

Fig. 1-1. Distribution of important metal ions in living cells.

that by active metabolic processes it can maintain concentration gradients from inside to outside the cell. In this way cells tend to exclude sodium and calcium, and tend to accumulate potassium and magnesium. They also accumulate other metals, but usually in much smaller amounts.

Outside the cell, calcium is involved in maintaining the structure of parts of the cell membrane, muscle contractility, and hard structures such as bones, teeth, and shells. Sodium is involved in both osmotic and pH buffering equilibria as well as in nerve impulse propagation. Inside the cell, potassium and magnesium are often associated with proteins and deoxyribonucleic acid (DNA), and many enzymes require these ions either for activation or activity enhancement. Trace metals found inside the cell are almost always associated with catalytic rather than stoichiometric function. They are required by metalloenzymes and redox proteins, either for formation of their active center or as cofactors which form a bridge between enzyme and substrate.

The other generalization about metals in biology concerns how firmly they are held in place and how this relates to their functioning (Tables 1-1 and 1-2). Sodium and potassium function largely as highly mobile charge carriers. When a cell is put into a solution of radioactive sodium ions, the radioactive

TABLE 1-1

Mobility and Function of Biologically Important Metals

Na, K	Fast exchange	Highly mobile charge carriers
Ca, Mg	Moderate exchange	Coupling nervous and hormonal stimuli to effector cells; contractility; structure
Mn, Zn	Slow exchange	Lewis acids in metalloenzymes
Fe, Co, Cu	No exchange	Redox processes

TABLE 1-2

The Biological Metals: Their Occurrence and Some Presently Known Functions

Na, K, Ca, Mg	Osmotic and acid-base equilibria
Na, K	Nerve impulse propagation, electrical potential of cell membrane
Mg, Ca	Muscle tone and contractility
Mg	Chlorophyll
	Activates enzymes which utilize ATP
Ca	Bones, teeth, shells, exoskeletons, kidney stones, blood clotting mechanisms, serum complement reactions, cell membrane responses to stimuli
Mn	Enzyme activator; e.g., phosphoenol pyruvate carboxylase
	Metalloenzymes; e.g., pyruvate kinase, pyruvate carboxylase, arginase
	Protein conformation; e.g., concanavalin A
V	Plant growth factor, porphyrin derivatives occur in petroleum, oxygen transport in tunicates
Fe	Hemoproteins; e.g., myoglobin, hemoglobin, cytochromes
	Enzymes; e.g., aconitase, catalase, peroxidase
	Hemerythrin, an oxygen transport protein in seashore worms
Co	Vitamin B_{12} and B_{12} coenzymes
Ni	Integral component of the enzyme urease
Cu	Oxidase enzymes; e.g., dopamine-β-hydroxylase, tyrosinase, amine oxidase, diamine oxidase, cytochrome oxidase, ascorbate oxidase, prostaglandin synthetase
	Oxygen transport: hemocyanin in insects
	Hydrolytic enzyme: uricase
	Ceruloplasmin, a copper storage protein
Mo	Oxidoreductase enzymes; e.g., nitrate reductase, hydrogenase, nitrogenase, liver aldehyde oxidase, xanthine oxidase
Zn	Yeast aldolase
	Hydrolytic enzymes; e.g., carbonic anhydrase, carboxypeptidase, alkaline phosphatase
	Dehydrogenase enzymes; e.g., alcohol dehydrogenase, glutamic dehydrogenase, lactic dehydrogenase, malic dehydrogenase
	Enzyme activation; e.g., arginase, various peptidases, histidine deaminase, enolase
Cr	Deficiency a suspected factor in diabetes

"labeled" sodium very rapidly crosses the membrane in both directions, even though the concentration gradient across the membrane is not affected. Calcium and magnesium also exchange, but not quite as rapidly as sodium or potassium. Zinc and manganese are usually bound to proteins inside the cell and undergo still slower exchange. They often function as Lewis acids in metalloenzymes which carry out isomerization or hydrolysis reactions, and their tighter binding reflects their appreciable Lewis acidity. Iron in its 3+ oxidation state is a strong Lewis acid. In the body it constitutes an immobile storage form of iron. In its 2+ oxidation state iron is more mobile, and it is in this form that iron is absorbed from the gut and distributed throughout the body. Copper, cobalt, and molybdenum are involved in redox processes and are highly resistant to exchange.

Metals in Organic Chemistry

Three of the most important types of reactions in modern synthetic organic chemistry are oxidation, reduction, and carbon–carbon coupling. In considering the range of reagents and catalysts commonly used to effect these transformations one is immediately struck by the ubiquitous involvement of metals and their derivatives. This point is amply illustrated by the listings in Tables 1-3, 1-4, and 1-5. Although these listings are rather selective and far from complete, their main purpose is to provide some illustration of the *diversity* of reaction types within each category as well as the *selectivity* and *specificity* attainable with some of the individual metal reagents. Similarly, as these and other reactions are discussed in the succeeding chapters, the main point of emphasis will be the general features of metal-catalyzed and metal-directed reactions and the reaction of coordinated ligands. Whenever possible references will be given to more thorough reviews of individual reactions.

In analyzing the effects of metal ions on the course of a chemical reaction a clear distinction must be made between *kinetic* effects and *thermodynamic* effects. The latter always depend on a *stoichiometric* amount of metal ion and the formation of a metal complex as a primary reaction product. If the metal ion exerts its influence at the kinetic level, only a *catalytic* amount of a metal ion should be necessary. However, if once formed the reaction product binds the metal strongly, a stoichiometric amount will still be needed, thus obscuring the fact that the main effect was kinetic rather than thermodynamic.

Other factors must also be considered in analyzing metal-catalyzed or metal-directed reactions. Frequently the metal center acts as a "template" or collector of ligands prior to their reaction, thus transforming the uncatalyzed intermolecular reaction into one which is "intramolecular." However, if the metal center is to function catalytically it is important that it be "labile," i.e.,

TABLE 1-3

Metal-Dependent Reduction Reactions

1. Hydrogenation of multiple bonds $C\equiv C$, $C=C$, $C=N$, $C\equiv N$, etc.
   a. Heterogeneous: Ni, Pd, Pt, Rh, Ru, etc.
   b. Homogeneous: $(Ph_3P)_3RhCl$ and analogs

2. Hydrogenolysis of C—Cl, C—Br, C—O, C—N, N—O, etc., generally Pd or Pt

3. Metal hydride reagents
   a. Boron: $NaBH_4$, $NaBH_3(CN)$, $Me_2S\cdot BH_3$, B_2H_6, others
   b. Aluminum: $LiAlH_4$, $NaH_2Al(OCH_2CH_2OCH_3)_2$, $(iBu)_2AlH$, $Al[OCH(CH_3)_2]_3$
   c. Magnesium:

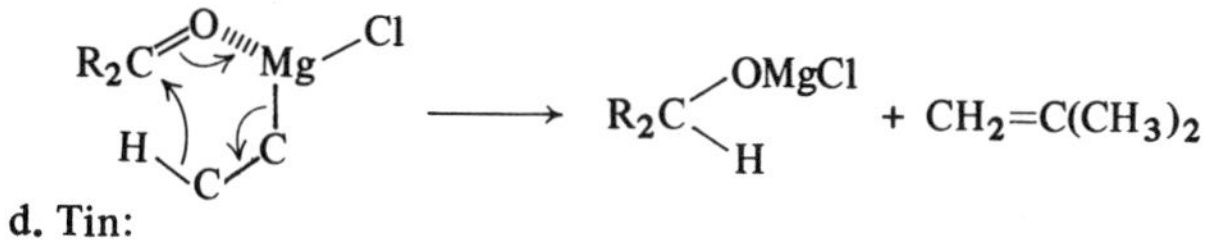

   d. Tin:

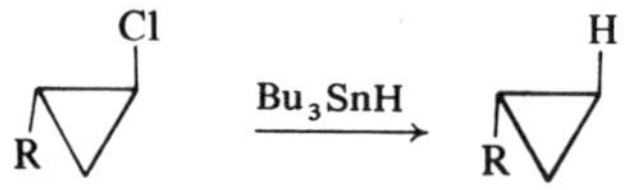

4. Dissolving metal reductions
   a. Fe, Sn, or Zn in acid

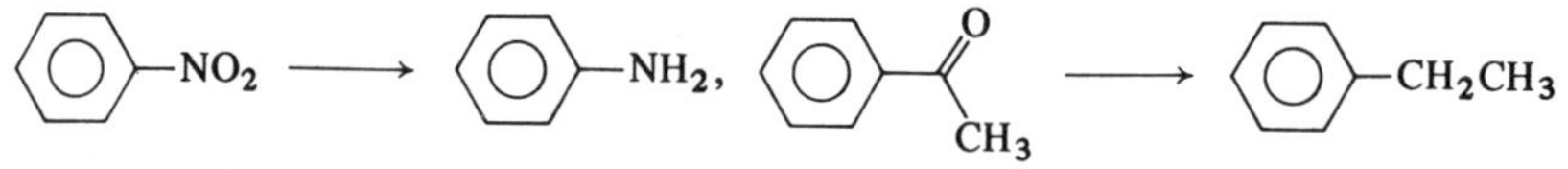

   b. Li or Na in NH_3 or amine solvents

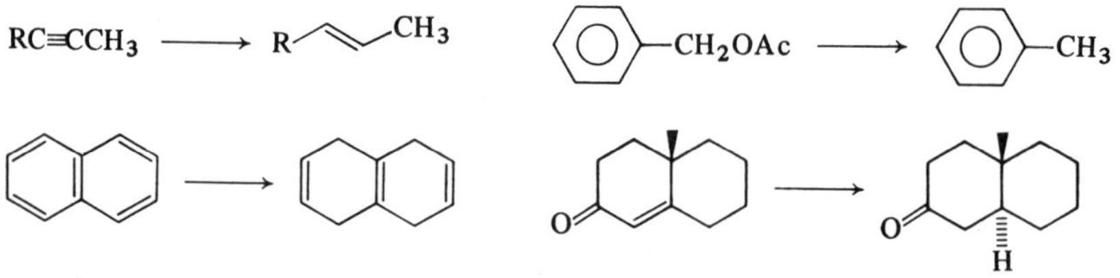

---

undergo facile ligand exchange, and that the product not bind the metal too strongly. On the other hand, multiple bonding between ligand and metal strengthens their association and increases the probability of the ligand reacting while under the influence of the metal.

Steric effects can be quite important in complex formation. If some of the ligands around a metal are chiral, or if their steric bulk is dissymmetrically distributed about the metal center, the complex may be able to discriminate between diastereomers or enantiomers when adding another ligand to fill

TABLE 1-4

Oxidations of Organic Compounds with Metal Reagents

Reagent	Oxidation reaction
$Ag(NH_3)_2OH$	$ArCHO \longrightarrow ArCO_2H$
$(NH_4)_2Ce(NO_3)_6$	$ArCH_3 \rightarrow ArCH_2OH \longrightarrow ArCHO$
MnO_2  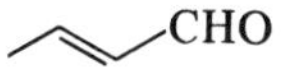	
K_2FeO_4 or $CrO_3 \cdot pyridine/CH_2Cl_2$	$RCH_2OH \longrightarrow RCHO$
H_2CrO_4/acetone or RuO_4/CCl_4	$R_2CHOH \longrightarrow R_2C=O$
$KMnO_4$ or OsO_4  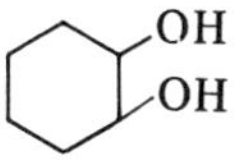	
$Mo(O)(O_2)_2 \cdot HMPA$ or $tBuOOH/VO(AcAc)_2$  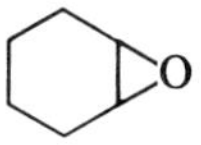	
RuO_4 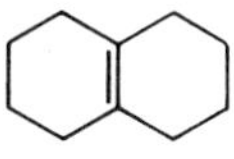 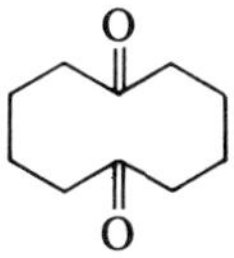	
Air, V_2O_5 catalyzed 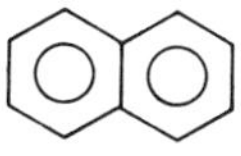 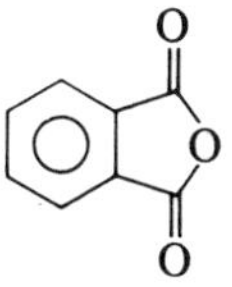	
Air, Ag catalyzed	$CH_2=CH_2 \longrightarrow CH_2-CH_2$ (epoxide) 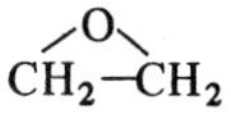
Air, Pd^{2+}/Cu^{2+} catalyzed	$CH_2=CH_2 \longrightarrow CH_3CHO$
H_2O_2/Fe^{2+}  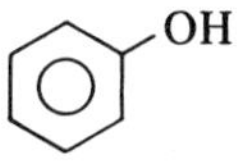	
$[Fe(DMF)_3Cl_2][FeCl_4]$ 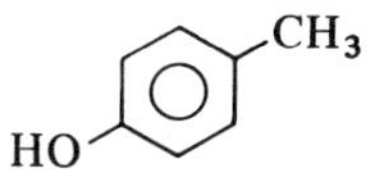 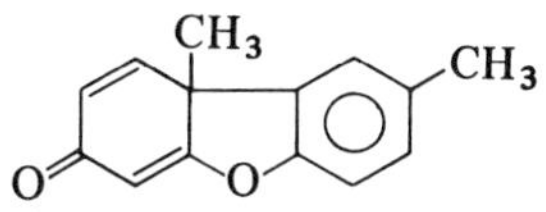	

(continued)

TABLE 1-4–continued

Reagent	Oxidation reaction
$Pb(OAc)_4$	 cyclohexane-1,2-dicarboxylic acid (CO_2H, CO_2H) $\longrightarrow$ cyclohexene $+ CO_2$
$Hg(OAc)_2/EtOH$	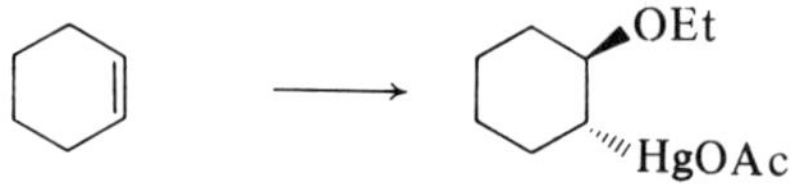 cyclohexene $\longrightarrow$ cyclohexane with OEt and $HgOAc$ substituents
$Mn(OAc)_3$	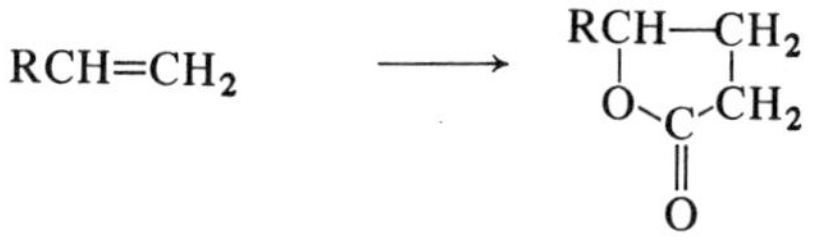 $RCH{=}CH_2 \longrightarrow$ $RCH{-}CH_2$ dioxolanone ($O{-}C{-}CH_2$, $\underset{O}{\parallel}$)
$Tl(OCOCF_3)_3$	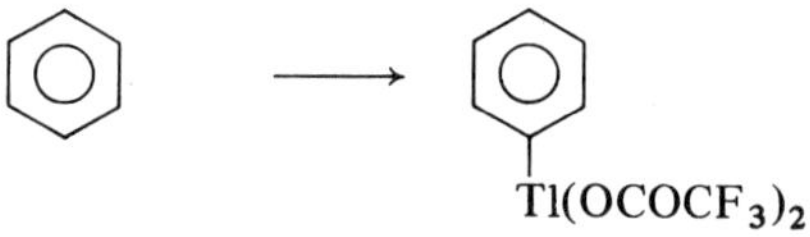 benzene $\longrightarrow$ benzene–$Tl(OCOCF_3)_2$

TABLE 1-5

Some Carbon–Carbon Couplings via Organometallic Reagents

$$RLi + R'CO_2H \longrightarrow RCOR'$$

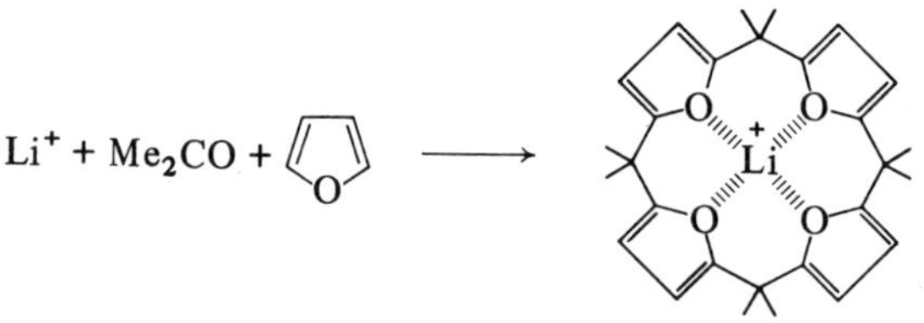

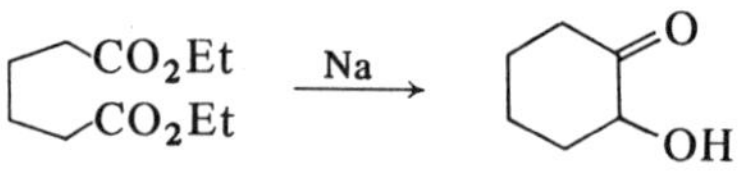

$$Li^+ + Me_2CO + \text{(furan)} \longrightarrow \text{lithium–tetrafuran macrocyclic complex}$$

$$\text{cyclohexane-1,1-diester} \; (CO_2Et, CO_2Et) \xrightarrow{\;Na\;} \text{2-hydroxycyclohexanone} \; (O, OH)$$

$$RMgCl + R'CO_2H \longrightarrow R_2R'COH$$

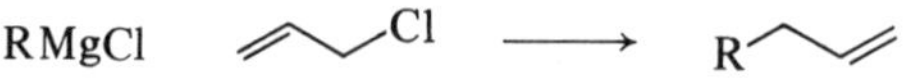

$$RMgCl + \text{allyl chloride} \longrightarrow R\text{–allyl}$$

$$R_2CO \xrightarrow{\;Mg\;} R_2C{-}CR_2 \; (OH, OH)$$

TABLE 1-5-continued

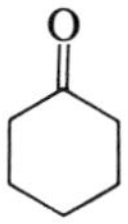

cyclohexanone $\xrightarrow{\text{Mg(OCO}_2\text{CH}_3)_2}$ 2-oxocyclohexanecarboxylic acid (CO_2H)

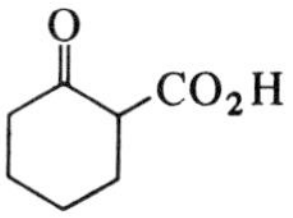

$CH_2=CH_2 \xrightarrow{\text{Et}_3\text{Al} \cdot \text{TiCl}_4}$ polyethylene

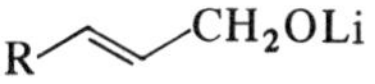

$R\text{—CH=CH—CH}_2\text{OLi} \xrightarrow{\text{TiCl}_3 \cdot \text{MeLi}} R\text{—CH=CH—CH}_2\text{—CH}_2\text{—CH=CH—}R$

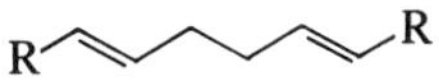

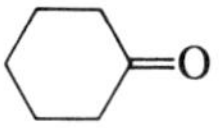

cyclohexanone (C=O) $\xrightarrow{\text{TiCl}_3 \cdot \text{LiAlH}_4}$ dicyclohexylidene

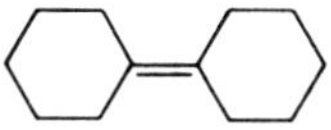

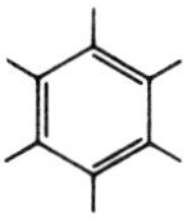

$CH_3C{\equiv}CCH_3 \xrightarrow{\text{Ph}_3\text{Cr(THF)}_3}$ hexamethylbenzene

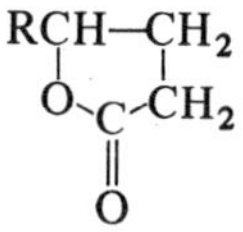

$RCH=CH_2 \xrightarrow{\text{Mn(OAc)}_3}$ RCH—CH$_2$ / O—C(=O)—CH$_2$ (lactone)

$Na_2Fe(CO)_4 \xrightarrow[\text{(2) R'COCl}]{\text{(1) RI}} RCOR'$

$H_2 + C_2H_4 + CO \xrightarrow{\text{Co}_2\text{(CO)}_8} HCH_2CH_2CHO$

$Ni(CO)_4 \xrightarrow[\text{DMF}]{R\text{—CH=CH—CH}_2\text{Br}} R\text{—CH=CH—CH}_2\text{—CH}_2\text{—CH=CH—}R$

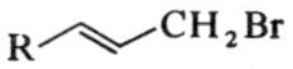

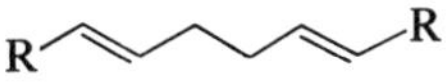

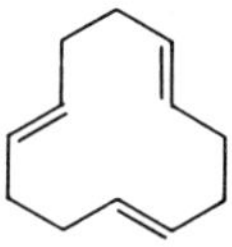

$Ni(CO)_4 \xrightarrow[-40°C]{\text{butadiene}}$ cyclododecatriene

$PhBr + NaCN \xrightarrow[\text{catalyzed}]{\text{Ni(Ph}_3\text{P)}_3} PhCN + NaBr$

$[\text{allyl NiBr}]_2 \xrightarrow[\text{DMF}]{\text{PhBr}} Ph\text{—CH}_2\text{—CH=CH}_2$

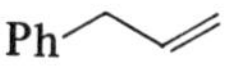

$PhN_2BF_4 \xrightarrow{\text{CuBr}} PhBr$

(continued)

TABLE 1-5-continued

Some Carbon–Carbon Couplings via Organometallic Reagents

$$\text{Ar}-\text{Br} \xrightarrow{\text{Cu, 200°C}} \text{Ar}-\text{Ar}$$

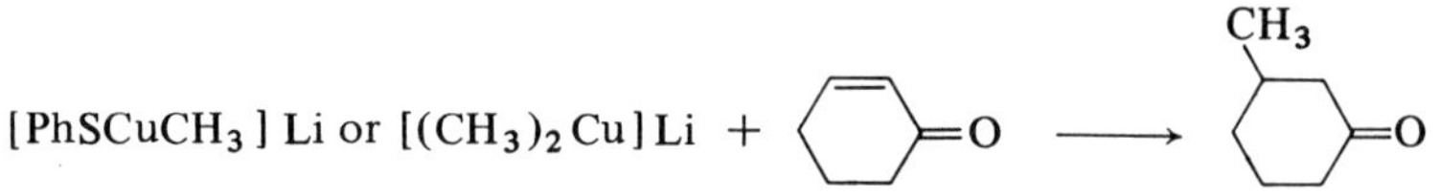

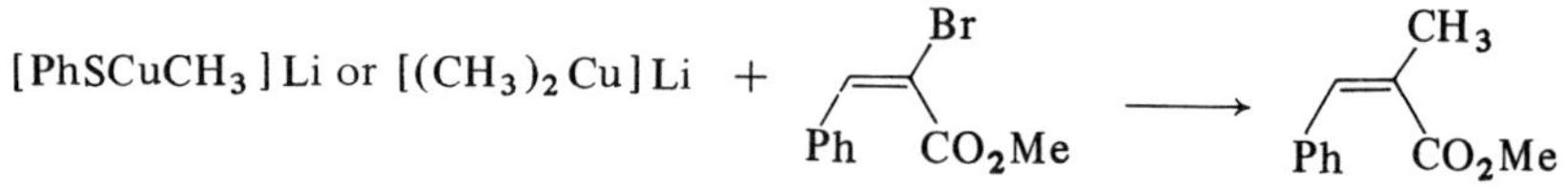

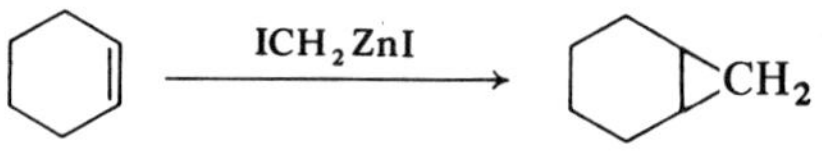

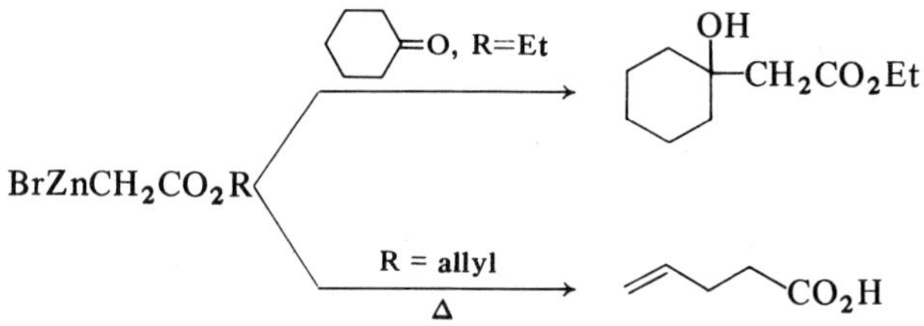

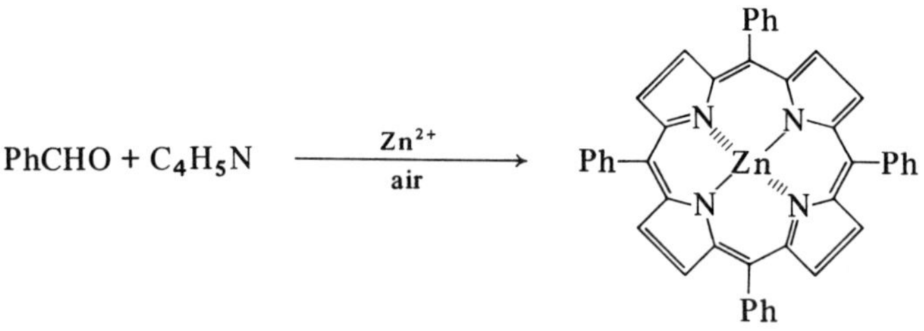

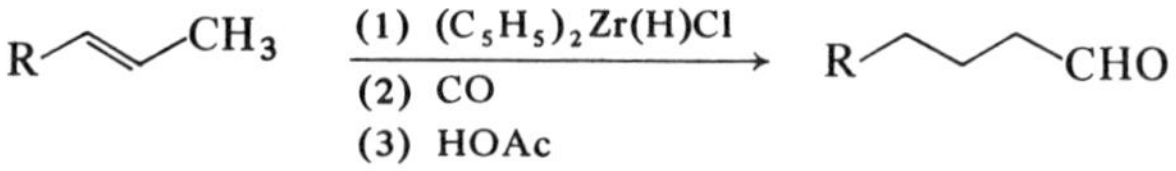

$$[\text{PhSCuCH}_3]\,\text{Li or}\,[(\text{CH}_3)_2\,\text{Cu}]\,\text{Li} \;+\; \text{(cyclohexenone)} \longrightarrow \text{(3-methylcyclohexanone)}$$

$$[\text{PhSCuCH}_3]\,\text{Li or}\,[(\text{CH}_3)_2\,\text{Cu}]\,\text{Li} \;+\; \underset{\text{Ph}\quad\text{CO}_2\text{Me}}{\overset{\text{Br}}{\diagup}} \longrightarrow \underset{\text{Ph}\quad\text{CO}_2\text{Me}}{\overset{\text{CH}_3}{\diagup}}$$

$$\text{(cyclohexene)} \xrightarrow{\text{ICH}_2\text{ZnI}} \text{(bicyclic)}\diagdown\text{CH}_2$$

$$\text{BrZnCH}_2\text{CO}_2\text{R} \begin{cases} \xrightarrow{\text{=O, R=Et}} \text{(cyclohexane)}\overset{\text{OH}}{\underset{}{-}}\text{CH}_2\text{CO}_2\text{Et} \\[2em] \xrightarrow[\Delta]{\text{R = allyl}} \diagup\diagup\diagdown\text{CO}_2\text{H} \end{cases}$$

$$\text{PhCHO} + \text{C}_4\text{H}_5\text{N} \xrightarrow[\text{air}]{\text{Zn}^{2+}} \text{(Zn porphyrin with Ph substituents)}$$

$$\text{R}\diagup\diagdown\text{CH}_3 \xrightarrow[\substack{(2)\ \text{CO} \\ (3)\ \text{HOAc}}]{(1)\ (\text{C}_5\text{H}_5)_2\text{Zr(H)Cl}} \text{R}\diagup\diagdown\diagup\text{CHO}$$

$$\text{RhCl(CO)(Ph}_3\text{P)}_2 \xrightarrow[(2)\ \text{R'COCl}]{(1)\ \text{RLi}} \text{RCOR}'$$

TABLE 1-5-continued

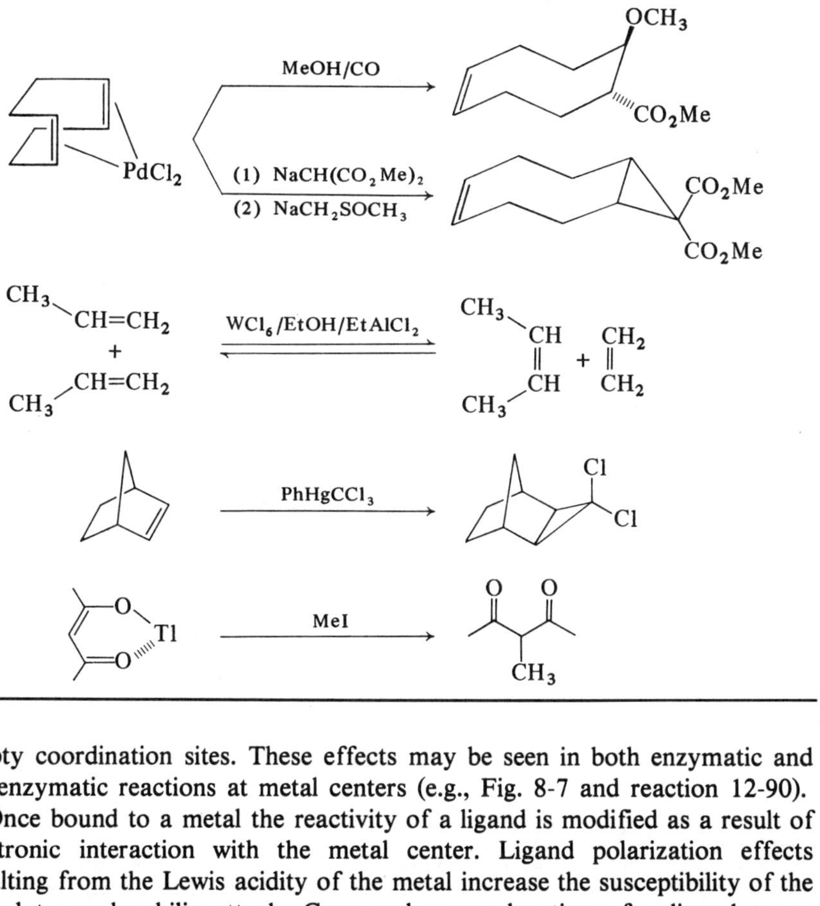

empty coordination sites. These effects may be seen in both enzymatic and nonenzymatic reactions at metal centers (e.g., Fig. 8-7 and reaction 12-90).

Once bound to a metal the reactivity of a ligand is modified as a result of electronic interaction with the metal center. Ligand polarization effects resulting from the Lewis acidity of the metal increase the susceptibility of the ligand to nucleophilic attack. Conversely, complexation of a ligand to an electron-rich metal may enhance its susceptibility to electrophilic attack or its tendency to undergo solvolysis reactions. Metal ions with partially filled d orbitals may provide a convenient "bridge" for electron transfer between the ligands in its coordination sphere. The multiplicity of spin states for many metals allows them to interact with or bridge between ligands of differing spin states, notably triplet molecular oxygen. In other cases the availability of several stable oxidation states allows the metal to catalyze electron transfer between ligands via a chain mechanism.

The various factors which enter into the analysis of the effects of metals on chemical reactions are discussed in greater detail in later chapters, particularly

Chapters III, VIII, and XII. In proceeding through the remaining chapters it should become apparent that there is a vast potential to be tapped through the judicious use of metal ions to catalyze or alter the course of chemical reactions.

GENERAL REFERENCES

Inorganic Chemistry

F. A. Cotton and G. Wilkinson, "Advanced Inorganic Chemistry," 3rd ed. Wiley, New York, 1973.

M. C. Day, Jr. and J. Selbin, "Theoretical Inorganic Chemistry," 2nd ed. Van Nostrand-Reinhold, Princeton, New Jersey, 1969.

H. J. Emeleus, ed., "M. T. P. International, Review of Science, Inorganic Chemistry Series One," Vols. 1–9 (particularly Vols. 6 and 9). Univ. Park Press, Baltimore, Maryland, 1972.

J. J. Lagowski, "Modern Inorganic Chemistry." Dekker, New York, 1973.

Metals in Biological Systems

I. J. T. Davies, "The Clinical Significance of the Essential Biological Metals." Thomas, Springfield, Illinois, 1972.

R. Dessy, J. Dillard, and L. Taylor, "Bioinorganic Chemistry," *Advan. Chem. Ser.,* Vol. **100**, 1974.

J. M. Diamond and E. M. Wright, Biological membranes: The physical basis of ion and non-electrolyte selectivity. *Annu. Rev. Physiol.* **31**, 581 (1969).

G. Eichhorn, ed., "Inorganic Biochemistry," Vols. 1 and 2. Elsevier, Amsterdam, 1973.

M. N. Hughes, "The Inorganic Chemistry of Biological Processes." Wiley (Interscience), New York, 1973.

E. T. Kaiser and F. J. Kezdy, eds., "Progress in Bioorganic Chemistry," Vol. 1. Wiley (Interscience), New York, 1971.

H. Sigel, ed., "Metal Ions in Biological Systems," Vols. 1–6. Dekker, New York, 1973–1976.

D. Williams, "The Metals of Life." Van Nostrand-Reinhold, Princeton, Jersey, 1971.

R. J. P. Williams, The biochemistry of sodium, potassium, magnesium, and calcium. *Quart. Rev., Chem. Soc.* **24**, 331 (1970).

Metals in Organic Chemistry

R. L. Augustine, "Catalytic Hydrogenation." Dekker, New York, 1965.

R. L. Augustine, ed., "Reduction." Dekker, New York, 1968.

D. H. Busch, Metal ion control of chemical reactions. *Science* **171**, 241 (1971).

L. F. Fieser and M. F. Fieser, "Reagents for Organic Synthesis," Vols. 1–3. Wiley, New York, 1973–1976.

D. Forster and J. F. Roth, eds., "Homogeneous Catalysis" *Advan. Chem. Ser.* Vol. **132**, (1974).

R. F. Heck, "An Introduction to Organotransition Metal Chemistry." Academic Press, New York, 1974.

B. R. James, "Homogeneous Hydrogenation." Wiley, New York, 1973.

M. M. Jones, "Ligand Reactivity and Catalysis." Academic Press, New York, 1968.

A. W. Langer, ed., Polyamine-chelated alkali metal compounds. *Advan. Chem. Ser. Vol.* **130** (1974).

A. E. Martell, Catalytic effects of metal chelate compounds. *Pure Appl. Chem.* **17**, 129 (1968).

A. E. Martell, Artificial enzymes. *In* "Metal Ions in Biological Systems" (H. Sigel, ed.), Vol. 2, pp. 208–262. Dekker, New York, 1973.

G. N. Schrauzer, ed., "Transition Metals in Homogeneous Catalysis." Dekker, New York, 1971.

J. M. Swan and D. St. C. Black, "Organometallics in Organic Synthesis." Chapman & Hall, London, 1974.

M. M. Taqui Kahn and A. E. Martell, "Homogeneous Catalysis by Metal Complexes." Academic Press, New York, 1974.

II

Inorganic Chemistry of Group Ia and IIa Metals

We will soon be examining the chemical reactivity of metals and metal ions in many different kinds of chemical systems. Chemical reactivity is of course defined in terms of changes in the chemical bonding present in a system undergoing reaction; some existing bonds are broken and some new bonds are formed. Chemical bonding is based on electronic interactions which give rise to attractive forces tending to hold two or more nuclei together. Therefore it behooves us first to review some basic aspects of the electronic structure of atoms. To begin with we will consider the elements in groups Ia and IIa of the periodic table, the so-called alkali and alkaline earth metals. In Chapter 4 we will return to consider the electronic structures of atoms and ions in a more general way.

Chemistry of Ia and IIa Metals

The alkali metals are extremely reactive and are found in nature only as cations. Their outstanding chemical property is perhaps their ability to form salts, nearly all of which are water soluble, but there are some important chemical uses for the metals, mainly as very powerful reducing agents. Considering either the metals or their cations in sequence by atomic number, one finds as a general rule only gradual changes in a given physical or chemical property.

14

The heaviest alkali metal, francium, is unique in that all its isotopes are radioactive. It is also unusual in being liquid at room temperature, an attribute shared by only four other elements: the alkali metal cesium, and mercury, gallium, and bromine. Rubidium is a solid at room temperature but melts very easily a few degrees higher. It is more abundant in nature than either cesium or the very rare francium. Cesium and rubidium are the most electropositive of all the elements; they react violently with water and ignite spontaneously in air.

Sodium and potassium are quite common in nature, and many uses have been found for their salts. As metals they too react vigorously with water, and potassium metal may ignite if a fresh surface is exposed to humid air. Sodium–potassium alloy (1:5) is a pyrophoric liquid used as an extremely powerful reducing agent in organic synthesis. It is also used as a heat transfer liquid in closed systems such as nuclear reactors because of its great thermal stability. Sodium amalgam, Na–Hg, is a similar but less powerful reducing agent and can be used in aqueous solvents.

Lithium is the lightest member of the alkali metals. In nature it is less abundant than sodium and potassium, but more abundant than the others. Lithium metal melts at $180°C$, and it is much harder than the other alkali metals. It reacts not only with oxygen but also with nitrogen, forming a coating of lithium nitride, Li_3N. Lithium metal has many uses in organic chemistry, either as a reducing agent or in preparation of organolithium analogs of Grignard reagents. The lithium ion has no important role in biology as do sodium or potassium. However, lithium carbonate is used as a depressant drug in the management of manic and severe psychotic states. Its mechanism of action is not clearly understood, but it involves more than simple substitution for sodium or potassium in or around nerve cells.

Like the alkali metals, the alkaline earth metals are rather reactive and occur in nature only as salts. Unlike the alkali metals, however, one finds more examples of water-insoluble salts for the alkaline earths. The heaviest alkaline earth, radium, is radioactive and rare in nature. Barium and strontium are often found in nature as their insoluble sulfates. Calcium occurs commonly as chalk, marble, and limestone, all of which are forms of calcium carbonate, $CaCO_3$. Calcium oxide reacts with carbon at high temperatures to give calcium carbide, CaC_2, which reacts further with water to form acetylene, H_2C_2, and calcium hydroxide. Calcium metal reacts with oxygen and nitrogen, and is used to scavenge air from inside vacuum tubes since either oxygen or nitrogen would destroy a hot tungsten filament. Magnesium is abundant in various types of silicate rocks and also in seawater from which it is extracted commercially. Magnesium metal is combined with zinc, manganese, and aluminum, to produce strong lightweight alloys. Both beryllium metal and its compounds are extremely poisonous. The metal has a very high specific heat and has been used as a lightweight heat sink on aircraft brakes because it will absorb a great amount of heat but undergo only a small temperature change.

As with the alkali metals, the chemistry of the alkaline earths is dominated by the chemistry of their cations. Differences in chemical properties between members in the series again tend to vary in sequence with atomic numbers, but the jumps are a bit larger than with the alkali metals.

Structures of Ia and IIa Atoms and Ions

One of the more important factors in maintaining the structure of an atom is simply electrostatic attraction between the positively charged nucleus and the negative electrons. The formation of atoms from a random collection of nuclei and electrons releases a large amount of energy, and conversely the removal of an electron from an atom requires an input of energy. The amount of energy required to separate one electron from an atom in the gas phase is known as the first ionization potential for that atom. Since most of the chemistry of the Ia and IIa elements is that of their ions, we should not be surprised that much of this chemistry can be correlated with the ionization potentials of the atoms.

TABLE 2-1

Properties of the Group Ia Elements

Property	Li	Na	K	Rb	Cs
Atomic number	3	11	19	37	55
Configuration	$2s^1$	$3s^1$	$4s^1$	$5s^1$	$6s^1$
Ionization energy, kcal, I_1	124	118	100	96	90
Ionization energy, kcal, I_2	1744	1091	734	634	579
Atomic radius, Å	1.33	1.57	2.03	2.16	2.35
ΔH_{sub}, kcal	38.4	25.9	21.5	19.5	18.7
Ionic radius, M^+, Å	0.68	0.98	1.33	1.48	1.67
ΔH_{hyd}, kcal	121	95	76	69	62
$E_0(M^+, M)$, V	−3.02	−2.71	−2.92	−2.99	−2.99
$D(M_2)$, kcal	25	17	12	11	10.4
Lattice energy, MF, kcal	243	216	192	185	176

The relative magnitudes of the sequential ionization potentials of the Ia and IIa atoms and ions (see Tables 2-1 and 2-2 and Fig. 2-1) can be rationalized by assuming a simple electrostatic model in which energy is inversely proportional to the distance between the charges ($E \propto e^2/r$). As expected the first ionization potential of the atoms decreases as the atomic number increases, and correspondingly the difference between I_1 and I_2 decreases with increasing atomic number. That the second ionization potential of an element is always greater than its first ionization potential is the result of the increased effectivity of the

TABLE 2-2

Properties of the Group IIa Elements

Property	Be	Mg	Ca	Sr	Ba
Atomic number	4	12	20	38	56
Configuration	$2s^2$	$3s^2$	$4s^2$	$5s^2$	$6s^2$
Ionization energy, kcal, I_1	214	175	140	132	120
Ionization energy, kcal, I_2	429	345	274	253	230
Atomic radius, Å	0.89	1.36	1.74	1.91	1.98
ΔH_{sub}, kcal	77.9	35.6	42.2	39.1	42.5
Ionic radius, M^{2+}, Å	0.30	0.65	0.94	1.10	1.29
ΔH_{hyd}, kcal	570	460	395	355	305
$E_0(M^{2+}, M)$, V	−1.70	−2.34	−2.87	−2.89	−2.90
Lattice energy, MF_2, kcal	—	702	623	592	557

nucleus in binding the remaining electrons; for example in the case of lithium I_2 is more than 13 times greater than I_1, whereas for beryllium I_2 is only about double I_1. This comparison may simply reflect the fact that the second ionization of a berylium atom leaves an ion which has a very stable electronic configuration similar to that of the rare gas helium, while the corresponding second ionization of a lithium atom destroys this favorable electronic configuration. The influence of net charge on ionization potential can best be seen by comparing the isoelectronic species H^-, He, and Li^+ (Fig. 2-1). Although the size of the helium atom is not known, it is probably in between the hydride ion, radius 2.08 Å, and the lithium ion, radius 0.68 Å. The isoelectronic beryllium ion, with a radius of 0.30 Å, is severely contracted by the strong net charge of 2+.

In summary, the important features of ionization processes are the effect of positive charge in increasing ionization potential, and the observation that the electron in the least stabilized orbital will be lost first. Later we will see that these features are important in the ionization of molecules and metal complexes as well. Another feature illustrated by the ionization of Ia and IIa metals, which is of very general importance in chemistry, is the observation that these atoms easily achieve the electronic configuration of a rare gas by losing one or two electrons. The concept of atoms or ions tending to undergo chemical reaction in order to achieve a rare gas configuration is a very useful, although not infallible, predictive device. It works very nicely with elements in the A groups of the periodic table, such as sodium, oxygen, or chlorine. In these cases it is known simply as the "octet rule." Among the B group elements, especially the transition metals, it is called the "18-electron rule." The latter is not as rigid as the octet rule, and many exceptions are known. The 18-electron rule is discussed in greater detail in Chapter 5.

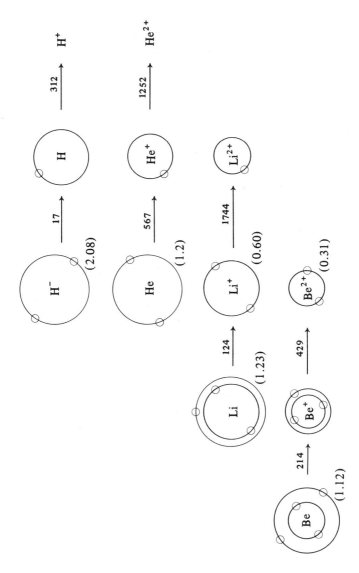

Fig. 2-1. Relationships of charge, size, electronic configuration, and gas phase ionization potential (kcal/mole) for some light atoms. Figures in parentheses give the radius in angstroms when known; circles are roughly proportional. Wilkinson.[12]

Influence of Ionization Potential on Chemical Properties

The formation of molecules from atoms, as well as their reformation by inter- or intramolecular chemical reactions, is a function of the outer or valence electrons of the constituent atoms. Since most of the chemistry of the Ia and IIa elements is that of their ions, we should expect to be able to correlate ionization potentials with the basic chemical properties of these elements. Two such important correlations involve heats of hydration of cations and lattice energies for fluoride salts. In both cases the proportionality is obviously direct although it is nonlinear. A simple rationalization of these correlations is that the movement of a negative charge, such as an electron or a fluoride ion, down the positive electrostatic field gradient of a lithium cation releases energy. Combining an electron with a lithium ion, the reverse of ionization, releases 124 kcal/mole. Similarly, the interaction of lithium and fluoride ions produces a solid three-dimensional network or lattice and is accompanied by the release of 243 kcal/mole. The interaction of a permanent dipole, such as that of a water molecule, with an electrostatic gradient also causes a release of energy. For example, the energy released by hydration of a lithium ion in dilute solution is 122 kcal/mole.

The best correlations with the chemistry of Ia and IIa ions can be obtained with their corresponding electrostatic field strengths, but for practical purposes it is convenient and reasonably accurate to think of this simply as being inversely proportional to the *size* of the ion. Many examples of this will be apparent in later portions of this and the next chapter, but even from Tables 2-1 and 2-2 it can be seen that the smaller the cation of a given charge the larger are its corresponding lattice and hydration energies. These correlations, more than any other factors, control the chemistry and biochemistry of the Ia and IIa ions.

Dissolution of Crystal Lattices and Hydration of Cations

In the previous section we developed an analogy between the formation of a crystal lattice or a hydrated cation and the *reversal* of the process of ionization of an atom, i.e., that these and related processes represent the acquisition of an electron or electron density by the cation, with the result that energy is released and a more stable system is achieved. To illustrate this in somewhat greater detail let us consider the sodium chloride ion pair, the sodium chloride crystal lattice, and the hydration of sodium and chloride ions.

The formation of a sodium chloride ion pair from the separated ions releases energy mounting to 140 kcal/mole. As shown in Fig. 2-2 there is a compromise between attractive forces and repulsive forces, the latter being significant only

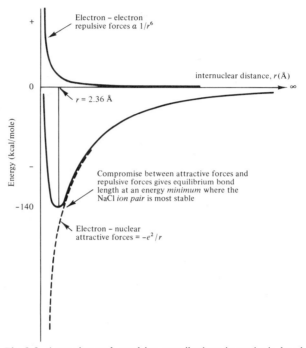

Fig. 2-2. Attractive and repulsive contributions in an ionic bond.

over short distances, which gives rise to the equilibrium bond length and bond strength characteristic of the ion pair. The ion-pair bond strength is the energy required to completely dissociate an ion pair in the gas phase, just as the ionization potential of a sodium atom is the energy required to remove an electron to an infinite distance.

In the sodium chloride crystal lattice the ion–ion interactions extend in three dimensions rather than one, and the lattice energy is therefore greater than the ion-pair dissociation energy. In order to calculate the energy released in forming an ionic lattice the basic approach of calculating attractive and repulsive forces is simply extended in three dimensions. For the rock salt lattice (Fig. 2-3) the following expression is used:

$$\text{Lattice energy} = U = \frac{-Az_1z_2e^2N}{r} \left(1 - \frac{1}{n}\right)$$

In this formula the lattice energy is given as a negative number because energy is released by a system during lattice formation. The contributing factors are z, the charge on the ions; e, the electrostatic unit of charge; N, Avogadro's number; and r, the internuclear distance. The term A is the Madelung constant,

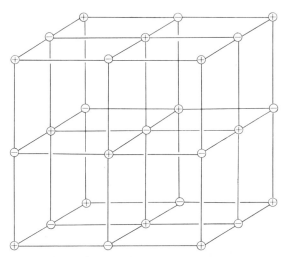

Fig. 2-3. The rock salt lattice.

which is an approximate sum of an infinite series of terms taking into account the fact that each sodium is surrounded by six neighboring chlorides at distance r, 12 more chlorides at distance $\sqrt{2}\ r$, and so on. The value of n is calculated from the mechanical compressibility of the crystal and takes repulsive interactions into account. Using this method the calculated lattice energy of NaCl is 179.2 kcal/mole with a Na–Cl distance of 2.82 Å. This is much greater than the energy formed by a single Na–Cl ion-pair bond, which the electrostatic model gives as 140 kcal/mole for a bond length of 2.36 Å.

Since the lattice energy, in theory, represents the energy required to convert one mole of crystalline substance to gas phase ions separated from each other at infinite distances, it is not an easily measured quantity. However, an indirect method for determining lattice energies can be used. It is based on a theoretical thermochemical cycle known as the Born–Haber cycle. The basic assumption of this method is that in considering two thermodynamic states (for example, elemental sodium and chlorine versus sodium chloride) energy differences between the states are independent of the processes which convert one state to the other. A Born–Haber cycle for NaCl is shown in Fig. 2-4. Any single unknown energy term, which may be one that is difficult to measure such as electron affinity, may be calculated if all the others are known.

The Born–Haber cycle in Fig. 2-4 compares two ways in which solid NaCl can be formed from one mole of solid sodium metal and one-half mole of chlorine gas. Their direct combination will produce NaCl, and 98.3 kcal of heat energy will be evolved. In this process sodium has lost an electron, chlorine has gained an electron, and a crystal lattice has formed. As an alternative theoretical reaction pathway we might first vaporize the sodium at an

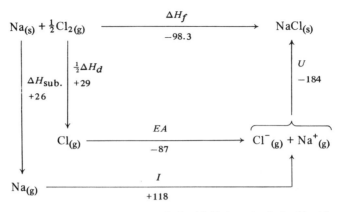

Fig. 2-4. The Born–Haber cycle for NaCl (energies in kcal/mole).

expense of +26 kcal and then ionize it at the further expense of +118 kcal. The chlorine molecule must first be dissociated to chlorine atoms, requiring 29 kcal per half mole of Cl_2. The electrons obtained by ionizing sodium combine with chlorine atoms to form chloride ions, still in the gas phase. This releases 87 kcal of energy, which represents the electron affinity of the chlorine atom. Finally, formation of solid NaCl from gaseous Na^+ and Cl^- releases 186 kcal. This term is called the lattice energy and represents the stability of the lattice formed. Adding these terms gives us $(+26) + (+118) + (+29) + (-87) + (-184) = -98$ kcal/mole, which is very close to the experimentally determined value of −98.3 for the heat of formation of NaCl. Table 2-3 gives lattice energies of Ia and IIa halides in kilocalories per mole.[1,2]

TABLE 2-3

Lattice Energies of Ia and IIa Halides (in kcal/mole)[a,b]

Element	Fluoride	Chloride	Bromide	Iodide
Li	242.8	201.7	191.0	178.4
Na	216.6	183.9	175.5	164.8
K	191.8	168.3	160.7	151.5
Rb	184.6	162.8	157.1	147.9
Cs	176.0	157.2	151.2	143.7
Mg	702.3	—	573	547.1
Ca	623.4	532.1	509.4	487.4
Sr	591.6	508.3	487.4	463.6
Ba	557.1	484.4	465.6	441.0

[a] Data from Ketelaar.[1]
[b] Data from Brackett and Brackett.[2]

In addition to releasing energy by attracting an electron or an anion to itself, a sodium ion can attract dipolar molecules and release energy. Dipolar molecules such as H_2O and NH_3 have asymmetric distributions of charge such that a permanent electrostatic dipole is developed. Water dipoles align

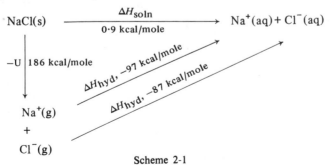

themselves with their negative end toward the center of the positive electrostatic field generated by a nearby sodium cation. As the dipole draws close to the cation, potential energy is released, and an ion–dipole bond is formed. These bonds are not as strong as covalent or ion–ion bonds, but they are very important in the process of dissolving ionic crystals. In solution, where ion–dipole bonds can form in large numbers around a spherical ion like Na^+, their sum can represent a very substantial energy release. Ion–dipole bonding is also very important in the formation of complexes of group Ia and IIa ions with organic compounds, most of which have only become known during the last decade or so.

The heat of hydration of an ion refers to the energy that would be released by the bare gas phase ion upon dissolving to form an infinitely dilute aqueous solution. For the sodium ion this energy term amounts to about 95 kcal/mole. Thus, hydration releases much of the ionization energy (118 kcal) of the sodium atom. Heats of hydration for individual ions cannot be measured directly since it is impossible to obtain ions as individual species. In other words, it is impossible to hydrate a metal cation without simultaneously hydrating its accompanying anion. However, by studying a matrix of compounds such as the halides of the alkali metals, and by using a thermochemical analysis like a Born–Haber cycle, one can obtain reasonably good estimates for the hydration of many individual ions.

TABLE 2-4

Heats of Hydration of Anions[a]

Anion	Unit (kcal/mole)
F^-	120.8 ± 3.5
Cl^-	86.8 ± 3.1
Br^-	80.3 ± 3.5
I^-	70.5 ± 3.5
ClO_4^-	57 ± 4
HO^-	110 ± 8

[a] Data from Halliwell and Nyburg.[3]

For NaCl the cycle has the form shown in Scheme 2-1. Again the assumption is made that energy differences between thermodynamic states are independent of the process which connects the states. The direct dissolution of solid NaCl evolves 0.9 kcal/mole of heat energy. The indirect theoretical approach first requires 186 kcal/mole to disrupt the NaCl lattice before the separated ions are hydrated. The *sum* of their hydration energies is then the difference between 186 and 0.9 or 185.1 kcal/mole. How to apportion this total between ΔH_{hyd} for Na^+ and ΔH_{hyd} for Cl^- depends on making some approximations and estimates. The values given in Tables 2-1 and 2-2 are good to within ± 3 kcal/mole; those in Table 2-4[3] have larger uncertainties.

Water Structure

Water is unique among solvents in that it is so well suited as a milieu for many kinds of chemical reactions, be they inorganic, organic, or biochemical. It is the most abundant compound on the earth's surface, and this is reflected in the fact that most living systems contain from 60 to 90% water. In light of the ubiquity of water and its suitability as a medium for chemical reactions, it is worthwhile if not imperative to consider the *mutual* effects of solutes and water on each other. For example, water is well known to disrupt the NaCl lattice by dissolving and separating the ions, but what is the effect of the ions on the properties of water, and why will water not dissolve non-polar organic compounds? To answer these and many other important questions we must consider the effects of solutes on water structure. Then we can better understand the formation of complexes of Ia and IIa ions in aqueous solution.

After considering ionic crystal lattices it is difficult to conceive of liquid water as having any kind of structure. However, it certainly has structural characteristics, mainly by virtue of hydrogen bonding between individual molecules.[4] We can approach the structure of liquid water by starting with the crystalline state of ice I, which occurs at atmospheric pressure just below the freezing temperature of 273°K. In ice I the oxygen atoms occupy a diamondlike lattice (Fig. 2-5), and the hydrogen atoms lie on the axes between the oxygens. Hydrogen bonding between individual molecules is substantial, as indicated by the large heat of fusion of water, 79.7 cal/gm. For ethyl alcohol, which is less able to self-associate by hydrogen bonding, the heat of fusion is 24.9 cal/gm.

When ice melts its crystal lattice is disrupted and there remains no extensive rigid network of hydrogen bonded molecules. However, that there is still exten-

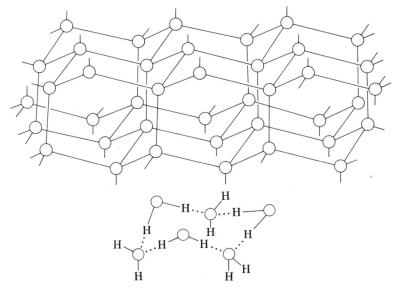

Fig. 2-5. The ice I lattice. Oxygens are shown as open circles in a diamondlike lattice. The inset shows the hydrogen bonded hexameric unit which repeats throughout the lattice.

sive hydrogen bonding throughout the liquid is evident from a comparison of the heats of vaporization for water and ethanol, which are 539 and 204 cal/gm, respectively. There are currently two theories on how these hydrogen bonds are organized throughout the bulk of the liquid, and they have come to be known as the continuum theory and the mixture theory. The mixture theory states that at any given instant there are regions in the liquid where hydrogen bonding gives rise to localized "icebergs" with very short lifetimes on the order of the vibrational frequency of chemical bonds, about 10^{-12} to 10^{-13} sec. Thus, the bulk of water is visualized as a constantly equilibrating mixture of monomeric water and various aggregates from dimers on up to "molecular icebergs." Because it has not been possible to obtain evidence for the existence of icebergs, the continuum theory arose, postulating that hydrogen bonding is developing and breaking down uniformly and randomly throughout the entire bulk of the water. In either case however, on a time scale slower than 10^{-12} sec, every water molecule would experience the same average environment.

Solute Effects on Water Structure[5-7]

Solutes may be classified into three types according to their effect on the structure of water. The first of these, and the easiest to understand, are the *electrostatic structure makers*. Examples of this kind of solute include Be^{2+}, Mg^{2+}, Li^+, HO^-, F^-, H^+, and probably ions such as Ca^{2+}, SO_4^{2-}, and Na^+ as well. They all share a common characteristic known as "hardness," which refers to the fact that these ions have a rather intense electrostatic field about them which is not easily perturbed. In fact, it is strong enough to perturb the electrostatic field of other ions and molecules instead. For example, the "hard" lithium ion and the "soft" iodide ion are both spherical, but as they approach each other the intense electrostatic field around the small lithium cation deforms or polarizes the large diffuse electron cloud of the iodide ion. Thus, polarizability or softness is associated with large diffuse electron clouds and weak electrostatic fields while hardness is associated with small dense electron clouds and intense electrostatic fields.

Another characteristic of some of the structure making ions is their ability to participate in hydrogen bonding with water molecules. Thus, a solute such as sodium fluoride imparts a significant amount of permanent structure to water in two ways. When the sodium ions undergo hydration, the oxygens of water are attracted toward the cation leaving the hydrogens relatively fixed, pointing outward, and relatively more available for hydrogen bonding as compared to a solution of pure water. With fluoride ions the reverse obtains; hydration occurs with water hydrogens pointing in and hydrogen bonding to the fluoride ion, leaving the oxygens pointing outward, relatively available for hydrogen bonding.

In general the electrostatic structure makers fit well into a space lattice like ice I because they are small, and their hardness allows them to bond to water strongly enough to maintain permanent structure as far away as two or three layers of water molecules. The results of the increased hydrogen bonding and increased net structure in an aqueous solution are manifested as increases in heat capacity, thermal conductivity, and viscosity. The maximum density of the solution as a function of temperature and the diffusion rate of water molecules decrease with increasing structure in solution.

The second category of solutes which affect the structure of water is the *hydrophobic structure makers*. Detergents and soaps are very good examples of this category, as are fatty acids, tetraalkylammonium salts, the tetraphenyl-boron anion $(C_6H_5)_4B^-$, certain amino acids, and "hydrophobic" portions of

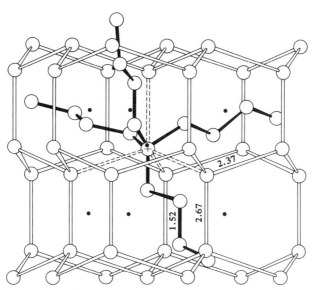

Fig. 2-6. Proposed model for structured water induced by a hydrophobic structure-making solute, tetrabutylammonium fluoride. The possible orientation of the butyl groups in cavities in an ice I lattice is shown. Dots indicate centers of cavities; open circles indicate oxygens or a fluoride ion; hydrogens not shown.[8]

proteins. These compounds have in common a small or zero ionic charge, and a portion of their structure is hydrocarbon-like or lipophilic. Everyone knows that oil and water don't mix, but the reason they don't is that when a hydrocarbon chain enters an aqueous solution it induces around itself a layer of ice, or in other words, highly structured water (Fig. 2-6).[8] This requires an increase in the free energy of the system mainly as a result of the large decrease in entropy as the water changes from a disordered to a highly ordered state. Thus, hydrocarbons tend to be excluded from water into a separate liquid phase. A detergent is a molecule which has a water-soluble portion and a lipophilic or water-insoluble portion.

Detergents in solution often form micelles (Fig. 2-7) in which the lipophilic or hydrophobic portions self-associate in a central region leaving their polar

Fig. 2-7. A micelle. Circles represent polar or ionic end groups such as CO_2^- or NMe_3^+; wavy lines represent hydrocarbon chains (cf. Fig. 3-4).

water-soluble portions exposed to the water on the outside. Detergents can associate with molecules of oily greasy dirt in a similar way. The two are "squeezed together" by the structured water surrounding them, and so are held together in aqueous solution by the hydrophilic properties of the detergent molecule.

In the case of proteins in solution, hydrogen bonding and ionic interactions between amino acid side chains make large contributions to the stability of the tertiary folded structure. However, hydrophobic interactions are important as well, not only in maintaining protein tertiary structure but also in the binding of small molecules such as hormones or drugs to proteins. These "hydrophobic bonds" are mainly because the association of nonpolar solute molecules with water is thermodynamically unfavorable. Consequently, nonpolar molecules, or fragments of molecules, are disassociated from water regions and are left with self-association as the only alternative. Hydrophobic structure-making solutes have basically the same gross effects as the electrostatic structure makers on the properties of aqueous solutions. Again, any favorable enthalpy change resulting from increased structure and increased hydrogen bonding in the solution is more than offset by a large decrease in entropy, most of which results from the extensive ordering of solvent water molecules. Table 2-5[9,10] gives some data which illustrate the relative importance of enthalpy and entropy changes for the transfer of small nonpolar molecules from benzene to water.

The third category of solutes are the *electrostatic structure breakers*. These

TABLE 2-5

Thermodynamic Changes Accompanying Transfer of Small Nonpolar Molecules from a Nonpolar Phase to Water at 25°C[a]

Molecule	Phases	ΔS (eu)	ΔH (cal/mole)	ΔG (cal/mole)
He	Benzene/water	−16.9	−3550	+1486
Ne	Benzene/water	−19.5	−4440	+1371
N_2	Benzene/water	−17.7	−3100	+2174
O_2	Benzene/water	−18.5	−3370	+2143
CH_4	Benzene/water	−18.3	−2820	+2633
C_2H_6	Benzene/water	−19.5	−2180	+3631
C_2H_4	Benzene/water	−15.2	−1610	+2914
C_3H_8	C_3H_8/H_2O	−23	−1800	+5050
C_4H_{10}	C_4H_{10}/H_2O	−23	−1000	+5850
Benzene	Benzene/H_2O	−14	0	+4070
Toluene	Toluene/H_2O	−16	0	+4650
Ethylbenzene	C_8H_{10}/H_2O	−19	0	+5500

[a] Data from Kauzmann[9] and Frank and Evans.[10]

TABLE 2-6

Effects of Solutes on the Properties of Aqueous Solutions

	Solute type[a]	
Solution property	Hydrophobic or electrostatic structure-making	Electrostatic structure-breaking (chaotropic)
Heat capacity	+	−
Thermal conductivity	+	−
Viscosity	+	−
Water diffusion rate	−	+
Surface tension	−	+
Total entropy	−	+
Maximum density	−	+

[a] The plus indicates an increase; the minus a decrease.

solutes and their aqueous solutions (Table 2-6) have properties which are opposite to those of the structure-making solutes. They are generally large soft ions with small charges, and they do not fit well into an ice I type of hydrogen bonding scheme. Therefore they tend to disrupt and disorganize water structure in the several layers of water surrounding them. Examples of structure-breaking solutes include

$$Cs^+ > Rb^+ > K^+ \approx NH_4^+$$
$$SCN^- > ClO_4^- \approx IO_3^- > NO_3^- > I^- > Br^- > Cl^-$$

Their effects on aqueous solutions can be striking; for example, most solutes increase the structure of water and thereby increase the viscosity of the solution over that of pure water. Cesium iodide however actually decreases the viscosity of water because it disrupts the structure which gives water its inherent viscosity. Structure-breaking salts also enhance the water solubility of organic compounds in water, making water more lipophilic by disrupting the icebergs that tend to form around hydrocarbon chains and exclude them from solution. Quite the opposite behaviour is seen when sodium chloride is used to salt out organic compounds for extraction from aqueous solutions.

Complexation of Ia and IIa Cations

Perhaps the most important "chemical property" of the Ia and IIa cations is their size. In the formation of ionic crystal lattices, the relative sizes of cation and anion are of prime importance in determining the lattice packing arrangements. In solution, the size of the cation or anion greatly affects the

TABLE 2-7

Common Coordination Number and Charge/Size Ratios for Alkali and Alkaline Earth Ions

Ion	C.N.[a]	1/r	Ion	C.N.[a]	2/r
Li	4, 6	1.47	Be	3,4	6.60
Na	4, 6	1.02	Mg	6	3.06
K	8, 10	0.75	Ca	6, 8	2.13
Rb	10	0.67	Sr		1.82
Cs		0.60	Ba		1.55

[a] Coordination number.

way in which it interacts with water structure. Similarly, in the formation of hydrated ions or complex ions, the size of the ion again plays by far the major role.

When a gas phase cation is transferred to dilute solution it attracts one or more layers of water molecules to itself. The innermost layer is held most tightly, and is called the inner or first coordination sphere. In most cases there are a definite number of positions in the first coordination sphere, and this number is known as the coordination number. Table 2-7 gives the usual coordination numbers for the Ia and IIa cations.

The smaller the ion, the less room there is on its surface for coordinated molecules, and the smaller its coordination number will be. However, the smaller ions have more intense electrostatic fields about them and hence are better able to attract several additional layers of water for their outer coordination spheres. This is nicely illustrated by comparing the specific or equivalent conductances of ions having the same charge but different sizes. Since electrical conductance depends on the ability of an ion to move through solution between two electrodes, the fastest moving and best conducting ions will be those whose effective size, including hydration layers, is smallest (Table 2-8).[11, 12] This behavior is also shown by the ammonium ions NH_4^+, Me_4N^+, Et_4N^+, and Bu_4N^+ whose equivalent conductances in water at 25°C are 74.3, 45, 33, and 19, respectively. The large Bu_4N^+ ion is a poor conductor because it cannot migrate through solution toward an electrode as rapidly as the smaller ions. The proton and hydroxide ion are exceptions, for although they are highly hydrated, an alternative mechanism is available to explain their high conductance values (Fig. 2-8).

The formation of complex ions is often an equilibrium process in which solvent molecules compete with other molecules for sites in the first coordination sphere of the metal ion. The general term for this process is ligation (Latin, ligare, to bind) and the molecules being bound are called ligands. Solvation is merely a specific type of ligation. The complex chemistry of the Ia and IIa cations was very slow to develop, compared to the transition metal complexes

TABLE 2-8

Comparison of Ionic Equivalent Conductance with Ionic Size and Other Parameters[a]

Property	Li+	Na+	K+	NH4+	Rb+	Cs+	F-	Cl-	I-	NO3-	ClO4-
Ionic radius, Å	0.6	0.95	1.33	1.44	1.48	1.69	1.36	1.81	2.16		
Hydrated radius, Å	3.40	2.76	2.32		2.28	2.28					
Total hydration number	25	16	10		10	10					
ΔH_{hyd}, kcal/mole	124	95	76		69	62	121	87	71		57
λ, 25°C, ohm^{-1} [b]	39	50	73.5	73.4		78.1					
λ, 18°C, ohm^{-1} [b]	33	43	64		67	68	46.6	65.4	67.4	61.7	58.4
Binding strength[c]											

[a] Data from Gurney[11] and Cotton and Wilkinson.[12]
[b] Here λ stands for equivalent conductance in water.
[c] Binding to cation exchange resin increases from Li+ to Cs+.

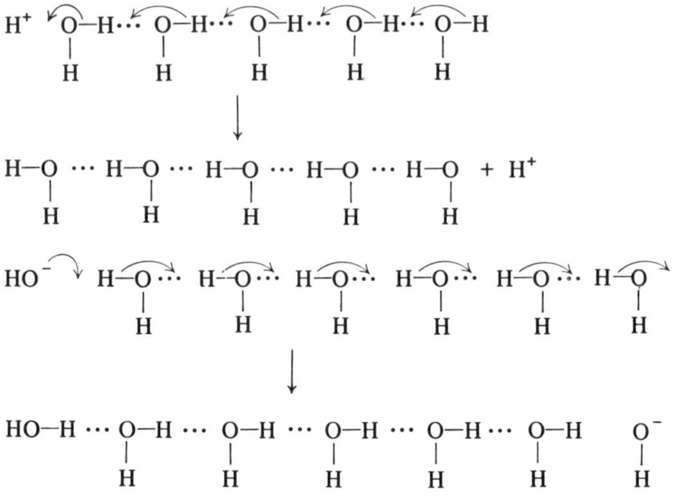

λ, ohm^{-1}	H$^+$	OH$^-$
25°	350	197
18°	316	177

Fig. 2-8. Schematic representation of rapid transference mechanisms for protons and hydroxide ions.

which have been under study since the nineteenth century. Prior to the 1925–30 era the only known "complexes" of Ia and IIa cations were their ionic crystals and a few insoluble salts. Some of the earliest indications that ions such as sodium would form complexes were observations made by Brewer and Sidgwick, who reasoned that if stable neutral complexes of alkali ions formed, they should be soluble in organic solvents. They then proceeded to show that the sodium salt of 1-phenylbutane-1,3-dione would dissolve in toluene

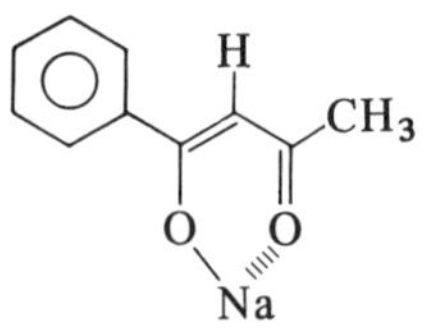

Sodium enolate of 1-phenylbutane-1,3-dione

if ethylene glycol was added. Moreover, they found that these solutions were nonconducting, which meant that there were no free ions. Eventually the structure of this complex was elucidated by X-ray diffraction (Fig. 2-9).[13] In this case the sodium ions show the uncommon coordination number 5 rather than 6 as shown by the rock salt lattice or the complex illustrated in Fig. 2-10.[13]

The next big step forward in the chemistry of Ia and IIa complexes was ac-

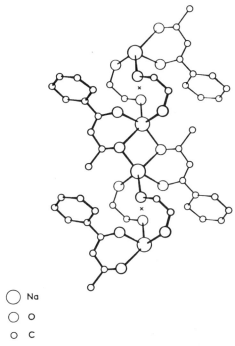

○ Na

○ O

○ C

Fig. 2-9. The structure of (PhCOCHCOCH$_3$)$^-$Na$^+$, HOCH$_2$CH$_2$OH. The glycol molecules form double bridges between centrosymmetrically related sodium ions. The resulting dimers are held by sharing of one carbonyl oxygen to give a one-dimensional polymer. Sodium is coordinated by an irregular square pyramid. Crosses represent centers of symmetry.[13]

tually two simultaneous discoveries: the crown ethers and the ionophorous antibiotics. In the remainder of this chapter we will examine the properties of these unusual ligands and the interesting and important complexes they form. The applications will be discussed in the next chapter.

Nonactin, the first member of a large series of diverse natural products known as the ionophorous or ion-carrying antibiotics,[14] was isolated in 1955. Other members of this rapidly expanding group include valinomycin, gramicidin A, nigericin, X-537A, monensin, enniatin B, monactin, dinactin, and trinactin. Some of their structures, or those of their complexes, are given in Figs. 2-11 to 2-13.[13,15-18] All of these antibiotics either possess, or can fold up to form, a large ring of 14 or more members, furnishing a more or less symmetrical arrangement of —CO—, —NH—, or —O— groups which act as ligands toward the central metal ion. The ionophorous antibiotics have no human use because of their toxicity to mammalian cells, although monensin is used in chicken feeds. Their main action is to cause cells to leak potassium ions and ultimately to die. For this reason, however, they have been intensely

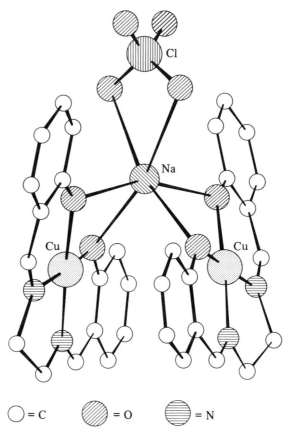

$\bigcirc$ = C   = O   = N

Fig. 2-10. One molecule of the sodium perchlorate complex with bis[*NN'*-ethylenebis-(salicylideneiminato)copper(II)].[13]

studies as a "model" system for ion transport through biological membranes (see Chapter 3).

These antibiotics are closely paralleled by the crown ethers[19,20] in ability to form complexes, particularly with potassium ion, except that the crowns are slightly less cation-selective. The first crown ether ligand was discovered by C. J. Pedersen in 1967 as a curious byproduct of an organic synthesis. This by-product appeared as a white fibrous solid which was almost insoluble in methanol, but which dissolved readily if sodium chloride was added. In Pedersen's nomenclature this compound was called dibenzo-[18]-crown-6 in reference to its 18-membered ring with 6 oxygen donors. It is shown as a rubidium thiocyanate complex in Fig. 2-14.[21] Before considering the chemistry of these fascinating complexes, some comments on their synthesis are appropriate. The unsubstituted prototype of the series can be synthesized in

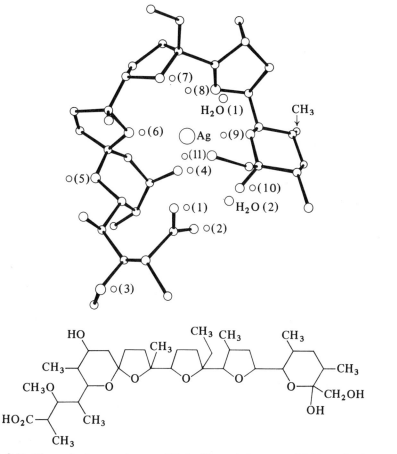

Fig. 2-11. Silver salt of monensin, an antibiotic. Monensin is a potent inhibitor of cation transport in rat liver mitochondria and has broad anticoccidial activity. Note that the carboxyl does not coordinate to the metal but instead ties the ligand into a macrocyclic ring by hydrogen bonding. The free ligand is also shown[15] (for a related example, see Jones *et al.*[16]). (Reprinted with permission from the *Journal of the American Chemical Society.* Copyright by the American Chemical Society.)

high yield as shown in Scheme 2-2.[22] When the reaction solvent is glyme $(CH_3OCH_2CH_2OCH_3)$ the yield of crown complex is 93%. However, the reaction is very specific for potassium ion. If tetrabutylammonium hydroxide is substituted for *t*BuOK, the reaction is very slow and only polymeric products are formed, even at high dilution. This observation leads to the postulate that potassium ion serves as a template upon which the reaction proceeds. The tetrabutylammonium ion is far too large to act in a way similar to that shown for potassium in Scheme 2-2. The "template effect" has been very useful in

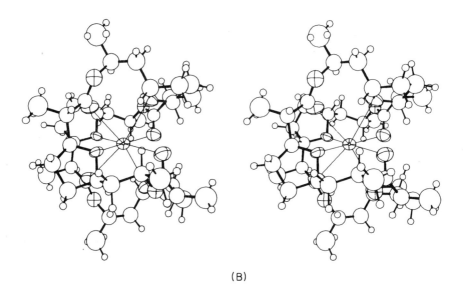

(A)

(B)

Fig. 2.12. The actins and the K$^+$-nonactin structure. (A) Diagram of the macrotetrolide, non-actin, showing the oxygen atoms which make contact with potassium in the KNCS complex. The cyclic molecule is wrapped round the potassium "like the seam of a tennis ball" to give approximately cubic coordination.[13] (B) Stereoscopic view of nonactin-K$^+$ along the crystallographic b axis.[17] From Helvetica Chimica Acta, with permission.

In nonactin, all four R's are methyls. Monactin has one ethyl and three methyls; dinactin, two each; trinactin, three and one; tetranactin, one of the most potent of the four (nonactin is inactive), has four ethyl groups. The R's are outside in the folded structure—look for them in the stereoview diagram.

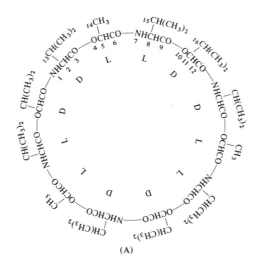

(A)

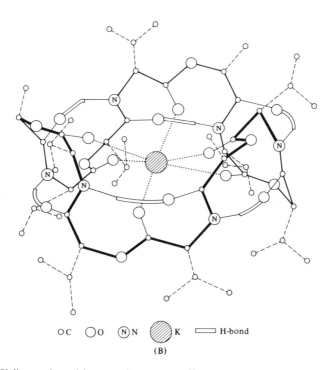

OC ◯ O Ⓝ N ⬡ K ▭ H-bond

(B)

Fig. 2-13. Valinomycin and its potassium complex:[18] (A) structure of valinomycin and (B) conformation of the K+ complex of valinomycin.

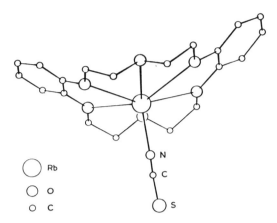

Fig. 2-14. RbNCS complex of dibenzo-[18]-crown-6.[21]

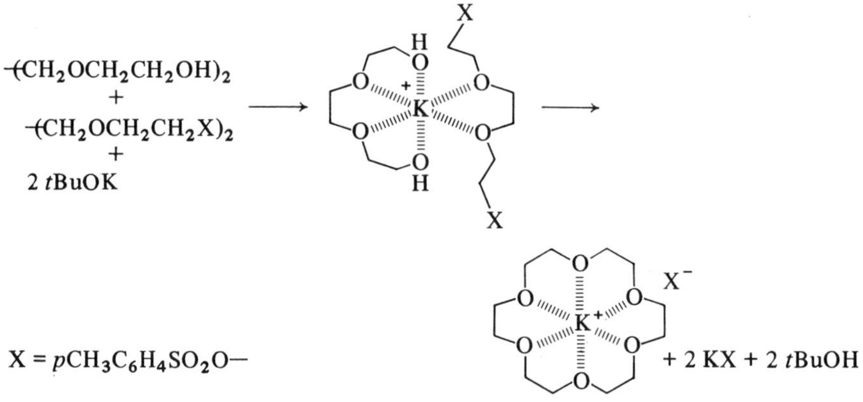

$$+CH_2OCH_2CH_2OH)_2$$
$$+$$
$$+CH_2OCH_2CH_2X)_2$$
$$+$$
$$2\ t\text{BuOK}$$

$$X = p\text{CH}_3\text{C}_6\text{H}_4\text{SO}_2\text{O}-$$

$$+ 2\ KX + 2\ t\text{BuOH}$$

Scheme 2-2

organic and inorganic synthetic reactions and is undoubtedly used to advantage by some metalloenzymes as well. This effect also provides a very clear example of the way in which entropy considerations can affect the course of a chemical process.

In reexamining the synthesis of [18]-crown-6, one observes that the base tBuOK is dissolved in glyme and initially has a structure as shown below, with the potassium ion solvated by glyme. When triethylene glycol is added it undoubtedly displaces most or all of the glyme from the potassium ion. The reason is that this action produces a favourable (positive) entropy change, such that the equilibrium lies to the right (Scheme 2-3).

The reason the entropy change is favorable is that in forming the intermediate complex, a number of solvent molecules are released. They are then free to diffuse at will, increasing the randomness and disorder of the system.

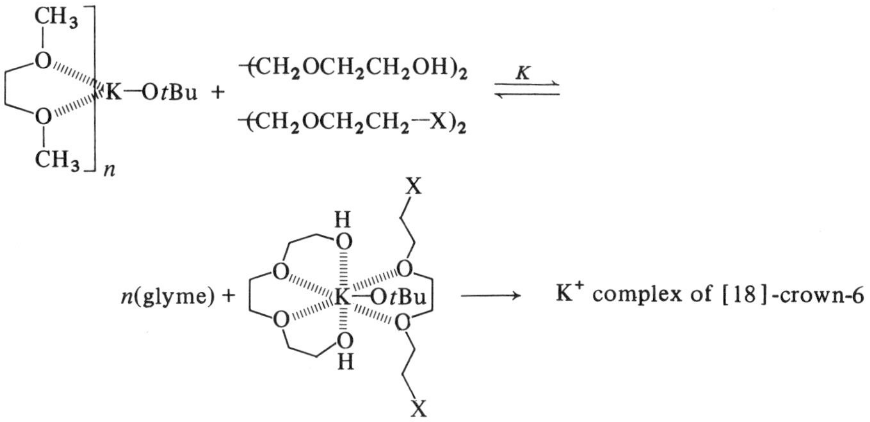

$$X = p\text{CH}_3\text{C}_6\text{H}_4\text{SO}_2\text{O}-$$

Scheme 2-3

Of course, some entropy is lost because of the restrictions placed on the random flopping and wiggling of the long-chain reactant molecules, but this is more than offset by the change from a three-particle system (left side of K) to a higher multiparticle system (right side of K). The effect of this is to increase the concentration of the intermediate complex by increasing K, and this in turn gives a faster rate of product formation with less opportunity for formation of linear polymeric by-products.

This argument might be rather unconvincing were it not supported by quantitative data for an analogous case. The analogy involves the competition reaction between a monodentate (single donor ligand) solvent tetrahydrofuran (THF), and a series of glymes, $\text{CH}_3\text{O}-(\text{CH}_2\text{CH}_2\text{O})_n-\text{CH}_3$, ($n = 1, 2, \ldots, 5$). The glymes are polydentate chelating (Greek, *chele*, claw) ligands competing with THF and the 9-fluorenyl anion (F), for binding to lithium or sodium cations, (M) (Scheme 2-4).

The absorption spectrum of the fluorenyl metal compound F–M differs from that of the solvent-separated ion pair, and this enables their ratio to be determined conveniently. By adjusting the experimental conditions so that the ratio

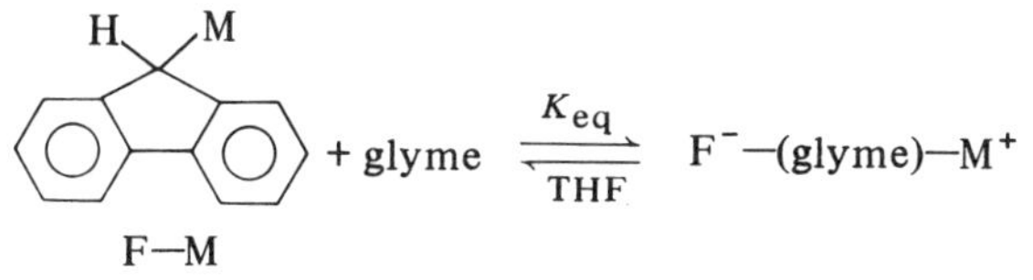

$$K_{eq} = \frac{[\text{F}^-][\text{M}^+]}{[\text{F}-\text{M}]} \cdot \frac{[\text{free glyme}]}{[\text{total glyme}]}$$

Scheme 2-4

TABLE 2-9

Equilibrium Constants for Formation of a Series of Glyme Separated
9-Fluorenyllithium and Sodium Ion Pairs at 25°C[23]

Glyme $CH_3(OCH_2CH_2)_nOCH_3$	Equilibrium constant, K	
	Li	Na
$n = 2$	3.1	1.4
3	130	9.0
4	240	170
5		450
6		800

of free glyme to total glyme is approximately unity, it is then easy to determine
K photometrically. The K values obtained for lithium and sodium with a series
of glymes appear in Table 2-9.[23] It is important to note that along with this
calorimetric studies of the reactions indicated negligible enthalpy changes.
Thus it is clear that the greater tendency to separate ion pairs with increasing n
values results from a progressively smaller loss in entropy for the process as
compared to solvent-separated ion-pair formation in tetrahydrofuran.
Mathematically the argument takes the following form:

$$\Delta H - T\,\Delta S = \Delta G = -RT\,\ln K$$

ΔH is determined to be virtually zero calorimetrically; therefore, the large
entropy increase obtained when one multidentate glyme displaces many uni-
dentate THF molecules causes a favorable decrease in free energy (ΔG) in the
system, and this in turn increases the equilibrium constant K. Entropy and
chelation effects are similarly important in transition metal coordination
chemistry, and further examples are given in Chapters 6 and 8.

With this background on the crown ethers, we now return to consider their
metal complexation chemistry. It should be no great surprise that charge-size
considerations again predominate among the controlling factors. The stability
constants for a series of crown ethers and alkali ions in methanol are given in
Table 2-10. Note that the complexes are most stable when the size of the cation
is just a little less than the size of the crown, as judged by the largest sphere
which molecular models of the crown can accommodate without being distorted. For maximum stability the ion-size–crown-size ratio for univalent
cations is around 0.9. For the divalent cations, the stability constant is considerably greater than for monovalent ions because of the greater polarizing
power of these ions. Furthermore, the stability constant is more critically
dependent on the ion-size–crown-size ratio as shown by the steeper curves in
Fig. 2.15. The fact that the optimum ratio for divalent ions is 0.8 rather than

TABLE 2-10

Stability Constants for Crown Alkali Complexes[a]

Complex	$\log_{10} K$ in MeOH, 25°C			Inside diameter of crown cavity (Å)
	Na[+]	K[+]	Cs[+]	
[14]-Crown-4	2.2	1.3		1.2–1.5
[15]-Crown-5	3.7	3.6	2.8	1.7–2.2
[18]-Crown-6	4.3	6.1	4.6	2.6–3.2
[21]-Crown-7	2.4	4.3	4.2	3.4–4.3
Tetraglyme	1.5	2.2		
Diameter of ion, Å	1.94	2.26	3.34	

[a] Data from Pedersen and Frensdorff[19] and Christensen et al.[20]

0.9 or 1.0 probably reflects the much larger enthalpy changes for the ligation of divalent ions than for monovalent ions. (Compare ΔH_{hyd} for the Ia and IIa ions, Tables 2-1 and 2-2.) In other words, the divalent ions bind so tightly that they can cause the crown ligand to bend in a little.

The ammonium ion has about the same size as the potassium ion, and ammonium ions form crown complexes that are about as stable as the potassium complexes. Monoalkylammonium ions are only weakly complexed by the crown ethers, and the di-, tri-, and tetraalkylammonium ions do not form

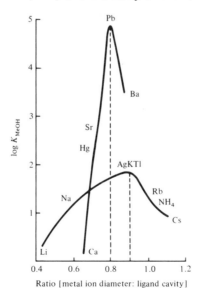

Fig. 2-15. Metal-crown ether complex stability as a function of metal: ligand size ratio.[20] (Copyright 1971 by the Association for the Advancement of Science.)

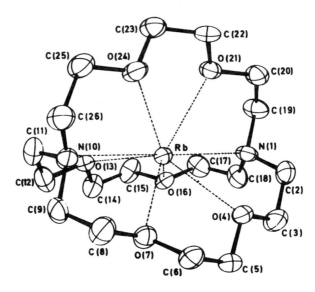

Fig. 2-16. Structure of a rubidium cryptate complex.[24]

TABLE 2-11

Stability Constants of Cryptate Complexes[a]

$CH_2CH_2(OCH_2CH_2)_n$					Ligand	m	n
					I	1	2
$N—CH_2CH_2(OCH_2CH_2)_m—N$					II	2	1
					III	2	2
$CH_2CH_2(OCH_2CH_2)_m$					IV	2	3

Cryptate complexes					$\log_{10}$ stability constant in water, 25°C				

Cation	Li	Na	Ag	Tl	K	Rb	Cs	Ca	Sr	Ba
Ion radius, Å	0.86	1.12	1.26	1.40	1.44	1.58	1.84	1.18	1.32	1.49
Ligand (size)[b]										
I (0.80)	4.30	2.80			[c]	[c]		2.80	[c]	[c]
II (1.15)	2.50	5.40	10.6		3.95	2.55		6.95	7.35	6.30
III (1.40)	[c]	3.90	9.6	6.30	5.40	4.35		4.40	8.00	9.50
IV (1.80)	[c]	[c]			2.2	2.05	2.20	[c]	3.4	6.00

[a] Data from Lehn and Sauvage.[25]

[b] Size measured as radius of a sphere accommodated within a CPK space-filling model of ligand with nitrogens endo-endo.

[c] Value $\leqslant 2.0$.

crown complexes. Their decreasing complexation tendency results from the increasing size of the ions, combined with the decreasing opportunity for hydrogen bonding of the N—H protons to the ether oxygens. In this context it is interesting to note that the activity of many enzymes is enhanced in the presence of potassium ions. Other monovalent alkali ions have relatively little stimulatory effect, but the thallous ion, Tl^+, and the ammonium ion, both of which are nearly the same size as potassium, have nearly the same stimulatory effect on the enzyme. At least for thallium it has been possible in some cases to show that it competes with potassium for the same binding site on the enzyme.

It is important to note that the above discussion of crown ether complexes refers to their behavior in methanol solutions. If similar experiments are run in aqueous solution one finds that much less cation discrimination or selectivity is displayed by the crown ethers. The reason is that solvation enthalpies (ΔH) for the ions are much larger in water than in methanol. Therefore, the crowns compete much better with methanol than with water for the metal cations. Methanol has another more curious effect on the properties of crown complexes. Because the crown ligands are large lipophilic molecules, their metal ion or ion-pair complexes are appreciably soluble in organic solvents like toluene. However, if a very small amount of methanol is present the solubility increases greatly. One explanation for this effect is that the metal ion would like to see a more spherically symmetrical distribution of ligands and the addition of traces of methanol fills this need.

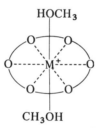

It has been possible to synthesize ligands which in fact have a more nearly spherical array of ligand atoms, and indeed they do form very stable complexes. These ligands are referred to as cryptate ligands or football ligands

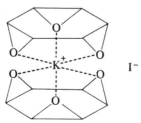

Fig. 2-17. A novel type of "crown ether" complex. bis(*cis,cis*-benzenetrioxide)potassium iodide.[26]

because of their shape. The structure of a rubidium cryptate has been determined by X-ray diffraction and is shown in Fig. 2-16.[24] Note that the configuration of the two apical nitrogens is *endo-endo*, which means they too are bonding to the metal. The stability constants for a series of cryptate complexes are given in Table 2-11.[25] These ligands are powerful complexing agents and even in water they are more cation-selective than the crown ethers are in methanol. Figure 2-17 shows a novel type of "crown ether" complex.[26]

REFERENCES

1. J. A. A. Ketelaar, "Chemical Constitution," 2nd ed. Elsevier, Amsterdam, 1958.
2. T. E. Brackett and E. B. Brackett, *J. Phys. Chem.* **69**, 3611 (1965).
3. H. F. Halliwell and S. C. Nyburg, *Trans. Faraday Soc.* **59**, 1126 (1963).
4. H. S. Frank, *Science* **169**, 635 (1970).
5. D. T. Warner, *Annu. Rep. Med. Chem.* **4**, 256 (1969).
6. T. S. Sarma and J. C. Ahluwalia, *Chem. Soc. Rev.* **2**, 203 (1973).
7. C. Tanford, "The Hydrophobic Effect." Wiley, New York, 1973.
8. A. H. Narten and S. Lindenbaum, *J. Chem. Phys.* **51**, 1108 (1969).
9. W. Kauzmann, *Advan. Protein Chem.* **14**, 1 (1959).
10. H. S. Frank and M. W. Evans, *J. Chem. Phys.* **13**, 507 (1945).
11. R. W. Gurney, "Ionic Processes in Solution." Dover, New York, 1962.
12. F. A. Cotton and G. Wilkinson, "Advanced Inorganic Chemistry." Wiley (Interscience), New York, 1967.
13. M. R. Truter, *Chem. Brit.,* 203 (1971).
14. B. C. Pressman, in "Inorganic Biochemistry" (G. Eichhorn, ed.), Vol. 1, p. 203. Elsevier, Amsterdam, 1973.
15. A. Agtarap, J. W. Chamberlain, M. Pinkerton, and L. Steinrauf, *J. Amer. Chem. Soc.* **89**, 5737 (1967).
16. N. D. Jones, M. O. Chaney, J. W. Chamberlain, R. L. Hamill, and S. Chen, *J. Amer. Chem. Soc.* **95**, 3399 (1973).
17. M. Dobler, J. D. Dunitz, and B. T. Kilbourn, *Helv. Chim. Acta* **52**, 2573 (1969).
18. M. M. Shemyakin, Yu. A. Ovchinnikov, V. T. Ivanov, V. K. Antonov, E. I. Vinogradova, A. M. Shkrob, G. G. Malenkov, A. V. Evstratov, I. A. Laine, E. I. Melnik, and I. D. Ryabova, *J. Membrane Biol.* **1**, 402 (1969).
19. C. J. Pedersen and H. K. Frensdorff, *Angew. Chem., Int. Ed. Engl.* **11**, 16 (1972).
20. J. J. Christensen, J. O. Hill, and R. M. Izatt, *Science* **174**, 459 (1971).
21. D. Bright and M. R. Truter, *Nature (London)* **225**, 176 (1970).
22. R. N. Greene, *Tetrahedron Lett.* p. 1793 (1972).
23. L. L. Chan, K. H. Wang, and J. Smid, *J. Amer. Chem. Soc.* **92**, 1955 (1970).
24. B. Metz, D. Moras, and R. Weiss, *Chem. Commun.* p. 217 (1970).
25. J. M. Lehn and J. P. Sauvage, *Chem. Commun.* p. 440 (1971).
26. R. Schwesinger and H. Prinzbach, *Angew. Chem.* **84**, 990 (1972).

III

Complexes of Ia and IIa Cations in Organic and Biological Chemistry

Complex Cations in Organic Chemistry

Some of the organic reactions and processes involving alkali and alkaline earth cations have already been mentioned in Chapter I. Because of the recent developments in complexation of these ions by new types of ligands, it is reasonable to expect that new modifications of old familiar reactions and reagents will be forthcoming.

SOLUBILIZATION OF INORGANIC REAGENTS

One area for such application is the use of crown ethers for rendering inorganic salts soluble in organic solvents (see reaction 3-1). For example,

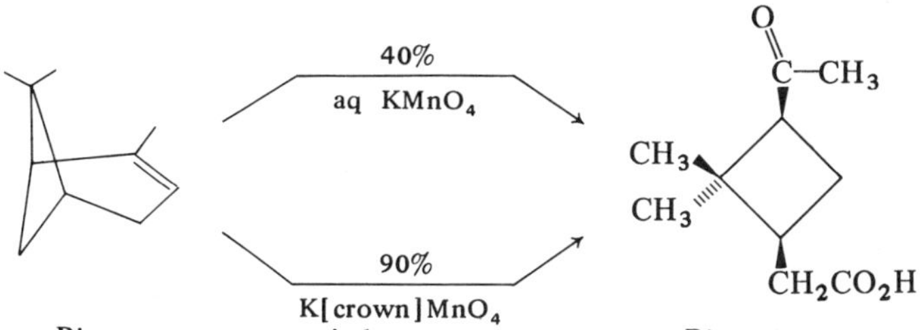

α-Pinene   Pinonic acid   (3-1)

potassium permanganate dissolves in benzene, if dicyclohexyl-[18]-crown-6 is present, to give a purple solution which cleanly oxidizes alcohols, olefins, and hydrocarbons such as toluene. The same reactions can also be run using a two-phase system with aqueous potassium permanganate, but the reactions in benzene are faster and cleaner, as illustrated in the accompanying tabulation.[1]

Substrate	Product	Yield (%)
t-Stilbene	Benzoic acid	100
Cyclohexene	Adipic acid	100
α-Pinene	Pinonic acid	90
Benzyl alcohol	Benzoic acid	100
1-Heptanol	Heptanoic acid	70
p-Xylene	Toluic acid	100

Dibenzo-[18]-crown-6 has been used to solubilize silver nitrate in alcohols in order to carry out the synthesis of glycosides from α-bromo sugars under homogeneous conditions (reaction 3-2).[2] Note that while the binding of the

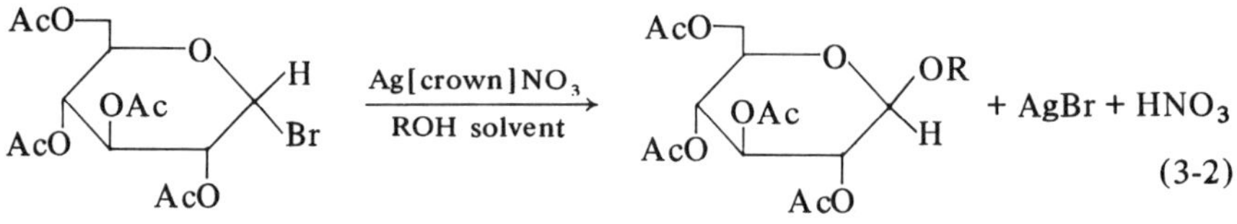

$$R = CH_3, CH(CH_3)_2, C(CH_3)_3, C_6H_{11}$$

silver ion to the crown ligand is strong enough to form a stable soluble complex, it is not as strong as the AgBr lattice. If it were, the driving force for the reaction would have been lost. Since the silver ion is roughly the same size as the potassium ion it is natural to assume that both ions are bound by the crown ligand in the same way. However, it is also possible that the silver ion forms a π-complex with the benzene rings of the ligand. It is because of such metal-π interactions that, for example, silver fluoroborate is readily soluble in toluene. Crown ethers will also render sodium and potassium hydroxides soluble in benzene, and if traces of methanol are present, concentrations up to 1 M may be achieved. These solutions are powerful reagents for the saponification of highly hindered water-insoluble esters such as those of mesitoic acid (reaction 3-3[3]).

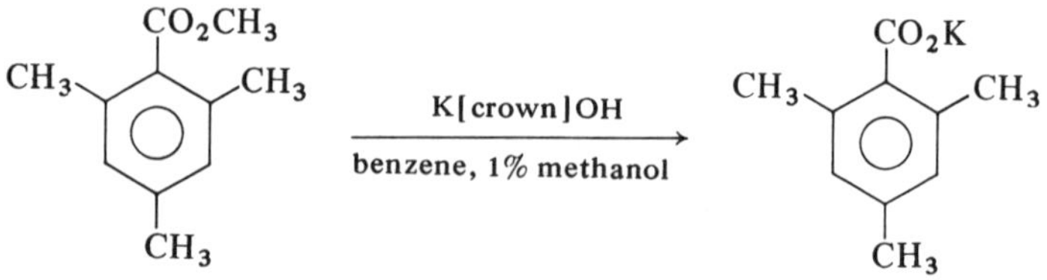

Similar effects can also be achieved in heterogeneous systems by the use of large organic cations, usually quaternary ammonium ions or cationic detergents, which form ion pairs with anionic reagents like permanganate or hydroxide. The ion pairs are somewhat soluble in organic solvents so that small amounts of the quaternary ion can shuttle back and forth transferring anionic reagents into an organic phase and anionic products back to the aqueous phase; hence, the term "phase transfer catalysis."[4] Such processes are thought to be very important in the passage of ionic compounds, both organic and inorganic, through lipoidal cell membranes (see later). The synthetic utility of this method is illustrated by reactions 3-4[5] and 3-5.[6]

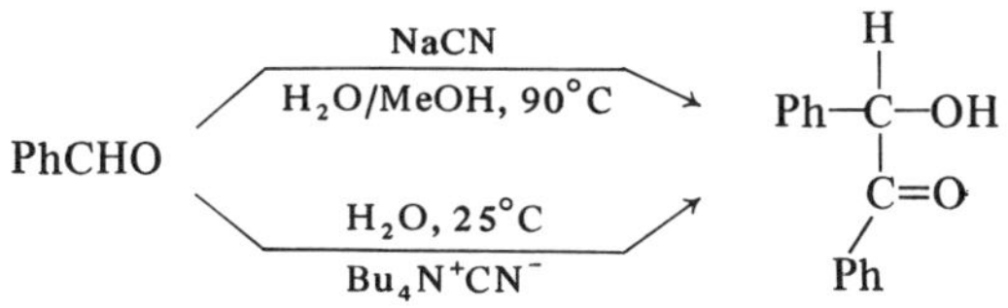

In the benzoin condensation the use of the quaternary ion allows the reaction to proceed quickly and efficiently at room temperature to give a 71% yield in 1 hr. With aqueous NaCN, in the absence of the quaternary ion, no reaction is observed even after 20 hr. Similarly, the permanganate oxidation of olefins proceeds very efficiently in heterogeneous systems using quaternary ions, and the results compare very favorably with those obtained by the use of K[crown]MnO$_4$ in benzene. Many other examples of the uses of phase transfer catalysis and crown ethers may be found in the reviews[4, 7] listed at the end of this chapter as well as those listed at the end of Chapter 2.

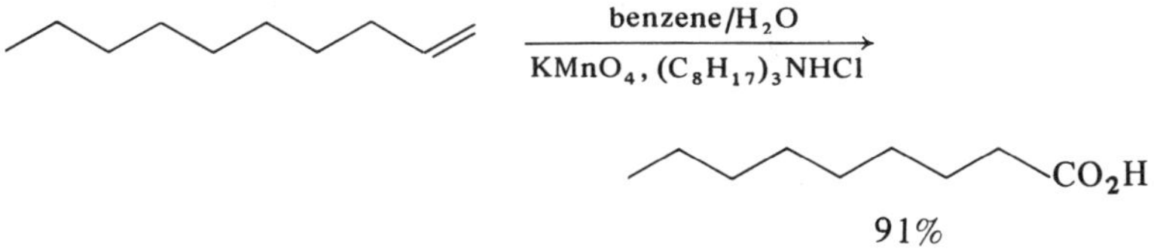

DISSOLVING METAL REDUCTIONS

The alkali metals themselves are very useful in organic chemistry because they dissolve in polar solvents like liquid ammonia to produce solvated metal cations and "solvated electrons" according to the equilibrium

$$M \rightleftharpoons M^+ + e^-$$

$$M^+ + ligands \rightleftharpoons complex$$

Stabilization of M^+ by solvation forces the equilibrium to the right. The electrons produced can be used for various kinds of reductions such as the Birch reduction of aromatic hydrocarbons with sodium in liquid ammonia. Since ether and THF do not complex metal cations as strongly as ammonia, these solvents are generally not useful for Birch or other types of dissolving metal reductions. However, in the presence of the crown ethers both potassium and cesium metals dissolve in THF and ether forming the characteristic blue solution of "solvated electrons," which have been identified by their characteristic electron spin resonance signals.[8] This finding is likely to add a new dimension in versatility and control to the dissolving metal reductions.

CARBONIC REAGENTS

Organolithium reagents serve many roles in organic synthesis, often either as a source of nucleophilic carbanions complementary to the Grignard reagent, or as very strongly basic carbanions for proton abstractions. Their utility is sometimes limited by the tendency of the lithium–carbon bond to have a large amount of covalent character. The lithium ion has such strong electron-attracting properties that the R—M bond does not develop as much ionic character as the analogous R^-, M^+ organosodium or organopotassium compounds. In order to promote reactivity of the carbanion, ligands are sometimes added to complex the lithium and begin to separate it from the carbanion; for example, one ligand commonly used for this purpose is N,N,N',N'-tetramethylethylenediamine (TMED).

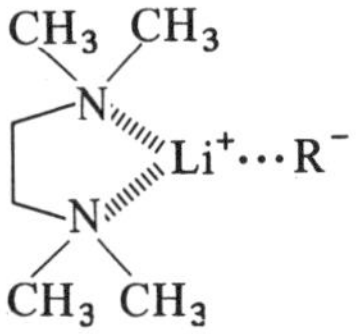

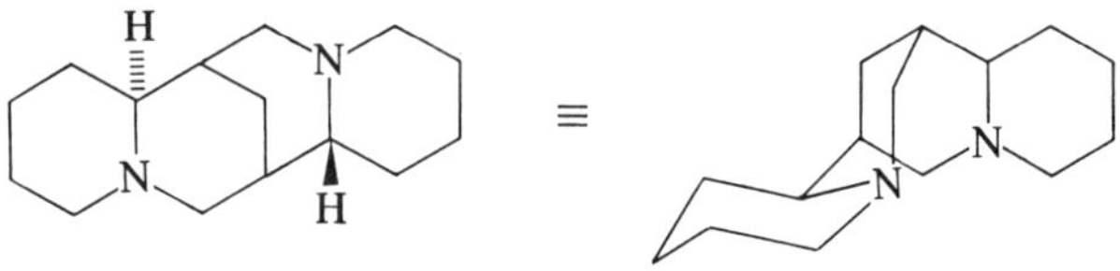

(−)Sparteine

Another related complexing agent, (−)sparteine, is interesting in that it is optically active and its chirality results in asymmetric induction when it is used to modify reactions of Grignard and organolithium reagents (reactions 3-6 and 3-7).[9-11] Similar effects but lower optical yields had been previously reported

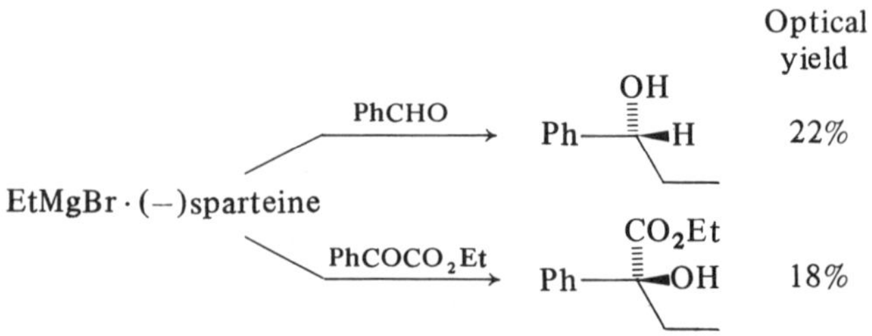

(3-6)

$$PhCH_2CH_3 \xrightarrow[\text{(2) } CO_2]{\text{(1) } (-)\text{sparteine} \cdot \text{BuLi}} Ph-\overset{CO_2H}{\underset{CH_3}{\overset{|}{\text{---}}}}H \quad 30\% \quad (3\text{-}7)$$

for Grignard reactions run in (+)(2R, 3R)-2,3-dimethoxybutane. The effects of crown ethers on the reactions of organolithium and Grignard reagents have not been explored, but there is a large potential for such application, particularly for the chiral crown ethers discussed below.

REACTIONS OF METAL ENOLATES

The chemical reactions of resonance-stabilized anions such as enolates can also be greatly affected by the nature of the metallic counterion. Those metal ions which are strong hard Lewis acids, i.e., lithium, magnesium, zinc, and aluminum, tend to bind at oxygen rather than at the carbon end of the enolate ion because oxygen is a hard Lewis base, whereas carbon anions are soft Lewis base (see Table 6-6). With the hard metal ions the bonding in O-metallated

Soft   Hard

enolates tends to have considerable covalent character, especially in nonpolar solvents like ether and benzene in which contact ion pairing is strongly favored over dissociation. This effect can be seen in the NMR spectrum of the enolate of phenylacetone. The positions of the aromatic protons are shifted to higher field as the cation is varied from Li^+ or Mg^{2+} to Na^+ and as the solvent is varied from ether to THF. The shift in the NMR absorption correlates with the increased amount of electron density on the enolate ion as the enolate-metal bond becomes more ionic and less covalent with the weaker Lewis acid sodium and the more polar solvent THF.

It has been appreciated for some time that the alkylation of enolates can occur at either oxygen or carbon, and that alkylating agents with soft leaving groups like bromide or iodide give primarily C alkylation while those with hard leaving groups like tosylate, sulfate, or chloride give primarily O alkylation.[11a] From this one might anticipate that the use of magnesium or lithium enolates with alkyl iodides should give the best results for C alkylation, while sodium or potassium enolates with alkyl chlorides or tosylates should be best for O alkylation. In one reported case a very large stereochemical effect was observed upon changing from potassium to lithium as the counterion in an alkylation of a nitrile (Scheme 3-1).[12]

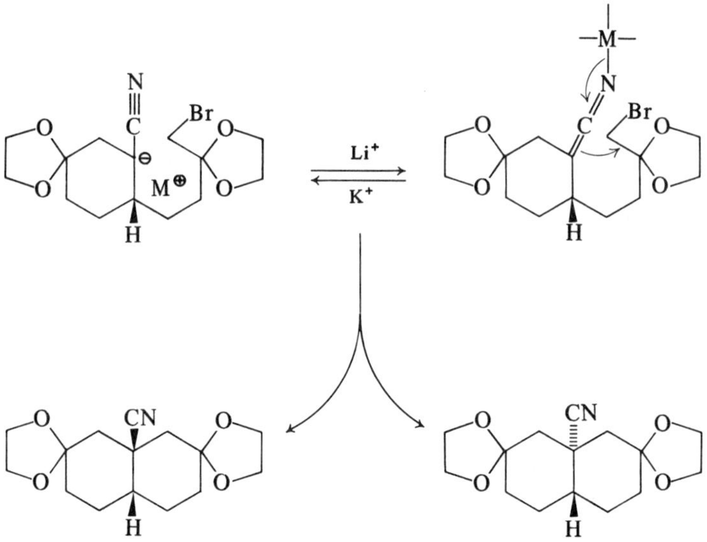

$$M = K^+ \quad 95\% \qquad\qquad 5\%$$
$$M = Li^+ \quad 5\% \qquad\qquad 95\%$$

Scheme 3-1

The factors which cause such a large stereochemical dependence are not immediately obvious, but among them must be considered the greater extent of C=C=N—M covalency with lithium and perhaps the greater size of the partly solvated or aggregated lithium ketimine intermediate.

In the acylation of enolates, the ratio of C:O acylation can be affected significantly by the nature of the metal counterion. Lithium, magnesium, and zinc favor metal-oxygen covalent character and tend to give the best yields of C acylation, while sodium and potassium, which are more highly ionic, favor O acylation. The mercuric ion, which is covalently bound to the enolate carbon, gives almost exclusively O acylation. With thallium(I) salts of β-diketones, either C or O acylation may be cleanly obtained depending on the acylating agent used,[13, 14] as shown by the examples in Table 3-1.[13, 15]

An interesting extension of this reasoning is the recently reported use of MgI_2 for the conversion of alkyl tosylates into alkyl iodides,[16] and the direct

$$R{-}OSO_2C_6H_4CH_3 \ + \ MgI_2 \ \longrightarrow \ R{-}I \ + \ Mg(OSO_2C_6H_4CH_3)_2$$

Soft–hard   Hard–soft   Soft–soft   Hard–hard

high yield conversion of cyclopropylcarbinols to homoallylic halides with ether solutions of anhydrous magnesium halides (see reaction 3-8 and accompanying tabulation).[17]

TABLE 3-1

Effects of Metal Ions on the Acylation of Enolates[a]

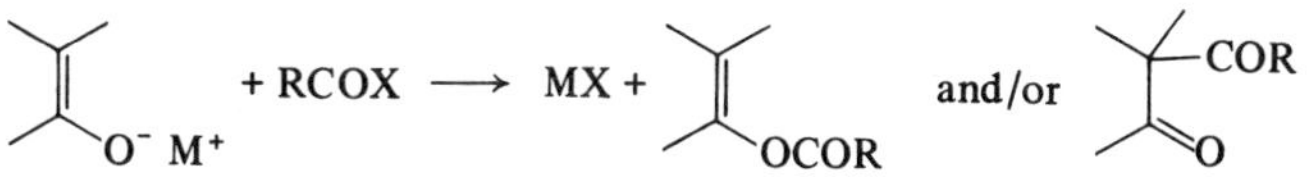

				Yield (%)	
Enolate	Metal ion	Solvent	RCOX	O Acylation	C Acylation
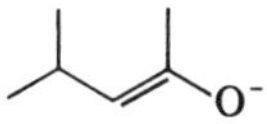	Li	Et$_2$O	Ac$_2$O	68	4
	MgBr	Et$_2$O	Ac$_2$O	37	34
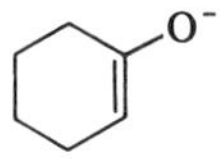	Li	DME	AcOEt	1	40
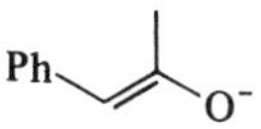	Na	DME	Ac$_2$O	92	1
	Na	Et$_2$O	Ac$_2$O	75	8
	Na	Et$_2$O	AcBr	12	46
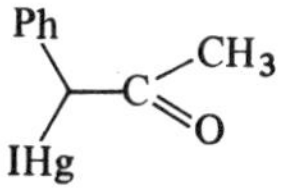		DME	AcCl	100	0
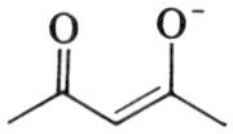	Tl$^+$	Et$_2$O	AcCl	95	
			AcF		90

[a]Data from Taylor et al.[13] and House et al.[15,15a]

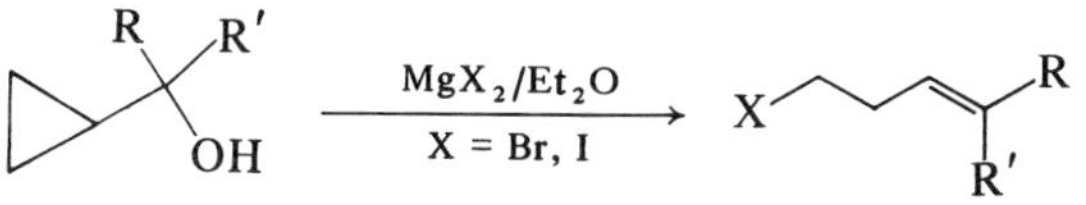

$$(3-8)$$

R	R'	MgX$_2$	Yield (%)	(E:Z ratio)
Me	Me	MgI$_2$, MgBr$_2$	100	—
Me	H	MgI$_2$	95	(96:4)
Me	H	MgBr$_2$	70	(97:3)
Ph	H	MgI$_2$	100	(95:5)
H	H	MgI$_2$, MgBr$_2$	0	—

RESOLUTION OF AMINES AND AMINO ACIDS

Primary alkylammonium ions form reasonably stable complexes with the same crown ethers that bind potassium ion well. By cleverly designing asymmetric crown ethers with the proper cavity size, it has been possible to completely resolve racemic primary amines and α-amino acids. The basis of the ligand design is the restricted rotation about the 1–1' bond in the binaphthol unit, which permits the separation of dissymmetric forms of the ligand. By using the optically pure crown ether (3-9), Cram and co-workers[18] were able to resolve α-phenylethylamine by placing it (as the HPF_6 salt) on top of a column containing $NaPF_6$ solution adsorbed on silica and eluting with a chloroform solution of the optically active crown ether (3-9). The R-isomer of

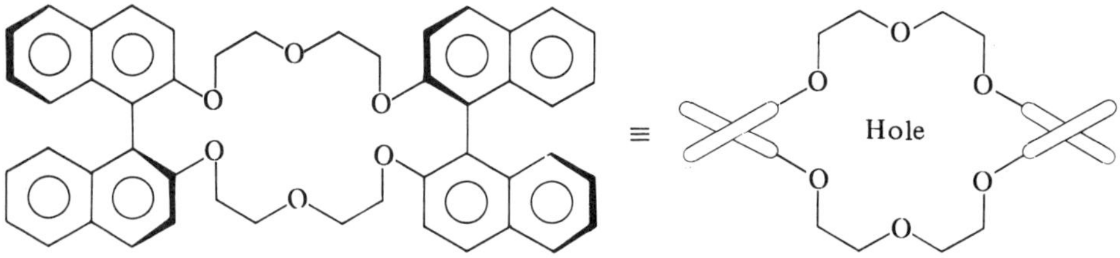

(3-9)

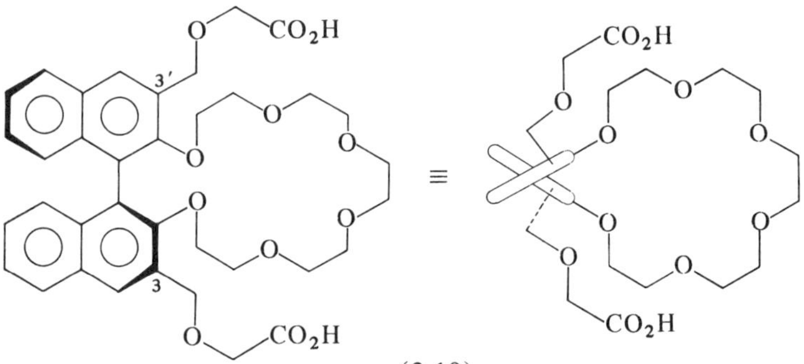

(3-10)

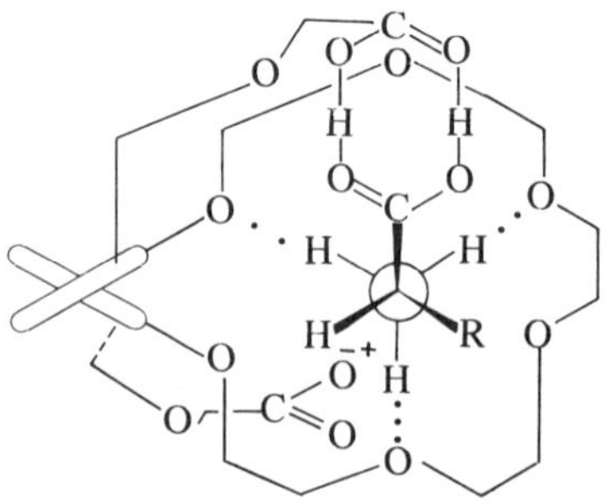

the amine eluted first as the complex with HPF_6 and the crown ether. This experiment involves the same type of partitioning equilibria as that illustrated in Fig. 3-6, except that there are many serial equilibrations occurring along the column. No doubt the large "hydrophobic" PF_6^- anion is important to the success of the equilibration since it can form nice stable ion pairs which readily partition into chloroform.

In a similar way pure S-valine has been used to resolve the crown ether (3-10) into its R- and S-isomers.[19] In this case the carboxylic acid side chains were added to hydrogen bond to the valine carboxyl and ion pair with the valine ammonium group. When a benzene solution of racemic (3-10) was passed through a column of S-valine in acetic acid–water adsorbed on silica, the S-ether was retained preferentially in the aqueous phase and the R-ether eluted first. These experiments provide an interesting basis for a large-scale separation of amino acids on columns containing crown ethers like (3-10) convalently bound to an inert support. They also provide insight into possible mechanisms for enantiomeric selectivity during absorption, transport, or metabolism of amino acids by living systems. These results should be compared to the metal-ion transport models discussed below, and to the optical resolution of amino acids discussed in Chapter 8.

Metal–Macromolecule Interactions in Biology

Biological macromolecules such as proteins, nucleic acids, and carbohydrates are polymers of basic building block units. The most chemically diverse of these are the amino acids, and their derived polypeptides and proteins. Not only the functional groups on amino acid side chains but also the oxygen and nitrogen atoms of the peptide bond are excellent donor groups for forming metal complexes. Here we will consider the effects of the metal on the protein. In other chapters we will discuss the effects of the protein on the properties of the metal and cooperative catalytic effects in metalloenzymes.

PROTEIN STRUCTURE

Metal ions can greatly affect the conformational stability of proteins, and many proteins undergo conformational changes upon complexation with metal ions. Proteins are usually multidentate toward metal ions, and the tighter the complexation of the metal, the greater the stability of the metalloprotein complex. In many cases the complexation is so strong that attempts to remove the metal actually destroy the protein. In other cases the metal can be reversibly removed but the metal-free apoprotein is much more sensitive to denaturation. One example of this, which is even more philosophically intriguing, is provided by the zinc metalloprotease enzymes from bacteria. These enzymes contain zinc at their active site and calcium at several other sites in the molecule. If calcium but not zinc is removed by a strong selective calcium complexing agent, the proteolytic enzyme now acts upon *itself* in solution. It literally digests itself by cleaving peptide bonds until the fragments no longer have any enzymatic activity![20]

Concanavalin A (con-A) is a very interesting metalloprotein for a number of reasons. Concanavalin A is a member of a group of proteins called lectins, which have the property of binding strongly and specifically to carbohydrates and polysaccharides, in the fashion of an antigen–antibody complex. More interesting is the fact that con-A agglutinates red blood cells and binds to glycoproteins on the surfaces of other cells which have been "transformed" by DNA tumor viruses or carcinogenic chemicals. Thus con-A is finding use as an important laboratory tool in immunology, cancer research, and fundamental studies of cell membranes. These special properties of con-A depend critically upon the presence of manganous and calcium ions. Recent X-ray diffraction studies[21] have revealed the structure of the tetrameric aggregate of the metalloprotein, and the crystal structure of one of the monomers (MW 27,000) is shown in Fig. 3-1.

The X-ray data show that the metal ions are not *directly* involved in carbohydrate binding since they are located about 22 Å away from the carbohydrate binding site. Since apo-con-A crystallizes in a different form from MnCa-con-A and does not bind carbohydrates in solution, the binding of metal ions must cause a conformational change in the protein such that a carbohydrate binding site is formed. These effects are called *allosteric* effects, and they can be caused by small organic molecules as well as metal ions. A detailed view of the unique double binding site is shown in Fig. 3-2. Apo-con-A will

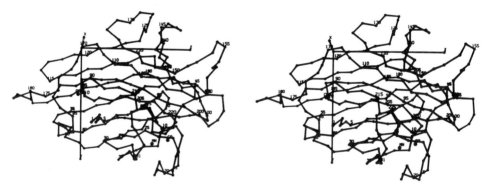

Fig. 3-1. Stereogram of the α-carbon positions of the 231 amino acid residues in the con-A monomer. The positions of the Mn^{2+}, Ca^{2+}, carbohydrate binding site, and the amino and carboxyl termini are labeled MN, CA, CHO, N, and C, respectively.[21] (Reprinted with permission from *Biochemistry*. Copyright by the American Chemistry Society.)

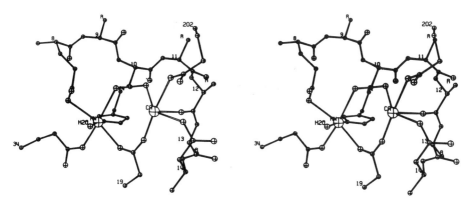

Fig. 3-2. Stereogram of the Mn^{2+} and Ca^{2+} double ion site. The amino acid residues are labeled at the α-carbon position. The Mn^{2+} and Ca^{2+} ions are hexa- and penta-coordinated, respectively, and both carboxyl groups of residues 10 and 19 donate oxygen ligands to both ions.[21] (Reprinted with permission from *Biochemistry*. Copyright by the American Chemistry Society.)

bind neither carbohydrates nor calcium ion unless it has first bound manganous ion, although Ni(II) can substitute for Mn(II). The binding constant for nickel ion is 1.3×10^5, and that for calcium is 3×10^3, which suggests that the binding of the transition metal ion helps form the binding site for the calcium. Note that two carboxyl ligands actually bond to *both* metal ions. In the absence of the metals the negatively charged carboxyl groups would repel each other into a more open configuration. The metal cations neutralize the excess negative charge and bind the groups causing the protein to change shape. Stated another way, some of the energy released by metal binding is used to perform the work of deforming the apoprotein from its preferred solution conformation. Figure 3-2 also highlights the fact that transition metal ions almost always demand certain rather regular coordination geometries, while the Ia and IIa ions are not unhappy with irregular geometries. Another example of this is the observation that transition metal ions do not readily form crown ether complexes, although they do form complexes with tetradentale macrocyclic ligands (see Chapter 6).

ISOMORPHOUS SUBSTITUTION

For a variety of reasons it is often desirable to substitute another metal ion for the one which is normally present in a given complex or metalloprotein. It is known from mineralogy that when cations are substituted for one another in a crystal lattice, their relative sizes are more important than their net charges; hence, the better substitution is *isomorphous* rather than *isoelectric*. We have already seen two examples of the application of this procedure. One was the substitution of Ni(II) for Mn(II) in con-A, the other was in the use of silver ion rather than sodium or potassium for the X-ray structure determination of monensin (Fig. 2-11). In the latter case silver acts as a "heavy atom" point of reference for solving the crystal structure from the X-ray diffraction data. There are many examples of the use of isomorphous substitution, especially in

TABLE 3-2

Isomorphous Exchanges of Probe Ions[a]

Ion	(radius, Å)	Properties
Zn^{2+}	(0.74)	Colorless, diamagnetic
Co^{2+}	(0.78)	Visible spectra, ESR, NMR line broadening
Fe^{3+}	(0.64)	Mössbauer, ESR, visible spectra
Ga^{3+}	(0.62)	Diamagnetic
Ru^{3+}	(0.69)	Visible spectra, redox properties
K^+	(1.33)	Colorless, diamagnetic
Tl^+	(1.40)	Fluorescent ion, NMR line broadening
Ag^+	(1.26)	Useful for X-ray studies
NH_4^+	(1.45)	Stimulates K-sensitive enzymes *in vitro* (also Tl^+)
Na^+	(0.95)	^{23}Na NMR
Li^+	(0.60)	Poor substitute for Na^+
Ca^{2+}	(0.99)	^{43}Ca NMR
Eu^{2+}	(1.12)	Mössbauer, NMR contact shifts
Mn^{2+}	(1.12)	NMR line broadening
Lanthanides	(0.93–1.15)	Visible spectra
UO_2^{2+}	(1.10)	Visible spectra
Mg^{2+}	(0.65)	
Mn^{2+}	(0.80)	NMR line broadening
Ni^{2+}	(0.78)	Visible spectra, ESR, NMR line broadening

A metal ion with probe (spectral, magnetism) properties can often be used to replace a "silent" metal in a protein or a mineral lattice. The silent metals such as Zn^{2+} or K^+ do not give any useful spectroscopic clues as to their molecular environments.

studies involving catalysis by metal complexes or metalloenzymes, and some of these will be mentioned in later chapters. Table 3-2 is intended to illustrate some of the choices which can be made, and some of the properties of the substitute ions which justify their use in studies of binding sites and reaction mechanisms. By substitution of the appropriate ion, its environment may be monitored by physical methods such as spectroscopy so that the ion is a sensitive "probe" of its environment.

Metal Ions and Biological Membranes

The types of membranes which enclose living cells appear to be as varied as the cells themselves, yet there are some generalizations which can be made. One of the most universal features of membranes is that they all contain a large amount of lipid material in addition to proteins. Quite generally small molecules enter and leave cells by dissolving into the membrane lipid, diffusing through it, and dissolving out on the other side, although a few special molecules such as the vitamins and essential amino acids are transported in and out by specific processes coupled to cellular metabolism. Since metal ions are not very soluble in lipid media, various theories have arisen to explain the extremely facile flow of ions through membranes. We will introduce two of these theories in the light of some membrane "models" and the "real" membrane of nerve cells. The nerve cell membrane has been a favorite to study because of the involvement of transmembrane ion fluxes in transmitting nervous impulses.[22] Across the nerve cell membrane there occurs a concentration gradient of sodium and potassium ions which gives rise to a transmembrane electrical potential of about 60–90 mV inside (negative) with respect to the outside of the membrane. If a microelectrode (+) inserted inside the membrane of a living nerve cell and another (−) is grounded outside the membrane, the transmembrane potential can be decreased by the application of an opposite potential from the electrodes. If the perturbation of the normal resting potential is small the cell quickly returns to its normal potential, but if the perturbation raises the potential above a critical threshold, a dramatic series of events takes place. Above the threshold potential the permeability of the membrane to sodium and potassium increases considerably and there is a large flow of sodium into and potassium out of the cell. This further raises the membrane potential, producing a "spike" or "action potential." The propagation of the action potential down the long slender axon of the nerve cell occurs because the change in transmembrane potential in one area causes surrounding areas of membrane to begin to leak ions and reverse polarity. Thus the nervous impulse is really like a wave on the ocean: The water (ions) moves up and down and the wave (nerve impulse) travels horizontally. After the impulse metabolically coupled processes restore the original resting ion distribution by pumping out sodium and pumping in potassium in exchange. The energy which goes into the production of the cation gradients comes from the hydrolysis of ATP by the so-called "sodium pump" or Na/K activated adenosinetriphosphatase (ATPase), but the mechanism of this process is poorly understood.

REAL MEMBRANES AND MODEL MEMBRANES

In order to understand how ions pass through lipid membranes it is helpful to know what kinds of lipids are in the membrane. The major constituents of

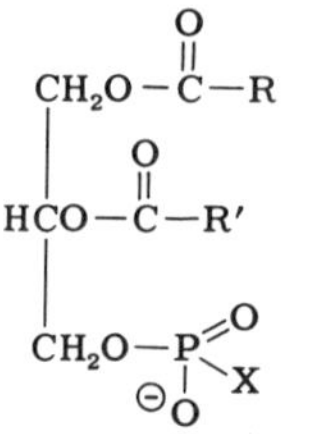

CH$_2$O$-$C$-$R
          ‖
          O

HCO$-$C$-$R'
        ‖
        O

CH$_2$O$-$P$\overset{O}{\underset{\ominus O}{\diagdown}}$X

Phosphatide structure 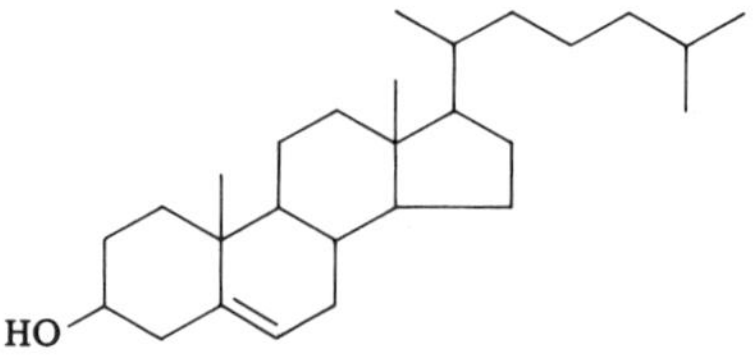   Cholesterol

HO

Figure. 3-3. Lipid constituents of membranes; (a) phosphatide structure and (b) cholesterol. R and R' are C$_{16}$ or C$_{18}$ fatty acids. X $= -$OCH$_2$CH$_2$N(CH$_3$)$_3{}^+$, phosphatidylcholine (lecithin); X $= -$OCH$_2$CHCO$_2{}^-$, phosphatidylserine; X $= -$OCH$_2$CH$_2$NH$_2$, phosphatidylethanolamine.
            |
            NH$_3{}^+$

mammalian cell membranes are cholesterol and phospholipids. Their structures are shown in Fig. 3-3. The phospholipids generally align themselves in double layers with their polar groups exposed at the outer surfaces as shown in Figs. 3-4[23, 24] and 3-5.[25] Since these lipid bilayers can be made in the laboratory, they have come to be used as membrane models. However an even simpler membrane model, if it may be called that, consists of a glass U-tube with a layer of chloroform representing the lipid membrane in the bottom and aqueous layers in the side arms representing the aqueous inside and outside of the cell (Fig. 3-6).[26]

The free energy of an ion as it approaches and enters a lipid phase from an aqueous phase is given in Fig. 3-7. As the ion approaches the membrane its electrostatic field interacts with the membrane's surface charge from phosphates, inorganic cations, amines, and acids present on the polar portion of the lipid constituents. If the membrane has a relatively positive surface it will repel approaching cations, while a negative surface will attract them. Regardless of the surface charge, there is a large increase in free energy when

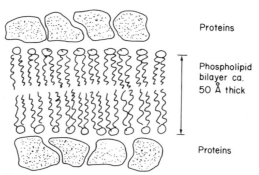

Proteins

Phospholipid bilayer ca. 50 Å thick

Proteins

Fig. 3-4. Davson–Danielli model for biological membranes. The small circles represent the polar ends of the phosphatides; the wavy lines represent fatty acid side chains.[23, 24]

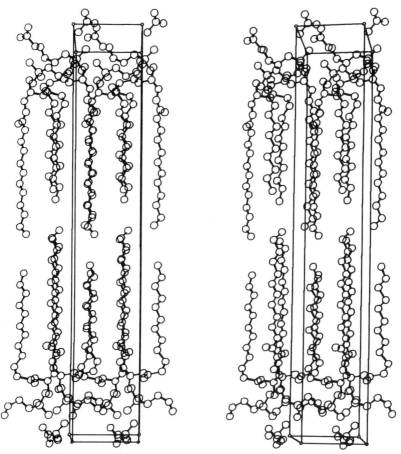

Fig. 3-5. Stereoscopic view of the ordered structure in a bilayer membrane composed of 1,2-dilauroyl-(±)-phosphatidylethanolamine-acetic acid.[25]

the ion leaves the aqueous phase and enters the lipid phase, especially if the hydration layers are lost. Experiments with phosphatidylserine bilayers as model membranes illustrate this nicely. It is found that when magnesium, calcium, strontium, or barium ions are bound to the surface of the bilayers, the conductance of the K^+-nonactin complex across the membrane *decreases* but the conductance of the I_3^- complex ion across the membrane *increases*.[27]

Pain perception depends on conductance of nervous impulses to the brain, but under certain conditions analgesic drugs or pain killers can block the perception of pain. The analgetic potency of a series of salicylates, including aspirin, is directly related to the lipid-water solubility properties of the compound; the more lipid soluble, the better the analgetic action. Recent studies

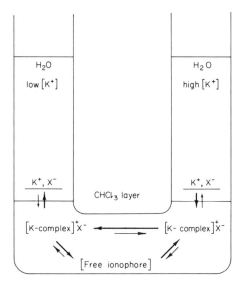

Fig. 3-6. U-tube demonstrating transport of K^+ ions. X^- is an anion such as picrate, and the carrier is an ionophorous antibiotic.[26]

have shown that salicylates increase the K^+ conductance and decrease the Cl^- conductance of nerve membranes,[28] which led to the suggestion that the salicylate ions dissolve into the membrane making its surface more negatively charged, and that this effect may be the basis for the analgesia caused by salicylates. Much more work will be needed to substantiate such a hypothesis.

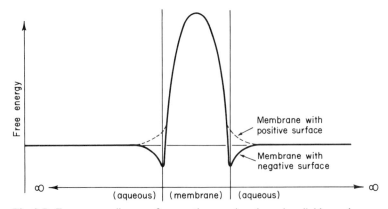

Fig. 3-7. Free-energy diagram for a cation passing through a lipid membrane.

MODELS FOR MEMBRANE CONDUCTANCE

We have considered the structure of membranes and the approach of ions to a membrane surface from an aqueous environment. To explain the facile conductance of ions through membranes, we need to find a mechanism which provides a pathway of lower energy than the one indicated in Fig. 3-7. Two such hypothetical mechanisms have been entertained for quite some time, and although neither one has been verified in a real membrane, both have good support from studies in model membrane systems.

The Pore Model

The pore model for ionic conductance conceives of the cell membrane as being perforated by channels or pores through which ions can pass without actually dissolving in the lipid layers.[29] A laboratory system which fulfills this model consists of two aqueous electrolyte phases separated by a lipid bilayer. The conductance across the membrane, as measured with a very sensitive ammeter, is very small because of the reluctance of ions to dissolve into it and out the other side. When gramicidin A is dissolved into the bilayer, the conductance across the membrane increases very dramatically. The effect of gramicidin A on biological membranes is also to cause them to leak ions. Gramicidin A is a protein of molecular weight 1880 containing both D- and L-amino acids. In solution it adopts a helical conformation with a hollow center about 4 Å in diameter and about 35 Å long. The relative efficiency with which ions pass through this pore in the membrane is $NH_4^+ > Cs^+ > Rb^+ > K^+ > Na > Li^+$, which is approximately the *reverse* of the size of the *hydrated ions* (Table 2-8). The conductance of membranes containing gramicidin A is far too efficient to fit the "carrier model."

The Carrier Model

The carrier model postulates that ions form complexes with carriers at the membranes surface and that the complexes diffuse through the membrane to release the ion at the other side.[29] This shuttlebuslike process is illustrated in Fig. 3-8. Although the hypothetical carrier molecules which function physiologically have never been identified, the toxic ionophorous antibiotics discussed in Chapter 2 certainly perform such a function in a passive way, uncoupled to metabolism, in both real and model membranes. The simplest model experiments are performed in U-tubes as shown in Fig. 3-6. The antibiotic is dissolved in the chloroform layer, and the side arms are filled with solutions of potassium ions. An equilibrium is set up at each phase boundary, and eventually the potassium ion concentrations in the two arms will be equal. Anions

$$K^+_{aq} + \text{ligand}_{CHCl_3} \rightleftharpoons K^+ \text{complex}_{CHCl_3}$$

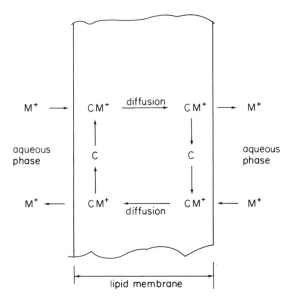

Fig. 3-8. The carrier model for ion transport through lipid membranes: C is the carrier molecule or ionophore, and M^+ the metal cation such as K^+.

are also dragged along and "transported" through the chloroform, and large soft anions such as I^-, $AuCl_4^-$, or picrate are transported most easily. Cation transport occurs down a concentration gradient, from highest to lowest, but since this process is coupled to anion transport, it can be used to *generate* an anion concentration gradient or vice versa. A similar system based on crown ethers which extracts amino acids with enantiomeric selectivity was described earlier.

More sophisticated model systems utilize a lipid bilayer stretched across a hole in a partition which separates two compartments of a plastic tank filled with electrolyte. An electrode is immersed in each compartment to measure the current generated by cations passing through the membrane. When valinomycin is dissolved into the membrane the conductivity increases in direct proportion to the valinomycin concentration. The efficiency of conduction of ions through the membrane is $Li^+ < Na^+ < K^+ < Rb^+ > Cs^+$, which is the same as the order of stability constants for the valinomycin–cation complex. As the potassium ion concentration increases on one side of the membranes, the conductance increases up to a point at which the carrier is saturated with potassium and cannot possibly carry ions any faster. This saturation phenomenon is an important characteristic not only of biological ion transport but also of enzyme systems in general. The transport process with valinomycin in lecithin bilayers occurs with rates approaching 10^4 K^+ ions per carrier per second, which is *faster* than most enzyme reactions.

Much remains to be learned about the mechanisms of ion transport across biological membranes such as those of nerve, heart muscle, or kidney. In these areas alone the potential for control or alleviation of human disease by manipulation of ionic membrane phenomena is truly enormous.

REFERENCES

1. D. J. Sam and H. F. Simmons, *J. Amer. Chem. Soc.* **94**, 4024 (1972).
2. A. Knochel, G. Rudolph, and J. Theim, *Tetrahedron Lett.* p. 551 (1974).
3. C. J. Pedersen, *J. Amer. Chem. Soc.* **89**, 7017 (1967).
4. J. Dockx, *Synthesis* p. 441 (1973).
5. J. Solodar, *Tetrahedron Lett.* p. 287 (1971).
6. C. M. Starks, *J. Amer. Chem. Soc.* **93**, 195 (1971).
7. D. J. Cram and J. M. Cram, *Science* **183**, 803 (1974).
8. J. L. Dye, M. G. DeBacker, and V. A. Nicely, *J. Amer. Chem. Soc.* **92**, 5226 (1970).
9. N. Nozaki, T. Aratani, T. Toraya, and R. Noyori, *Tetrahedron* **27**, 905 (1971).
10. G. Fraenkel, B. Appleman, and J. G. Ray, *J. Amer. Chem. Soc.* **96**, 5113 (1974).
11. A. W. Langer, ed., "Polyamine-Chelated Alkali Metal Compounds," Advan. Chem. Ser. No. 130. Advan. Chem. Ser., Washington, D.C., 1974.
11a. W. J. LeNoble, *Synthesis* p. 1 (1970).
12. G. Stork and R. K. Boeckman, *J. Amer. Chem. Soc.* **95**, 2016 (1973).
13. E. C. Taylor, G. H. Hawke, and A. McKillop, *J. Amer. Chem. Soc.* **90**, 2421 (1968).
14. A. McKillop and E. C. Taylor, *Chem. Brit.* p. 4 (1973).
15. H. O. House, R. A. Auerbach, M. Gall, and N. P. Peet, *J. Org. Chem.* **38**, 574 (1973).
15a. H. O. House, D. S. Crumrine, A. Y. Teranishi, and H. O. Olinstead, *J. Amer. Chem. Soc.* **95**, 3310 (1973).
16. J. Gore, P. Place, and M. L. Roumestant, *J. Chem. Soc., Chem. Commun.* p. 821 (1973).
17. J. P. McCormick and D. L. Barton, *J. Chem. Soc., Chem. Commun.* p. 303 (1975).
18. E. B. Kyba, K. Koga, L. R. Sousa, M. G. Siegel, and D. J. Cram, *J. Amer. Chem. Soc.* **95**, 2692 (1973).
19. R. C. Helgeson, K. Koga, J. M. Timko, and D. J. Cram, *J. Amer. Chem. Soc.* **95**, 3021 (1973).
19a. R. C. Helgeson, J. M. Timko, and D. J. Cram, *J. Amer. Chem. Soc.* **95**, 3023 (1973).
20. J. Feder and L. R. Garrett, *Biochem. Biophys. Res. Commun.* **43**, 943 (1971).
21. K. D. Hardman and C. F. Ainsworth, *Biochemistry* **11**, 4910 (1972).
22. B. Katz, *Sci. Amer.* **205** (9), 209 (1961).
23. J. F. Danielli and H. Davson, *J. Cell. Comp. Physiol.* **5**, 495 (1935).
24. P. B. Hitchcock, R. Mason, K. M. Thomas, and G. G. Shipley, *J. Chem. Soc. Chem. Commun.* p. 539 (1974).
25. P. B. Hitchcock *et al., J. Chem. Soc. Chem. Commun.* p. 539 (1974).
26. M. Pinkerton, L. K. Steinrauf, and P. Dawkins, *Biochem. Biophys. Res. Commun.* **35**, 512 (1969).
27. D. A. Haydon and S. B. Hladky, *Quart. Rev. Biophys.* **5**, 187 (1972).
28. H. Leviton and J. L. Barker, *Science* **178**, 63 (1972).
29. P. Lauger, *Science* **178**, 24 (1972).

IV

Atomic Structure and Structure–Activity Correlations

In the preceding chapters we have seen that most of the chemistry of the alkali and alkaline earth metals could be correlated with predictions based on electrostatic or coulombic interactions between their spherical cations and negative ions or dipoles. Unfortunately, the application of this approach to transition metal chemistry has been only partly successful at best. In order to develop a general basis for the qualitative and quantitative description of bonding interactions between atoms, we begin by first considering the nature of atoms. Our current concept of atomic structure began to evolve during the nineteenth century, but in more recent times the study of atoms has centered around two issues, the nature of the nucleus of the atom, and the arrangement of the electrons surrounding the nucleus. The study of the composition and structure of atomic nuclei is an important endeavor, but for purposes of describing chemical bonding and chemical reactions, it is adequate to characterize the nucleus of an atom simply by its mass and its net charge, or atomic number. As chemists we are more concerned with the layers of electrons around the atom, and the interactions of these electrons, attractive or repulsive, both within an atom and between different atoms.[1]

The Origins of Quantum Theory of Atomic Structure

During the early twentieth century, while the details of atomic structure were being revealed by the experiments of Millikan, Thompson, and Rutherford, others like Bohr and Schroedinger were trying to explain the stability of the atom. Classical physics was clearly inadequate since it predicted that the negative electrons should collapse in toward the positive nucleus and emit light or some other form of electromagnetic radiation.

Niels Bohr

In 1913, Bohr developed the first successful quantum theory for atomic structure. He used ideas generated by the work of Max Planck and Albert Einstein, who earlier had deduced and demonstrated that light energy was "quantized" in units whose energy was proportional to the frequency of the light, independent of the amplitude of the light wave. Bohr reasoned that since light energy is quantized, the energy levels of electrons in atoms must also be quantized, and that transitions of the electron between these energy levels must be accompanied by absorption or emission of light with the proper quantized energy or frequency. These concepts of Bohr's are retained in modern quantum theory, but his postulate of circular trajectories for the electron about the nucleus has been abandoned. The initial virtue of the Bohr theory was that it beautifully explained the lines in the hydrogen atom spark emission spectrum as arising from transitions of the electron between various energy levels. After only 12 years this feature became the ultimate shortcoming of the Bohr theory since it only explained the spectra of one-electron or "hydrogenlike" systems such as H, He^+, and Li^{2+}.

Erwin Schroedinger

The Schroedinger equation, developed in 1926, became the basis of modern quantum mechanics and atomic structure. Schroedinger applied DeBroglie's hypothesis of the dual particle-wave nature of light to the electrons in Bohr's quantized orbitals. Thus, electrons around a nucleus could be regarded as energy waves with a characteristic wavelength and amplitude rather than as "negative particles of matter" in Newtonian trajectories as Bohr postulated. In a three-dimensional form the Schroedinger equation is

$$E\Psi = V\Psi - \frac{h^2}{8\pi^2 m} \nabla^2\Psi$$

In this equation E represents the quantized total energy levels allowed the electron, Ψ is a *wave function* for the wave character of the electron, and V is the potential energy of the electron as a function of its location around the nucleus. The ∇^2 symbol represents a mathematical *operator* which takes the second derivative of the wave function.

In principle the Schroedinger equation is solved to obtain the wave functions *and* the energies E corresponding to stable states for the atom, although it is difficult to solve if the atom contains more than one electron. When solved for the hydrogen atom or for one-electron hydrogenlike atoms, the values of E are found to be related by integers. The wave functions Ψ can be solved for the amplitude of the wave, but they have no real physical interpretation. However, the value of Ψ^2 does have a tangible physical significance. It gives the *probability density* for the electron being in various regions of space around the nucleus. Plots of Ψ^2 on a coordinate system centred on the nucleus pictorially give the shapes of *orbitals*, regions where electron density is likely to be high.

The wave functions Ψ contain several terms, and the coefficients of these terms are called *quantum numbers*. The quantum numbers of an electron can furnish a complete description of the energy and distribution of an electron in space. According to the *Pauli exclusion principle* no two electrons in an atom may have all four quantum numbers identical.

The *principal* or *radial* quantum number n is related to the total energy of an electron in a hydrogenlike orbital. The values of n are the integers 1, 2, 3, 4, ..., n.

The *azimuthal* or *angular momentum* quantum number l gives the number of angular nodes in the wave function and thus is related to the angular kinetic energy of the electron, which is always less than or equal to the total energy. The allowed values of l include 0, 1, 2, ..., $n-1$.

The *magnetic* quantum number m relates to the magnetic field generated by the orbital angular momentum of the electron. The allowed values of m are $-l, -(l-1), \ldots, -1, 0, +1, \ldots, +(l-1), l$.

The *spin* quantum number s describes the magnetic moment of the electron itself, apart from orbital angular magnetic moments. This number is a consequence of the mathematical treatment and does not really represent a particulate electron spinning on its own axis as is sometimes claimed. Such a picture would have the same problems as Bohr's circular orbitals. The allowed values of s are $+\frac{1}{2}$ or $-\frac{1}{2}$.

Quantum Numbers and Atomic Orbitals

The wave functions and quantum numbers obtained from the Schroedinger equation can be put to a number of interesting uses; for example, we can graphically plot the Ψ functions on cylindrical coordinates to obtain the shapes of atomic orbitals, and we can "synthesize" atoms from nuclei and electrons using the "*aufbau*" approach in combination with the Pauli exclusion principle.

Considering the radial distribution of electron density we might ask: "What is the probability of an electron being around the nucleus in a layer between two concentric spheres of radius r and $r + dr$?" The answer, of course, depends on the quantum numbers n and l which appear in the radial part of the

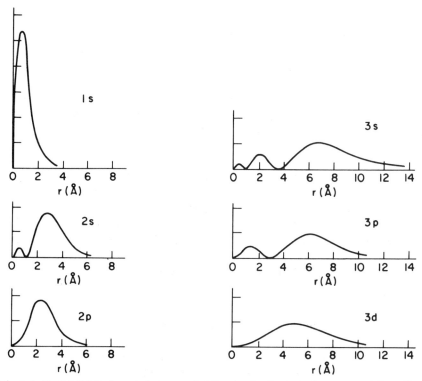

Fig. 4-1. Radial distribution of electron probability density for the hydrogen atom. The ordinate scale is proportional to $4\pi r^2 \Psi^2$.

Ψ functions, and is given by the plots of $4\pi r^2 \Psi^2$ vs r shown in Fig. 4-1. There are several things to notice in these plots. First, the maximum of the $1s$ distribution curve occurs at exactly the Bohr radius of the hydrogen atom. Second, for certain orbitals there are particular values of r at which the electron density is zero. These surfaces, called nodes, are the three-dimensional analog of the nodes in a vibrating violin string; the higher the energy the shorter the wavelength and the greater the number of nodes.

For atomic orbitals the total number of nodes is always $n - 1$, since the principal quantum number is related to the total energy of the electron wave. The $2p$, $3p$, and $3d$ orbitals have less than $n - 1$ radial nodes, but these orbitals have l angular nodes; thus, the number of total nodes is still given by $n - 1$.

$$n - l - 1 = \text{number of radial nodes}$$
$$\underline{l = \text{number of angular nodes}}$$
$$n - 1 = \text{total number of nodes}$$

The third feature to note in Fig. 4-1 is that for a given value of n, the s electrons can approach the nucleus more closely than p electrons, which in turn approach more closely than d electrons. The different abilities of s, p, and d electrons to penetrate toward the nucleus are related to the detailed chemical behavior of elements whose valence electrons are in such orbitals, and consequently to the structure of the periodic table. The quantum number l plays an important part in determining both the energy and geometry of an orbital, and special designations are used to indicate this.

If the angular dependence of the wave function is plotted, the orbitals take the shapes indicated in Fig. 4-2. The s orbital is always spherical, having only radial nodes. Each of the three degenerate p orbitals lies on one of the coordinate axes. (The term "degenerate" refers to the fact that all three have identical energies.) Its dumbbell shape reflects its one angular node, and the arithmetic signs for each lobe derive from the wave equation. The d orbitals have two angular nodes and hence, four lobes. The d_{z^2} orbital looks unusual because it results from a combination of two dependent solutions to the equation ($d_{z^2-y^2}$ and $d_{z^2-x^2}$), but in energy it is equivalent to the other four d orbitals.

Quantum Numbers and Chemical Periodicity

In the *aufbau* approach to the elements (synthesis of atoms with protons and electrons) we must keep the following points in mind:

1. An entering electron always seeks the lowest energy, or in other words, the lowest unoccupied orbital.

2. No two electrons in an atom may have the same values for all four quantum numbers (Pauli principle).

If one then considers electron pairs, with one electron having $s = +\frac{1}{2}$ and the other having $s = -\frac{1}{2}$, a table can be constructed which gives the energies of the orbitals in a *hydrogenlike* atom (see Fig. 4-3 and Table 4-1). Figure 4-4 shows the lifting of orbital degeneracy in a multi-electron atom and Fig. 4-5 gives the variation of orbital energy with atomic number. An expansion Table 4-1 would

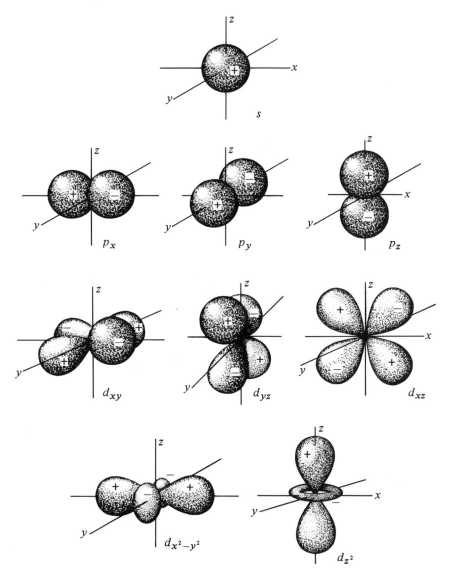

Fig. 4-2. The shapes of the s, p, and d orbitals. The orbitals increase in size considerably as the principal quantum number n increases. This property is better indicated by the radial distribution curves of Fig. 4-1.

show that filling an $n\,s$ orbital with two electrons generates an alkaline earth metal and that filling all the $n\,p$ orbitals generates a rare gas. This periodic repetition of chemical properties as a function of the quantum numbers of

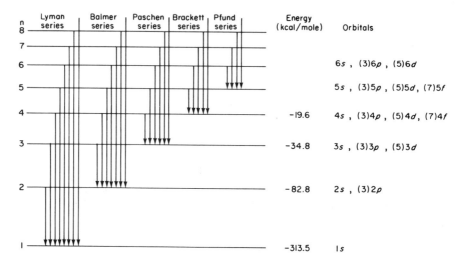

Fig. 4-3. Energies of orbitals for the hydrogen atom. The arrows represent spectral transitions between each energy level.

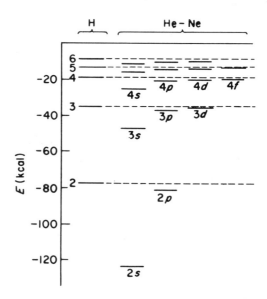

Fig. 4-4. Lifting of orbital degeneracy in a multi-electron atom. This diagram is approximately correct for the "light" atoms helium through neon. For heavier atoms the levels cross as shown in Fig. 4-5.

TABLE 4-1

Quantum Numbers, Orbitals, and Chemical Periodicity

Quantum numbers				Hydrogen like orbitals	Max. No. electrons	Atomic No.	Atom
n	l	m	s				
1	0	0	$\pm\frac{1}{2}$	$1s$	2	2	He
2	0	0	$\pm\frac{1}{2}$	$2s$	2	4	Be
2	1	$-1, 0, 1$	$\pm\frac{1}{2}$	$2p$	6	10	Ne
3	0	0	$\pm\frac{1}{2}$	$3s$	2	12	Mg
3	1	$-1, 0, 1$	$\pm\frac{1}{2}$	$3p$	6	18	Ar
3	2	$-2, -1, 0, 1, 2$	$\pm\frac{1}{2}$	$3d$	10	20	Ca
4	0	0	$\pm\frac{1}{2}$	$4s$	2	30	Zn
4	1	$-1, 0, 1$	$\pm\frac{1}{2}$	$4p$	6	36	Kr
4	2	$-2, -1, 0, 1, 2$	$\pm\frac{1}{2}$	$4d$	10	—	—
4	3	$-3, -2, -1, 0, 1, 2, 3$	$\pm\frac{1}{2}$	$4f$	14	—	—

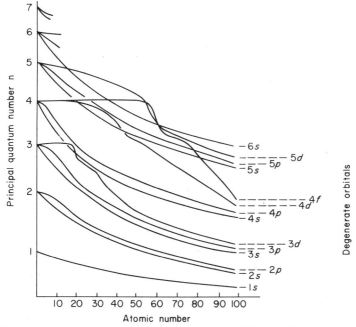

Fig. 4-5. Variation of orbital energy with atomic number. Note that in some cases there are inversions in the *relative* energies of two particular orbitals. Examples include the $4s$ and $3d$ orbitals for atomic numbers 20 to 30 (the first transition series), the $5s$ and $4d$ orbitals for atomic numbers 37 to 48 (the second transition series), and the $5d$, $4f$, and $6s$ orbitals for atomic numbers 55–80 (third transition series and the lanthanides).

atomic orbitals is not a coincidence. It is a *consequence* of the assumptions and conditions utilized in the quantum mechanical *aufbau* approach to atomic structure. In Table 4-1 calcium and zinc need to be inverted to keep the proper sequence, but this is because the table was constructed for the simple case of hydrogenlike atoms. These discrepancies do not mean that quantum mechanics is not a valid description of atomic structure, rather they show that calcium and zinc are not sufficiently hydrogenlike to be approximated by the simplest solutions to the Schroedinger equation. Allowing for this it is reasonable to say that one of the greatest triumphs of the quantum theory of atomic structure is that *the periodic chemical properties of atoms derive naturally from the solutions to the wave equations.* Thus the periodic table strongly reflects the type of orbitals containing valence electrons that are responsible for the chemical properties of the elements. This is emphasized by the block diagram of the periodic table in Fig. 4-6.

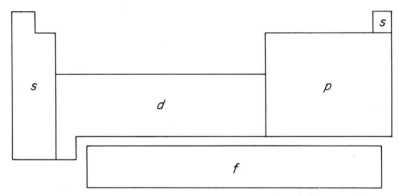

Fig. 4-6. Division of the periodic table into blocks of *s, p, d,* and *f* electrons. Compare this table to the periodic table in the front matter of this book.

Electronic Configurations and Hund's First Rule

We have already discussed the electronic configurations of the *s* block or Ia and IIa atoms. For atoms other than the alkali or alkaline earth metals, several electronic configurations are possible because of the degeneracy of *p, d,* and *f* orbitals. The problem of choosing the configuration of lowest energy is solved by the application of Hund's first rule, which states that for an atom (or a molecule) with electrons in degenerate orbitals, the ground state is the one with the highest multiplicity. Thus, for a gas phase carbon atom the ground state is a triplet, and for the first-row transition metals the same situation arises, as shown in Fig. 4-7. Particularly for the transition metals the chemical properties of both the atom, its complexes, and its ions are related to their electronic con-

(A) Carbon Atoms

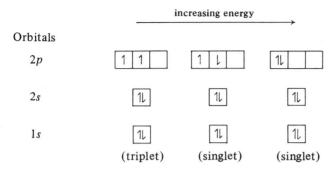

(B) First-row Transition Metal Ions

Atom	Outer electron configuration		Ionization energies			Common oxidation states
	$4s$	$3d$	I_1	I_2	$I_1 + I_2$	
K	1		100	–	–	+1
Ca	1⎸		141	274	415	+2
Sc	1⎸	1	151	297	448	+3
Ti	1⎸	1 1	158	312	470	+4 (3)
V	1⎸	1 1 1	155	338	493	+5 (4, 3, 2)
Cr	1	1 1 1 1 1	156	380	536	+3 (6, 3, 2)
Mn	1⎸	1 1 1 1 1	171	360	531	+2 (7, 6, 4, 3)
Fe	1⎸	1⎸ 1 1 1 1	182	373	555	+3 (2)
Co	1⎸	1⎸ 1⎸ 1 1 1	181	393	574	+2 (3)
Ni	1⎸	1⎸ 1⎸ 1⎸ 1 1	176	418	594	+2
Cu	1	1⎸ 1⎸ 1⎸ 1⎸ 1⎸	178	468	646	+2 (1)
Zn	1⎸	1⎸ 1⎸ 1⎸ 1⎸ 1⎸	216	414	630	+2

Fig. 4-7. Application of Hund's rule to (A) a carbon atom, and (B) first-row transition metal atoms.

figurations. Note in Fig. 4-7 the peculiar irregularities in configuration at V → Cr and Ni → Cu. These result from (1) the very close spacing of $4s$ and $3d$ energy levels around atomic number 25 (Fig. 4-5); (2) the fact that the most stable arrangement for a set of orbitals is either empty, half-filled, or filled; and (3) the fact that the repulsive interaction energy of two electrons in the same orbital is about the same as the energy difference between the $4s$ and $3p$ orbitals (i.e., pairing energy ≈ promotion energy).

Formation and Ligation of Transition Metal Ions and Implications for Organic and Biochemical Systems

The ionization of the first-row transition metals is such that the resultant dipositive gas phase cations have the configurations indicated in Table 4-2. As

TABLE 4-2

Gas Phase Electronic Configurations of First-row Metal Ions[a]

Ion	Configuration	Ion	Configuration
Ca^{2+}	$3d^0$	Mn^{2+}	$3d^5$
Sc^{2+}	$3d^1$	Fe^{2+}	$3d^6$
Ti^{2+}	$3d^2$	Co^{2+}	$3d^7$
V^{2+}	$3d^3$	Ni^{2+}	$3d^8$
Cr^{2+}	$3d^4$	Cu^{2+}	$3d^9$
		Zn^{2+}	$3d^{10}$

[a] All of these ions have the basic argon core $[1s^2\ 2s^2\ 2p^6\ 3s^2\ 3p^6]$ as well.

was the case with the neutral atoms, the configurations of the ions are determined by the relative stabilities of the $3d$ and $4s$ orbitals, but for the ions this is such that the $4s$ orbitals are always of higher energy and thus are always vacant of metal electrons. The ionization energy increases with increasing atomic number under the first row, but the progression is not smooth, as shown in Fig. 4-8. The dip in the curve mainly results from the special stability of the half-filled set of d-orbitals in the $d^5 Mn^{2+}$ ion and from the cumulative repulsive interactions of the paired d electrons in the series from manganese to zinc.

For the alkali and alkaline earth ions, we were able to correlate the ionization energy of an atom with the tendency of its ion to undergo such chemical

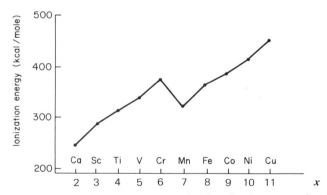

Fig. 4-8. Ionization energies (I_2) for first-row metals. The values are corrected so that they all correspond to the process $3d^{x-1} \rightarrow 3d^{x-2} + e^-$ for $M^+ \rightarrow M^{2+} + e^-$.

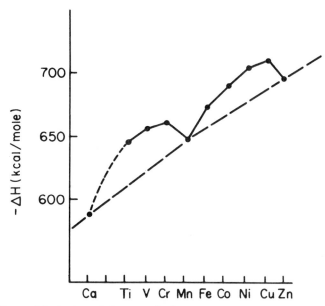

Fig. 4-9. Heats of hydration of first-row divalent cations. Dashed line is the behavior predicted by a simple electrostatic model.

bond-forming processes as hydration, ligation, and lattice formation. Figure 4-9 shows that for the first-row transition metals, the heats of hydration of the divalent cations do not increase *regularly*, as expected from a simple electrostatic model. Only Ca^{2+}, Mn^{2+}, and Zn^{2+} fall on a reasonable line. Later we will see that the discrepancy of observed and expected values depends on the fact that the d orbitals are highly directional and hence only d^0, d^5, and d^{10} transition metal ions approximate the spherical symmetry of the alkali and alkaline earth ions. Similar trends can be seen in the lattice energies and ionic radii for the first-row transition metals (Figs. 4-10 and 4-11). The right-hand portion of Figs. 4-10 and 4-11, which includes the series manganese to zinc, are of particular interest and importance in organic chemistry and biochemistry.

In 1953, Irving and Williams pointed out[2] that for a very broad range of complexes of divalent transition metal complexes the stability of the complexes was almost always in the order $Mn^{2+} < Fe^{2+} < Co^{2+} < Ni^{2+} < Cu^{2+} > Zn^{2+}$. Since then it has been widely recognized that for many nonredox organic reactions catalyzed by metals or metalloenzymes, the order of catalytic effectivity is also given by the Irving–Williams series $Mn^{2+} < Fe^{2+} < Co^{2+} < Ni^{2+} < Cu^{2+} > Zn^{2+}$. Not all of the effects of metals on enzymes or living systems are beneficial. Observations on the toxicity of metal cations to living organisms

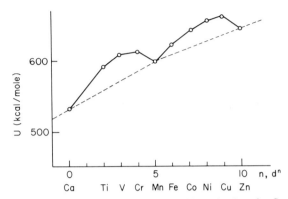

Fig. 4-10. Lattice energies of first-row divalent chlorides. The data for fluorides, bromides, iodides, and oxides all form similar curves with two distinct maxima compared to the gradual increase predicted by a simple electrostatic model (dotted line).

and on the ability of metal cations to inhibit enzymatic reactions date back to the nineteenth century. In 1961, Shaw pointed out[3] that despite the numerous types of ligands present in living systems, and the numerous steps in the chain of events which leads to the death of a cell or an organism, the toxicity of transition metal ions to aquatic organisms nicely parallels the ability of these ions to inhibit several metal-free enzymes, and that *overall, both processes follow the Irving–Williams sequence.* Some of Shaw's data are reproduced in Fig. 4-12. The striking similarity of his enzyme inhibition and toxicity curves to those in Figs. 4-9 through 4-11 is an impressive example of how *a wide variety of complex biochemical and biological processes can be correlated with some very basic chemistry that is well founded on theory.*

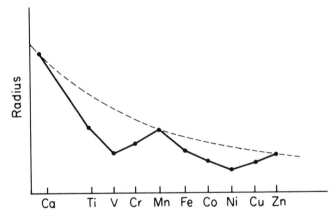

Fig. 4-11. Ionic radii of first-row divalent cations. Sc(II) is nonexistent, and the values for Cr(II) and Cu(II) are estimated.

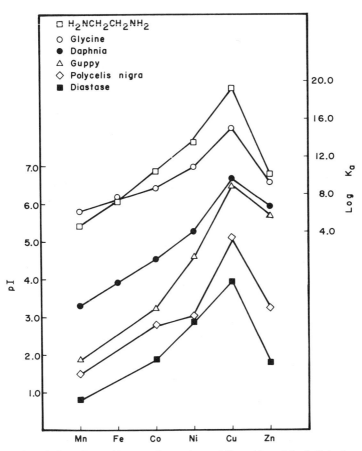

Fig. 4-12. Correlation of transition metal complex stability with toxicity.[3] K_a is the stability constant for the ethylenediammine or glycine chelate of the divalent ion. *pI* is the negative log of the concentration required to kill 50% of the organisms indicated (LD_{50}), or to inhibit the enzyme *diastase* to the 50% level. Similar toxicity curves were observed with tadpoles, stickelbacks, *Paramecia*, and *Fundulus* eggs.

REFERENCES

1. M. C. Day, Jr. and J. Selbin, "Theoretical Inorganic Chemistry," 2nd ed. Van Nostrand-Reinhold, Princeton, New Jersey, 1969; H. B. Gray, "Electrons and Chemical Bonding." Benjamin, New York, 1964.
2. H. Irving and R. J. P. Williams, *J. Chem. Soc., London* p. 3192 (1953).
3. W. H. R. Shaw, *Nature (London)* **192**, 754 (1961).

V

Bonding in Ligands and Metal Complexes

All chemical and biochemical reactions involve the formation and/or disruption of chemical bonds. Therefore, an understanding of the chemical bonding phenomenon itself is prerequisite for the analysis and understanding of dynamic, reacting chemical and biochemical systems. According to Pauling,[1] "a chemical bond exists between two atoms whenever the bonding force between them is of such strength as to lead to an aggregate of sufficient stability to warrant their consideration as an independent molecular species." While it is common practice to use a straight line drawn between the symbols of two atoms to *represent the existence* of a bond and to *indicate the spatial relationship* of the atoms, this representation gives us no clue as to the *nature* of the bond. For this it is necessary to consider the interactions of the participating atoms in terms of their electronic structures.

Concepts in Chemical Bonding[2,3]

THE ELECTRON PAIR BOND

The first theory of chemical bonding to have lasting impact was the valence theory proposed by Lewis[4] in 1916. In essence, Lewis conceived of a chemical bond as arising from the sharing of a pair of electrons by the two bonded atoms. Only the outer or "valence" electrons of an atom were considered to be involved in this process. This idea greatly reinforced the traditional line representation of a chemical bond because the line could now be identified with the physical reality of a pair of electrons in the bond. Moreover, within the framework of this theory, different types of bonding could be envisioned according to the origin of the electrons used to form the bonding pair and their distribution after bond formation. Thus two atoms A· and B·, each having one electron to contribute to a bond, could form a *covalent* bond by symmetrical sharing of the electron pair or an *ionic* bond by unsymmetrical sharing of the electron pair. Alternatively one atom could furnish both electrons and thus form a *coordinate covalent* bond to an electron-deficient atom.

$$A \cdot + B \cdot \qquad A^+ B:^- \qquad \text{Ionic bond}$$
$$\qquad\qquad\quad A-B \qquad \text{Covalent bond}$$
$$A: + B \qquad A \rightarrow B \qquad \text{Coordinate (covalent) bond}$$

The Lewis concept of acids and bases is built upon the notion of electron deficiency and the coordinate bond, and is in turn the basis for interpreting a large part of the bonding in transition metal complexes. According to Lewis,[5] "a base is any species that is capable of donating a pair of electrons to the formation of a chemical bond, and an acid is any species that is capable of accepting a pair of electrons to form a covalent bond." This definition can be used to explain not only the neutralization of proton acids by hydroxide, but the formation of molecular adducts and metal complexes as well. The silver ion

Lewis acid		Lewis base		Adduct
H^+	+	$HO:^-$	$\longrightarrow$	HOH
F_3B	+	$F:^-$	$\longrightarrow$	BF_4^-
Ag^+	+	$2 H_3N:$	$\longrightarrow$	$H_3N:Ag:NH_3^+$

and boron trifluoride are Lewis acids by virtue of their unfilled $5s$ and $2p$ orbitals, respectively.

ELECTRONEGATIVITY

The existence of dipole moments in such "covalent" compounds as CH_3Cl implies the existence of bond polarity, or partial ionic character, as a result of an uneven sharing of the bonding pair of electrons. In fact, most bonds in chemical compounds appear to behave as mixtures of partial ionic and covalent character. To account for this Pauling[6] introduced the term "electronegativity" to reflect "the ability of an atom *in a molecule* to attract electrons to itself." Although this term still defies precise definition, Pauling devised a relative electronegativity scale based on the observation that the bond dissociation energy of a heteronuclear diatomic molecule AB was greater than the average bond energies of the two corresponding homonuclear molecules AA and BB.

$$D_{AB} = \tfrac{1}{2}(D_{AA} + D_{BB}) + \Delta_{AB}$$

The Δ_{AB} term was thus a measure of the *increased* stability of the heteronuclear molecule attributable to the existence of "ionic resonance" contributions to the net bonding scheme, i.e., $A-B \leftrightarrow A^+B^-$. By arbitrarily defining the electronegativity of hydrogen $H = 2.1$, it was possible to evaluate the *relative* electronegativity of many other atoms with the expression $\Delta_{AB} = 23.06\,(\chi_A - \chi_B)^2$ kcal/mole. Other electronegativity scales have been devised, but all show that the most electronegative elements are found in the upper right-hand portion of the periodic table around fluorine, while the least electronegative are found in the lower left corner around cesium.

Since the electronegativity of an atom *in a molecule* is likely to depend on its molecular environment, Hinze and Jaffe[7] introduced the concept of "orbital electronegativity." According to this concept the electronegativity of a σ-molecular orbital is always greater than that of a π-molecular orbital, and the electronegativity of a hybrid atomic orbital is linearly related to the amount of s character present in the hybrid (e.g., sp, sp^2, or sp^3) orbital. As an example of this relationship we may consider the relative acidities of ethane, ethylene, and acetylene (reactions 5-1, 5-2, and 5-3, respectively). Acetylene is the strongest

$$H_3C-CH_3 \; \underset{}{\overset{K_1}{\rightleftharpoons}} \; H^+ + H_3C-CH_2^- \qquad (5\text{-}1)$$

$$H_2C=CH_2 \; \underset{}{\overset{K_2}{\rightleftharpoons}} \; H^+ + H_2C=CH^- \qquad (5\text{-}2)$$

$$HC\equiv CH \; \underset{}{\overset{K_3}{\rightleftharpoons}} \; H^+ + HC\equiv C^- \qquad (5\text{-}3)$$

$$K_1 < K_2 < K_3$$

acid (and the acetylide anion is the weakest conjugate base) because its sp hybrid orbital has more "s character" than either sp^2 or sp^3 orbitals. Consequently, the sp orbital is the more electronegative of the three and is best able to stabilize negative charge (i.e., an electron). As the ionic resonance form of the $C-H$ bond becomes relatively more stabilized, the ionization becomes easier and the molecule becomes a stronger acid. In an analogous manner, the relative base strengths of N-methylpiperidine, pyridine, and acetonitrile may be compared.

$$sp^3 \qquad\qquad\qquad\qquad sp^2 \qquad\qquad\qquad\qquad sp$$

Valence Bond Theory and Covalent Bonding

The valence bond approach to chemical bonding stems directly from the Lewis picture of the covalent bond as an interactive phenomenon which is *localized* between the two participating atoms, and its popularity derives from the fact that it lends itself so easily to graphical representation. Computationally, valence bond theory requires application of the Schroedinger equation to a wave function which is generated from a combination of atomic orbital wave functions suitably chosen to reflect the orbitals used in the bonding process. A general form for such a combination would be

$$\psi_{AB} = a\psi_A(1)\psi_B(2) + b\psi_A(2)\psi_B(1) + c\psi_{A+}\psi_{B-} + d\psi_{A-}\psi_{B+}$$

The first two terms of this equation take into account the fact that the electrons in a bond lose their identity with respect to a particular atom of origin. This is sometimes referred to as resonance, or exchange, and accounts for a major portion of the energy of the bond. The remaining terms take into account bond polarity resulting from electronegativity differences, while the coefficients a, b, c, d weight the terms according to their relative importance.

The phenomenon of exchange resonance has no physical interpretation, and indeed does not necessarily even exist if other approaches to the covalent bond, such as molecular orbital theory, are used. A somewhat different concept of resonance is more readily interpretable, however. This is the resonance which is found among the canonical valence bond isomers of symmetrical species with "*delocalized*" bonding systems such as the carbonate ion. Valence bond theory

can only treat each form separately, although each would have the same calculated energy. However, the observed energy would be lower than that calculated for one of the canonical forms by an amount known as the *resonance energy*. Similar arguments can be made for benzene, whose resonance energy amounts to 35–40 kcal/mole.

HYBRIDIZATION OF ATOMIC ORBITALS

The use of atomic orbitals to construct valence bond wave functions for a polyatomic molecule leads to the expectation of bond angles which reflect the geometry of the bonding orbitals. Thus bonds formed from pure "p" or "d" atomic orbitals would generate bond angles of 90°, and this angle is observed in many cases. To explain the cases where the bond angle is other than 90°, the concept of *hybridization* of atomic orbitals is used. In mathematical terms, a set of n hybrid orbitals is generated by combining the wave functions of a total of n independent atomic orbitals, each of which has a different orientation and possibly a different energy as well. The hybrid atomic orbitals *each* have the *same* energy but a different orientation which reflects the average orientation of the pure atomic orbitals. This is more easily seen graphically in Fig. 5-1, which illustrates the formation of hybridized orbitals such as those of carbon atoms as observed in organic compounds.

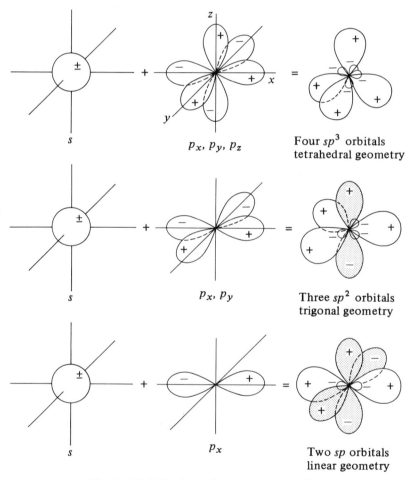

Four sp^3 orbitals
tetrahedral geometry

Three sp^2 orbitals
trigonal geometry

Two sp orbitals
linear geometry

Fig. 5-1. Hybridizations of s and p atomic orbitals.

Hybridization can involve *d* orbitals as well, and this treatment is often used in pictorial or valence bond representations of transition metal complexes. Here the most common geometries are octahedral, square planar, and tetrahedral, in that order (Fig. 5-2). For the idealized symmetrical octahedral complex there are six equivalent metal/ligand bonds directed along the coordinate axes from the central metal. The atomic orbitals which have components on the coordinate axes are the *s*, p_x, p_y, p_z, and the $d_{x^2-y^2}$ and d_{z^2} orbitals. For example, the 4*s*, 4*p*, and 3*d* set are combined to form six d^2sp^3 hybrid orbitals each of which has a major component directed alone one of the six arms of the coordinate axes. Similarly, for a symmetrical complex with four

Coordination number	Orbital hybridization	Geometry	Examples
2	*sp*	Linear	$HC{\equiv}CH$, $[H_3N{-}Ag{-}NH_3]^+$
3	sp^2	Trigonal planar	H_2CO, CH_3^+, BCl_3
4	sp^3	Tetrahedral	CH_4, $Zn(Cl)_4^{2-}$, BCl_4^- $N(CH_3)_4^+$, $Ni(CO)_4$
4	dsp^2	Square planar	$Pt(Cl)_4^{2-}$, $Cu(Cl)_4^{2-}$
5	dsp^3	Trigonal bipyramid	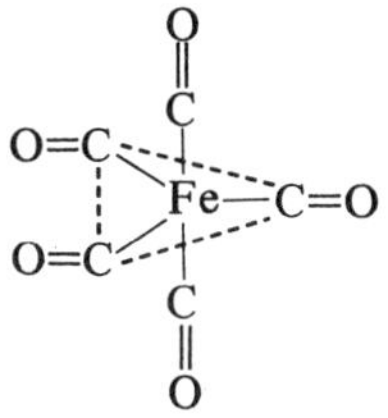 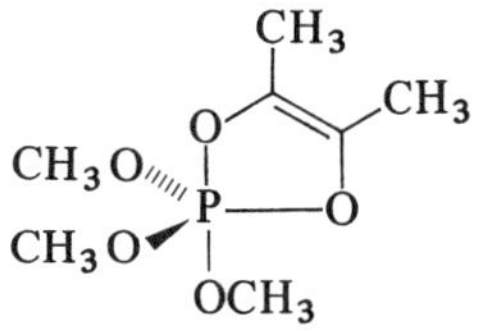
5	d^2sp^2	Square pyramid	Fe in deoxyhemoglobin 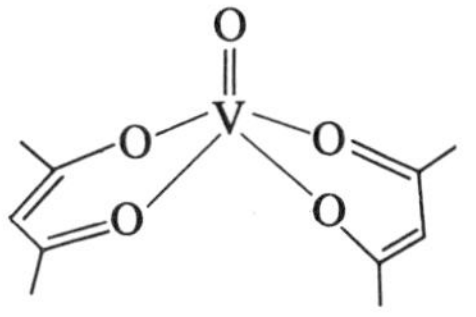 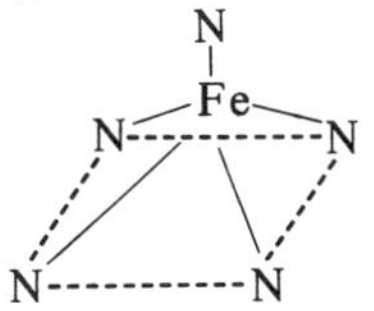 Vanadyl bis(acetylacetonate)
6	d^2sp^3	Octahedral	PF_6^-, $Co(NH_3)_6^{3+}$ Mn^{2+} in concanavalin A (see Fig. 3-2)

Fig. 5-2. Examples of molecules formed using hybridized atomic orbitals.

ligands on the axes in the x–y plane, the $4s$, $4p_x$, $4p_y$, and $3d_{x^2-y^2}$ orbitals are mixed to form four dsp^2 hybrids. These and other more or less common types of hybridized geometries are illustrated in Fig. 5-2.

TRANSITION METAL COMPLEXES

The analysis and description of bonding in transition metal complexes by valence bond theory was developed by Pauling. The transition metal atom or ion was considered to be a Lewis acid by virtue of unfilled s, p, and/or d orbitals. Thus, the complex formed as a result of *coordinate covalent* bonding between Lewis base donors or ligands and the Lewis acid metal center. Through the use of hybridized metal orbitals the coordination number, geometry, and magnetic properties (number of unpaired electrons) of the complex could be explained in many cases. The way in which valence bond theory is used for metal complexes may be seen from the examples given in Diagram 5-1. The orbitals being considered are indicated as boxes which may contain zero, one, or two (paired) electrons. Thus, the free Nickel(II) ion has eight $3d$

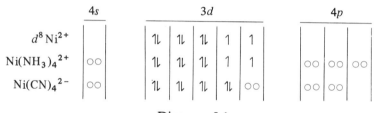

Diagram 5.1

electrons, two of which remain unpaired (Hund's rule). The tetraamminenickel(II) ion utilizes the empty $4s$ and $4p$ orbitals to form a paramagnetic sp^3 hybridized tetrahedral complex, while the tetracyanonickelate ion uses a $3d$ orbital, with a $4s$ and two $4p$ orbitals to form a diamagnetic dsp^2 hybridized square planar complex. The metal d electrons in this case are all paired because cyanide is a very strong ligand and its metal-binding strength is greater than the energy required to force electrons to pair in an orbital.

The Co(III) ion has six $3d$ electrons, four of which are unpaired in the free ion. Because of its greater charge, Co(III) is a very strong Lewis acid and binds most ligands very tightly. Thus in the diamagnetic hexaamminecobalt(III) ion, all the cobalt d electrons are paired leaving six empty orbitals to form six equivalent d^2sp^3 hybrid orbitals using $3d$, $4s$, and $4p$ orbitals. The situation is very similar for the ferric ion and the familiar ferricyanide ion. However, analysis of the corresponding fluoride complexes of Co(III) and Fe(III) reveals one of the shortcomings of valence bond theory. Magnetic measurements show that the octahedral $Fe(F)_6^{3-}$ and $Co(F)_6^{3-}$ ions have five and four unpaired electrons, respectively. Thus with the valence bond approach one is forced to use two empty $4d$ orbitals to form a d^2sp^3 set, but since the $4d$ orbitals are much higher in energy than $3d$, $4s$, or $4p$ orbitals, it is probably not realistic to use them in this way (see Diagram 5-2).

	4s	3d					4p			4d	
$d^6\,Co^{3+}$		↑↓	↑	↑	↑	↑					
$Co(NH_3)_6{}^{3+}$	oo	↑↓	↑↓	↑↓	oo	oo	oo	oo	oo		
$d^5\,Fe^{3+}$		↑	↑	↑	↑	↑					
$Fe(CN)_6{}^{3-}$	oo	↑↓	↑↓	↑	oo	oo	oo	oo	oo		
$Fe(F)_6{}^{3-}$	oo	↑	↑	↑	↑	↑	oo	oo	oo	oo	oo
$Co(F)_6{}^{3-}$	oo	↑↓	↑	↑	↑	↑	oo	oo	oo	oo	oo

Diagram 5.2

The metal–ligand bonding described above can be thought of as essentially σ-type covalent or coordinate covalent bonding. It is also possible to involve the metal d orbitals in π-type ligand bonding interactions if suitable p or d ligand orbitals are available. For example, a ligand having a filled σ-donor orbital and an empty π^*-acceptor orbital (see later) could form a very strong bond to a metal having an empty σ-type acceptor orbital and a filled π-type donor d orbital. A case in point is the ferricyanide ion, in which the excessive

$$[(CN)_5Fe\!-\!C\!\equiv\!N]^{3-} \quad\longleftrightarrow\quad [(CN)_5Fe\!=\!C\!=\!N]^{3-}$$

accumulation of negative charge is drained away from the central metal and distributed over the cyanide ligands through their empty π^* orbitals. For simplicity this is shown for only one of the six cyanide ligands in Fig. 5-3. Other types of π bonding with d orbitals are illustrated in Fig. 5-4.

Despite the advantages of valence bond theory lending itself to easy pictorial representation of bonding and predicting (in some cases) the stereochemistry, magnetic properties, and coordination numbers for various complexes of transition metal ions, it has some important shortcomings. For example, it gives no clue as to the relative energies of various geometrical isomers of a given complex. More important, it provides no way to explain or predict the spectral properties of various complexes. It is largely for these reasons that valence bond theory has been replaced by other more sophisticated theories for treating transition metal complexes.

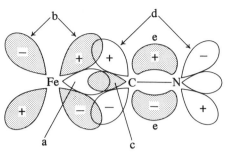

Fig. 5-3. Generalized valence bond picture of M → L $p\pi$–$d\pi$ back-bonding in a Fe(CN) unit of ferricyanide: a, one of six empty d^2sp^3 hybrids (σ acceptor); b, filled d_{xz} orbital (π donor); c, filled sp hybrid on CN (σ donor); d, empty π_{xz}^* orbital on CN (π acceptor); e, filled π_{xz} orbital on CN (bonding only on ligand). Analogous M → L $p\pi$–$d\pi$ bonding occurs between electron-rich metal centers, usually later transition elements in low oxidation states, and ligands with empty π-antibonding orbitals. Representative ligands include CO, NO_2^-, NO^-, RNC (isocyanides), CH_2 (carbene), and unsaturated organic ligands.

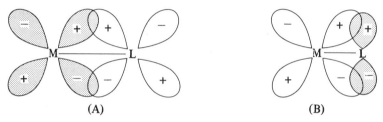

(A)   (B)

Fig. 5-4. Generalized valence bond pictures of (A) $d\pi$–$d\pi$ (M → L) back-bonding and (B) $p\pi$–$d\pi$ (L → M) donation. (A) M = metal with high electron density, usually later transition elements in lower oxidation states or with many anionic ligands. L = donor atom with low-lying empty d orbitals, usually an organic sulfide R_2S, phosphine R_3P, or arsine R_3As, possibly a mercaptide RS^-. (B) M = metal with few d electrons, usually earlier transition elements in high oxidation states such as Cr(VI), V(V), or W(VI). L = first row nonmetal donor such as F^-, alkoxides and β-diketonates, H_2N^-, N_3^-, O^{2-} as in VO^{2+} or the linear $[Fe-O-Fe]^{4+}$ system.

Crystal Field Theory

The application of a simple electrostatic bonding model to transition metal complexes does not lead to very satisfactory results. It consistently *under*estimates the stability of ionic complexes, it predicts that complexes should have *regular* geometries (e.g., tetrahedral instead of square planar), and it is useless for complexes with neutral ligands such as ammonia. As an alternative, crystal field theory (CFT) became popular with chemists in the 1950's for treating the magnetism and spectral properties of transition metal complexes, although it had been used by physicists since about 1920. Crystal field theory supposes metal–ligand bonding of an ion–ion or ion–dipole type, but departs from simple electrostatic models in considering the effects of the *intensity* and *symmetry* of the electrostatic field of the ligands on the energies of metal *d* orbitals. In CFT, electrons in *d* orbitals are regarded as *nonbonding*, which precludes consideration of their involvement in π bonding. Calculations of energies of complexes using CFT still require use of the Schroedinger equation but the assumption of an external electrostatic field resulting from the ligands introduces a complicating factor. Therefore, as we have done previously, we shall omit the details of the calculations and emphasize the qualitative results, their implications, and their applications.

CRYSTAL FIELD SPLITTINGS OF *d* ORBITAL ENERGY LEVELS

The directionality of the *d* orbitals with respect to Cartesian coordinates is shown in Fig. 4-2. Under the influence of a spherical external electrostatic field the energy of the five *d* orbitals would increase, but they would retain their degeneracy. However, if the external electrostatic field resulted from six equal point charges or dipoles equidistant from the nucleus and located on the coordinate axes, those orbitals lying off the axes would be destabilized less than those lying on the axes, as shown in Fig. 5-5. In this case the individual orbitals are either more stable or less stable than their *average* energy. If their energy difference is called Δ or 10Dq, electrons in the d_{xy}, d_{xz} or d_{yz}, orbitals will each experience a stabilization of -0.4Δ or -4Dq relative to the average, while electrons in the $d_{x^2-y^2}$ and d_z orbitals will each be destabilized or raised

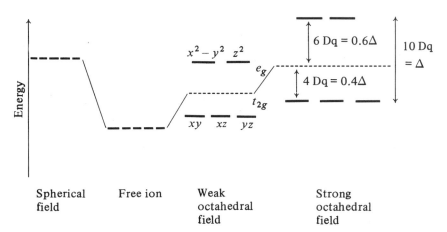

Fig. 5-5. Effects of external electrostatic fields on *d* orbital energy levels. Dotted lines give the *average* energy in a given field.

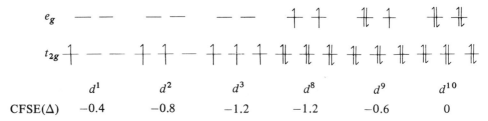

	d^1	d^2	d^3	d^8	d^9	d^{10}
CFSE(Δ)	-0.4	-0.8	-1.2	-1.2	-0.6	0

Fig. 5-6. d Electron configurations which are independent of octahedral field strength and Δ.

in energy by $+0.6\Delta$ or $+6Dq$ from the average. The formerly degenerate d orbitals are thus divided into a lower energy set called t_{2g} (from group theory) and a higher energy set called e_g. The cumulative effect of preferential filling of the lower t_{2g} orbitals is known as the *crystal field stabilization energy* (CFSE) of a complex, but the presence of electrons in the higher energy e_g orbitals tends to cancel this effect, as shown in Figs. 5-6, 5-7, and 5-8.

The energy difference Δ is a function of the *strength* of the external electrostatic field, but it also depends on the nature of the ligands and the type of bonding to the metal. Dipolar ligands such as NH_3 produce stronger fields and larger splittings than diffuse nondipolar ligands such as the halides. Ligands capable of π-bonding interactions with the metal (see Fig. 5-3) form strong bonds and produce large splittings. Note, however, that CFT does not acknowledge π bonding; one of its basic postulates is that d electrons are nonbonding!

The magnitude of Δ also depends on the oxidation state of the metal. Thus, for divalent first-row transition ions Δ usually varies from about 22 to 34

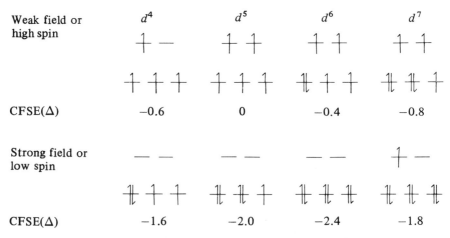

Weak field or high spin	d^4	d^5	d^6	d^7
CFSE(Δ)	-0.6	0	-0.4	-0.8
Strong field or low spin				
CFSE(Δ)	-1.6	-2.0	-2.4	-1.8

Fig. 5-7. d Electron configurations which depend on octahedral field strength.

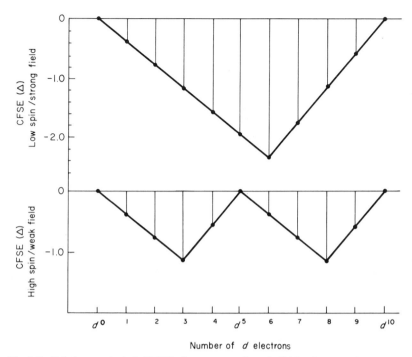

Fig. 5-8. Relative octahedral CFSE's for strong and weak fields. Compare lower curve to Figs. 4-9 through 4-12.

kcal/mole (7500–12000 cm^{-1}) while for the trivalent ions Δ varies from 40 to 72 kcal/mole (14,000–25,000 cm^{-1}). Such generalizations are less applicable to second- and third-row transition metals, but it is generally true that bonding strength increases down a triad (e.g., Fe, Ru, and Os). The magnitude of Δ is also dependent on the symmetry and geometry of the ligand electrostatic field; for example, Δ is less for tetrahedral than for octahedral complexes.

The lifting of the degeneracy of the d orbitals has some important consequences for the electronic configuration of the metal center. For a transition metal ion with 1, 2, 3, 8, 9, or 10 d electrons there is a single preferred d electron configuration regardless of whether there is a weak or a strong octahedral field present. These configurations are illustrated in Fig. 5-6. However, for an ion with 4, 5, 6, or 7 d electrons, two configurations are possible, depending on the strength of the field, or more precisely, on the magnitude of Δ (Fig. 5-7). For weak fields the energy Δ required to promote an electron from the t_{2g} to the e_g level is less than the energy required to pair two electrons in one orbital. The latter value is about 51–57 kcal/mole. Although this is greater than the strength of some chemical bonds, it is considerably less than, e.g., the heat of hydration of a divalent transition metal ion. Since weak field complexes have fewer paired electrons than do the corresponding strong field complexes, the former are often referred to as high spin or spin free complexes and the latter as low spin or spin paired complexes.

In the last section of Chapter 4, Figs. 4-9 to 4-12 were presented to introduce a consistently recurring trend in the properties of first-row transition metal ions. All of these plots show the same type of curvature or deviation from the behavior predicted by a simple electrostatic model. It is more than a coincidence that a plot of CFSE calculated for the high spin ions displays the same bimodal character as those mentioned above (Fig. 5-8). *The fact that the same qualitative patterns are found in systems varying from the crystalline state to enzymes to biological organisms is indicative of the fundamental importance of crystal field effects in determining the chemical properties of a metal ion.* The occasional variations in the relative positions of nickel, copper, and zinc arise from the fact that copper and zinc are seldom octahedral and therefore do not always fit correlations with octahedral CFSE's. Zinc ions tend to be four-coordinate and tetrahedral, although they can form octahedral complexes. Copper, on the other hand, often tends to form square planar four-coordinate complexes.

THE JAHN–TELLER EFFECT

Even when copper is six-coordinate it seldom has regular octahedral geometry. Instead it has a distorted octahedral geometry in which four ligands

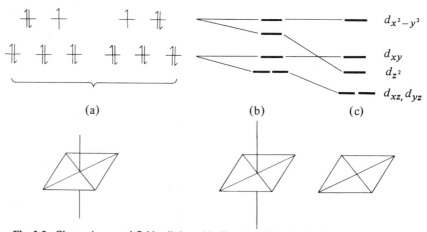

Fig. 5-9. Change in crystal field splitting with distortion of octahedral to square planar fields. (a) Degenerate ground states for a d^9 Cu(II) ion. The unpaired electron may be in either the $d_{x^2-y^2}$ or the d_{z^2} orbital. (b) Elongation of two trans ligands by tetragonal distortion lowers the energy of orbitals with components on that axis. (c) Extreme elongation leaves only a square planar ligand field.

lie in a square planar arrangement while the other two lie at a greater distance on the axis perpendicular to the plane. This is predicted by the Jahn–Teller theorem which states that any nonlinear molecule with a degenerate electronic state will tend to undergo a distortion to lower its symmetry and relieve the degeneracy of its electronic state. Figure 5-9 shows the degenerate electronic ground states of the Cu(II) ion in an octahedral field and the effect of elongating the metal–ligand distance on the z axis only. This results in a lessening of the interaction of the field with orbitals containing a z component relative to those that do not. If this process is carried to the extreme two trans ligands are totally removed and only a square planar complex remains. By analogous reasoning one obtains the qualitative crystal field splittings for square pyramidal and tetrahedral geometry shown in Fig. 5-10. The tetrahedral case is opposite to the octahedral case since the greatest intensity of a tetrahedral field is located off rather than on the coordinate axes. Hence the e_g set ($d_{x^2-y^2}$ and d_{z^2}) is lower in energy than the t_{2g} set (d_{xy}, d_{xz}, d_{yz}).

One of the greatest advantages of crystal field theory over valence bond theory or simple ionic models of bonding in transition metal complexes is that it can explain some features of their absorption spectra. In fact, the simplest way to estimate Δ for a given complex is from its absorption spectrum, where Δ is the energy of the photon absorbed when an electron is promoted from a lower to a higher energy state (i.e., a d–d transition). For a given metal center and geometry it is possible to arrange various ligands in order of their ability to cause a large crystal field splitting in a complex. For most transition metal ions in their lower common oxidation states this series, known as the spectrochemical series, is

$$CN^- \sim CO \sim NO > NO_2^- > bipy > en > NH_3 > SCN^- > pyridine$$

$$C_2O_4^{2-} \sim H_2O > HO^- > F^- > RS^- > Cl^- > Br^- > I^-$$

where bipy stands for 2,2'-bipyridyl and en for ethylenediamine. It is interesting that most of the strongest ligands (those causing the greatest splittings) are unsaturated ligands capable of π-bonding interactions with the metal. Yet CFT is modeled on ionic and ion-dipole bonding and assumes that the d orbitals are nonbonding. This clearly is one of the shortcomings of simple CFT. Others are that charge transfer spectral transitions cannot be explained and that no covalent character in the bonding can be accommodated.

Molecular Orbital Theory

In contrast to the *localized* bonding pairs pictured by VBT, molecular orbital theory (MOT) assumes that the electrons in a molecule experience the

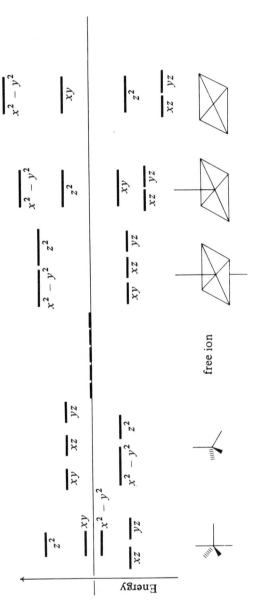

Fig. 5-10. Crystal field splittings for various geometries.

delocalized field of *all* the nuclei in the molecule. Molecular orbital theory further assumes that at any given instant an electron will be relatively close to one nucleus and distant from the others and that it will then behave as if it were in an atomic orbital localized on that particular atom. In turn this leads to the idea that a molecular orbital wave function ψ_{MO} could be formed by taking a *linear combination of atomic orbitals* (LCAO). In this method one forms a number of molecular orbitals equal to the number of atomic orbitals used to start with. The molecular orbitals are formed in pairs consisting of a *symmetric* and an *antisymmetric* MO which are called *bonding* and *antibonding* MO's, respectively. The molecule is then constructed by placing electron pairs in MO's according to Hund's rule and the Pauli principle, and filling first the MO's of lowest energy, the bonding MO's. Thus, just as in the *aufbau* approach to atoms using atomic orbitals, MOT is an *aufbau* approach to molecules using polycentric molecular orbitals constructed from atomic orbitals.

MOLECULAR ORBITAL ENERGY LEVEL DIAGRAMS

Rather than develop MOT from a mathematical base, we will take a more practical approach and illustrate how to construct and interpret MO energy level diagrams. In this process the atomic orbitals to be used are plotted on a vertical orbital energy scale at opposite sides of the diagram to represent the energy states of the separated atoms or molecular fragments. In the center, the MO's which result from the "mixing" of atomic orbitals are plotted on the same energy scale. Figure 5-11 shows a MO diagram for a diatomic molecule

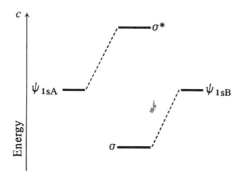

Fig. 5-11. Molecular orbital diagram for hydrogen or helium dimers. Interaction of two approaching degenerate 1s atomic orbitals splits their energies producing a bonding σMO and an antibonding σ^*MO. Placing electrons in orbitals gives four different possible molecular species.

Species	H_2^+	H_2	He_2^+	He_2
No. of electrons	1	2	3	4
Configuration	σ^1	σ^2	$\sigma^2\sigma^{*1}$	$\sigma^2\sigma^{*2}$
Bond order	$\tfrac{1}{2}$	1	$\tfrac{1}{2}$	0

which uses one AO from each atom to form the bond. If we specify that the two atoms be identical, their AO energy levels are identical or degenerate. This condition allows for maximum efficiency of interaction of the two AO's and is reflected by the large difference or splitting between the mean of the AO energies and the MO energies.

The diagram in Fig. 5-11 is appropriate for considering the two simplest homonuclear diatomic molecules, H_2^+ and H_2. The hydrogen ion molecule has only one electron and it would necessarily reside in the lowest energy or bonding MO. The hydrogen molecule has two electrons, and both of them can occupy the bonding MO if their spins are paired (Pauli principle). Using the diagram of Fig. 5-11, placed on a different energy scale, we can also consider

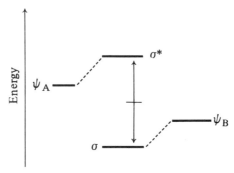

Fig. 5.12. Typical MO diagram for a heteronuclear diatomic molecule A—B; for example, this diagram could suffice qualitatively for A—B = He—H or for A—B = H—Br.

the bonding in the molecules He_2^+ and He_2. The He_2^+ molecule has three electrons, and therefore one of them must occupy an antibonding orbital. Electrons in antibonding orbitals tend to cancel the effects of electrons in bonding orbitals; thus, the bond in He_2^+ is not particularly strong. We can also do the same for the neutral three-electron system HeH, which is known to exist in the vapor phase. However, this is necessarily an approximation since the He $1s$ orbital and the H $1s$ orbital are not of equal energies. To accommodate a heteronuclear molecule a slightly different energy diagram is constructed, as in Fig. 5-12. The He_2 molecule (Fig. 5-11) would have equal numbers of bonding and antibonding electrons (2 + 2), or in other words zero net bonding, and this explains why the rare gases are monatomic. At this point we can introduce the concept of *bond order*. We say that the single bond of the hydrogen molecule has a bond order of 1. Therefore the bond orders for H_2^+, He_2^+ or HeH, and He_2 may readily be seen to be 0.5, 0.5, and 0, respectively. As the order of a bond between two atoms increases, the bond becomes shorter and stronger. Bond shortening in a series of compounds can be observed by precision X-ray crystallographic studies, and the strengthening can be observed as increases in the force constant of the bond as measured by infrared spectroscopy. Examples of this can be found in Chapters 10 and 11.

The molecular orbital *aufbau* approach can be extended to include homonuclear diatomic molecules formed from larger atoms up to fluorine, as shown by the MO energy level diagrams of Fig. 5-15. When, as in this diagram, both *s* and *p* orbitals are being used, it is necessary to consider the various ways that these types of orbitals can combine, or in mathematical terms, the ways in which their wave functions *overlap*. Overlap is easily represented pictorially using plots of the angular portions of the wave equations; for example, two *s* orbitals can combine in a symmetric or anti-symmetric (bonding or antibonding) fashion. With *p* orbitals several types of overlaps are possible, and overlap of *s* and *p* orbitals can occur in two ways. However, only one of these will produce a bonding interaction (Fig. 5-13).

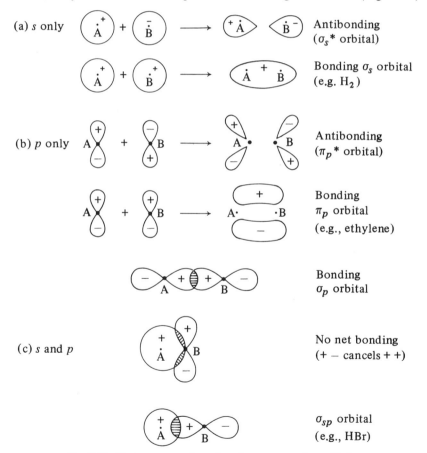

Fig. 5-13. Various types of overlaps between *s* and *p* orbitals.

We can now construct a MO energy level diagram which includes $1s$, $2s$, and $2p$ AO's. The relative energies of the AO's are indicated (not to scale) at the sides of the diagram in Fig. 5-14. Atomic orbital sharing common symmetry *elements*, but not necessarily having similar *total* symmetries, can be mixed to form MO's. Thus the $1s$ and $2s$ orbitals, which have both axial and spherical symmetry elements, combine to form σ and σ^* MO's just as in Fig. 5-11. The $2p$ atomic orbitals combine as indicated in Fig. 5-13 to form both σ- and π-type MO's. The scheme leads to the MO energy levels of Fig. 5-14a,

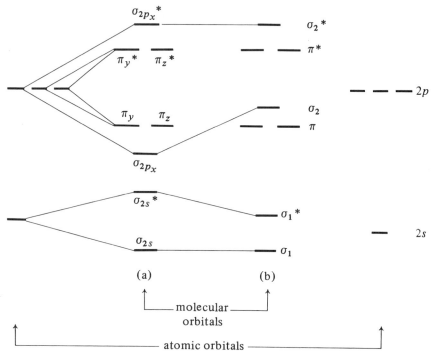

Fig. 5.14. Molecular orbital energy level diagram for homonuclear diatomic molecules. The $1s$ orbitals are not shown. See text for explanation of splittings in (b).

which can be modified further to acknowledge the fact that the σ_{2s}^* and σ_{2p_x} orbitals interact strongly and mix because of their great overlap in both symmetry and space. The diagram of Fig. 5-14b is thus more correct, and gives good predictions of bond order and magnetic properties for molecules such as N_2 and F_2. The real power of MOT is shown by its straightforward prediction of bond order and paramagnetism for the oxygen molecule. Thus, while VBT treatment of oxygen leads to the following predictions,

$$:\!O\!=\!O\!:\qquad \text{Bond order 2, diamagnetic}$$

or

$$\cdot \overset{..}{\underset{..}{O}} \!-\! \overset{..}{\underset{..}{O}} \cdot \qquad \text{Bond order 1, paramagnetic diradical}$$

MOT correctly predicts a bond order of 2 with paramagnetism (Fig. 5-15).

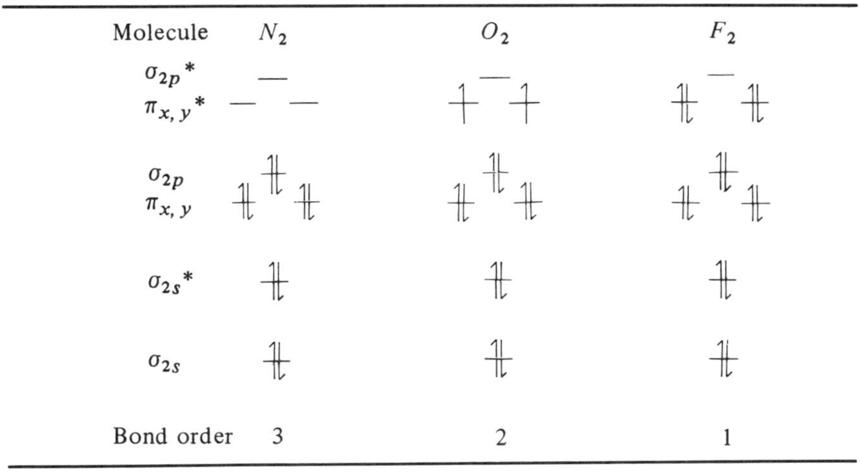

Molecule	N_2	O_2	F_2
σ_{2p}^*	—	— —	— —
$\pi_{x,y}^*$	— —	↑ ↑	↑↓ ↑↓
σ_{2p} $\pi_{x,y}$	↑↓ ↑↓ ↑↓	↑↓ ↑↓ ↑↓	↑↓ ↑↓ ↑↓
σ_{2s}^*	↑↓	↑↓	↑↓
σ_{2s}	↑↓	↑↓	↑↓
Bond order	3	2	1

Fig. 5-15. Electron configuration of N_2, O_2, and F_2.

MOLECULAR ORBITAL THEORY FOR TRANSITION METAL COMPLEXES

We will discuss the application of MOT to transition metal complexes for two different situations, namely, without π bonding and with π bonding included, following the same qualitative approach adopted earlier for the construction of MO energy level diagrams. One important point to remember is that the number of MO's produced must equal the number of metal and ligand orbitals used. For example, in the octahedral case without π bonding there are six σ orbitals from the ligands and a total of nine s, p, and d orbitals on the metal. Generally the ligand σ orbitals are lower in energy than any of the metal orbitals, as shown in Fig. 5-16. As in the examples discussed earlier, the metal atom and ligand orbitals which have similar symmetries can interact or "mix." Thus, one ligand σ orbital interacts with a metal s orbital, three more ligand σ orbitals interact with three metal p orbitals, and the remaining two ligand σ orbitals interact with the d_{z^2} and the $d_{x^2-y^2}$ metal orbitals. The d_{xy}, d_{yz}, and d_{xz} metal orbitals have π-type symmetry (see Fig. 5-3, for example) and do not mix or interact with ligand σ-type orbitals. Therefore they remain unchanged and are *nonbonding* in the molecular complex.

Several important points emerge from the MO diagram of Fig. 5-16. One is that the bonding orbitals are more ligandlike than metallike (i.e., closer in

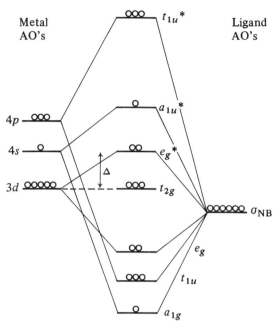

Fig. 5-16. Molecular orbital scheme for octahedral complexes with σ-bonding ligands. Examples of such complexes include FeF_6^{3-} and $Co(NH_3)_6^{3+}$.

energy to ligand than to metal orbitals) and thus may have considerable ionic character. Another important point is that if there are more than six metal d electrons in an octahedral complex they must go into the antibonding e_g^* level. As a consequence we see the d^7–d^{10} ions tending to form five- or four-coordinate complexes frequently whereas the d^0–d^6 ions rarely do so.

To consider both σ and π bonding only one change needs to be made. This is shown in Fig. 5-17, which is identical to Fig. 5-16 except for the addition of ligand π and π^* orbitals. For comparison to the VBT handling of π bonding we can again consider the ferricyanide complex ion. The MO energy level diagram utilizes six cyanide σ orbitals, but only three cyanide π^* orbitals, in addition to the metal orbitals. The σ orbitals are nonbonding on cyanide but bonding in the complex, and the π orbitals are bonding on cyanide but nonbonding in the complex since they do not interact with the metal orbitals. The three π^* orbitals are antibonding on cyanide but have the requisite π symmetry and orientation to interact with the t_{2g} metal orbitals. The result of this interaction is that the t_{2g} orbitals are now *bonding* instead of only nonbonding as before. The valence bond analogy of this interaction was given previously in Fig. 5-3. Such $p\pi$–$d\pi$ bonding or "back-bonding" is possible between a transition metal and *any* ligand with a π-type orbital, atomic or molecular, originating at the donor atom of the ligand, such as CN^-, CO, N_2, $HC{\equiv}C^-$, Schiff bases, and related

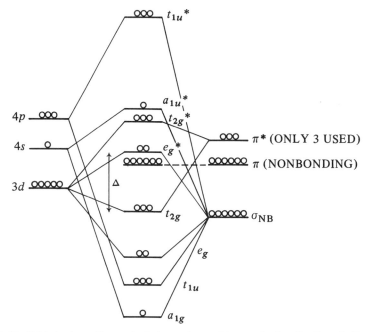

Fig. 5-17. Orbital scheme for octahedral complexes with σ- and π-bonding ligands. Examples of such complexes include $Cr(CO)_6$ and $Fe(CN)_6^{3-}$.

ligands. Back-bonding from filled metal d orbitals to vacant ligand d orbitals is also possible for ligands such as phosphines R_3P, and the Cl_3Sn- group.

CHEMICAL CONSEQUENCES OF BACK-BONDING

Back-bonding is not without precedent or importance in organic chemistry; for example, the acidity of silanols compared to alcohols results from the greater stabilization of the oxyanion by d orbitals on silicon. The d orbitals on carbon are of too high an energy to contribute to stabilizing an alkoxide anion.

$$
\begin{array}{cc}
CH_3 & CH_3 \\
| & | \\
CH_3-C-OH & CH_3-Si-OH \\
| & | \\
CH_3 & CH_3 \\
pK_a \geqslant 16 & pK_a \approx 11\text{--}12
\end{array}
$$

Similarly, while diazoalkanes are thermally unstable, the silicon-substituted diazoalkane can be distilled at atmospheric pressure. It enhanced stability

$$(CH_3)_3 Si-C=N=N$$
$$|$$
$$H$$

results from $p\pi$–$d\pi$ bonding over the entire Si–C–N–N system, and similar arguments may be made for the greater stability of phosphorous ylids (Wittig reagents) over nitrogen ylids. As a final example we may consider the effect of silicon d orbitals on the absorption spectrum of simple ketones. The Si–C–O

$$\begin{array}{ccc}
\text{CH}_3 & \text{(CH}_3)_3\text{Si} & \text{(CH}_3)_3\text{Si} \\
\diagdown\text{C=O} & \diagdown\text{C=O} & \diagdown\text{C=O} \\
\text{CH}_3\diagup & \text{CH}_3\diagup & \text{(CH}_3)_3\text{Si}\diagup
\end{array}$$

| λ_{max}, Å: | < 3000 | 3800 | 5000 |
| Color: | (colorless) | yellow | violet |

overlap considerably lowers the energy required for the n–π^* transition thus producing the above spectral shifts.

Back-bonding is very important in many ways in transition metal chemistry. It helps drain excess negative charge away from the metal center in complexes like $Fe(CN)_6^{4-}$. It helps stabilize metals in low oxidation states, such as Co(I) in vitamin B_{12}, and is an important component of many organometallic complexes containing metal-carbon σ and/or π bonds. Various back-bonding ligands can also be arranged in a spectrochemical series, according to the strength of their interaction with the metal:

$$NO > CO > PF_3 > (PhO)_3P > Ph_3P > R_3P > \text{pyridine}$$

Because the d orbital used in $p\pi$–$d\pi$ back-bonding extends in two directions from the metal, a trans or linear L–M–L arrangement of two back-bonding ligands is usually unfavorable since both ligands then compete for the same d orbital. Therefore the smaller stronger ligands such as NO, CO, and CN^- tend not to occupy trans positions in a complex whenever possible. Pyridines and phosphines however are often forced into trans positions because of their steric bulk. Note in Fig. 5-17 that only three of the twelve π^* orbitals on the six cyanide ligands were used in constructing the MO's.

The pK_a, spectral, and redox data for ruthenium complexes in Table 5-1[8] provide one excellent illustration of how back-bonding affects the properties of metal complexes, and also in this case, how *complexation to a metal can affect the properties of an organic ligand*. The Ru(II) cation exerts two influences on the heterocyclic compound to which it is bound at a nitrogen donor. There is a strong electron withdrawing influence through the σ-bond system of the complex and ligand, and there is simultaneously a tendency to donate electron density into the π^* system of the ligand from filled d orbitals on the d^6 metal. The former effect should be greatest at short distances from the metal, while the latter effect can extend over considerable distances as a result of resonance and π-overlap effects. Upon complex formation the pK_a of the pyridazine ring drops 2.3 units because of the negative inductive effect which overwhelms the back-donation effect. In the pyrimidine case there is a smaller but similar

TABLE 5-1

Spectral, pK_a, and Redox Properties of Free and Complexed Heterocyclic Nitrogen Compounds[a]

Ligand	pK_a		λ_{max}(nm)		$E_{1/2}$(V)
	LH^{+b}	MLH^{+b}	MLH^{+b}	ML^b	L^b
 Pyridazine	2.33	0.03	556	467	1.60
 Pyrimidine	1.3	0.0	463	445	1.78
 Piperazine	0.6	2.5	529	472	1.57
 Pyridine	—	—	—	407	2.09

[a] Data from Ford et al.[8]
[b] L = free ligand, LH^+ = protonated ligand, ML = $[(H_3N)_5RuL]^{2+}$, and MLH^+ = $[(H_3N)_5RuLH]^{3+}$.

decrease in pK_a. However, in the piperazine complex there is actually a pK_a increase upon complexation. Evidently back-bonding increases electron density at the para position through favorable resonance interactions, which can be represented by reaction 5-4, whereas the para nitrogen is too far from the

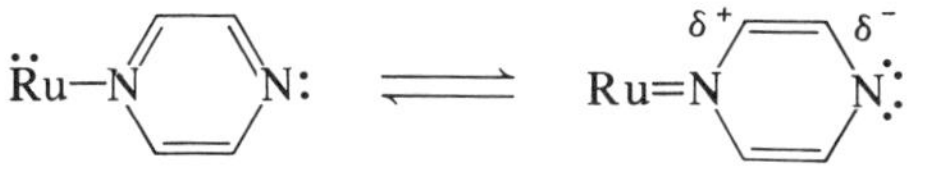

$$\ddot{R}u-N\diagup\hspace{-0.3em}\diagdown N: \rightleftharpoons Ru=N\overset{\delta^+}{\diagup}\hspace{-0.3em}\overset{\delta^-}{\diagdown}\ddot{N}: \qquad (5\text{-}4)$$

metal center to feel much of the negative inductive effect. However, the analogous Ru(III) piperazine complex has a pK_a of −0.8, indicative of only an inductive effect with no significant back-donation by the metal in its higher oxidation state.

The redox potential of the free heterocyclic ligand is a measure of the ease with which an electron can be added to the lowest empty π^*-molecular orbital on the ligand. In this series piperazine is most easily reduced to an anion radical while pyridine is the least easily reduced. This trend is exactly paralleled by the energy of the absorption band of the complex. Because of the intensity of the absorptions and the linear correlation of their energies with the reduction potential of the ligand, the absorptions are considered to be charge-transfer processes in which the absorption of a photon of proper energy promotes a metal d electron into a ligand π^* orbital. The protonated ligands are even better electron acceptors so the protonated complexes (MLH^+) absorb at correspondingly lower energies. It is easy to imagine that similar effects could be important in governing the reactivity of biological redox systems and oxygen carriers such as hemoglobin, and such effects are indeed observed (see Chapter 10).

THE EIGHTEEN-ELECTRON RULE

Based on the observation that many transition metal complexes contain a total of 18[metal + ligand] electrons, a special stability is usually ascribed to this situation, in which the metal appears to have the configuration of the next highest rare gas. The 18-electron rule predicts that a given metal center would accept ligands in order to attain 18, and only 18, electrons in its outer orbitals. But there are more than a few exceptions to the rule as can be seen from the accompanying tabulations. Figures 5-16 and 5-17 can be used to rationalize

Complexes Involving σ-Bonding Ligands or Metals with Few d Electrons

d^2	$W(CN)_8^{4-}$	(18)	d^6	$Pt(F)_6^{2-}$	(18)
d^3	$Cr(CN)_6^{3-}$	(15)	d^7	$Co(H_2O)_6^{2+}$	(19)
d^4	$Mn(CN)_6^{3-}$	(16)	d^8	$Ni(en)_3^{2+}$	(20)
d^5	$Fe(C_2O_4)_3^{3-}$	(17)	d^9	$Cu(NH_3)_6^{2+}$	(21)
d^6	$Co(NH_3)_6^{3+}$	(18)	d^{10}	$Zn(en)_3^{2+}$	(22)

Complexes of Strong π-Bonding Ligands

d^6	$Fe(CN)_6^{4-}$	(18)	d^8	$Fe(PF_3)_5$	(18)
d^6	$Cr(CO)_6$	(18)	d^8	$HCo(CO)_4$	(18)
d^6	π-$C_5H_5Mn(CO)_3$	(18)	d^9	$Co_2(CO)_8$	(18)
d^7	$Mn_2(CO)_{10}$	(18)	d^{10}	$Ni(1,5\text{-}COD)_2$	(18)

why a particular complex obeys or disobeys the rule.[9] For complexes where π bonding is weak or absent, Fig. 5-16 shows that the first six metal d electrons are nonbonding but that any additional d electrons must enter antibonding orbitals. For this reason complexes with more than 18 electrons such as $Cu(NH_3)_6^{3+}$ or $Zn(en)_3^{2+}$ are unstable and tend to lose ligands easily. The $Ni(en)_3^{2+}$ complex is less unstable because of the high charge/radius ratio for Ni(II) and the effects of chelation. It is of no special consequence if there are less than 18 electrons in these kinds of complexes since electrons in nonbonding orbitals neither contribute to nor detract from the overall stability of the complex. However, this is not the case when there are strong π-bonding ligands since their interaction with the t_{2g} orbitals causes them to become *bonding*. It thus becomes important to have the t_{2g} orbitals filled but to keep the $e_g{}^*$ orbitals empty, and this requires exactly 18 electrons. For the case of a metal with an odd number of d electrons a metal–metal bond often forms so that each metal can effectively have 18 electrons. Complexes with strong π-bonding ligands usually involve metals in low oxidation states since this ensures maximum availability of the metal d electrons for back-donation to the ligand. Cases of complexes with less than 18 electrons are known to exist in solution and may be quite important in catalytic cycles (see Chapter 11).

REFERENCES

1. L. Pauling, "The Nature of the Chemical Bond," 3rd ed. Cornell Univ. Press, Ithaca, New York, 1960.
2. M. C. Day, Jr. and J. Selbin, "Theoretical Inorganic Chemistry," 2nd ed. Van Nostrand-Reinhold, Princeton, New Jersey, 1969.
3. H. B. Gray, "Electrons and Chemical Bonding." Benjamin, New York, 1964.
4. G. N. Lewis, *J. Amer. Chem. Soc.* **38**, 762 (1916).
5. Lewis, "Valence and the Structure of Atoms and Molecules." Chem. Catalog Co. (Tudor), New York, 1923.
6. L. Pauling and D. M. Yost, *Proc. Nat. Acad. Sci. U.S.* **14**, 414 (1932); L. Pauling, *J. Amer. Chem. Soc.* **54**, 3570 (1932).
7. J. Hinze and H. H. Jaffe, *J. Amer. Chem. Soc.* **84**, 540 (1962).
8. P. Ford, DeF. P. Rudd, R. Gaunder, and H. Taube, *J. Amer. Chem. Soc.* **90**, 1187 (1968).
9. P. R. Mitchell and R. V. Parish, *J. Chem. Educ.* **46**, 811 (1968).

VI

Ligand Exchange Reactions and Factors in Complex Stability

The topics of complex stability and ligand–metal exchange rates are crucial to understanding the functioning of metals in both chemical and biological systems. Consider the metalloenzymes for example. For efficient catalysis the enzyme–metal association must be tenacious, yet there must be facile exchange of substrate and product ligands. Why aren't these restrictions mutually exclusive, and why does metal substitution or interchange produce an active enzyme modification in some cases but not in others? In this chapter we examine some of the factors which govern the reactivity and stability of complexes of both the transition metals and the alkali and alkaline earth metals.

At the outset it is well to distinguish between the terms "stable," "unstable," "inert," and "labile." The latter terms refer to the *kinetic tendency* of a complex to undergo ligand exchange, while the former terms refer to a ground state *thermodynamic* or *equilibrium* situation involving particular reactants *under specified conditions*. For example, the hexaamminecobalt(II) and hexaamminecobalt(III) complexes are both thermodynamically *unstable* in strong acid solution, yet the latter can exist in strong acid for long periods of time because there is no readily available *mechanism* for its conversion to the more stable products. Thus, kinetic inertness masks its instability in acid (see reactions 6-1).

Unstable state $\longrightarrow$ stable state

$$Co(NH_3)_6{}^{2+} + acid \xrightarrow[\text{very fast}]{\text{labile complex}} 6\,NH_4{}^+ + Co^{2+} \qquad (6\text{-}1)$$

$$Co(NH_3)_6{}^{3+} + acid \xrightarrow[\text{very slow}]{\text{inert complex}} 6\,NH_4{}^+ + Co^{3+}$$

Ligand Exchange Mechanisms[1-13]

There is a good deal of overall similarity in mechanism between ligand exchange at metal centers in complexes and substitution reactions at carbon centers in organic molecules. Both processes are studied using the same basic approaches: kinetics and rate laws, and stereochemistry. The mechanisms of both processes may be grouped roughly into associative processes, where in the transition state bond making with the entering ligand is more advanced than bond breaking with the leaving group, and *dissociative* processes, where the reverse situation obtains. In our discussion we shall consider only *heterolytic* processes, or those in which a ligand enters or departs the coordination sphere as a nucleophile or Lewis base with a pair of electrons. Homolytic or radicallike reactions at metal centers are also very important but are more logically grouped with metal redox chemistry.

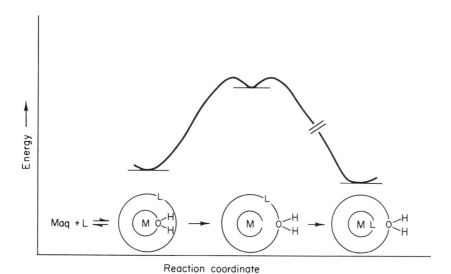

Fig. 6-1. Reaction diagram for the dissociative ligand exchange mechanism for d^0 cations. Circles represent inner and outer coordination spheres. Abbreviations: Maq, hydrated metal ion; M, metal; and L, ligand.

Ia and IIa Cations

In aqueous solution the Ia and IIa cations are hydrated, and their rates of solvent or ligand exchange are extremely fast, almost approaching the diffusion-controlled limit. They fall into the same familiar sequence as the ionic radii.[13] This is because the rate limiting step in the exchange process is the dissociation of a coordinated ligand or solvent molecule to leave a vacancy in the first coordination sphere (Fig. 6-1). The smaller ions with the more intense electrostatic fields and the higher heats of hydration will therefore have the largest activation energy requirements and the slowest rates of ligand dissociation (Table 6-1 and Fig. 6-2). Very small highly charged ions such as Be^{2+} and

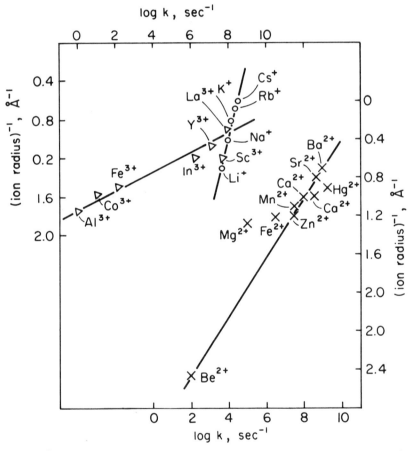

Fig. 6-2. First-order rate constants for solvent exchange as a function of ionic radius[1]: $\triangle$, trivalent ions; $\times$, divalent ions; $\bigcirc$, monovalent ions.

TABLE 6-1

First-Order Rate Constants for Complex Formation for Various Metal Ions[1]

Ion	$10^{-7} k$ (sec⁻¹)	Ion	k (sec⁻¹)	Ion	k (sec⁻¹)	Ion	k (sec⁻¹)	Ion	k (sec⁻¹)	Ion	$10^{-7} k$ (sec⁻¹)	Ion	$10^{-7} k$ (sec⁻¹)
Li^+	4.7	Be^{2+}	10^2	V^{2+}	30	Al^{3+}	1	Ti^{3+}	4.0×10^3	Ce^{3+}	9.5	Tb^{3+}	3.0
Na^+	8.8	Mg^{2+}	10^5	Cr^{2+}	10^8	Sc^{3+}	5×10^7	V^{3+}	1.1×10^2	Pr^{3+}	8.6	Dy^{3+}	1.7
K^+	15	Ca^{2+}	10^8	Mn^{2+}	3×10^7	La^{3+}	7×10^7	Cr^{3+}	1.8×10^{-6}	Nd^{3+}	9.3	Ho^{3+}	1.4
Rb^+	23	Sr^{2+}	5×10^8	Fe^{2+}	3×10^6	Y^{3+}	1.3×10^7	Fe^{3+}	1.3×10^2	Sm^{3+}	9.6	Er^{3+}	1.0
Cs^+	35	Ba^{2+}	9×10^8	Co^{2+}	2×10^6	In^{3+}	2×10^6	Co^{3+}	10	Eu^{3+}	8.2	Tm^{3+}	1.1
		Cd^{2+}	5×10^9	Ni^{2+}	2×10^4			Mn^{3+}	5.0×10^4	Gd^{3+}	5.2	Yb^{3+}	1.1
		Hg^{2+}	3×10^9	Cu^{2+}	2×10^8							Lu^{3+}	1.3
		Pb^{2+}	6×10^8	Zn^{2+}	3×10^7								

Al^{3+} would be extremely slow to undergo ligand exchange were it not that they undergo hydrolysis. Their extreme Lewis acidity renders coordinated water molecules quite acidic (reaction 6-2). Deprotonation of a coordinated water produces a coordinated hydroxide and lowers the net charge of the complex, which in turn "labilizes" the remaining water molecules.

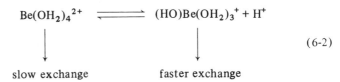

$$Be(OH_2)_4{}^{2+} \rightleftharpoons (HO)Be(OH_2)_3{}^{+} + H^{+}$$

(6-2)

slow exchange   faster exchange

Labile Transition Metal Complexes

Compared to the nontransition metals, there is much more variety in terms of mechanism, stereochemistry, and rate of ligand exchange at transition metal centers. The particular metal center involved plays a large part in determining the rate of ligand exchange, but the effects of one ligand on the exchange of another and the nature of the ligand being lost are important considerations. Again one finds that dissociative mechanisms are generally more common, especially for first-row metals, six-coordinate, and 18-electron complexes. Associative or Sn2-like mechanisms are less uncommon for the larger second- and third-row metals which can expand their coordination spheres, for square planar complexes, and for complexes of metals with few d electrons such as Ti(III) and V(III).

Early investigations of mechanisms and stereochemistry of ligand substitution reactions were of necessity confined to complexes of Pt(II), Pt(IV), Co(III), and Cr(III). This was because the kinetic inertness of these complexes made possible the manipulation and isolation of complexes that was necessitated in the days before rapid reaction monitoring techniques such as stopped-flow, NMR, ultrasound absorption, and pressure- or temperature-jump relaxation methods. These modern techniques have made possible the measurement of extremely fast ligand dissociation and association processes. Results obtained for the divalent first-row transition metal ions indicate (1) an overall similarity between their rates of solvent exchange and rates of complex formation, and (2) a familiar characteristic pattern in their relative reactivities (cf. Figs. 4-8 to 4-12) as shown in Fig. 6-3. The latter tendency is reflective of crystal field stabilization energy (CFSE) effects, superimposed on charge-radius electrostatic effects, on the dissociation of a coordinated group. With the trivalent first-row ions the crystal field splitting is larger, as is the charge-size ratio; thus, ions such as Co(III) and Cr(III) can be extremely inert toward ligand dissociation and exchange.

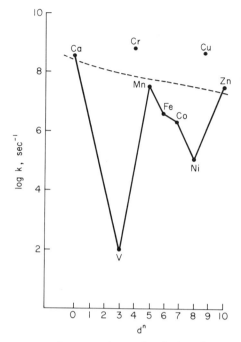

Fig. 6-3. Approximate rates of water exchange for divalent first-row transition metal ions.[1]

The electronic configuration of the metal center per se also has an effect on ligand dissociation rates. Complexes with filled or half-filled t_{2g} levels such as Cr(III), V(II), and the low spin forms of Co(III) and Ru(II), tend to be extremely inert. On the other hand, lability is found among those ions which have less than three d electrons, or which have enough electrons to populate e_g^* antibonding orbitals. The extreme lability of high spin d^4 and d^9 systems such as Cr(II) and Cu(II) results from Jahn–Teller distortion effects, which are equally distributed to all six ligand positions by metal–ligand vibrational stretching modes.

Inert Transition Metal Complexes

From the remainder of this section it should become evident that there is a great deal of similarity between dissociative reactions of metal complexes and Sn1 reactions at carbon in organic compounds. The differences between the metal and organic systems usually result from differences in bond strengths and in the solvation requirements of reaction intermediates or transition states. Some stereochemical differences are inherent in the different coordination numbers and initial geometries of metals compared to carbon centers.

LEAVING AND ENTERING GROUP EFFECTS

A rate limiting dissociative step in the ligand exchange process implies that the effects of varying the entering group should be small in comparison to the effects of varying the leaving group. This is convincingly demonstrated by the

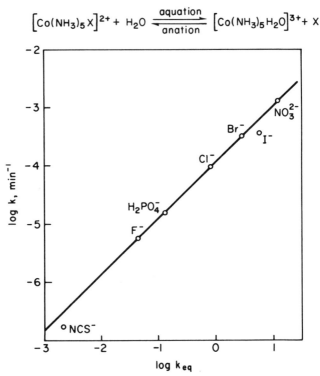

$$[Co(NH_3)_5 X]^{2+} + H_2O \underset{\text{anation}}{\overset{\text{aquation}}{\rightleftharpoons}} [Co(NH_3)_5 H_2O]^{3+} + X$$

Fig. 6-4. Linear free energy relationship for $(NH_3)_5CoX^{2+}$ aquation rates and equilibria.[14] (Reprinted with permission from *Inorganic Chemistry.* Copyright by the American Chemical Society.)

existence of a linear free energy correlation (Fig. 6-4)[14] between the rate constant and equilibrium constant for the aquation of a series of pentaammine-cobalt(III) derivatives. The fact that the slope of this correlation (reaction 6-3)

$$H_2O + (H_3N)_5 CoX^{2+} \underset{k', \text{ anation}}{\overset{k, \text{ aquation}}{\rightleftharpoons}} (H_3N)_5 CoOH_2^{3+} + X^- \qquad (6\text{-}3)$$

is unity indicates that ground state changes in the Co—X bond strength, and in the solvation of X^-, are similar in the transition state for the aquation reaction. This leads to the conclusion that in the transition state the X group resembles free X^-. Other linear free energy relationships supporting this conclusion include

1. A correlation of log k with the pK_a of HX for the aquation of $(H_3N)_5CrX^{2+}$ [15]
2. A correlation of activation entropies, $\Delta S^{\ddagger}$, and ground state solvation entropies, ΔS°, for the X group in $(H_3N)_5CrX^{2+}$ [15]
3. A correlation of the rate of loss of substituted pyridines (PyX) in the complex $(H_3N)_5RuPyX^{2+}$ with the Hammett σ constant for the particular X group [16]
4. A correlation of the frequency (energy) of the first ligand field absorption band with the activation energy for dissociation in a series of both Rh(III) and Co(III) complexes

Substitution rates are not entirely independent of the nature of the entering group, and the extent of this dependence is usually regarded as an indication that some bond making is occurring before complete loss of the leaving group. Entering group effects are more noticeable for second and third row transition metals, which are much less labile than their first row analogs. Anation rates greater than water exchange rates are also suggestive of associative mechanisms. The second- and third-row transition metals are also "softer" than those in the first row, and their substitution rates are often sensitive to the "softness" of the entering ligand (see below).

SOLVENT EFFECTS[2, 8]

Replacing solvent water by less polar organic solvents can enhance the rate dependence on the entering group by promoting enhanced ion pairing if the complex and entering group are oppositely charged, but the effects are not great. On a log scale the rates of solvolysis of alkyl halides are 2- to 4-fold more sensitive to solvent changes in aqueous organic solvent systems such as alcohols, acetone, dioxane, or formic acid. This reflects the much greater demand for solvation by carbonium ions intermediates than by metal complexes with coordination numbers reduced by one and agrees with the previous conclusion that the transition state for aquation of $(NH_3)_5CoX^{2+}$ resembles free X^- and $(NH_3)_5Co^{3+}$.

ACID CATALYSIS AND METAL ASSISTANCE IN SOLVOLYSIS

Both at carbon and at metal centers, conjugate bases of strong acids such as chloride, sulfate, or p-toluenesulfonate are good leaving groups. Conjugate bases of weaker acids such as HO^-, F^-, NO_2^-, N_3^-, CN^-, CO_3^{2-}, and $CH_3CO_2^-$ are not particularly good as leaving groups from saturated carbon, but their reactivity can be considerably enhanced by protonation or acid catalysis. That the same is true for their elimination from metal centers is shown by the rate law for hydrolysis of complexes containing these ligands, which usually has the form

$$-d(\text{complex})/dt = (k_0 + k[\text{H}^+])[\text{complex}]$$

indicating the presence of both acid-catalyzed and acid-independent pathways. A certain amount of caution is required in the application of this generalization, however. For instance, oxyanions such as CO_3^{2-}, $CH_3CO_2^-$, or ONO^- can react by two different pathways (reactions 6-4 and 6-5).[17–20]

$$
\begin{array}{c}
\text{M—O—X} \longrightarrow \text{M} + \text{O—X} \\
+ \qquad\qquad \;\, | \quad\; | \\
\text{O—H} \qquad\quad \text{OH} \;\; \text{H} \\
| \\
\text{H}
\end{array}
\qquad (6\text{-}4)
$$

$$X = SO_3^-, NO_2, PO_3^{2-}, \text{oxalate}$$

$$
\begin{array}{c}
\text{M—O—X} \longrightarrow \text{M—O} + \text{X} \\
+ \qquad\qquad\;\; | \quad\;\; | \\
\text{H—O} \qquad\quad \text{H} \;\;\; \text{OH} \\
| \\
\text{H}
\end{array}
\qquad (6\text{-}5)
$$

$$X = CH_3CO, CO_2^-, SO_2^-, NO$$

Studies with $^{18}OH_2$ have shown that these reactions often proceed via reaction 6-5, which is an acid-catalyzed reaction of a coordinated ligand rather than an acid-catalyzed dissociation of a ligand (reaction 6-4). Other parallels between solvolysis at metal and at carbon centers include

1. Departure of halide leaving groups assisted by metal ions such as Ag^+ and Hg^{2+},

$$Ph_2S + CH_3I + AgBF_4 \longrightarrow Ph_2SCH_3^+BF_4^- + AgI$$

$$
en_2Co\begin{array}{l} \diagup H_2NCH_2CO_2^- \\ \diagdown Br \end{array} + Hg^{2+} \longrightarrow en_2Co\begin{array}{l} \diagup NH_2\text{—}CH_2 \\ \diagdown O\text{——}C \end{array}\underset{O}{\overset{\|}{\;}} + HgBr^+
$$

2. Chemical generation of a small stable molecule such as CO_2, N_2O, or N_2 as a leaving group

$$PhCONH_2 \xrightarrow{\text{HONO}} PhCO_2H + N_2$$

$$(H_3N)_5CoN_3^{2+} \xrightarrow{\text{HONO}} N_2 + N_2O + (H_3N)_5CoOH_2^{3+} \quad [21]$$

$$(H_3N)_5CoOCONH_2 \xrightarrow{\text{HONO}} N_2 + CO_2 + (H_3N)_5CoOH_2^{3+} \quad [22]$$

The stereochemical outcome of spontaneous as well as acid-assisted aquation reactions such as those above is usually complete or at least predominant *retention* of configuration.[23, 24] This suggests that the reaction may proceed *via* a square pyramidal five-coordinate intermediate (6-6). In contrast, solvolytic reactions at asymmetric carbon centers usually give complete randomization

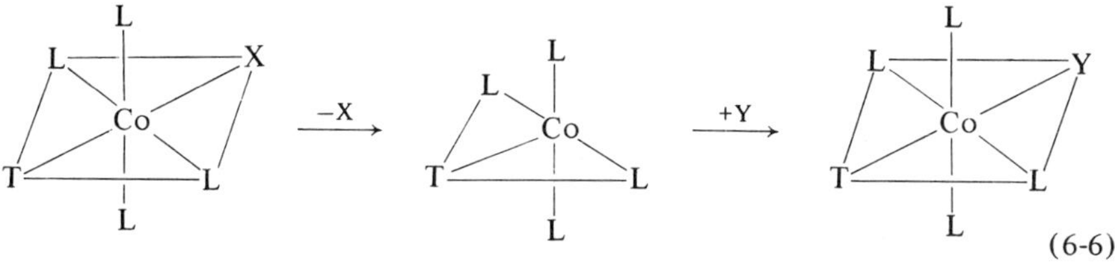

$$(6-6)$$

of configuration rather than retention, except where anchimeric assistance, steric hindrance, or other special effects prevail.

BASE-CATALYZED SOLVOLYSIS AND ANCHIMERIC ASSISTANCE

The solvolysis of alkyl halides or tosylates is greatly facilitated by the presence of a neighboring group which can donate a pair of electrons to stabilize the developing positive charge on carbon. This type of anchimeric assistance can have profound effects on the stereochemical outcome as well as the rate of the reaction because the transition state is no longer a planar carbonium ion (Scheme 6-1).

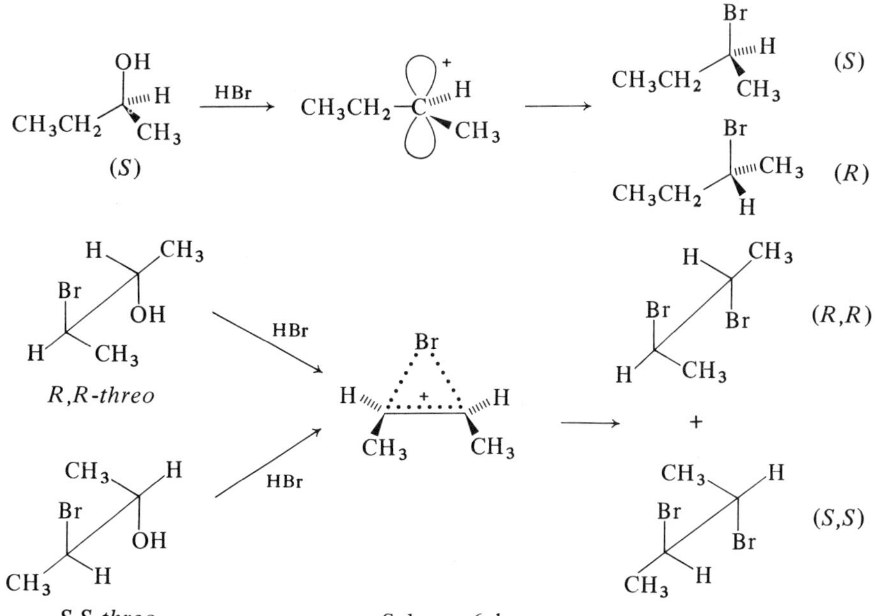

Scheme 6-1

A somewhat analogous process of internally assisted dissociation is common, if not the norm, in basic solutions of inert complexes having ligands capable of deprotonation, such as water, ammonia, or organic amines (reactions 6-7, 6-8, and 6-9, respectively). This process is somewhat analogous to

$$(H_3N)_5CoCl^{2+} + HO^- \rightleftharpoons H_2O + (H_3N)_4Co(NH_2)Cl^+ \tag{6-7}$$

$$(H_3N)_4Co(NH_2)Cl^+ \xrightarrow{slow} Cl^- + [(H_3N)_4CoNH_2]^{2+} \tag{6-8}$$

$$[(H_3N)_4CoNH_2]^{2+} \nearrow^{fast,\ H_2O} (H_3N)_5CoOH_2{}^{3+} \searrow_{fast}^{H_2O,\ X^-} (H_3N)_5CoX^{2+} \tag{6-9}$$

the hydrolysis of Be^{2+} or Al^{3+} complexes mentioned earlier. Since the dissociation is facilitated by the formation of a *conjugate base* of an amine ligand, the process is referred to as an Sn1cb mechanism. Protons on the nitrogen trans to the leaving group are observed by NMR to undergo base-catalyzed deuterium

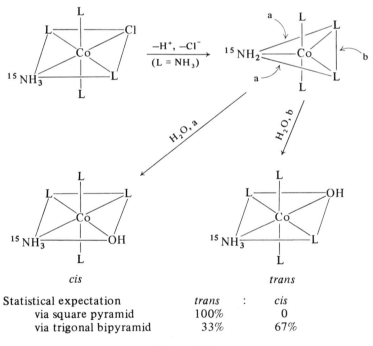

Statistical expectation	*trans*	:	*cis*
via square pyramid	100%		0
via trigonal bipyramid	33%		67%

Scheme 6-2

exchange 100–300 times faster than protons on *cis* nitrogens,[25] suggesting that the *trans* nitrogen is preferentially deprotonated. The labilization of the leaving group is thought to reflect its weakened bonding because of the strong σ- *and* π-*donor* character of the trans amido group.

The steric course of base hydrolysis usually leads to mixed product stereochemistry rather than predominant retention. An intermediate with geometry somewhere between square pyramidal and trigonal bipyramidal could explain the formation of both cis and trans products from a trans parent complex. This geometry would also be favorable for $p\pi$–$d\pi$ bonding from the labilizing amido group (or hydroxo group) to the metal. In this connection an experiment by Buckingham, Sargeson, and their co-workers is worthy of mention for its elegance of design, if not the generality of the result. They prepared the complex *trans*-$[(^{15}NH_3)(NH_3)_4CoCl]^{2+}$ and found[23] that base hydrolysis gave a product which was 40% *cis* and 60% *trans*-$[(^{15}NH_3)(NH_3)_4CoOH]^{2+}$. However, in the presence of competing azide ion, the base hydrolysis gave 70% *cis*- and 30% *trans*-$[(^{15}NH_3)(NH_3)_4CoN_3]^{2+}$, which closely approximates the 67:33 ratio expected for statistical attack on the edges of the equatorial plane of a trigonal bipyramidal intermediate (Scheme 6-2).[26]

LIGAND EXCHANGE IN NONOCTAHEDRAL COMPLEXES[1, 2, 12, 12a]

Square planar complexes of d^8 metal ions such as Pt(II), Ni(II), Pd(II), and Au(III) are relatively inert and hence have been extensively studied using classical techniques. In contrast to octahedral complexes, square planar complexes undergo ligand exchange predominantly via an associative mechanism involving a five-coordinate transition state or intermediate. Consequently, the rate law for ligand exchange has the form

$$-d(\text{complex})/dt = (k_1 + k_2[\text{Y}])[\text{complex}]$$

when Y is the entering ligand. The rate constant k_1 is usually much less than k_2, and is a *pseudo*-first-order rate constant for a pathway in which the leaving group is first replaced by a solvent molecule which is in turn displaced by Y (reaction 6-10). The k_1 depends on the solvent nucleophilicity (e.g., k_{DMSO} >

(6-10)

S = solvent

k_{MeOH}) indicates that even this pathway is an associative process. The five-coordinate intermediate is generally felt to have trigonal bipyramidal geometry, rather than square pyramidal, since the former would involve less steric repulsion between the ligands and would be more favorable for metal–ligand π bonding (reaction 6-11). Further evidence for the trigonal pyramid intermediate

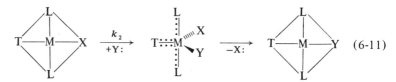

$$T\!-\!M\!-\!X \xrightarrow[+Y:]{k_2} T\vdots\!M\raise1ex\hbox{$\cdots X$}\lower1ex\hbox{$\searrow Y$} \xrightarrow{-X:} T\!-\!M\!-\!Y \quad (6\text{-}11)$$

is the general observation that the activation entropies for square planar substitution (k_2) are negative and rather large.

The effects of one ligand on the exchange or substitution of another are particularly pronounced in some square planar complexes, and this phenomenon has become known as the *trans* effect.[12a] Various ligands T may thus be ranked in order of their labilizing effect on a trans ligand X, as indicated in the pathway above. The general order is not much different whether considering kinetic or equilibrium data, and runs something like

$$C_2H_4 \sim NO \sim CO \sim CN > R_3P \sim H \sim thiourea > CH_3 >$$

$$C_6H_5 > SCN > NO_2 > I > Br > Cl > NH_3 > OH > H_2O$$

The large trans effect of ligands such as H^- and CH_3^- undoubtedly result in large part from their extreme "softness" and the strength of their σ bonding to soft d^8 metals such as Pt(II), which then somewhat weakens the bonding of the trans ligand. The large effect of most of the other ligands can be attributed to the π-bonding stabilization of the five-coordinate intermediate. At the bottom of the order are the small hard ligands which are the most labile, rather than labilizing, at soft metal centers.

For ligand substitution in tetrahedral complexes a five-coordinate intermediate is also suggested by both the negative entropy of activation observed for such processes, and by the fact that typical second-order kinetics are observed. Examples of such processes are the exchange[27] of radioactive α-picoline into the complex $Co(\alpha\text{-picoline})_2Cl_2$, and phosphine exchange in the complexes ML_2X_2 [where M = Co(II) or Ni(II), L = phosphine, and X = chloride or bromide].[28] These complexes have less than 18 electrons in their valence shell as well as being only four-coordinate, and both of these properties favor an associative pathway. For five-coordinate complexes with 18 electrons a dissociative mechanism for ligand exchange might be predicted to be more likely,[13] and this has been found to be the case for displacement of a carbonyl in $Fe(CO)_4(Ph_3P)$ by another phosphine.

Factors Influencing the Stability of Metal Complexes[29-35]

Having emphasized the distinction between kinetic inertness and thermo-dynamic stability, we may now analyze the factors which contribute to the stability of metal complexes. For this purpose we will use as a measure of stability K_f, the equilibrium constant for formation of a given complex. For a simple equilibrium such as $M + L \rightleftharpoons ML$ (charges omitted for convenience) the *stoichiometric* equilibrium constant K is defined as $K = [ML]/[M][L]$. Since M and L are often charged species, K is sensitive to the ionic strength of the measuring medium. Therefore, in order to generate a consistent basis for measurement and comparison of K's, one of two methods is usually adopted. One is to measure K as a function of ionic strength and extrapolate a plot of K vs $\mu^{1/2}$ to zero ionic strength in order to obtain a *thermodynamic* value for K. More commonly a large and constant concentration of an innocuous electrolyte such as $NaClO_4$ or Na_2SO_4 is used as a "swamping salt" to main-tain a "constant" ionic strength. Few metal complexation equilibria are as simple as the one above, however, and frequently a stepwise series of K's are

$$M + L \rightleftharpoons ML \qquad K_1 = \frac{[ML]}{[M][L]} \qquad (6\text{-}12)$$

$$ML + L \rightleftharpoons ML_2 \qquad K_2 = \frac{[ML_2]}{[ML][L]} = \frac{[ML_2]}{K_1[M][L]^2} \qquad (6\text{-}13)$$

$$ML_2 + L \rightleftharpoons ML_3 \qquad K_3 = \frac{[ML_3]}{[ML_2][L]} = \frac{[ML_3]}{K_1 K_2 [M][L]^3}, \text{ etc.} \quad (6\text{-}14)$$

involved (reactions 6-12, 6-13, and 6-14). For such multistep equilibria it is not uncommon to use *cumulative* equilibrium constants β, defined as

$$\beta_1 = K_1 = \frac{[ML]}{[M][L]} \qquad \beta_2 = K_1 K_2 = \frac{[ML_2]}{[M][L]^2} \qquad \beta_n = K_1 K_2 \ldots K_n = \frac{[ML_n]}{[M][L]^n}$$

Metal Ion Properties

The successive values of K generally decrease as n increases (Fig. 6-5), es-pecially for negatively charged ligands which are not good π bonders. This results from a combination of the statistical effect of fewer sites not already occupied by the ligand, from the effects of steric crowding of ligands in the first coordination sphere, and from the electrostatic effects of accumulation of negative charge at the metal center. To some extent the stability of the complex is also determined by the particular metal ion. This is often explained

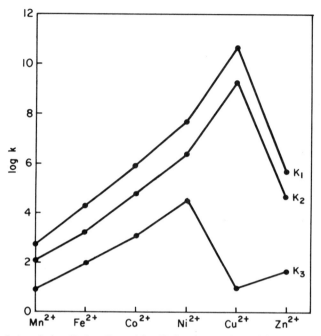

Fig. 6-5. Variations in log K_1, log K_2, and log K_3 for the reactions of first-row transition metal ions (M^{2+}) with ethylenediamine ($NH_2CH_2CH_2NH_2$) at 30°C.[31]

qualitatively in terms of charge-radius effects and the hard–soft compatibilities of specific metal–ligand combinations (see below). The importance of crystal field splitting effects has frequently been cited (cf. Chapter 4), but it must be remembered that octahedral CFSE values are somewhat smaller than the observed enthalpies of complexation for many first-row ions, that partial covalent character is important in many bonds, that the preferred geometry of various metals may not always be octahedral, and that the enthalpy and entropy of *ligand solvation* can play a significant part in complexation equilibria.

Ligand Properties

One of the ligand properties which is important in determining its metal complexing ability is its proton complexing ability, i.e., its pK_a. In correlating metal- and proton-binding abilities one must be careful to work within a family of Lewis bases having the same type of basic function, such as carboxylates, phenolates, aromatic amines, aliphatic amines, or pyridines. Then for a given metal ion a good correlation of K with ligand pK_a, of the form log $K = a(pK_a)$ + b can be established. Examples of systems for which such a correlation is known are aromatic amines with Ag(I), simple carboxylates with Pb(II), phenolates with Fe(III), and β-diketonates with Cu(II). While such correlations are useful in predicting the relative stabilities of complexes or the relative amounts of two different ligands that would be complexed at a given pH, the actual amount of metal complexed is determined by the relative concentrations of metal and hydrogen ions which compete for the ligand. The relative effectiveness of metal ion and hydrogen ion in this competition is measured by the a term in the correlation, which is usually in the range $0.5 \leqslant a \leqslant 1.5$. For ligands such as hydride, alkyl, and carbon monoxide this correlation is unimportant or even indeterminable, and π bonding and metal softness (see below) are much more significant factors.

Bidentate Ligands and the Chelate Effect

A ligand molecule possessing a second potential donor suitably situated with respect to the first, i.e., a *bidentate* ligand, may form two bonds to a metal ion. This multiple bonding, called chelation, greatly enhances the overall stability of the complex in comparison to a complex formed from similar donor sites each on a monodentate ligand. One example of the chelate effect involving an alkali metal was discussed in Chapter 2, where the importance of entropy effects was noted. For divalent first-row transition metal ions it appears that the relative importance of entropic and enthalpic effects is determined by the nature of the donor atoms and the charge on the bidentate ligand. Chelation by strong donors such as bidentate amines is accompanied by large negative enthalpies and small entropies. Chelation by weak donors like carboxylates is accompanied by small enthalpy changes, but the entropies are large and positive as a result of the release of many solvent molecules from both metal and ligand when a complex of lower total charge forms. At least for first-row metals from Mn to Zn, enthalpy effects become less important along the series $-SH > -NH_2 > -CO_2^- > -OH$, while entropy effects are most important with ligands of high negative charge.

The size of a chelate ring is an important factor in determining its stability, with the usual preferred sequence being $4 \ll 5 > 6 > 7$ or larger. Four-membered rings are found mainly with ligands having two large donor atoms, such as sulfur in $Et_2NCS_2^-$ or $(EtO)_2PS_2^-$. The difference in stability between analogous five- and six-membered chelates usually amounts to a factor of 20- to 60-fold, and most of this difference is attributed to ring strain effects. Just as in organic systems, this includes bond angle strain and steric repulsion energies for eclipsing of ring hydrogens or other substituents. The difference in stability of the Ni(II) complexes of the two diastereomers of 2,3-diaminobutane, which differ only in the cis-trans relationship of the two

TABLE 6-2

Influence of Ligand Type and Ring Size on Stability of Ni(II) Complexes[a]

Ni^{2+} + ligand $\rightleftharpoons$ 1:1 complex

Ligand	log K (25°C)	ΔH (kcal/mole)	$-T\Delta S$ (kcal/mole)
Oxalate	5.2	0.15	7.21
Malonate	4.1	1.88	7.45
Succinate	2.3	—	—
Glycine	6.2	−4.14	4.30
β-Alanine	4.7	−3.81	3.04
Thioglycolate	7.0	—	—
Mercaptoethylamine	11.0	—	—
Ethylenediamine	7.7	−9.05	1.19
1,3-Propylenediamine	6.3	−7.7	0.89
meso-2,3-Diaminobutane	7.04	—	—
dl-2,3-Diaminobutane	7.71	—	—

[a] Data from Sillen and Martell[31] and Nancollas.[36]

methyls on the chelate ring, illustrates the importance of steric effects (Table 6-2).[31,36] Another parallel comparison between inorganic and organic systems, which applies in *both* a kinetic and a thermodynamic sense, is the relative stability and reactivity of the following series of ketals, ranked in order of ease of hydrolysis.

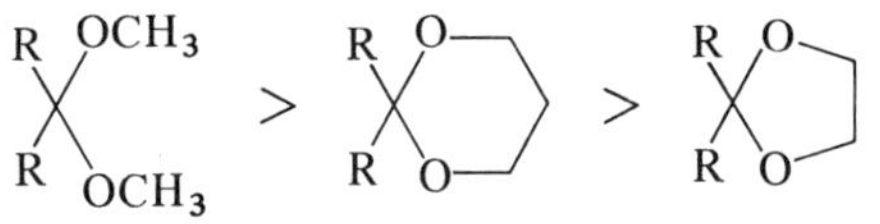

CHELATE FORMATION, OPENING, AND STABILITY

The formation of a metal chelate usually involves a series of discrete ligation steps, the first of which is an *inter*molecular reaction. The succeeding steps are *intra*molecular reactions, and consequently are often very fast. Thus a general mechanism for chelate formation, based on the Eigen–Tamm mechanism for solvent exchange, is the following:

(a) Formation of a solvent-separated ion pair or outer sphere complex at essentially diffusion-limited rates

$$M_{aq} + A{-}B_{aq} \xrightarrow{K_{ip}} (H_2O)M(H_2O)A{-}B$$

(b) Dissociation of a solvent molecule from the inner coordination sphere of the metal followed by rapid monoligation

$$(H_2O)M(H_2O)A-B \underset{k_{-1}}{\overset{k_1}{\rightleftharpoons}} (H_2O)M-A-B + H_2O$$

(c) Dissociation of another solvent molecule followed by chelation

$$(H_2O)M-A-B \underset{k_{-2}}{\overset{k_2}{\rightleftharpoons}} M\begin{pmatrix} A \\ B \end{pmatrix} + H_2O$$

For the formation of five-membered chelates of first-row transition metal ions, the rate of ring closure k_2 is usually greater than k_1. This is suggested by the observation that the overall rate of formation of many *chelates* is approximately the same as the rate of complexation of the same metal by *mono*dentate ligands. In some important cases however k_1 *and* k_{-1} may be greater than k_2. For example, if a ligand must deprotonate to reveal a second donor site for use in chelation, the rate of chelate formation may be limited, sometimes severely, by the rate of proton transfer from ligand to solvent. Eigen, who has contributed greatly to the study of "fast" reactions in general and to the study of ligand exchange reactions in particular, has shown that the rate of transfer of a proton from a weak base (strong conjugate acid) to a strong base (to form a weak conjugate acid) is very rapid.[37] The reverse transfer is very slow however, in accordance with the predicted equilibrium for such a competition ($K_{eq} = k_f/k_r$). The rates of closure of six-membered chelates may also be reduced in comparison to five-membered systems. Examples of these effects may be found in the data of Table 6-3.[4,38] The basis behind the chelate effect is also applied in many attempts to synthesize medium-ring organic compounds. The high dilution technique used disfavours intermolecular reactions, i.e., polymerization, without affecting the rate of intramolecular ring closure.

Multidentate and Other Special Ligands

The formation of chelates of multidentate ligands is qualitatively the same as with bidentate ligands. Table 6-4[31,39] shows that the complex formed generally increases in stability as the number of rings increases. Although such direct comparisons of equilibrium constants are not strictly proper owing to the different units encountered for various K_n values, the log numbers do reflect the relative stabilities observed under equilibrium conditions. Ring strain effects in multidentate chelates are dependent on the individual rings present as well as the strain of the fused ring system *as a whole*. For this reason the incorporation of some six-membered rings can be beneficial to stability in spite of

TABLE 6-3

Rate Constants for Formation and Dissociation of Labile Chelate Complexes [a]

Reaction [b]	k_f [c] $(M^{-1} sec^{-1})$	k_d [c] $(M^{-1} sec^{-1}$ or $sec^{-1})$
$Mg_{aq}^{2+} + Ox^- \rightleftharpoons MgOx^+$	3.8×10^5	7
$Mg^{2+} + HOx \rightleftharpoons MgOx^+ + H^+$	1.1×10^4	—
$Mn^{2+} + Ox^- \rightleftharpoons MnOx^+$	1.1×10^8	140
$Mn^{2+} + HOx \rightleftharpoons MnOx^+ + H^+$	4.4×10^5	—
$Mn^{2+} + \alpha\text{-alanine}$	6.0×10^7	—
$Mn^{2+} + \beta\text{-alanine}$	5×10^4	—
$Co^{2+} + Gly^- \rightleftharpoons CoGly^+$	1.6×10^6	34
$Co^{2+} + \alpha\text{-alanine}$	6.0×10^5	—
$Co^{2+} + \beta\text{-alanine}$	7.5×10^4	—
$(bipy)Co^{2+} + Gly \rightleftharpoons (bipy)CoGly^+$	1.6×10^6	55
$(Gly)Co^+ + Gly^- \rightleftharpoons CoGly_2$	2×10^6	330
$Co^{2+} + \alpha\text{-ABA}$	2.5×10^5	—
$Co^{2+} + \beta\text{-ABA}$	2.0×10^4	—
$Ni^{2+} + bipy \rightleftharpoons Ni(bipy)^{2+}$	1.6×10^3	—
$(en)Ni^{2+} + bipy \rightleftharpoons (en)Ni(bipy)^{2+}$	5.0×10^3	—
$(en)_2Ni^{2+} + bipy \rightleftharpoons (en)_2Ni(bipy)^{2+}$	8.4×10^3	—
$Ni^{2+} + {}^-O_2CCH_2CO_2^- \rightarrow Ni^{\pm}OCOCH_2CO_2^-$	$k_1 = 7.5 \times 10^3$	—
$Ni^+OCOCH_2CO_2^- \rightarrow chelate$	$k_2 = 1.1 \times 10^2 sec^{-1}$	—
$Cu^{2+} + Gly^- \rightleftharpoons Cu(Gly)^+$	4×10^9	22
$(bipy)Cu^{2+} + Gly^- \rightleftharpoons (bipy)Cu(Gly)^+$	1.6×10^9	19
$Cu^{2+} + \alpha\text{-alanine}$	1.3×10^9	—
$Cu^{2+} + \beta\text{-alanine}$	2.0×10^8	—
$(bipy)Cu^{2+} + \alpha\text{-Ala}^- \rightleftharpoons (bipy)Cu(\alpha\text{-Ala})^+$	1.0×10^9	10
$(bipy)Cu^{2+} + \beta\text{-Ala}^- \rightleftharpoons (bipy)Cu(\beta\text{-Ala})^+$	3.4×10^8	110
$Cu^{2+} + en \rightleftharpoons (en)Cu^{2+}$	3.8×10^9	0.1
$(bipy)Cu^{2+} + en \rightleftharpoons (bipy)Cu(en)^{2+}$	2.0×10^9	1.4
$Cu^{2+} + Hen^+ \rightleftharpoons H^+ + Cu(en)^{2+}$	1.4×10^5	—
$(bipy)Cu^{2+} + Hen^+ \rightleftharpoons H^+ + (bipy)Cu(en)^{2+}$	2.2×10^4	1.2×10^5

[a] Data from Sharma and Leussing[4] and Kustin and Swinehart.[38]

[b] Abbreviations: HOx, 8-hydroxyquinoline; en, ethylenediamine; bipy, 2,2'-bipyridyl; ABA, aminobutyric acid; Gly, glycine; and Ala, alanine.

[c] Most data for $\mu = 0.1$ and 25°C, k_f = observed chelate formation rate, k_d = observed rate of chelate dissociation.

the general $5 > 6$ stability order for single rings. Thus tet-a forms a planar Cu(II) chelate more stable by several powers of ten than the analogous trien chelate (Table 6-4).

If the ends of a multidentate ligand are joined to form a macrocyclic ligand an additional increment in stability is usually observed. It is difficult to measure the equilibrium stability constants of such complexes because they are

TABLE 6-4

Long K_f Values for Cu^{2+} Complexes of Some Multidentate Ligands at 25°C[a,b]

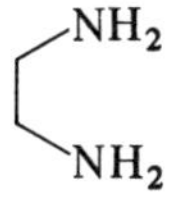

en, 10.7

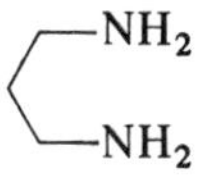

pn, 9.8

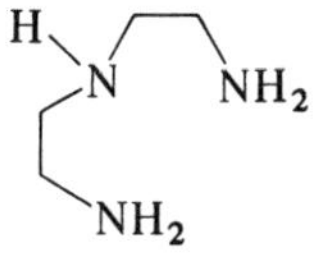

dien, 16.1

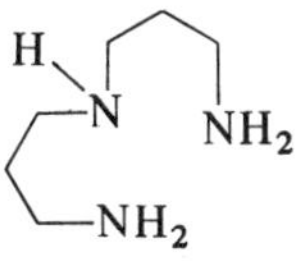

dipn, 14.2

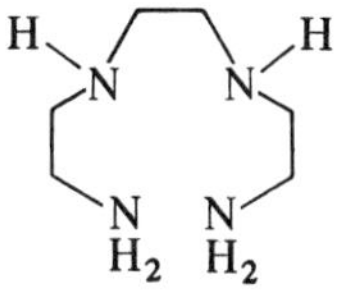

trien, 20.1

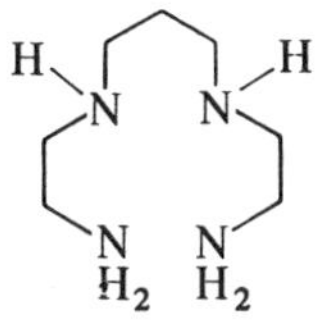

2,3,2-tet, 23.9

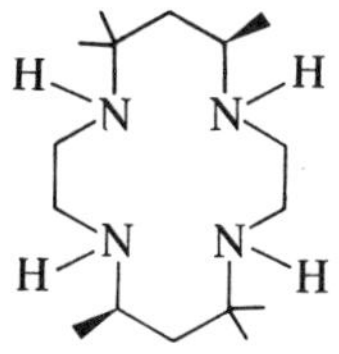

tet-a, 28

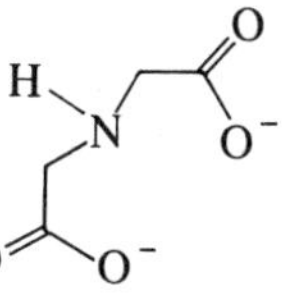

Gly, 8.3

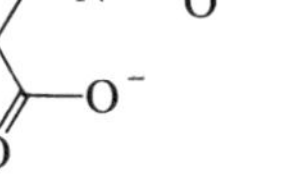

β-Ala, 7.1

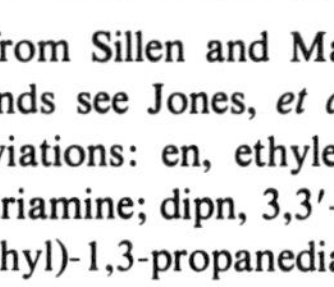

imda, 10.4

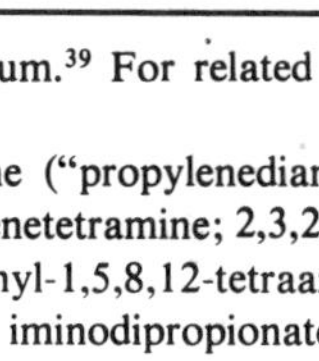

imdp, 9.43

---

[a]Data from Sillen and Martell[31] and Cabbiness and Margerum.[39] For related examples of sulfur ligands see Jones, *et al.*[39a]

[b]Abbreviations: en, ethylenediamine; pn, 1,3-diaminopropane ("propylenediamine"); dien, diethylenetriamine; dipn, 3,3'-iminodipropylamine; trien, triethylenetetramine; 2,3,2-tet, *N,N'*-di-(2-aminoethyl)-1,3-propanediamine; tet-a, 2,2,4,9,9,11-hexamethyl-1,5,8,12-tetraazacyclotetradecane; gly, glycine; β-ala, β-alanine; imda, iminodiacetate; imdp, iminodipropionate.

very large, but more importantly because the association and dissociation rates, and hence the rates of equilibration, are very slow (Table 6-5). This results from the geometric strains the ligand must undergo in order to form, or to dissociate from, a square planar complex. The origin of the "macrocyclic effect" is not completely understood, but it can be broken down for analysis just as was done for the chelate effect.

TABLE 6-5

Formation and Dissociation Rates for Cu^{2+} Complexes[a]

Ligand	k_f (M^{-1} sec^{-1}) (0.5 OH^-)[b]	k_d (sec^{-1}) (acid)[b]
trien	10^7	4.1
tet-a	1.6×10^3	3.6×10^{-7}
Hematoporphyrin(IX)	2×10^{-2}	—

[a] Data from Busch et al.[7]
[b] Reaction conditions.

Provided the macrocyclic ligand is reasonably flexible the rate of complex formation is limited by the first ligation step as in ordinary chelate formation. In the reverse direction, however, ligand dissociation requires that the ligand assume a strained conformation. Since large macrocycles can accommodate this strain more easily than small macrocycles, their complexes dissociate more readily and become less stable as the ring size increases.[39a]

A recent thermodynamic study of the complexation of Ni(II) by 2,3,2-tet and cyclam show that the 10^6-fold increase in stability for the cyclam complex arises from the fact that this ligand is less hydrogen bonded to solvent than 2,3,2-tet.[40] This results in a much more negative complexation enthalpy, since fewer ligand–water hydrogen bonds are broken, which more than offsets the slightly less positive entropy change as a result of fewer waters of solvation being released from the ligand during complex formation. Few studies have been done with the naturally occurring macrocyclic ligands, i.e., the porphyrins and corrins. Undoubtedly the extreme stability of their metal derivatives is the result of metal-ligand π bonding as well as macrocyclic effects.

Another recently recognized factor in the "stability" of certain multidentate chelates is the *rigidity* of the ligand. Actually this "stability" may only be an apparent stability more properly viewed as kinetic inertness. An example of this is the greatly reduced rates of chelate formation by phenanthroline H^+ (Hphen$^+$) compared to the more flexible bipyridyl H^+ (Hbipy$^+$). Another very interesting example of this is the unusual ligand tris(anhydro)aminobenz-aldehyde, first prepared via a template reaction by Eichhorn,[41] shown in Diagram 6-1 as an octahedral Ni(II) complex. This nickel complex is so inert kinetically that it can be resolved into dissymmetric isomers which are optically stable in solution for weeks.[7] The enormous rigidity of this complex, together

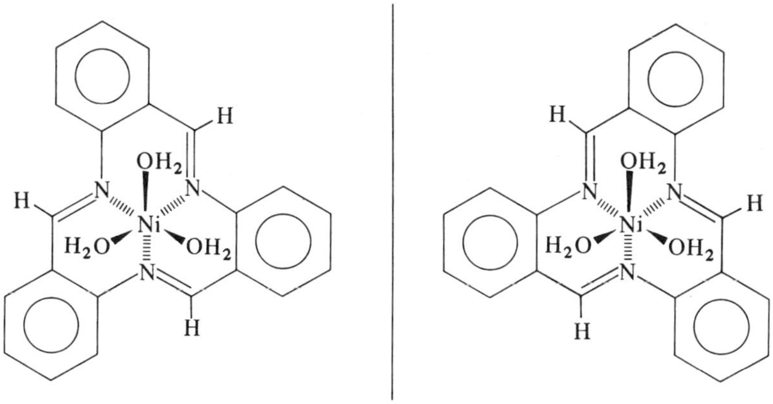

(mirror plane)

Diagram 6-1

with the nearly ideal fit of the ligand on one face of an octahedral Ni(II) ion, imposes an enormous energy (enthalpy?) of activation requirement, possibly equivalent to the simultaneous rupture of *two* bonds, for dissociation of this complex. The most remarkable feature of this complex, however, is that *the three coordinated water molecules undergo exchange at approximately the same rate as waters coordinated to a free Ni(II) ion*, 3.8×10^4 vs 3.2×10^4 sec^{-1}, respectively.[42] This important observation may provide insight to an answer for one of the questions posed at the outset of this chapter, namely, the paradox of tenacious metal–enzyme binding with simultaneous rapid exchange of substrate and product ligands.

Information on the binding of metal ions in metalloenzymes is difficult to obtain and consequently scarce, and information on the rates and equilibria of metal binding and dissociation in these systems is confined to only a few investigations.[6, 9] However, results with *apo*-carbonic anhydrase indicate that only minor changes in protein conformation occur upon reconstitution with Zn(II) ions. This reassociation is 10^2–10^3 times slower than the formation of "model" complexes, as might be predicted for a rigid apoenzyme ligand, but since the binding site is at the bottom of a pocket or fold in the enzyme, and may have a somewhat unusual geometry, ligand rigidity may not be the only factor in the slowness of metal exchange for this enzyme.

Mutual Metal–Ligand Effects in Complex Stability[43–47]

HARD AND SOFT ACIDS AND BASES

According to Pearson,[46, 47] an important consideration in determining the strength of a given metal–ligand interaction is their mutual compatibility in terms of hardness and softness. This parameter is qualitatively correlated with the charge-size ratio of the ion in that large ions of low ionic charge have easily polarizable or deformable, i.e., soft, electrostatic fields about them, while small highly charged ions with relatively intense electrostatic fields are hard. Table 6-6 gives some examples of hard and soft metals and ligands. On the basis that *hard metal ions prefer to bind hard ligands, and vice versa,* one can understand why Fe^{3+} binds halides in the order $F > Cl > Br > I$ while Hg^{2+} prefers the reverse order, and why first-row ions bind thiocyanate (NCS^-) at N while second- and third-row ions bind at S.

TABLE 6-6

Classification of Metal Ions and Ligands by the Hard–Soft Criteria

Hard	Borderline	Soft
	Metal Ions	
H^+, Li^+, Na^+, K^+	$Fe^{2+}, Co^{2+}, Ni^{2+}$	Cu^+, Ag^+, Au^+, Tl^+
$Be^{2+}, Mg^{2+}, Ca^{2+}, Sr^{2+}$	$Cu^{2+}, Zn^{2+}, Pb^{2+}$	$Hg_2^{2+}, Hg^{2+}, Pd^{2+}, Pt^{2+}$
$Mn^{2+}, Al^{3+}, Sc^{3+}, Ga^{3+}$	$Sn^{2+}, Sb^{3+}, Bi^{3+}$	Pt^{4+}, Tl^{3+}
$In^{3+}, La^{3+}, Gd^{3+}, Lu^{3+}$		
$Cr^{3+}, Co^{3+}, Fe^{3+}, Si^{4+}$		
$Ti^{4+}, Sn^{4+}, WO^{4+}, VO^{2+}$		
	Ligands	
H_2O, OH^-, F^-, Cl^-	$C_6H_5NH_2, C_5H_5N$	R_2S, RS^-, I^-
$CH_3CO_2^-, PO_4^{3-}, SO_4^{2-}$	N_3^-, Br^-, NO_2^-	$SCN^-, S_2O_3^{2-}$
$CO_3^{2-}, ClO_4^-, NO_3^-$	SO_3^{2-}, N_2	$R_3P, (RO)_3P, R_3As$
ROH, R_2O, NH_3		$CN^-, RNC, CO,$
RNH_2, N_2H_4		H^-, R^-, C_2H_4

The hard–soft dichotomy provides a rationale for many features of the behavior of metal systems in chemistry and biology. For example, the poisonous heavy metals lead, cadmium, and mercury exert many of their toxic effects by binding to important −SH groups in enzymes thus blocking their function. The usual antidote is a sulfur-containing ligand such as penicillamine or BAL (British Anti-Lewisite). These molecules complex the metals in a soluble form which can then be excreted. Acting as ligands, organic sulfides and phosphines rapidly "poison" platinum and palladium hydrogenation

$$\begin{array}{cc}
CH_3 \; CH_3 & \\
\diagdown\!\diagup & \\
\quad\;\; CH\!-\!CO_2H & CH_2\!-\!CH\!-\!CH_2OH \\
\;\; | \quad\;\; | & \quad | \quad\quad | \\
SH \quad NH_2 & \;\; SH \quad\;\; SH \\
\text{Penicillamine} & \text{BAL}
\end{array}$$

catalysts, but Raney nickel removes sulfur from many kinds of organic compounds. Toxic anions such as cyanide and fluoride exert their effects by binding respectively to Fe(II) in cytochromes and Cu(II) in cytochrome oxidase, and to Mn(II) in many other enzymes, thus preventing their function. A further example of hard–soft effects may be seen in the slopes of the plots of binding strength vs metal ion in Fig. 6-6. These slopes reflect the relative

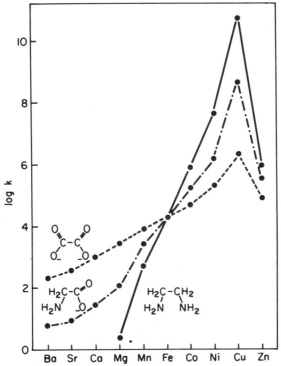

Fig. 6-6. Log K_1 values for some metal ions and ligands of biological interest.[45] The d^0 ions and d^5 Mn^{2+} are hard and hence prefer hard ligands. They bind oxalate > glycine > ethylenediamine, while the reverse order is displayed by the softer transition metal ions.

hard–soft compatibilities of the various metal–ligand combinations. Note that there is a crossover point at Fe^{2+} in the plot, implying that this biologically common and important metal, *in its lower oxidation state*, is very flexible in its hard–soft ligand preferences. This versatility is undoubtedly an important factor in iron absorption, storage, and transfer in living systems. The ferrous form is readily absorbed but it is stored as ferric iron in ferritin and phosvitin, the latter being a protein with many serine phosphate groups as potential hard donors for binding ferric iron.

Mixed Ligand Complexes

In all but a few special cases the equilibrium constant for binding a second ligand is lower than that for the first, for reasons outlined earlier in this chapter. However, there are a few cases in which K_2 is greater than K_1 for the same ligand, i.e., penicillamine and cysteine, and other cases in which one ligand increases the affinity of a metal ion for a second different ligand. These effects are caused by *mutual* ligand–ligand interactions mediated or translated through the central metal of a mixed ligand complex. They can be quantitated by use of the equilibrium expressions 6-15, 6-16, and 6-17. For the special case

$$M + B \rightleftharpoons MB \qquad K_{MB}^{M} = [MB]/[M][B] \qquad (6\text{-}15)$$

$$MA + B \rightleftharpoons MAB \qquad K_{MAB}^{MA} = [MAB]/[MA][B] \qquad (6\text{-}16)$$

$$\Delta \log K = \log K_{MAB}^{MA} - \log K_{MB}^{M} \qquad (6\text{-}17)$$

$A = B$, $\Delta \log K$ usually ranges from -0.5 to -0.8 for monodentate ligands and from -1 to -2 for bidentate ligands. Positive values for $\Delta \log K$ indicate enhanced binding of B as a result of the effects of A on the properties of the metal center. $\Delta \log K$ has been determined for a number of mixed systems with copper ions. The $\Delta \log K$ values for other metals are smaller (Irving–Williams series) but follow the same trends with respect to ligand–ligand compatibility. Representative data from a review by Sigel and McCormick[45] are listed in Table 6-7. Examination of this and similar data leads to several interesting conclusions. First, bipyridyl *increases* the stability of complexes of oxyanion (hard) ligands such as oxalate, phenolates, and phosphates, and at the same time *reduces* the metal's affinity for amine and amino acid ligands. The large difference in K_1 and K_2 for bipy results from steric interactions of the ligands as well as from the antagonistic or competitive π-bonding effects associated with two trans π-acceptor ligands. Second, exactly the opposite trend is observed for the enCu^{2+} and (NH$_3$)$_2$Cu^{2+} systems where the strong σ-donor effects of saturated amines reduces the Lewis acidity of the central metal. Amino acid ligands tend to have a similar effect, i.e., destabilization of σ-donor ligands, especially oxyanions, and stabilization of π-acceptor ligands. Third, there are some indications that other biologically important π-acceptor ligands such as histamine (found at many metal-binding sites of enzymes), benzimidazole and the corrin ring (found in vitamin B$_{12}$), porphyrin derivatives (found in hemoglobin and cytochromes), and sulfur groupings (CH$_3$S—R of methionine and HS—R of cysteine, found in hemoproteins and nonheme iron proteins) do show effects similar to although smaller than those of bipyridyl.

TABLE 6-7

Stability Data for Some Mixed Ligand Cu^{2+} Complexes[a,b]

$$MA + B \overset{K}{\rightleftharpoons} MAB$$

MA	B	$\Delta \log K$
bipyCu^{2+}	Oxalate	0.7
	AMP(ADP, ATP)	0.4–0.5
	HPO$_4^{2-}$	0.4
	Pyr^{2-}	0.4
	SSal^{3-}	0.5
	gly$^-$	−0.35
	β-ala$^-$	−0.6
	en	−1.3
	bipy	−1.9
enCu^{2+}	Oxalate	−1.1
	Pyr^{2-}	−0.8
	AMP	−0.45
(NH$_3$)$_2$Cu^{2+}	gly$^-$	−1.2
	SSal^{3-}	−1.2

[a] Data from Sigel and McCormick.[45]

[b] Abbreviations: bipy, 2,2'-bipyridyl; Pyr^{2-}, pyrocate-cholate; SSal^{3-}, 5-sulfosalicylate; gly$^-$, glycinate; β-ala$^-$, β-alaninate; en, ethylenediamine; AMP, adenosine-5'-monophosphate; and $\Delta \log K$ as defined in the text.

Not surprisingly there are kinetic effects corollary to these thermodynamic effects. Thus, strong σ-donor ligands like aliphatic amines, or ligands like acetylacetonate which are *both* σ- and π-donors, tend to *labilize* coordinated water. Strong π-acceptor ligands such as 1,10-phenanthroline (phen) or 2,2'-bipyridyl (bipy) do not labilize coordinated water, which implies that the σ-donor effect of the Lewis base ligand is compensated by its π-acceptor proper-ties, leaving the bipyCu^{2+} ion with substantial Lewis acid character.

The labilization of coordinated water is no doubt an important factor in the rapid closure of some five-membered chelate rings and may also be important to the catalytic functioning of some metal complexes and/or metalloenzymes. One special effect observable in mixed ligand complex formation is the ability of one chiral ligand to generate selectivity in the complexing of another. This effect is of obvious importance in enzymatic chemistry and is being successfully applied to various chemical systems, most notably in asymmetric hydrogenation via chiral catalysts.[48] Other examples of this effect will be dis-cussed in later chapters.

Although the demonstration that any of the mixed ligand or other special binding effects is used to advantage by metalloenzymes is still awaited, there is some reason to believe that these effects are important. During the 1960's many talented research teams devoted their efforts toward unraveling the mysteries of enzymatic catalysis. Most of these groups concentrated on the organic chemistry of the enzymes and substrates and "chemical model systems," and a few began the arduous task of X-ray structure determinations for purified crystalline enzymes. A relatively few groups[49] took the unorthodox approach of "inorganic functional group analysis" in the study of *metallo-enzymes*, predicated on the belief that the "probe properties" of a metal ion *which was already an integral part of an enzyme active site* could be monitored to get information about that active site, while perturbing it as little as possible. Using probe properties common to many transition metals[50] (see Table 3-2), and sometimes substituting one metal for another in the enzyme to take advantage of special probe properties, it rapidly became apparent that simple coordination compounds did not resemble metalloenzymes in many of these properties. In an important paper, Vallee and Williams summarized many of these differences and offered an interesting hypothesis which attempted to both explain the differences *and* link them to the catalytic properties of the enzymes. According to Vallee and Williams,[5]

> Present treatments of the mechanism of enzymatic catalysis postulate an active intermediate, and the rate of a reaction depends on the ease with which the related transition state is reached. In our terms this would be facilitated by *a state of entasis: the existence in the enzyme of an area with energy, closer to that of a unimolecular transition state than to that of a conventional, stable molecule, thereby constituting an energetically poised domain.* On this basis the differences between the physical properties of metals in metalloenzymes and models could be understood, since conventional metal complexes, which generally serve as models, do not have geometries of the proposed unstable intermediates of chemical reactions but relax to conventional, stable geometries. It appears that the future design of suitable models should incorporate the emerging structures of metal complexes in enzyme centers, which may deviate significantly from simpler systems currently known.

The "entatic state hypothesis" of geometrically strained or otherwise unusual binding may account for the unusual spectral properties of metals in metalloenzymes, but it almost seems to contradict the great stability or inertness to dissociation of metals at the active sites of most metalloenzymes. Busch *et al.* have pointed out one way by which the metalloenzymes might escape this paradox.[7] They suggested that while distortion from idealized geometry ("entasis") might well lead to weakened metal binding and reduced electron density (or heightened electrophilicity) at the metal center, the rigidity (or "multiple juxtapositional fixedness," MJF) of the protein ligand could compensate in terms of a kinetically inert metal–protein complex. In this way the enzyme might be able to bind its metal ion tightly yet entatically (see Scheme 6-3).

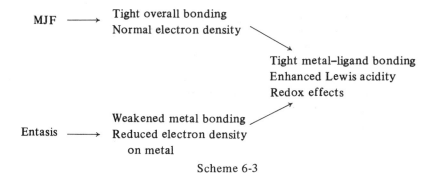

MJF $\longrightarrow$ Tight overall bonding
Normal electron density

Tight metal–ligand bonding
Enhanced Lewis acidity
Redox effects

Weakened metal bonding
Entasis $\longrightarrow$ Reduced electron density
on metal

Scheme 6-3

Mixed Ligand Effects on Metal-Catalyzed Reactions

There are an increasing number of examples which show clearly that various ligands can modulate the *catalytic* properties of metal ions. Several of these will be discussed in succeeding chapters, and thus are mentioned here only to illustrate the scope of the situation.

Redox: Ancillary ligands affect the rate of the Cu^{2+}-catalyzed disproportionation of H_2O_2 according to their expected effect on Cu^{2+} binding of an oxyanion (HOO^-); bipy > en > gly > pyrophosphate.[52]

Enantiomer selectivity: Racemic mixtures of amino acids have been resolved by passing them through a column containing Cu^{2+} bound to an optically active anionic resin.[53]

Preferred Binding Sites for Biologically Important Metals

Metal	Donors at site
Na^+	Neutral oxygen donors, crown ethers, peptide carbonyls, possibly one negative charge at site, small site
K^+	Large site of neutral donors, possibly one negative charge at site
Mg^{2+}	Prefers basic (hard) nitrogen donors > hard oxyanions (carboxylate, phosphate, phenols, and catechols)
Ca^{2+}	Hard oxyanions > nitrogen donors
Mn^{2+}	Similar to Mg^{2+}, often interchangeable but Mn^{2+} decidedly more octahedral
Fe^{2+}	$-SH > -NH_2 > -SCH_3 >$ carboxylates
Fe^{3+}	Serine OH, phosphate, hydroxamate, hydroxide, often polymeric with $-O-$ bridges
Cu^+	Soft donors, $-SH > -NH_2 \gg$ others
Cu^{2+}	Imidazole and amines, peptide $-CON^=$
Zn^{2+}	Imidazole, $-SH, >$ carboxylates

Hydrolytic reactions: The 1:1 Cu^{2+} complexes of bipy, phen, imidazole, and histidine are good catalysts for the hydrolysis of diisopropylfluorophosphate (DFP) but glyCu$^+$ is not.[54]

Decarboxylations: Oxalacetate is catalytically decarboxylated to pyruvate by a manganese-containing enzyme and by many metal cations. BipyCu^{2+} is twice as effective as Cu_{aq}^{2+} in this reaction. While Mn^{2+} is only 1/200th as active as Cu^{2+}, complexation with bipy increases its efficiency *tenfold*![55]

REFERENCES

1. A. McAuley and J. Hill, *Quart. Rev., Chem. Soc.* **23**, 18 (1969).
2. J. Burgess, ed., "Inorganic Reaction Mechanisms," Vol. 2. Chem. Soc., London, 1972.
3. J. O. Edwards, ed., "Inorganic Reaction Mechanisms," Progr. Inorg. Chem. No. 13. Progr. Inorg. Chem., New York, 1970.
4. V. S. Sharma and D. L. Leussing, *in* "Metal Ions in Biological Systems" (H. Sigel, ed.), Vol. 2, p. 127. Dekker, New York, 1973.
5. R. G. Wilkins, *Accounts Chem. Res.* **3**, 408 (1970).
6. J. E. Coleman, *Progr. Bioorg. Chem.* **1**, 159 (1971).
7. D. H. Busch, K. Farmery, V. Goedken, V. Katonic, A. C. Melnyk, C. R. Sperati, and N. Tokel, *Advan. Chem. Ser.* **100**, 44 (1971).
8. C. H. Langford and V. S. Sastri, "Mechanism and Steric Course of Octahedral Substitution," in *MTP Int. Rev. Inorg. Chem. Ser. 1* **9**, 203 (1972).
9. R. G. Wilkins, *Pure Appl. Chem.* **33**, 583 (1973).
10. A. M. Sargeson, *Pure Appl. Chem.* **33**, 527 (1973).
11. M. L. Tobe, *Accounts Chem. Res.* **3**, 337 (1970).
12. M. Cattalini, *MTP Int. Rev. Sci., Inorg. Chem. Ser. 1* **9**, 269 (1972).
12a. F. R. Hartley, *Chem. Soc. Rev.* **2**, 163 (1973).
13. M. Eigen and R. G. Wilkins, *Advan. Chem. Ser.* **49**, 55 (1965).
14. C. Langford, *Inorg. Chem.* **4**, 265 (1965).
15. T. P. Jones and P. K. Phillips, *J. Chem. Soc., A* p. 674 (1968).
16. P C. Ford, J. R. Kuempel, and H. Taube, *Inorg. Chem.* **7**, 1976 (1968).
17. D. J. Francis and R. B. Jordon, *J. Amer. Chem. Soc.* **91**, 6629 (1969).
18. L. Hin-Fat and W. C. E. Higginson, *J. Chem. Soc., A* p. 2836 (1970).
19. D. W. Carlyle and E. L. King, *Inorg. Chem.* **9**, 233 (1970).
20. V. S. Sastri and G. M. Harris, *J. Amer. Chem. Soc.* **92**, 2943 (1970).
21. A. Haim and H. Taube, *Inorg. Chem.* **2**, 1199 (1963).
22. D. A. Buckingham, I. I. Olson, A. M. Sargeson, and H. Satrapa, *Inorg. Chem.* **6**, 1027 (1967).
23. D. A. Buckingham, I. I. Olson, and A. M. Sargeson, *J. Amer. Chem. Soc.* **89**, 5129 (1967).
24. D. A. Buckingham, I. I. Olson, and A. M. Sargeson, *Austr. J. Chem.* **20**, 597 (1967).
25. D. A. Buckingham, P. A. Marzilli, and A. M. Sargeson, *Inorg. Chem.* **8**, 1595 (1969).
26. D. A. Buckingham, I. I. Olson, and A. M. Sargeson, *J. Amer. Chem. Soc.* **90**, 6539 (1968).
27. S. S. Zumdahl and R. S. Drago, *J. Amer. Chem. Soc.* **89**, 4319 (1967).
28. W. De W. Horrocks and L. H. Pignolet, *J. Amer. Chem. Soc.* **88**, 5929 (1966).
29. R. M. Izatt, J. J. Christensen, and J. H. Rytting, *Chem. Rev.* **71**, 439 (1971).
30. S. J. Ashcroft and C. T. Mortimer, "Thermochemistry of Transition Metal Co.nplexes." Academic Press, New York, 1970.

31. L. G. Sillen and A. E. Martell, *Chem. Soc. Spec. Publ.* **17**, (1964); **25**, (1970).
32. G. Beech, *Quart. Rev., Chem. Soc.* **23**, 410 (1969).
33. R. J. Angelici, in "Bioinorganic Chemistry," (G. Eichhorn, ed.), Vol. 1, p. 63. Elsevier, Amsterdam, 1973.
34. W. G. Bardsley and R. E. Childs, *Biochem. J.* **137**, 55 (1974).
35. L. F. Lindoy and D. H. Busch, *Prep. Inorg. React.* **6**, 1 (1971).
36. G. H. Nancollas, "Interactions in Electrolyte Solutions." Elsevier, Amsterdam, 1966.
37. M. Eigen, *Angew. Chem.* **75**, 489 (1963).
38. K. Kustin and J. Swinehart, *Progr. Inorg. Chem.* **13**, 107 (1970).
39. D. K. Cabbiness and D. W. Margerum, *J. Amer. Chem. Soc.* **91**, 6540 (1969).
39a. T. E. Jones, L. L. Zimmer, L. L. Diaddario, D. B. Rorabacher, and L. A. Ochrymowycz, *J. Amer. Chem. Soc.* **97**, 7163 (1975).
40. F. P. Hinz and D. W. Margerum, *J. Amer. Chem. Soc.* **96**, 4993 (1974).
41. G. Eichhorn and R. A. Latif, *J. Amer. Chem. Soc.* **76**, 5180 (1954).
42. J. E. Letter and R. B. Jordan, *J. Amer. Chem. Soc.* **93**, 864 (1971).
43. H. Sigel, ed., "Metal Ions in Biological Systems," Vol. 2. Dekker, New York, 1973.
44. H. C. Freeman, in "Bioinorganic Chemistry," (G. Eichhorn, ed.), Vol. 1, p. 121, Elsevier, Amsterdam, 1973.
45. H. Sigel and D. B. McCormick, *Accounts Chem. Res.* **3**, 201 (1970).
46. R. G. Pearson, *Science* **151**, 172 (1966).
47. T.-L. Ho, *Chem. Rev.* **75**, 1 (1975).
48. J. W. Scott and D. Valentine, *Science* **184**, 943 (1974).
49. B. L. Vallee and R. J. P. Williams, *Chem. Brit.* **4**, 397 (1968).
50. H. Sigel, ed., "Metal Ions in Biological Systems," Vol. 4. Dekker, New York, 1974.
51. B. L. Vallee and R. J. P. Williams, *Proc. Nat. Acad. Sci. U.S.* **59**, 498 (1968).
52. H. Sigel, *Angew. Chem., Int. Ed. Engl.* **8**, 167 (1969).
53. R. V. Snyder, R. J. Angelici, and R. B. Meck, *J. Amer. Chem. Soc.* **94**, 2660 (1972).
54. A. E. Martel, *in* "Metal Ions in Biological Systems" (H. Sigel, ed.), Vol. 2, p. 208. Dekker, New York, 1973.
55. R. W. Hay and N. K. Leong, *Chem. Commun.* p. 800 (1967); *J. Chem. Soc., A* p. 3639 (1971).

VII

Redox Potentials and Processes

Electrodes, Electromotive Forces, and Sign Conventions[1,2]

Many interesting and important chemical reactions involve oxidation or reduction. Actually the two processes never occur independently of one another, even though they may sometimes appear to do so. Since oxidation–reduction or redox reactions involve the transfer of electrons, many such reactions can be induced to occur at electrodes immersed in appropriate solutions. The electrons can then be transferred via an external electrical circuit, and their flow, or current, can be used to do electrical work. For this to occur there need only be a difference in potential between two electrodes. When they are connected in a circuit, electron flow will be spontaneous in the direction from the better reducing (negative) electrode to the better oxidizing (positive) electrode.

Under the European convention the sign of the electrode potential is invariant, and is either positive or negative as determined experimentally by comparing the electrode to the normal hydrogen electrode. Thus this system actually deals with *constant electrostatic potentials*. Under the American convention, an electrode potential is negative if the reaction *as written* tends to occur spontaneously, and positive if the *reverse* reaction tends to occur spontaneously. Under this system the sign of the potential is not invariant but changes according to the tendency of the reaction to proceed "to the right." Thus the term "electrode potential" is reserved for the European system, and the term "electromotive force," or emf, is used in the American system (see tabulation below, and the more extensive collection of redox potentials given at the end of this chapter in Tables 7-4, 7-5, and 7-6).

Reaction	Standard electrode potential (V)	Standard emf or E_0 (V)
$Zn = Zn^{2+} + 2\,e^-$	+0.763	+0.763
$Zn^{2+} + 2\,e^- = Zn$	+0.763	−0.763

The difference in potential between two electrodes can be measured with a potentiometer, but there is no way to measure the *absolute* potential of any single electrode. To create a uniform system for comparing electrode potentials, the potential of the normal hydrogen electrode (NHE) was arbitrarily

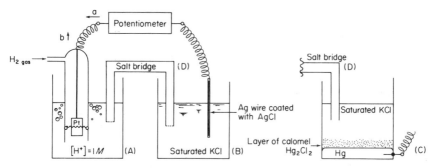

Fig. 7-1. The normal hydrogen, Ag/AgCl, and saturated calomel electrodes. (A) Normal hydrogen electrode (NHE) consists of a piece of Pt metal exposed to 1 atm H_2 and a solution of $H^+ = 1\,M$ (more precisely H^+ at unit activity). The half-reaction is $H^+ + e^- = \frac{1}{2}H_2$, and its potential, by definition, is zero. (B) The Ag/AgCl electrode in saturated KCl solution has a potential of +0.199 V with respect to the NHE. The half-reaction which occurs is $AgCl + e^- = Ag + Cl^-$. (C) The SCE consists of a pool of mercury under a layer of calomel, Hg_2Cl_2, in saturated KCl solution. The half-reaction $\frac{1}{2}Hg_2Cl_2 + e^- = Hg + Cl^-$ has a potential of 0.246 V vs NHE. (D) The salt bridge is a glass U-tube filled with an electrolyte gel (usually agar/KCl) so that electrical conductance can be established without diffusion or mixing of the two solutions from the electrode baths (A) and (B). In the ideal case the net current flow is zero, and the potentiometer measures the potential difference by applying an equal but opposite potential to the cell as shown by arrow a. In this way the electrical resistance of the cells and the salt bridge are *not* important. If the potentiometer were replaced by a light bulb, electrons would flow *from* the Pt electrode ($\frac{1}{2}H_2 \rightarrow H^+ + e^-$) *to* the Ag electrode ($AgCl + e^- \rightarrow Ag + Cl^-$) as arrow b indicates. This would constitute a discharging Galvanic cell. If the potentiometer supplied more than an equal opposite potential, we would have an *electrolytic* cell which oxidizes silver and reduces protons. This would correspond to charging a reusable battery if no hydrogen were lost.

defined as 0.000 V. The most reasonable basis for this choice is the experimental observation that the hydrogen electrode is near the center of the range of electrode potentials of many of the common elements and their ions. Many of the metallic elements are oxidized by the proton, while other metal ions and most nonmetals can be reduced by elemental hydrogen.

Since the normal hydrogen electrode is not very convenient to construct and use routinely, it is common practice to use other standard reference electrodes whose potentials, relative to the hydrogen electrode, are accurately known. Two such electrodes are the Ag/AgCl electrode and the saturated calomel electrode (SCE). These electrodes are compared in Fig. 7-1. Commercial pH meters frequently use the Ag/AgCl reference electrode while electrochemists often use the SCE. The potentials of the latter two electrodes are both positive with respect to the NHE, which means that their reactions, as written below, are spontaneous relative to the reaction $\frac{1}{2} H_2 \longrightarrow e^- + H^+$. Thus in the NHE/SCE electrode couple, the net spontaneous half-reactions are

$$\frac{1}{2} H_2 \longrightarrow H^+ + e^- \qquad \text{at the cathode}$$

$$e^- + \frac{1}{2} Hg_2Cl_2 \longrightarrow Hg + Cl^- \qquad \text{at the anode}$$

Net reaction: $\frac{1}{2} H_2 + \frac{1}{2} Hg_2Cl_2 \longrightarrow H^+ + Hg + Cl^-$

Similarly, the spontaneous half-reactions for the NHE/Ag/AgCl couple are

$$\frac{1}{2} H_2 \longrightarrow H^+ + e^- \qquad \text{at the cathode}$$

$$e^- + AgCl \longrightarrow Ag + Cl^- \qquad \text{at the anode}$$

Net reaction: $\frac{1}{2} H + AgCl \longrightarrow H^+ + Ag + Cl^-$

Changes in pH or chloride ion concentration will affect the potential difference between these electrodes; for example, the potential of the Ag/AgCl electrode is 0.290 V in 0.1 M KCl, but it is 0.199 V in saturated KCl. Since the equilibrium constant of a reaction can be related to an electrical potential difference, it is easy to see, for example, how one could use the Ag/AgCl reference in combination with a salt bridge and a hydrogen electrode to measure the pH of solutions.

To derive an expression for the potential of a hydrogen electrode as a function of the solution pH, we can begin with the basic equilibrium $\frac{1}{2} H_2 = H^+ + e^-$, for which the equilibrium constant $K = [H^+]/[H_2]^{1/2}$ and $\log K = \log[H^+] - \log(\text{constant})$. Using the important relationships $\Delta G = -nFE = -RT \ln K$, and the definition $pH = -\log [H^+]$, we may write

$$E = \frac{-2.303RT}{nF} [\log(\text{constant}) + pH]$$

Since the potential of the NHE equals 0.00 at pH 0, the value of the term [log(constant)] must also be zero. The log(constant) term is actually the potential of the NHE so that when the pH is not different from 0, the term $+$pH drops out and the equation then becomes an identity. Thus, at any other pH we may write

$$E = \frac{-2.303RT}{nF} \text{ pH} = -0.059\text{pH}$$

As the pH of a solution increases, the emf of a hydrogen electrode in that solution decreases relative to the NHE. This is in general agreement with the notion that as pH increases from zero, H^+ becomes a poorer oxidant and H_2 a better reductant. At the physiological pH of blood and tissues, pH 7.4, the potential of the hydrogen electrode has decreased to -0.43 V vs NHE.

Working with Redox Potentials

EQUILIBRIUM CONSTANTS

One of the main values of measuring emf's is that they can be related to equilibrium constants which might be difficult to obtain by other methods. For example, the pH equilibrium already discussed is an exceedingly important application. Another situation where redox potentials can be used advantageously is in the measurement of K_{sp}, the solubility product constant, for a salt such as silver chloride. By combining the half-reactions one determines

$$
\begin{array}{lll}
AgCl + e^- = Ag + Cl^- & E = 0.222 \text{ V} \\
Ag \quad\quad = Ag^+ + e^- & E = -0.799 \text{ V} \\
\hline
AgCl \quad = Ag^+ + Cl^- & E = -0.577 \text{ V}
\end{array}
$$

that ΔE for the solid-solution equilibrium is -0.577 V. Using the Nernst equation one then calculates that the K_{sp} for AgCl is 1.7×10^{-10}:

$$\log K = \frac{nFE}{2.303RT} = \frac{nE}{0.059} = \frac{(1)(-0.577)}{0.059} = -9.78 \quad \text{or } K = 1.7 \times 10^{-10}$$

The disproportionation of molecules or ions in unstable oxidation states into other more stable oxidation states can also be described quantitatively by relating the potential changes to the equilibrium constant for the process. For example, the cuprous ion is unstable in water, and disproportionates rapidly to copper metal and cupric ion. This can be represented as

$$Cu^+ + e^- \longrightarrow Cu^0 \qquad E = +0.52 \text{ V}$$

$$\underline{Cu^+ \longrightarrow Cu^{2+} + e^- \qquad E = -0.15 \text{ V}}$$

$$2\,Cu^+ \longrightarrow Cu^{2+} + Cu^0 \qquad E = +0.37 \text{ V}$$

From this the equilibrium constant can be calculated:

$$\log K = \log \frac{[Cu^{2+}]}{[Cu^+]^2} = \frac{(1)(+0.37)}{0.059} = 6.271 \text{ or } K = 1.87 \times 10^6$$

Therefore, the cuprous ion almost disappears in aqueous solution, especially if cupric is less than 1 M. This disproportionation is very rapid and is accompanied by a release of free energy calculated as

$$-(1)(96,500 \text{ Coul})(+0.37 \text{ V})(0.239 \text{ cal/Joule}) =$$

$$-8.53 \text{ kcal/mole} \qquad (1 \text{ V} \cdot \text{Coul} = 0.239 \text{ cal})$$

The disproportionation of hydrogen peroxide may be calculated similarly from the equations

$$\tfrac{1}{2} H_2O_2 \longrightarrow e^- + H^+ + \tfrac{1}{2} O_2 \qquad E_0 = +0.68 \text{ V}$$

$$\underline{\tfrac{1}{2} H_2O_2 + e^- + H^+ \longrightarrow H_2O \qquad E_0 = +1.77 \text{ V}}$$

$$H_2O_2 \longrightarrow \tfrac{1}{2} O_2 + H_2O \qquad E_0 = +2.45 \text{ V}$$

$$\log K = \frac{(1)(+2.45)}{0.059} = 41.5$$

and

$$\Delta G = -(1)(96,500)(2.45)(0.239) = -56.5 \text{ kcal/mole}$$

This disproportionation releases a tremendous amount of free energy and at equilibrium the concentration of hydrogen peroxide is infinitesimally small. Yet pure hydrogen peroxide is very stable in solution and shows almost no tendency to decompose "spontaneously" as did the cuprous ion.

These two disproportionation reactions are another illustration of the difference between kinetic and thermodynamic stability (cf. Chapter 6). Both cuprous ion and hydrogen peroxide are *thermodynamically* unstable in aqueous solution, but hydrogen peroxide enjoys *kinetic* stability, the reason being that it has no easily available mechanistic pathway leading to its thermodynamically more stable state, oxygen plus water. In contrast the cuprous ion has a readily available mechanism and thus reaches its equilibrium quickly. If one were to dip a rusty spatula into a solution of hydrogen peroxide, the disproportionation would proceed rapidly, almost explosively if the peroxide were concentrated.

The mechanism shown in Scheme 7-1 is generally accepted (but unproved) to be the rapid reaction pathway by which iron ions catalyze H_2O_2 decomposition. It is based on a mechanism originally proposed by Haber and Weiss in 1934 to explain the autoxidation of ferrous solutions exposed to air.[3] Its chief experimental support comes from kinetic studies and from the fact that if toluene is present, the intermediate hydroxyl radicals (HO·) are intercepted to give o-, m-, and p-cresol in a 59:13:28 ratio. Addition of hydrogen peroxide to acidic ferrous sulfate produces Fenton's reagent, a powerful oxidizing system which has often been compared to certain oxygenase enzymes which also oxidize toluene to form cresols.[4,5]

$$FeOH^{2+} + H_2O_2 \longrightarrow Fe^{2+} + H_2O + HO_2\cdot$$

$$Fe^{2+} + H_2O_2 \longrightarrow HO\cdot + FeOH^{2+}$$

$$HO_2\cdot + FeOH^{2+} \longrightarrow Fe^{2+} + H_2O + O_2$$

$$Fe^{2+} + HO\cdot \longrightarrow FeOH^{2+}$$

$$2\,H_2O_2 \longrightarrow 2\,H_2O + O_2$$

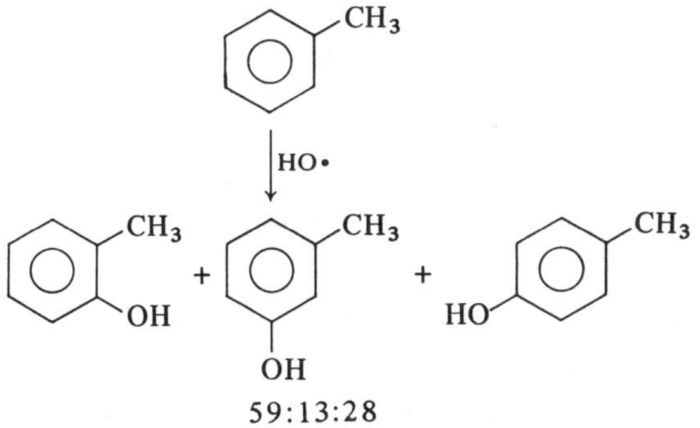

59:13:28

Scheme 7-1

STABILIZATION OF PARTICULAR OXIDATION STATES OF METALS

A preferential increase in the stability of one particular oxidation state of a metal over another will increase the energy difference and the redox potential difference between these states. The best example of this occurs in the formation of complex ions. By correlating the ligand structure with its effect on

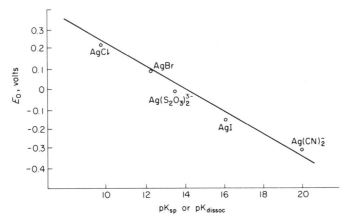

Fig. 7-2. Effect of anions on the emf of the Ag/Ag^+ couple. The points are plotted, the line is drawn to have slope -0.059, as predicted from the equation $E = (2.303RT/nF)\log K = -0.059pK$.

redox potential one can predict what types of ligands can be used to stabilize one particular oxidation state over another. As an example the effects of anions on the emf of the Ag/Ag^+ couple are shown in Fig. 7-2. A further example involving copper complexes and copper redox proteins is discussed in the section on electron transfer mechanisms.

Factors Influencing the Redox Potential of Metals

PERIODIC CORRELATIONS

Since the oxidation of a metal is somewhat similar to the ionization of a metal atom or ion, we might expect to observe periodic trends in redox potential similar to the trends in ionization potentials. Applying this reasoning to the alkali metals leads one to predict that lithium, with the highest ionization potential, would be the poorest reducing agent. In actual fact however, it is the best reducing agent. The reason the prediction was in error is that it neglected the anomalously high hydration energy for lithium. Just as chloride ion stabilizes Ag^+ by forming a stable lattice, water stabilizes Li^+ by forming an extremely stable hydrated cation. In both cases the metal becomes a better reducing agent, and the complexed cation, a correspondingly poorer oxidant.

With the exception of lithium there is a general trend in the redox potential exhibited by nontransition metals: The redox potential (emf) increases toward the *left* of a row and toward the *bottom* of a family in the periodic table. Among the transition metals, trends in redox potentials are complicated by crystal field stabilization energy effects on hydration energy and the problems associated with the changing of d orbital energy levels with atomic number. Among a vertical triad, however, the heavier atoms usually are more stable in higher oxidation states, and the lighter atoms, in their lower oxidation states. Compare, for example, Cr(III) vs WO_4^{2-}, Fe(III) vs OsO_4, and Ni(II) vs Pt(IV).

LIGAND EFFECTS

The lower the energy of an orbital, the more stable an electron in that orbital will be. Because of the way in which ligands split the energies of the d orbitals, the type of orbital an electron enters or leaves during redox will have an effect on the redox potential. This can be seen by considering the Fe(II)/Fe(III) couple for the complexes shown in Diagram 7-1.

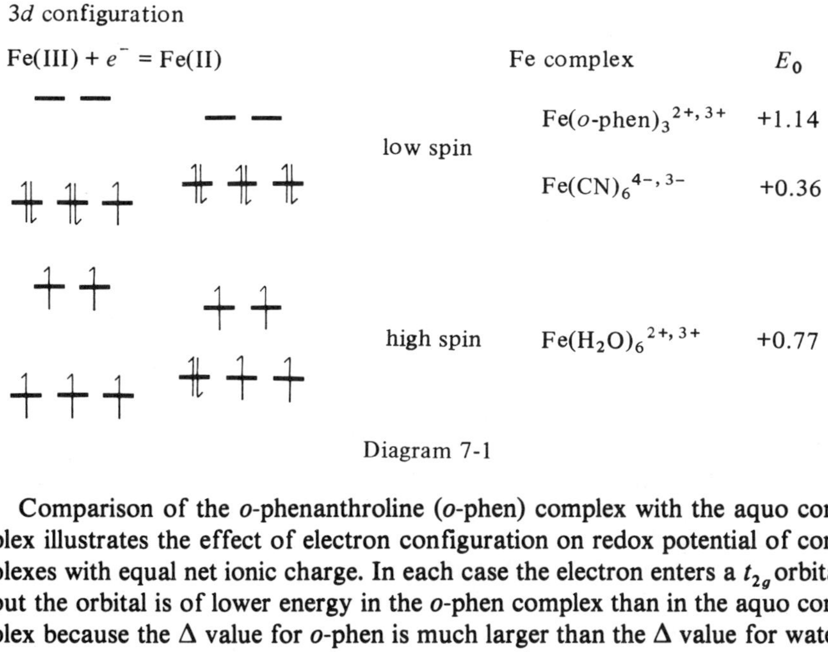

3d configuration

Fe(III) + e^- = Fe(II)   Fe complex E_0

low spin $Fe(o\text{-phen})_3^{2+,3+}$ +1.14

$Fe(CN)_6^{4-,3-}$ +0.36

high spin $Fe(H_2O)_6^{2+,3+}$ +0.77

Diagram 7-1

Comparison of the o-phenanthroline (o-phen) complex with the aquo complex illustrates the effect of electron configuration on redox potential of complexes with equal net ionic charge. In each case the electron enters a t_{2g} orbital, but the orbital is of lower energy in the o-phen complex than in the aquo complex because the Δ value for o-phen is much larger than the Δ value for water. Although the cyano complex has the same electronic configuration as the o-phen complex, its redox potential is far lower, mainly because it has a net charge of 3−, rather than 3+ as the others have, and this large accumulation of negative charge tends to resist the entrance of another electron. Arguments of this type can be useful in comparing or predicting the redox potentials of complexes, as long as it is realized that they may be oversimplifications of a complicated situation and therefore are only qualitative at best.

Redox Reactions of Water and Oxygen

Because of their ubiquity, the redox behavior of water and oxygen is uniquely important. We have already discussed the hydrogen electrode and its sensitivity to pH. A comparable electrode could be constructed for oxygen, but the situation is more complicated since there are intermediate oxidation states for oxygen. The pertinent data for oxygen, hydrogen peroxide, and water are given in Table 7-1[6,7] for the standard electrode conditions (pH = 0), and for

TABLE 7-1

Electrode Potentials of Species in the Water-Hydrogen-Oxygen System[a]

Electrode reaction	$[H^+] = 1$, E_0, (V)	pH 7.4, E_0', (V)
1. $HO\cdot + H^+ + e^- = H_2O$	+2.74	+2.31
2. $\frac{1}{2}H_2O_2 + H^+ + e^- = H_2O$	+1.77	+1.34
3. $HO_2\cdot + H^+ + e^- = H_2O_2$	+1.68	—
$O_2^- + 2H^+ + e^- - H_2O_2$	—	+0.96
4. $\frac{1}{4}O_2 + H^+ + e^- = \frac{1}{2}H_2O_2$	+1.23	+0.80
5. $H_2O_2 + H^+ + e^- = H_2O + HO\cdot$	+0.88	+0.37
6. $\frac{1}{2}O_2 + H^+ + e^- = \frac{1}{2}H_2O_2$	+0.68	+0.25
7. $O_2 + H^+ + e^- = HO_2\cdot$	+0.57	—
$O_2 + e^- = O_2^-$	—	+0.15
8. $H^+ + e^- = \frac{1}{2}H_2$	0.00	−0.43

[a] Data from George[6] and Hayon and Simic.[7]

physiological conditions (corrected to pH 7.4). Overall there is a fairly constant difference between E_0 and E_0' of −0.43 V, a consequence of the effect of pH on the hydrogen electrode (reaction 8). Reactions 3 and 7 are exceptions because the pK_a of the hydroperoxy radical (HOO·) is 4.8 while water and hydrogen peroxide are much weaker acids.

The species HO·, H_2O_2, and HOO· are, in that order, very powerful oxidizing agents. H_2O_2 is a toxic by-product of the autoxidation of some kinds of organic compounds found in cells, such as flavins and dihydropyridines (see below), and is also a convenient source for HO· and HOO· via the Haber–Weiss mechanism. It is therefore not surprising that nearly all aerobic organisms have an iron-containing enzyme known as catalase which removes hydrogen peroxide by catalyzing its disproportionation. The hydroperoxy radical HOO·, or at pH 7.4 the superoxide anion O_2^-, are also formed both enzymatically and nonenzymatically from oxygen present in cells. The cells of all aerobic organisms which have been tested are found to contain a metalloenzyme known as superoxide dismutase (SOD). This enzyme, which catalyzes the reaction

$$2O_2^{\overline{\cdot}} + 2H^+ \rightarrow O_2 + H_2O_2$$

contains at its active site either manganese, copper, or iron, depending on its source organism.[8,9] Molecular oxygen is essential to aerobic organisms but at higher concentrations, such as in oxygen therapy, oxygen causes tissue damage and cell destruction. Current evidence suggests that oxygen toxicity begins by *in vivo* formation of $O_2^{\overline{\cdot}}$ and H_2O_2 via autoxidation reactions. Neither of these is extremely destructive alone, but it is postulated that the two react via a Haber–Weiss type of mechanism to produce HO·, which is very destructive toward any organic cellular constituents.

Respiration: The Biological Reduction of Oxygen[2,10]

A physiologically more constructive use for oxygen depends on harnessing the energy released by its reduction to water. The energy released, per electron, is given by the difference between the potential of oxygen and the "effective" potential of biological reducing agents (i.e., reduced flavins and pyridines, Figs. 7-3 and 7-8) at pH 7.4: $(+0.80) - (-0.32) = +1.12$ V or approximately 26 kcal/mole. Measurements have shown that upon reduction of one-half mole of oxygen ($\frac{1}{2}O_2$), three moles of the energy storage compound adenosine triphosphate (ATP) are formed, corresponding to a capture of about 24 kcal/mole of the theoretically possible yield of 52 kcal/mole, or an efficiency of 46%. The remaining energy not captured by forming unstable (high energy) chemical bonds is wasted as heat, just as heat is wasted by the I^2R heating of a storage battery when it is used to power electrical equipment. This energy-capturing process is analogous to a waterwheel since energy released by the electron as it moves to lower potentials is used to drive the synthesis of high energy chemical bonds. The energy-capturing process is known as oxidative phosphorylation (Fig. 7-8). Agents which block phosphorylation of ADP without blocking substrate oxidation and electron transport are known as "uncouplers." One of the best known uncouplers, 2,4-dinitrophenol, was once used medicinally as a weight reducing agent but is now considered unsafe for this purpose. The idea was that it would lower the efficiency of ATP synthesis thereby wasting a lot of food and stored fat by nonproductive oxidation.

The process of biological electron transport is a magnificent merger of organic and inorganic redox processes involving the synchronization of one-, two-, and four-electron redox processes. Before examining the process we will examine the individual components. Some of them have complicated structures which are associated with enzymatic recognition or solubility roles, but in all cases it is only a very small portion of the molecule, often a single functional group, that becomes involved in the redox chemistry.

Although there are some extremely important biological oxidations in which a substrate is directly oxygenated by an enzyme utilizing molecular oxygen,[11] the biological oxidations important in energy production are actually dehydrogenations as far as the substrates are concerned. By alternating dehydrogenation and hydration reactions it can appear as if molecular oxygen has been introduced directly, thus masking an important distinction in mechanism (see Scheme 7-2). In any case, for these oxygenations and

Direct

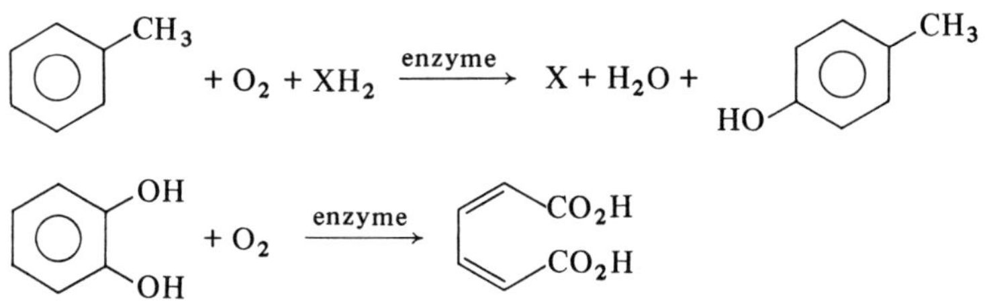

$$\text{[benzene-CH}_3] + O_2 + XH_2 \xrightarrow{\text{enzyme}} X + H_2O + \text{[HO-benzene-CH}_3]$$

$$\text{[benzene with OH, OH]} + O_2 \xrightarrow{\text{enzyme}} \text{[diene-CO}_2H, CO_2H]$$

Indirect

$$RCH_2CH_2CO_2H + X \xrightarrow{\text{enzyme}} XH_2 + RCH{=}CHCO_2H$$

$$RCH{=}CHCO_2H + H_2O \xrightarrow{\text{enzyme}} R\underset{|}{C}HCH_2CO_2H \\ OH$$

$$XH_2 + \tfrac{1}{2}O_2 \xrightarrow[\text{transport}]{\text{electron}} X + H_2O$$

Scheme 7-2

dehydrogenations, X and XH$_2$ are the oxidized and reduced forms of the coenzymes FAD, FMN, NAD, or NADP, whose structures are given in Fig. 7-3.

FLAVINS AND PYRIDINES

As indicated in Fig. 7-3, the chemical behavior of these coenzymes are different. The flavins generally react like free radical hydrogen abstractors, or like hydrogenation catalysts, and they often interconvert $-CH_2CH_2-$ and $-CH{=}CH-$ groups in substrate molecules. The pyridine derivatives, on the other hand, are often involved in oxidation of alcohols or reduction of ketones or aldehydes, and their chemistry is better represented as hydride donation to a carbonyl group or its reverse. One of the best studied enzymes which utilizes NADPH as a cofactor, alcohol dehydrogenase (ADH) has a zinc ion at its catalytic site (reaction 7-1). Evidence continues to accumulate suggesting that

Oxidized forms   Reduced forms

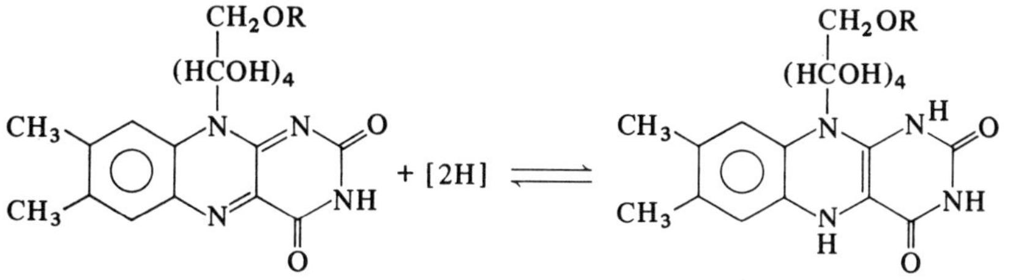

Flavin adenine dinucleotide (FAD)   (FADH$_2$)

 R = $-PO_2^-$$-O-$ribose$-$adenine

flavin mononucleotide (FMN)   (FMNH$_2$)

 R = PO_3^{2-}

Nicotinamide-adenine dinucleotide (NAD$^+$) (NADH)

 R = $-PO_2^-$$-O-$ribose$-$adenine

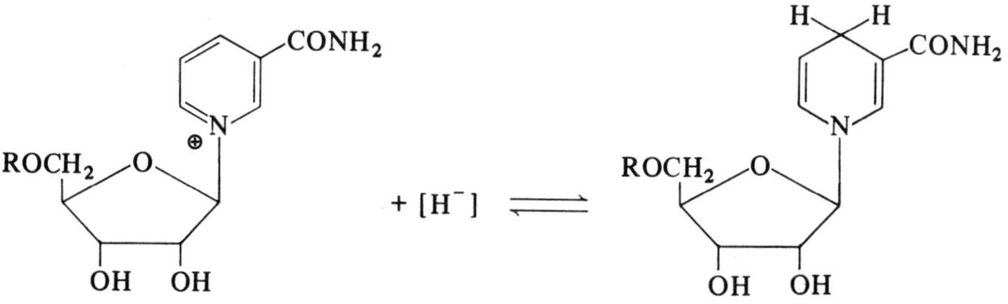

Nicotinamide-adenine dinucleotide phosphate (NADPH)
 (NADP$^+$)

 R = $-PO_2^-$$-$ribose$(PO_3^{2-})$$-$adenine

Fig. 7-3. Structures of hydrogen transfer coenzymes.

the mechanism of this enzyme may be similar to the Oppenauer oxidation and the Meerwein–Pondorff–Verely (MPV) reduction (reaction 7-2).[12]

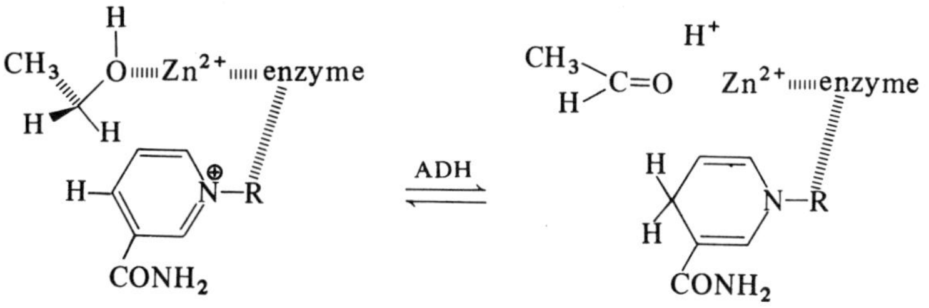

(7-1)

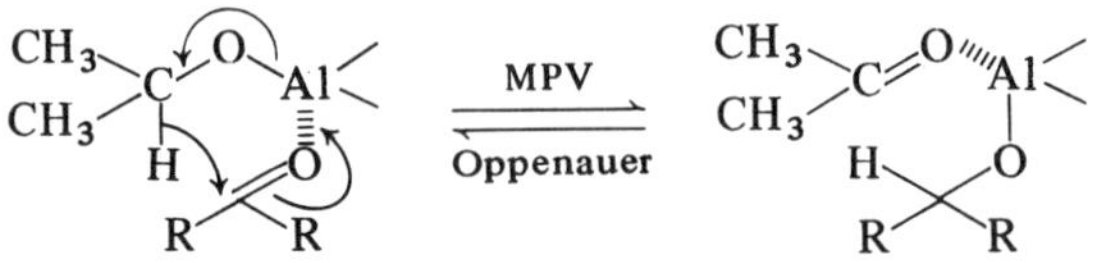

$$
\begin{array}{ccc}
\text{CH}_3\text{—C—O—Al} & \xrightleftharpoons[\text{Oppenauer}]{\text{MPV}} & \text{CH}_3\text{—C—O—Al} \\
\text{CH}_3 \quad \text{H} \quad \text{O} & & \text{CH}_3 \quad \text{H} \quad \text{O} \\
\text{R} \quad \text{R} & & \text{R} \quad \text{R}
\end{array} \qquad (7\text{-}2)
$$

QUINONES

Another class of components of the electron transfer chain which are organic in nature are the quinones and their corresponding reduced or hydroquinone forms. Various kinds of quinones have been found in all forms of life examined. In plants, plastoquinones are found in chloroplasts, the subcellar photosynthetic apparatus. In mammals, ubiquinone, also called coenzyme Q_{10} or CoQ, is found in the mitochondria, which is where electron transport and oxidative phosphorylation (ATP synthesis) occur. Vitamin K is a closely related quinone, but its primary function is not involved with redox. The impor-

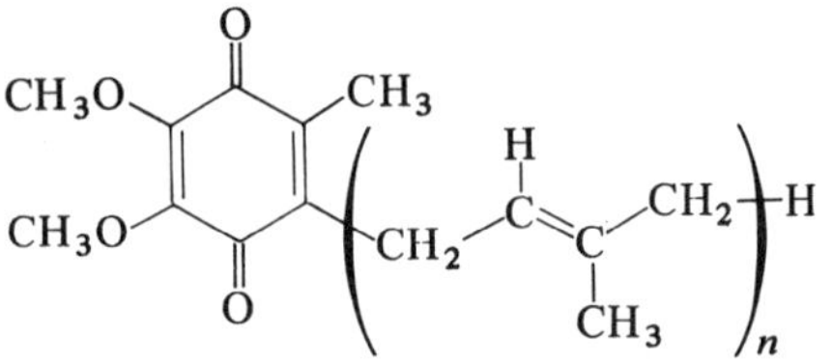

Ubiquinone

$$\left(\begin{array}{l}n = 10 \text{ in mammals} \\ n = 6\text{--}9 \text{ in other species}\end{array}\right)$$

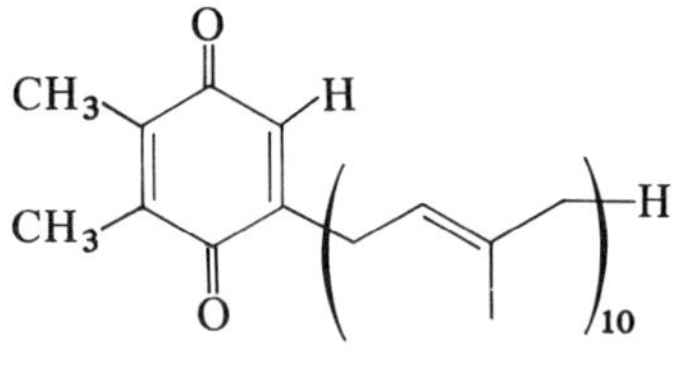

Plastoquinone

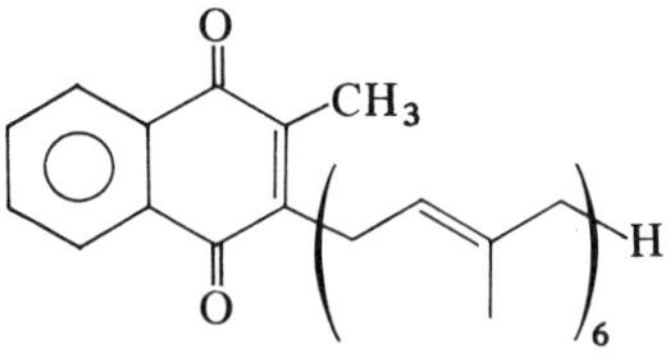

Vitamin $K_2(30)$

tant feature of these compounds from the redox point of view is their ability to undergo two-electron oxidations or reductions in one-electron steps. This flexibility allows them to act as couplers of redox processes which are obligate one- or two-electron changes. The intermediate form is the semiquinone, a free radical or anion radical which is stabilized by resonance.

Vitamin C, or ascorbic acid, is also an important reducing agent in cells of all kinds. The functional group involved is an enediol, which bears some resemblance in structure and chemical properties to the enediamine unit of the

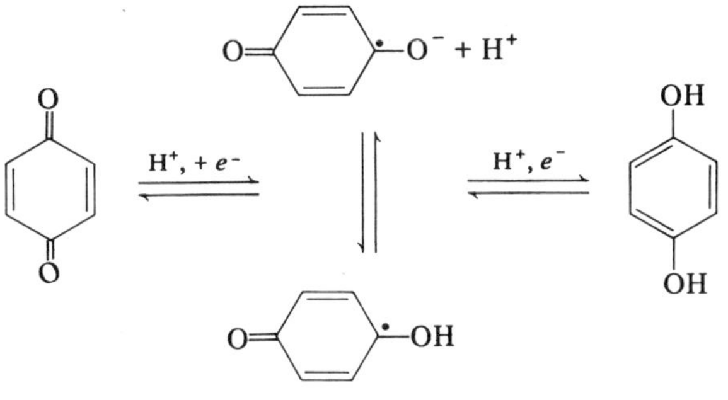

Quinone form   Semiquinone form   Hydroquinone form

flavins and to catechol and hydroquinone groups. Except for man and a few vegetarian mammals, vitamin C is not a dietary requirement for it is biosynthesized by all plants and all other animals. In living cells ascorbic acid functions as a hydrogen donor and is a cofactor for an important brain enzyme, dopamine-β-hydroxylase.

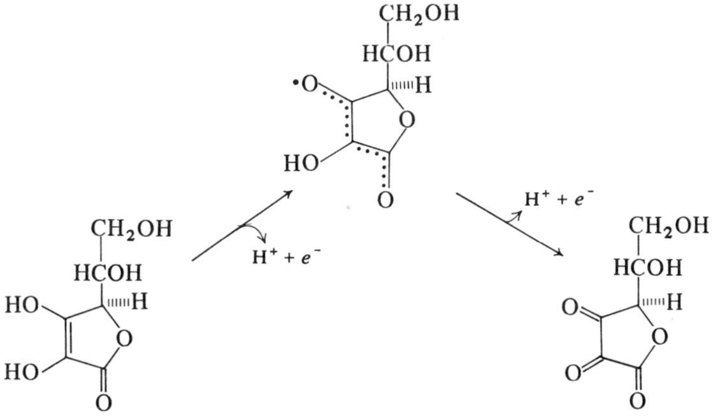

L-Ascorbic acid   Dehydroascorbic acid

HEMOPROTEINS

The cytochromes[13,14] are a group of mitochondrial respiratory redox proteins which contain iron in the form of a heme prosthetic group (see Table 7-2).* They are divided into three categories a, b, and c, according to their absorption spectra, although there are other differences. More than a dozen varieties of a-, b-, and c-type cytochromes have been isolated from diverse biological sources. Other hemoproteins include myoglobin and hemoglobin,

* A prosthetic group is a nonprotein unit which is tightly bound by and necessary to the functioning of a biologically active protein; coenzymes are less tightly bound.

TABLE 7-2

Properties of Representative Hemoproteins and Hemes[a]

Proteins[b]	$MW \times 10^{-3}$	Redox components	Ligands		CO/CN⁻ inhibition	Autoxidation	Potential (V)
			Fifth	Sixth			
Hemoglobin	64.5	4 Ferroprotoheme	His(N)	—	Yes	No	−0.13
Myoglobin	17	1 Ferroprotoheme	His(N)(protein)	—	Yes	No	+0.046
b-Cytochromes	13–40	1 Ferroprotoheme	Protein	—	No	No	−0.05 to +0.10
Cytochrome P_{450}	40	1 Ferroprotoheme	Protein	—	CO	Yes	−0.27 to −0.41
c-Cytochromes	12–37	1 Heme C	His(N)	Met(S)	No	No	+0.20 to +0.26
a-Cytochromes	100	2 Heme A + 2 Cu	Protein?	—	Yes	Yes	+0.25 to +0.35
Catalase	200	4 Ferroprotoheme	$-CO_2^-$ or His(N)	—			
Heme derivatives							
Protoheme IX • 2 pyr			Pyr	Pyr			+0.137
Heme A • 2 pyridine			Pyr	Pyr			+0.288

[a] Data from Mahler and Cordes.[2]

[b] Abbreviations: Pyr, pyridine; His, histidine; and Met, methionine.

which function in oxygen storage and transport, respectively, and catalase, which was mentioned earlier.[15] There are several types of hemes, each characteristic of a group of hemoproteins. Hemoglobin, myoglobin, catalase, and the *b*-type cytochromes all contain protoheme, or ferroprotoporphyrin IX, bound to the protein by ionic and hydrophobic interactions. The *a*-type cytochromes contain heme A, and the *c*-type cytochromes contain heme C. In the case of heme C, however, there are two covalent thioether bonds from the protein to the prosthetic heme unit.

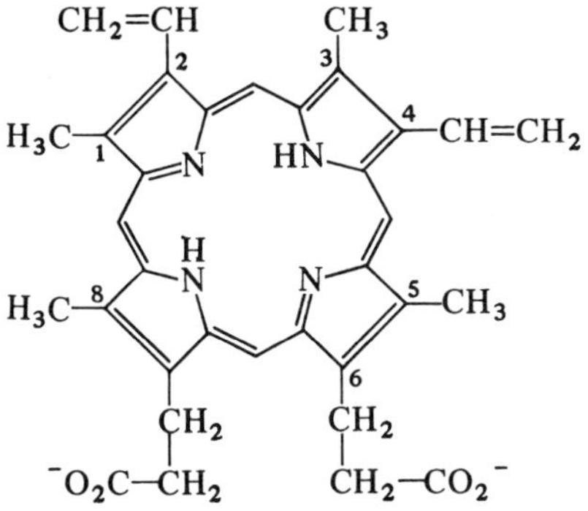

Protoporphyrin IX

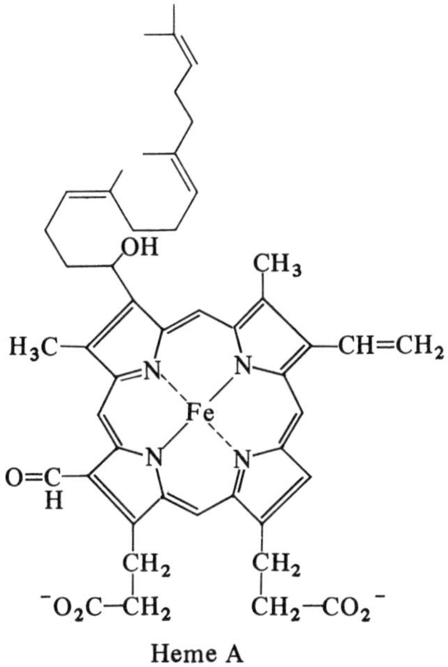

Heme A
Prosthetic group of
cytochromes of class A

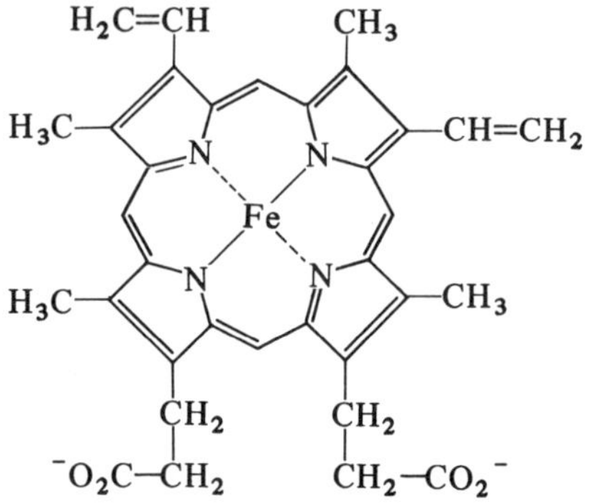

Ferroprotoporphyrin IX
(proto)heme (IX)

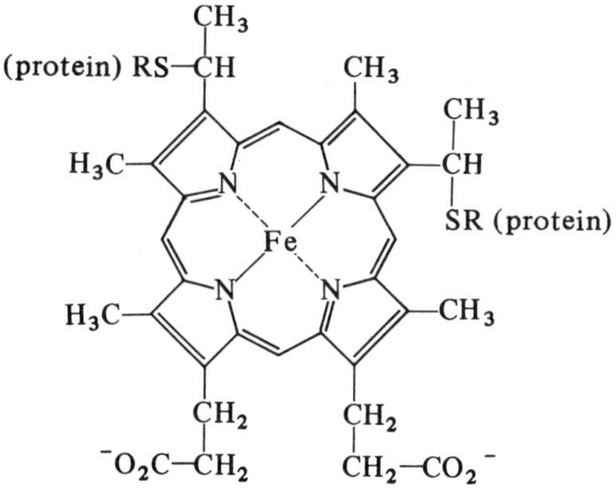

Heme C
Prosthetic group of
cytochromes of class C

In the ferroprotoporphyrin IX prosthetic group the iron is four-coordinate and sits slightly above the square plane of the porphyrin ligand. The iron in all hemoproteins is either five- or six-coordinate because the protein donates at least one ligand, often an imidazole-N from a histidine side chain. Whether or not the protein donates a sixth ligand to the iron has very important consequences for the chemical reactivity of the iron center. In hemoglobin and myoglobin, which are not redox proteins, the iron is five-coordinate. This allows room for an oxygen molecule to add reversibly and complete the coordination sphere of the iron. In the absence of the globin protein, ferroheme is rapidly oxidized to its Fe(III) form *hemin*. The nature of the iron–oxygen interaction in hemoglobin itself has eluded scientists for many years.

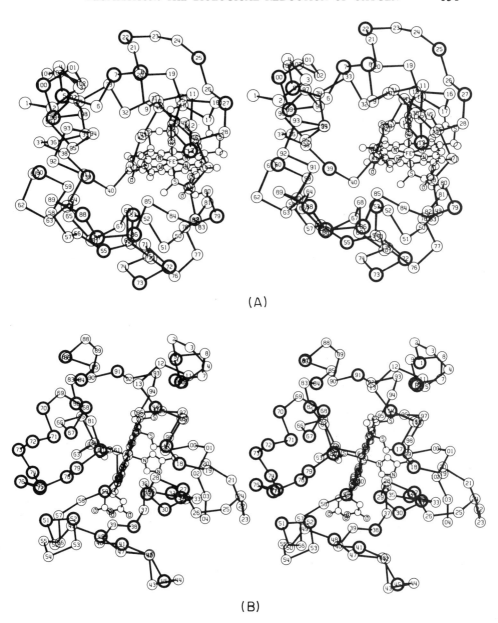

(A)

(B)

Fig. 7-4. Stereoscopic α-carbon diagrams of oxidized horse heart cytochrome c at 2.4 Å resolution. Note the heme crevice and the axial thioether and imidazole ligands. The apparently hollow center is packed with side chains of nonpolar or "hydrophobic" amino acids.[16] (A) and (B) are different views of the same structure.

Ferrohemoglobin is a paramagnetic, high spin complex, but oxyhemoglobin is diamagnetic [low spin d^6Fe(II)]. The change in magnetic properties has been attributed to the oxygen–iron interaction, which can be viewed either as side-on binding of the π-acid oxygen to the iron or as the formation of an anti-ferromagnetically coupled Fe(III)-$O_2^=$ ion pair or charge-transfer complex. Appropriate chemical models for both kinds of behaviour are described in Chapter 10. Figure 7-4 shows stereoscopic diagrams of oxidized horseheart cytochrome c.[16]

Iron is also five-coordinate in the a cytochromes. But rather than reversibly binding oxygen, the a-type ferrous cytochromes are *autoxidized* by oxygen to their ferric forms. Both hemoglobin and cytochrome a can bind cyanide or carbon monoxide as the sixth iron ligand. The fatal effects of cyanide poisoning are associated with inhibition not of hemoglobin but of the autoxidation of cytochrome a, thereby preventing its catalytic function in respiration. Indeed, for cyanide poisoning the usual treatment is inhalation of amyl nitrite or injection of sodium nitrite, which oxidizes some hemoglobin to methemoglobin [Fe(III)]. Cyanide is much more tightly bound to methemoglobin than to either hemoglobin or cytochrome a, and is removed as blood is filtered through the spleen, where erythrocytes containing large amounts of methemoglobin are selectively destroyed. Methemoglobin does not bind CO; thus, for carbon monoxide poisoning the only hope is to administer oxygen and displace the CO by an equilibrium process. The a-type cytochromes have molecular weights in the range of 100,000 to 200,000, and have two hemes and two copper ions per molecule. They are the only autoxidizable respiratory cytochromes, and they function at the high potential end of the electron transport chain in transferring electrons to oxygen.

In contrast to hemoglobin and the a-type cytochromes, the iron centers of the respiratory b and c cytochromes are sixth-coordinate. These cytochromes do not react with cyanide or CO and are not autoxidized by oxygen. The b cytochromes in general have lower potentials than the c types; thus, the overall flow of electrons in respiration is $b \rightarrow c \rightarrow a \rightarrow$ oxygen. All the respiratory cytochromes participate in electron transport by undergoing alternate one-electron oxidation and reduction. Their broad range of redox potentials is an elegant example of the fine control which can be exerted by subtle changes in the substituents on the porphyrin periphery and in the nature of the protein–iron interaction.

COPPER REDOX PROTEINS

Many nonheme metalloproteins have long been recognized to participate in biological redox reactions. The two main themes of interest surrounding these proteins, as compared to simple complexes of the same metals, have been their atypical absorption and electron spin resonance spectra and their abnormally high or low redox potentials. Most of the evidence suggests that the metal has a very unusual coordination geometry, compared to simple complexes, but more will be said about this later.

A number of redox proteins and oxidase enzymes require copper for activity. In some cases it is known that copper alternates between Cu(I) and Cu(II) as it functions, but in other cases this oxidation state change cannot be detected.[17,18] The redox potential of many of these proteins is in the range +0.3 to +0.4 V, compared to −0.5 to +0.2 V for ordinary copper complexes. The copper enzyme, ceruloplasmin, has attracted much attention, possibly because its deficiency in humans causes Wilson's disease, a rare degenerative disorder of the brain and liver. Ceruloplasmin has a molecular weight of 1.3 to 1.5 × 10^5 daltons, contains seven atoms of copper, and constitutes 0.5% of the protein and 98% of the copper in normal human blood plasma. One of its roles is simply to serve as a storage and transport form for copper. More recently it was discovered that it is an enzyme for oxidizing ferrous iron. In the body iron is absorbed and transported across membranes as Fe(II), but it is transported in plasma and stored in tissue as Fe(III)–protein complexes. Conversion of Fe(III) to Fe(II) takes place nonenzymatically with a variety of organic reducing agents such as ascorbic acid, catechols, sulfhydryl compounds, and phenols. Thus ceruloplasmin, a metalloenzyme, is an important factor in the metabolism of the metallosubstrate, iron, and has a direct effect on the synthesis of heme and iron proteins.

NONHEME IRON PROTEINS

During the 1950's and early 1960's several proteins were isolated which contained nonheme iron, and had the unusual property of producing hydrogen sulfide when acidified.[19-24] Further studies disclosed (1) that these proteins had very unusual spectral and redox properties compared to ordinary iron complexes; (2) that they could be found in many species of plants, animals, and microorganisms; (3) that they were involved in such diverse redox processes as photosynthesis, hydroxylation of steroid hormones, and nitrogen fixation; and (4) that they contained up to eight iron atoms and eight "labile sulfides." In addition, ESR and Mössbauer studies indicated that the iron was in a site of low symmetry. Other properties of some iron–sulfur proteins are given in Table 7–3.[19-24] During the early 1970's structure elucidation studies revealed new and unusual coordination sites for the iron in these proteins. In some cases the synthesis of iron complexes with structures closely similar to those of the iron–sulfur cores of the proteins has been achieved.

TABLE 7-3

Physicochemical Properties of Some Iron–Sulfur Proteins[a]

Protein	Source	Functions in	MW	Fe:S^{2-}	E_0' (mV)	EPR signal	
						Oxidized	Reduced
Rubredoxin	C. pasteurianum	?	6000	1:0	−60	+	+
	P. oleovorans	Hydroxylation	19,000	2:0			
Putidaredoxin	P. putida	Hydroxylation	12,000	2:2	−235	−	+
Adrenodoxin	Adrenal gland	Hydroxylation	13,000	2:2	−367	−	+
Ferredoxin	Chloroplasts of spinach	Photosynthesis	10,600	2:2	−430	−	+
HIPIP[b]	Chromatium	?	10,000	4:4	+350	+	−
Ferredoxin	B. polymyxa	Nitrogen fixation?	9000	4:4	−390	−	+
Ferredoxin	C. pasteurianum	Nitrogen fixation	6000	8:8	−410	−	+

[a] Data from references 19–24.
[b] High potential iron protein.

Rubredoxin

The simplest of the iron–sulfur proteins is rubredoxin, a red protein containing one Fe and no "labile sulfide." The iron is bound by four cysteine side chains in a site of distorted tetrahedral geometry. As shown in Fig. 7-5,[25] one

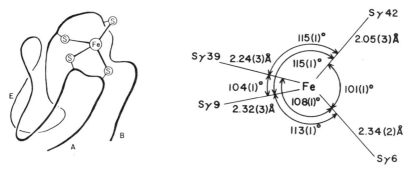

Fig. 7-5. Structure of rubredoxin from *C. pasteurianum*. A and B designate the N- and C-terminus, respectively. The geometry about the iron is also given.[25]

of the Fe–S bonds is unusually short. Rubredoxins from several bacterial species have been purified and sequenced. The peptides consist of 53 or 54 residues, giving a molecular weight of ~ 6000, and although there are variations in the sequences, the positions of the four cysteines are invariant at positions 6, 9, 39, and 42.

The distortion in the coordination site (entasis?) affects the properties of the protein in a variety of ways. The red oxidized form of the protein contains high spin ferric iron, while the colorless reduced form contains high spin ferrous iron. The redox potential (E_0', pH 7) for rubredoxin is -0.057 V for the transfer of a single electron. As yet there is no known metabolic role for the one-iron rubredoxins which cannot also be filled by ferredoxin. The rubredoxin from *Pseudomonas oleovorans* functions in aliphatic hydroxylation reactions, but this protein is larger (MW 19,000) and has two isolated one-iron sites.

2 Fe: 2 S²⁻ Ferredoxins

Proteins with $2\,Fe:2\,S^{2-}$ groups have been isolated from chloroplasts of many species of plants, several diverse species of bacteria, beef and pig adrenal gland, and beef heart mitochondria. They are characterized by a low redox potential (-0.2 to -0.4 V) for a one-electron change, intense visible absorption bands which decrease by about 50% in intensity upon reduction, paramagnetism in the reduced form ($S = \frac{1}{2}$), and diamagnetism in the oxidized form.

Although crystal structures are not yet available for these ferredoxins, ENDOR and Mössbauer spectroscopy indicate high spin iron in a roughly tetrahedral site,[22,26] as shown in Fig. 7-6. The apparent diamagnetism of the

$$\text{Cys}-\text{S} \diagdown \diagup \text{S} \diagdown \diagup \text{S}-\text{Cys}$$
$$\text{Cys}-\text{S} \diagup \text{Fe} \diagdown \text{S} \diagup \text{Fe} \diagdown \text{S}-\text{Cys}$$

Fig. 7-6. Geometry of the active site in the 2 Fe:2 S proteins.

oxidized form is accounted for by antiferromagnetic coupling. Each high spin Fe(III) has a net spin of 5/2, but because they are connected by sulfide bridges, the irons recognize each other's spin states and couple to give a net S of zero. The bridging sulfurs are reactive and can be titrated with triphenylphosphine, which is converted to Ph_3PS. Treatment with acid produces H_2S, which is detectable even at low levels by its familiar odor.

8 Fe: 8 S^{2-} and 4 Fe:4 S^{2-} Ferredoxins

Although both of these types of ferredoxins contain similar Fe:S cores, their properties are widely different. The 8 Fe:8 S^{2-} proteins have very low redox potentials ($E_0' = -0.25$ to -0.43 V), which ranks them among the most powerful of all biochemical reducing agents. *Clostridial* ferredoxin, for example, functions in nitrogen fixation (*q.v.*) as an electron transfer species. Bacterial ferredoxins are small proteins (MW $\sim$ 6000) having two 4 Fe:4 S^{2-} clusters, each of which undergoes a one-electron change (see Fig. 7-7). The oxidized forms are diamagnetic and are thought to contain equal amounts of ferric and ferrous iron. Evidently there is appreciable antiferromagnetic coupling of spins, and/or extensive spin delocalization through the covalent character of the Fe—S (and possibly Fe—Fe) bonds.

X-ray diffraction studies of the ferredoxin from *P. aerogenes* shows that the two 4 Fe:4 S^{2-} cores are each suspended by four cysteine residues in a sea of hydrophobic amino acid side chains.[27] No doubt this serves to insulate the delicate clusters from disruption by nucleophiles or solvent. The two clusters are *ca.* 12 Å apart in the protein, but located in this gap are two tyrosyl side chains whose phenolic rings are parallel to and in contact with one face of their nearest 4 Fe:4 S^{2-} cluster. One edge of each phenolic side chain is also exposed to solvent, and this may be important as an electron transfer pathway, much as with the exposed porphyrin edge in cytochrome *c*.

The 4 Fe:4 S^{2-} ferredoxin isolated from *Chromatium*, high potential iron protein (HIPIP), is an enigma. It has essentially the same 4 Fe:4 S^{2-} unit as found in the other 8 Fe:8 S^{2-} bacterial ferredoxins, yet the potential for its one-electron change is nearly 0.75 V higher ($E_0' = +0.35$ V). Unfortunately, its

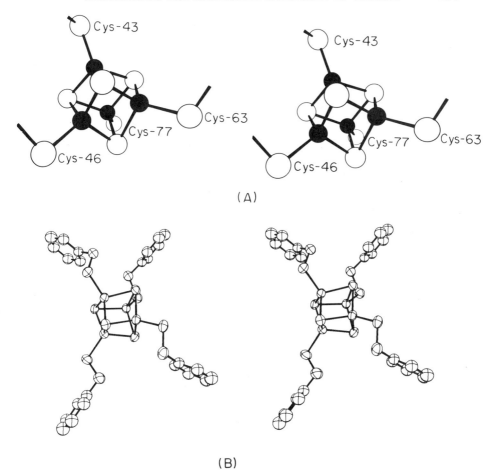

Fig. 7-7. Structures of 4 Fe:4 S^{2-} clusters. (A) Stereoscopic representation[28] of the cluster from *Chromatium* HIPIP, and (B) stereoscopic view of the [$Fe_4S_4(SCH_2Ph)_4$]$^{2-}$ ion.[32] (Reprinted with permission from *Accounts of Chemical Research*. Copyright by the American Chemical Society.)

physiological role is not yet known. There are two possible explanations for the great difference in redox potential.[28] It is possible that there are structural differences between the 4 Fe:4 S^{2-} clusters of the two kinds of proteins which are not detectable at the 2.0–2.5 Å resolution level at which the proteins have been examined. Alternatively, there may be a third as yet undetected redox state for HIPIP, possibly even lower than that for other reduced ferredoxins. Three oxidation states have been observed for model iron–sulfur cluster complexes (see below).

Synthetic Models for Iron–Sulfur Proteins

In designing synthetic models for Fe:S proteins perhaps the most important factor to consider is geometry; the absorption spectra, spin states, and perhaps even redox potentials should then follow in order. In the case of the ferredoxins and HIPIP, another important target property is the chemically reactive "labile sulfide" moiety. The various physical, chemical, and spectral properties whose composite describes the Fe:S proteins have been observed independently in a number of iron complexes of sulfur-containing ligands. However, relatively few ion–sulfur complexes display several of the requisite characteristics to make them good models, and thus far only one fully characterized complex which is truly a model for the iron–sulfur proteins is known.

The active site of rubredoxin has been approached by several iron chelates of $(S)_2$ ligands.[29,30] The complexes are air-sensitive, tetrahedral, and high spin, but the asymmetry of the rubredoxin site is lacking. The decapeptide

Off-white solid
SFeS = $100.5°$–$114.9°$
Fe—S = $2·34$–$2·38$ Å

yellow solid
tetrahedral

$(H_2N)Cys$-Thr-Leu-Cys-Gly-Cys-Pro-Leu-Cys-Gly(CO_2H) is claimed to form an iron complex with spectral properties similar to those of the rubredoxin.[31] However, since no complex was actually isolated, and no biological assay is available, the significance of this "model" is open to question.

The Fe_2S_2 proteins have not been modeled very closely although some interesting approaches have been made. Monomeric and dimeric Fe(III) complexes of alkylthioxanthates $(RSCS_2^-)$ have been prepared by Lippard and co-workers.[21] The monomeric complex $Fe(tBuSCS_2)_3$ is octahedral and has a redox potential $(E_{1/2})$ of -0.36 V. n-Alkyl analogs extrude CS_2 and form binuclear complexes with metal–metal bonds and much lower redox potentials. Although these are octahedral low spin complexes which do not have any "labile sulfide," their low redox potentials mimic one important aspect of the ferredoxins. In this case the low redox potential of $[Fe(PrS)(PrSCS_2)_2]_2$ is attributed to the population of the σ^* orbital of the metal–metal bond upon reduction of the dimer from the Fe(III) to the Fe(II) state. The reduced forms were not isolated, but a related *isoelectronic* Co(III) complex was found *not* to be *isostructural* and not to have a metal–metal bond. Thus in the Fe_2S_2

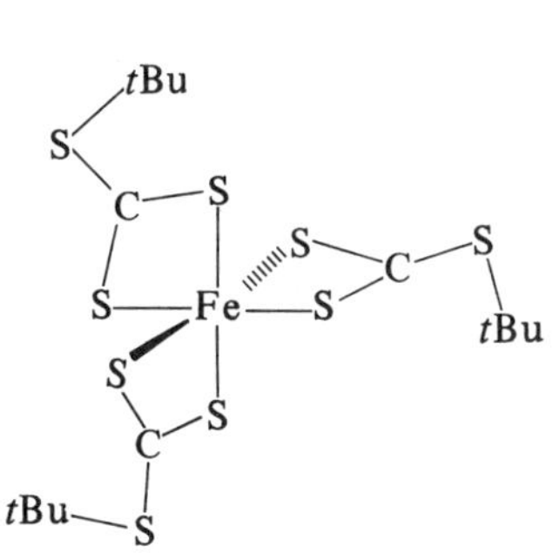

$E_{1/2} = -0.36$ V

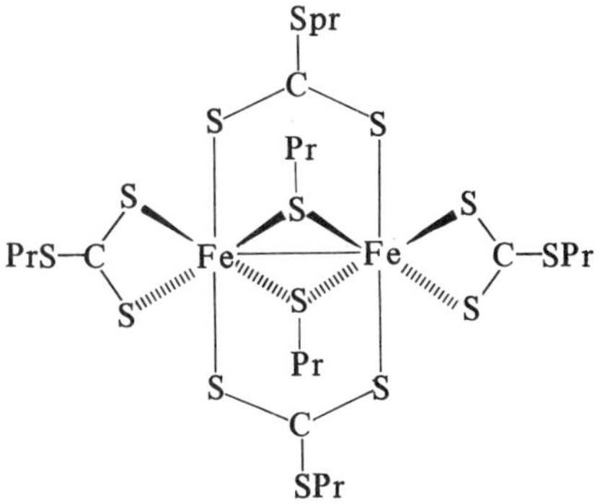

$E_{1/2} = -0.64$ V

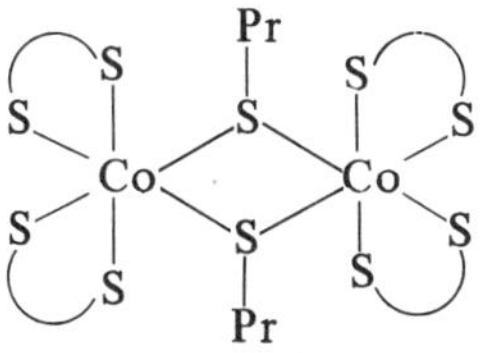

S—S = PrSCS$_2^-$

ferredoxins metal–metal bonds, together with rigid protein structure tending to resist demands for rearrangement of the coordination geometry, could be contributing factors in explaining the low reduction potentials.

A very good chemical model has been found for the iron–sulfur core of the Fe$_4$S$_4$ proteins. The complex [Et$_4$N]$_2$[Fe$_4$S$_4$(PhCH$_2$S)$_4$], consisting *formally* of 2 Fe(II), 2 Fe(III), 4 S^{2-}, and 4 PhCH$_2$S$^-$ units in addition to the quaternary cations has been prepared recently.[32,33] Detailed comparisons of this complex with oxidized ferredoxin from *M. aerogenes* and with reduced HIPIP from *Chromatium*, in terms of Mössbauer spectra, absorption spectra, magnetic susceptibility, and bond lengths and angles about the Fe$_4$S$_4$ unit, suggest that all three species possess virtually identical core structures. Since the reduction potentials of ferredoxin and HIPIP differ by nearly 0.75 V, the close similarity of their core structures is a paradox. The potential difference ($\sim$ 17.5 kcal/mole) is far too great to be accounted for by "entatic" effects resulting from the protein backbone. Carter *et al.* have rationalized this paradox in terms of a "three state" model[28] analogous to one proposed for chlorophyll. This model hypothesizes the existence of an additional reduced state of low potential for HIPIP, as well as an additional oxidized state of high potential for ferredoxin. This hypothesis also ties in the magnetic properties of the proteins and the model complex[32] (see accompanying tabulation). Although the Fe$_4$S$_4$

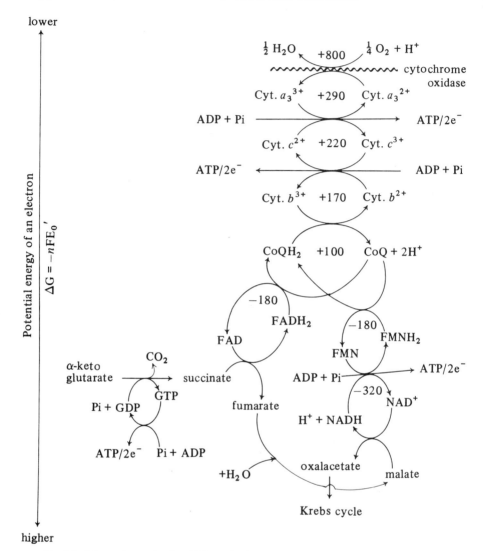

Fig. 7-8. Mammalian mitochondrial respiration. Malate, oxalacetate, fumarate, succinate, and α-ketoglutarate are substrates in the Krebs cycle; Pi is inorganic phosphate. Phosphorylation of ADP is coupled to the flow of electrons from a high potential energy level (substrates) to a low potential energy level (H_2O). In some nonmammalian systems iron–sulfur proteins couple between flavins and CoQ. The numbers give the approximate E_0' in millivolts. See also Table 7-7.

cores formally contain pairs of Fe(II) and Fe(III), no differences in the irons can be detected from X-ray studies or from photoelectron spectroscopy of the model.[33] X-ray studies of the protein cores are of lower resolution but also fail

Fe:S complex	Redox change[a] (C = Fe₄S₄ core)	Potential (V), E_0' or (vs SCE)	
		Observed	Hypothesized
Fd	$*C^+ + e^- = C$		+0.7?
HIPIP	$*C^+ + e^- = C$	+0.35	
Fd	$C + e^- = *C^-$	−0.43	
HIPIP	$C + e^- = *C^-$		−0.8?
$Fe_4S_4(PhCH_2)_4^{2-}$	$C + e^- = *C^-$	(−1.19)	
$Fe_4S_4(PhCH_2)_4^{2-}$	$*C^+ + e^- = C$	(−0.15)	

[a] Asterisk, Paramagnetic species.

to detect differences between the irons. The only significant difference observed is that upon oxidation the core in HIPIP contracts by about 0.1–0.2 Å, as expected for tighter bonding in higher oxidation states.[28]

ELECTRON TRANSPORT AND OXIDATIVE PHOSPHORYLATION

In aerobic bacteria and in the mitochondria of aerobic plant and animal cells the oxidation of organic molecules, with oxygen as the ultimate electron acceptor, is used to supply energy to the organism. The relationship between substrates, oxygen, and the members of the electron transport chain in mammalian systems is shown in Fig. 7-8. As the electrons from substrates flow through the pathway they release energy. Some of this energy is captured by the poorly understood process of oxidative phosphorylation in the form of polyphosphate bonds of ATP.[2,10] When ATP is hydrolyzed to ADP + PO_4^{3-}, the conserved energy is released. This basic reaction is coupled to all biological processes which do work, such as the generation of osmotic or electrostatic gradients, muscle contraction, or chemical synthesis.

Electron Transfer Mechanisms for Metals

Studies of redox reactions of transition metals have shown that there are basically two kinds of mechanisms for electron transfer, "outer sphere" mechanisms and "inner sphere" mechanisms.[34-37] In addition to this distinction, redox reactions are divided for study and discussion purposes into self-exchange reactions between two different oxidation states of the same metal, and redox reactions between complexes with different metal centers. In this section we will briefly illustrate the nature of the mechanisms and the factors which can influence the rates of redox reactions.

OUTER SPHERE REACTIONS

The simplest type of redox reaction is the outer sphere self-exchange reaction, which can be represented in terms of contact between the outer electron clouds of the reacting complexes. When the contact occurs the electron "jumps" from an orbital on the reducing complex to an orbital on the oxidizing complex (Scheme 7-3). In a typical outer sphere experiment, a Fe(II) complex is

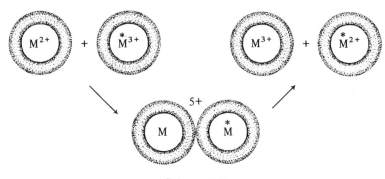

Scheme 7-3

mixed with its corresponding Fe(III) complex which contains *radioactive* iron (Fe*). At various times after mixing the Fe(II) and Fe(III) forms are separated, perhaps by ion exchange chromatography, and each is analyzed for radioac-

Complex	Electron configuration (II)	(III)	$k(M^{-1}\ sec^{-1})$	$T(^{\circ}C)$
1. $Cr(H_2O)_6$	$\begin{array}{c}+\ -\\+\ +\ +\end{array}$	$\begin{array}{c}-\ -\\+\ +\ +\end{array}$	$\leqslant 2 \times 10^{-5}$	25
2. $Fe(H_2O)_6$	$\begin{array}{c}-\ -\\+\!\!\!+\ +\!\!\!+\ +\!\!\!+\end{array}$	$\begin{array}{c}-\ -\\+\!\!\!+\ +\!\!\!+\ +\end{array}$	4.2	25
3. $Co(H_2O)_6$	$\begin{array}{c}+\ +\\+\!\!\!+\ +\!\!\!+\ +\end{array}$	$\begin{array}{c}-\ -\\+\!\!\!+\ +\!\!\!+\ +\!\!\!+\end{array}$	5	25
4. $Co(en)_3$	$\begin{array}{c}+\ -\\+\!\!\!+\ +\!\!\!+\ +\!\!\!+\end{array}$	$\begin{array}{c}-\ -\\+\!\!\!+\ +\!\!\!+\ +\!\!\!+\end{array}$	1.4×10^{-4}	50
5. $Fe(o\text{-}phen)_3$	$\begin{array}{c}-\ -\\+\!\!\!+\ +\!\!\!+\ +\!\!\!+\end{array}$	$\begin{array}{c}-\ -\\+\!\!\!+\ +\!\!\!+\ +\end{array}$	$> 10^5$	0
6. $Co(o\text{-}phen)_3$	$\begin{array}{c}+\ -\\+\!\!\!+\ +\!\!\!+\ +\!\!\!+\end{array}$	$\begin{array}{c}-\ -\\+\!\!\!+\ +\!\!\!+\ +\!\!\!+\end{array}$	1.1	0

Fig. 7-9. Second-order rate constants for electron exchange reactions ($M^{2+} + M^{3+} \rightarrow M^{3+} + M^{2+}$) and electronic configurations of the reactants.[36,37]

tivity. The rate at which radioactivity appears in the Fe(II) form, or disappears from the Fe(III) form, is a measure of the rate of electron exchange at an iron center. The rates of outer sphere electron transfer for some transition metal ions is given in Fig. 7-9 along with the d electron configurations of the reactants. By comparing the relative rates we can determine the effect of ligand structure and electron configuration on the exchange rate, just as we did with exchange equilibria and redox potentials. Before doing so, however, we must keep two things in mind, the Franck–Condon principle and the effect of oxidation state on the geometry of the complex.

Different oxidation states of a given transition metal have different preferences for bond lengths and bond angles, and any change in oxidation state must be accompanied by an appropriate change in the coordination sphere. If the coordination sphere does not adjust following a redox change, the resulting complex will not be in its most stable form. This line of reasoning has some interesting consequences for redox potentials as well as redox rates. Consider the ions Cu(I) and Cu(II). The former prefers tetrahedral coordination while the latter prefers square planar or tetragonal coordination, and a redox interconversion would require the reorganization shown in reaction 7-3.

$$\text{Tetrahedral Cu(I)} \qquad \xrightleftharpoons[{+e^-}]{{-e^-}} \qquad \text{Square planar Cu(II)} \tag{7-3}$$

Now suppose that the ligands X were tied together in such a way that they could only adopt one or the other geometry, but not both. This would certainly stabilize the oxidation state that preferred that particular geometry and, in turn, could move the redox potential for the reaction far outside the "normal" range. A good example of this is the copper chelate of the tridentate Schiff base tris(anhydro)-o-aminobenzaldehyde. This ligand can conveniently occupy one face of an octahedron or tetrahedron, but cannot satisfactorily bind to a metal requiring square planar geometry such as Cu(II). Thus when Cu(II) is used as a template for the condensation of o-aminobenzaldehyde, a Cu(I) complex of the trimer ligand is isolated.[38] In the blue copper proteins such as ceruloplasmin, a similar geometric redox potential effect may be operative. This line of reasoning led Williams and Vallee to formulate their "entatic state hypothesis" (Chapter 6). It is easy to see how it might be applied to the blue copper redox proteins and to the iron–sulfur proteins to explain their "abnormal" redox potentials.

Returning to the discussion of redox rates we may note that changes in bond length, if not bond angles as well, probably occur for every change in oxidation

Copper chelate of the tridentate Schiff base
tris(anhydro)-o-aminobenzaldehyde

state for all transition metals. Bonds from ligands to trivalent metal ions are usually shorter and tighter than bonds to divalent metal ions. If an electron is added to an Fe(III) complex, for example, the Fe–ligand bonds will be lengthen. Since redox is an electronic event, it must be subjected to the Franck–Condon principle. This principle states that electrons pass between orbitals faster than the stretching and bending motions of chemical bonds and that *during an electronic event* (absorption of $h\nu$ or redox) *the nuclei of the reacting molecule do not change their relative positions.* The implication of this principle is that if an electron is "pumped in" to a Fe(III) center it will enter so rapidly that the Fe–ligand bonds won't have time to elongate properly. This means the first-formed product would have to be an excited-state Fe(II) complex, and the reduction would be very endothermic overall. To avoid these complications it is usually considered that the metal center must undergo a geometrical reorganization *before* it can accept an electron. If the reorganization requires little input of "activation energy" the reduction may proceed rapidly, but if a large activation energy is required the metal will be hesitant to accept an electron.

The data of Fig. 7-9 show that second-order rate constants for electron exchange may vary over an extremely large range. If reactions 1 and 4 are compared to reaction 2 the effects of Franck–Condon restrictions on rates of electron transfer can be seen. The low spin $d^3\mathrm{Cr(H_2O)_6^{3+}}$ ion and the low spin $d^6\mathrm{Co(en)_3^{3+}}$ ion have very stable electron configurations. They require a large activation energy for metal–ligand bond elongation prior to accepting an electron, and therefore react slowly. In contrast the $\mathrm{Fe(H_2O)_6^{3+}}$ ion does not have such a stable configuration; consequently, it has a lower activation energy for reorganization and undergoes faster electron exchange. The same comparison may be seen in reactions 5 and 6. Unfortunately, the situation is not always this simple, and there is no satisfactory explanation for the high rate constant for reaction 3. Although hydrolysis could be a factor, high spin cobalt would also fit the arguments nicely. Comparisons of reactions 6 and 4 or 5 and 2 show that if there are π-bonding ligands on the metals, the rate of electron exchange may be increased by a factor of 10^4 or more. The effect of the π-bonding ligands is to provide a low energy pathway between the metal d orbitals and the outer reaches of the complex which contact each other during their redox encounter. The aquo and amine complexes not having a set of orbitals with such favorable energy and symmetry undergo slower electron exchange.

In the redox protein cytochrome c, the porphyrin is a π-bonding ligand system, and although most of the heme is buried in the protein, one edge of the porphyrin is relatively exposed at the surface of the protein. This can easily be seen in the crystal structure diagrams shown in Fig. 7-4. From this one can estimate that about 3% of the surface of a cytochrome c molecule is occupied by the exposed porphyrin edge, which means that the probability of two cytochrome c molecules colliding in solution at the porphyrin edges for an outer sphere transfer of an electron is $(0.03)^2$ or about 10^{-3}. Using the Fe(o-phen)$_3^{2+}$/Fe(o-phen)$^{3+}$ couple as a model for the π-bonding heme of the cytochromes, one can crudely calculate an expected cytochrome c electron self-exchange rate about 10^{-3} that of the Fe(o-phen)$_3$ system.[35] Since the exchange rate of the Fe(o-phen)$_3$ system approaches 10^8 M^{-1} sec^{-1} at 25°C, one could expect a cytochrome c electron self-exchange rate of about 10^5 M^{-1} sec^{-1}. The observed value is 5×10^4 M^{-1} sec^{-1}. Although this does not prove that outer sphere mechanisms are operative in cytochrome systems, it certainly supports such an hypothesis.

INNER SPHERE REACTIONS

For couples which react via electron transfer at rates faster than those at which they undergo ligand exchange, it is usually safe to postulate an outer sphere mechanism. However, for complexes which undergo ligand exchange as fast as or faster than electron transfer, an inner sphere mechanism may be involved. Taube, Meyers, and Rich[38a] first demonstrated this mechanism by taking advantage of the extreme substitution inertness of Co(III) and Cr(III) ions. In the reaction

$$[(NH_3)_5CoCl]^{2+} + Cr^{2+} \text{ aq} \longrightarrow [ClCr(OH_2)_5]^{2+} + Co^{2+} \text{ aq}$$

the only reasonable way to explain the appearance of chloride in the product is to assume it was retained by chromium upon breakdown of a *chloride-bridged transition state*:

$$[(NH_3)_5Co \ldots Cl \ldots Cr(OH_2)_5]^{4+}$$

Various anions and groups bridge with different efficiency, generally in the following order:

$$HO^- > F^- > Cl^- > Br^- > I^- > N_3^- > NCS^- > CN^- > H_2O > CH_3CO_2^-$$

The fact that fluoride is a highly effective bridge rules out the possibility that the bridge itself is sequentially oxidized by Co(III) and reduced by Cr(II). A transfer involving electron tunneling through s and p orbitals is more likely. Unsaturated π-bonding ligands can bridge, but in contrast to the outer sphere mechanism they are not the best bridging ligands for inner sphere mechanisms.

Redox Reactions of Coordinated Ligands

Metal ions may facilitate electron transfer or redox reactions in several ways. In what is perhaps the simplest of these, the metal brings the two reactants together as ligands and acts as a bridge between them. This process is roughly analogous to the bridging of two metal centres by one ligand during an inner sphere electron transfer process. In both cases the bridging system constitutes a low energy pathway, often involving π-bonding orbitals, by which the electron transfer can proceed. Metals can also catalyze redox reactions by undergoing alternate oxidation by one substrate and reduction by another. This type of pathway depends on the existence of two oxidation states of the metal which are not too widely separated in energy, as well as facile mechanisms for *both* the oxidation and reduction reactions of the metal ion. Another property unique to metals is the frequent existence of several spin states which are close in energy. This allows the metals to escape the spin restrictions which prevent, for example, the direct combination of singlet organic molecules with molecular oxygen in its triplet ground state. Several metal complexes reversibly form diamagnetic oxygen adducts, such as hemoglobin, myoglobin, and others which are mentioned in Chapter 10. It has been suggested by some that the latter may represent complexes of singlet oxygen, but hard evidence to support this idea is lacking. Still another device which metals may use in facilitating or catalyzing redox reactions is the stabilization of "unusual" oxidation states of organic compounds by virtue of their *complexation* to the metal. This appears to be involved in the formation of the superoxide complexes which are discussed in more detail in Chapter 10. The superoxide ion itself is rather unstable in aqueous solution, and its formation is *endo*thermic under acidic conditions; hence, it tends to disproportionate rapidly[8] ($k_2 \approx 10^8$ M^{-1} sec^{-1} in water).

OXIDATION OF OLEFINS BY ONE- AND TWO-ELECTRON PROCESSES

Because of the very factors which make transition metals such good catalysts for redox reactions, it is often difficult to determine the mechanisms of the reactions. When the oxygen molecule is one of the reactants it can become a very sticky problem to determine whether one-electron (free radical) or two-electron steps are involved. Under normal circumstances, free radical autoxidation is the most common mechanism for the aerobic oxidation of organic compounds. This process is a self-perpetuating chain reaction in many cases, so that only a little help from an initiator is needed to get things started. If the propagating steps are much faster than the terminating steps, the chain length will be very long and the process will be efficient in terms of moles of product per mole of initiator (In). Metal ions can "promote" the oxidation of $R-H$ by catalyzing the formation of new initiating and propagating species as shown in Scheme 7-4. Many transition metal ions can promote these reactions,

Initiation: $In \cdot + RH \longrightarrow InH + R \cdot$

Propagation: $R \cdot + O_2 \longrightarrow RO_2 \cdot$

$RO_2 \cdot + RH \longrightarrow R \cdot + RO_2H$

Termination: Radical species $\longrightarrow$ nonradical products

$(e.g., 2 R \cdot \longrightarrow R-R)$

"Catalysis" or
"promotion": $RO_2H + M^{2+} \longrightarrow M^{3+} + HO^- + RO \cdot \rightarrow$ more initiation

$RO_2H + M^{3+} \longrightarrow M^{2+} + H^+ + ROO \cdot \rightarrow$ more propagation

Scheme 7-4

but manganese, iron, and cobalt are particularly effective. In these reactions very little of the hydroperoxide can be detected, although it is certainly an intermediate.

The discovery of metal complexes which form stable *diamagnetic* adducts with molecular oxygen revived interest in metal-catalyzed oxygenation of organic compounds, particularly olefins because of the tremendous importance of ethylene oxide and propylene oxide for antifreeze and plastics. It was hoped that if oxygen and olefin could be brought together in the coordination sphere of a metal ion, a clean specific and facile reaction such as 7-4 or 7-5 might occur to give oxygenated organic products. Reaction 7-4 is reminiscent of the reaction of olefins with singlet oxygen, and it has been speculated that the com-

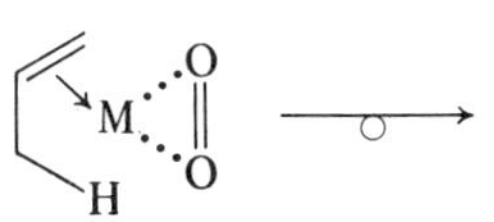

$$(7-4)$$

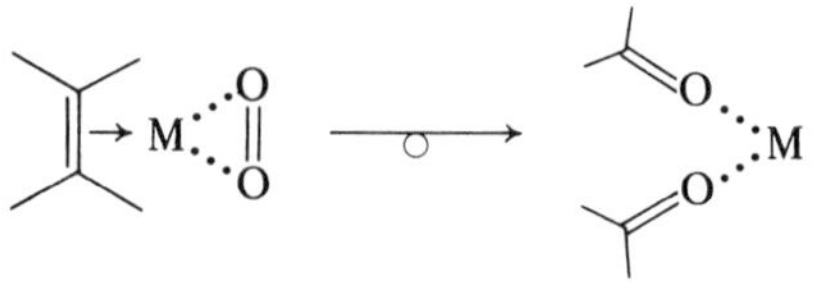

$$\text{(7-5)}$$

plexed oxygen is in fact in its singlet state since the complex is diamagnetic. There are now many accounts of such attempts in the chemical (and patent) literature,[15] but perhaps one of the most informative concerns the oxidation of tetramethylethylene with Vaska's compound $IrCl(CO)(Ph_3P)_2$ or its rhodium analog $RhCl(CO)(Ph_3P)_2$ to form[39] epoxide and allylic alcohol (see reaction 7-6). In this system it was shown that the metal complexes probably catalyzed

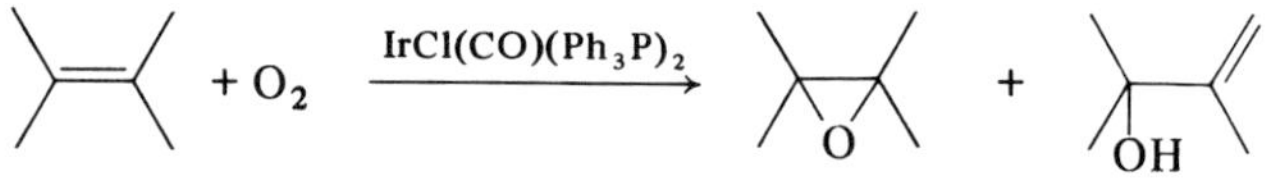

$$\text{(7-6)}$$

the formation of a hydroperoxide (reaction 7-7) *and* the subsequent reduction of this peroxide by the olefin to form the observed products, epoxide and alcohol (reaction 7-8).

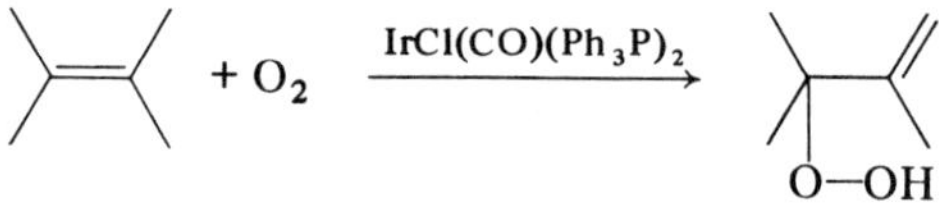

$$\text{(7-7)}$$

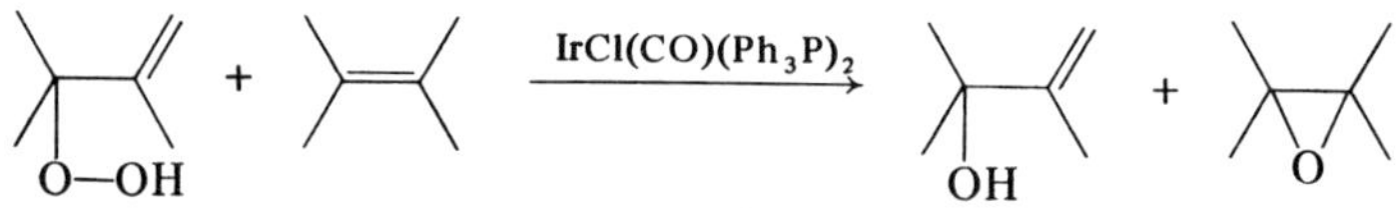

$$\text{(7-8)}$$

The rhodium- and iridium-catalyzed epoxidations of an olefin with a hydroperoxide are probably similar to the epoxidation of olefins catalyzed by various molybdenum and vanadium complexes. These reactions were first described in the patent literature, but some mechanistic studies have been carried out. With molybdenum catalysts the rate law is first order in Mo, peroxide, and olefin, and the epoxidation of *cis*- and *trans*-2-butene is stereospecific. This suggests that one function of the metal is to complex and polarize the peroxide unit in order to promote its *heterolytic* cleavage, much as the carbonyl group polarizes the peroxide unit in peracetic acid (reaction 7-9).[40]

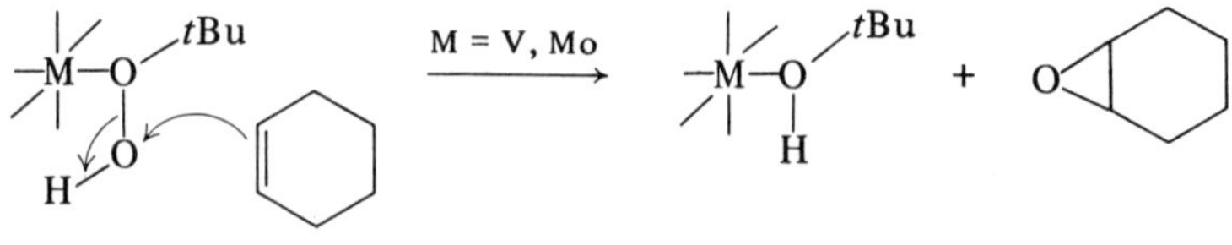

$$\text{(7-9)}$$

In contrast, metals which easily undergo one-electron oxidation and reduction, such as cobalt and manganese, promote the homolysis of peroxides. The peroxy radicals which are formed attack olefins and give epoxides but the process is nonstereospecific (Scheme 7-5).[41]

$$ROOH + M^{2+} \longrightarrow RO\cdot + M^{3+} + HO^-$$

$$ROOH + M^{3+} \longrightarrow M^{2+} + H^+ + ROO\cdot$$

$$ROOH + RO\cdot \longrightarrow ROH + ROO\cdot$$

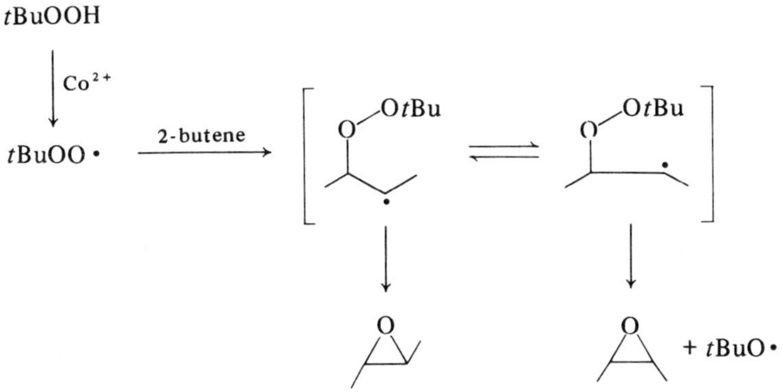

cis-2-Butene $\longrightarrow$	1.5:1.0
trans-2-Butene $\longrightarrow$	3.3:1.0

Scheme 7-5

Whether the hydroperoxide in reaction 7-7 is formed via the free radical route or a mechanism such as (7-4) is not known. However, the autoxidation of cyclohexene with complexes related to $IrCl(CO)(PH_3P)_2$ and $RhCl(CO)(Ph_3P)_2$ is strongly inhibited by hydroquinone, which suggests that the free radical route is most important. In addition attempts to trap singlet oxygen upon decomposition of the oxygen adducts of $IrCl(CO)(Ph_3P)_2$ and $RhCl(CO)(Ph_3P)$ have not succeeded (see below and also Chapter 10).

The term "singlet oxygen" usually refers to the oxygen molecule in its lowest excited state, $^1\Delta$, which lies 23.4 kcal above the triplet $^3\Sigma$ ground state. There is also another excited singlet $^1\Sigma$ state 37.5 kcal above the ground state, but this state is not involved in most of the chemistry[42] of singlet oxygen. The electron configurations of these states differ only in the π^* MO level (7-10); the complete MO diagram for $^3\Sigma O_2$ is given in Fig. 5-15. The characteristic reactions of singlet oxygen generated by either of these means, with organic compounds,

$$\pi^* \text{ orbitals} \quad \underset{}{\uparrow}\,\underset{}{\uparrow} \xrightarrow{\frac{23.4}{\text{kcal}}} \underset{}{\uparrow\downarrow} \; - \xrightarrow{\frac{14.1}{\text{kcal}}} \underset{}{\uparrow}\,\underset{}{\uparrow} \qquad (7\text{-}10)$$

$$\text{State} \qquad\qquad {}^3\Sigma \qquad\qquad {}^1\Delta \qquad\qquad {}^1\Sigma$$

are Diels–Alder Addition across *cisoid* dienes and formation of allylic hydro-peroxides via what seems to be a relatively concerted electrocyclic process (Scheme 7-6).

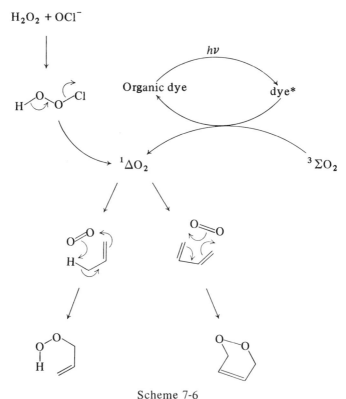

Scheme 7-6

Both the formation and reactions of singlet oxygen are good examples of the operation of the principle of spin conservation. Thus the reason that ${}^3\Sigma O_2$ and singlet organic molecules do not combine in a direct concerted fashion is that the product would necessarily be a triplet, which therefore makes the process highly endothermic and unlikely (reaction 7-11). Singlet oxygen, however, may and does combine readily as indicated. Note the striking similarity of these reactions to the Diels–Alder reactions of maleic anhydride with dienes and the "ene" reaction of olefins. Similarly, the decomposition of HOOCl formed from

$$^3\Sigma O_2 \quad \begin{matrix} O \overset{\uparrow\downarrow}{=\!=} O \\ + \\ C \diagdown \diagup C \\ \quad C \!-\! C \quad \end{matrix} \quad \xrightarrow{\;\;/\!/\;\;} \quad \begin{matrix} \downarrow\uparrow O \!-\! O \uparrow\downarrow \\ C \diagdown \qquad \diagup C \\ C \!=\! C \\ \uparrow\downarrow \end{matrix} \qquad (7\text{-}11)$$

Triplet excited state

H_2O_2 and OCl^- must produce singlet oxygen rather than triplet oxygen since the reactant is a singlet species and the decomposition is a two-electron process.

Free radicals are in a unique position with respect to spin conservation restrictions in that they can react with either triplet or singlet states of molecules. In this respect they are very similar to paramagnetic transition metal ions. Therefore the stepwise one-electron changes involved in free radical autoxidation of organic compounds are facile and subject to promotion by metal compounds which can undergo alternate oxidation and reduction. The superoxide anion radical has attracted a large amount of attention as a possible biological precursor to singlet oxygen. This possibility seems extremely unlikely however, since there is no apparent reason why in giving up an electron the more stable $^3\Sigma$ state should not be formed, i.e., only energetics and not spin conservation is involved (reaction 7-12).[43] However if electron transfer is very rapid, the singlet state can be formed.[44]

$$(7\text{-}12)$$

Singlet oxygen has attracted some attention as a possible intermediate in some enzymatic oxidations, and it has even been reported that enzymes such as the iron-containing lipoxygenases, which convert unsaturated fatty acids to allylic hydroperoxides, can convert "singlet oxygen traps" to compounds identical to those formed using chemically generated singlet oxygen. However, it is doubtful that results with highly unnatural substrates designed to trap singlet oxygen bear any relationship to the normal enzymatic reactions because the same products could possibly have arisen via mechanisms other than the trapping of singlet oxygen. Evidence against singlet oxygen as an intermediate in the action of lipoxygenase on fatty acids was provided by a deuterium labeling study which showed that the hydrogen atom abstraction and oxygen processes occurred from *opposite* sides of the plane of the carbon system.[45] In singlet oxygen reactions both events take place on the same side of the plane of the

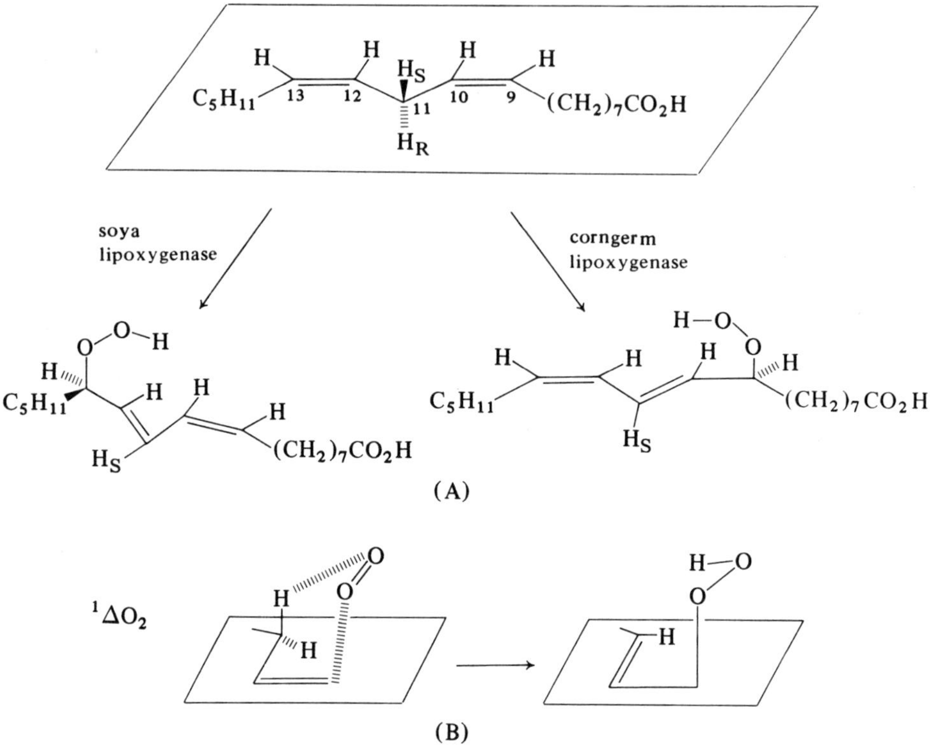

Fig. 7-10. Enzymatic (A) vs chemical (B) formation of allylic hydroperoxides. Part A is from Egmond *et al.*[45]

allylic system (Fig. 7-10). A similar conclusion was reached for the enzymatic hydroperoxidation of cholesterol.[46]

One case in which a metal–oxygen compound actually appears to furnish singlet oxygen upon decomposition is K_3CrO_8. This compound oxygenates olefins to give exactly the product distribution obtained from the reaction of the olefins with photochemically generated singlet oxygen.[47]

SOME REACTIONS WHICH FORM OR CONSUME HYDROGEN PEROXIDE

The reduction of molecular oxygen by ascorbic acid is strongly catalyzed by transition metal ions, and both ionic (two-electron) and free radical (one-electron) mechanisms may be involved, depending on the reaction conditions. The two pathways can be distinguished by their kinetic rate laws and the difference in hydrogen peroxide production (Fig. 7-11).[48] In the absence of extraneous ligands the reaction is first order in ascorbate, copper, and oxygen,

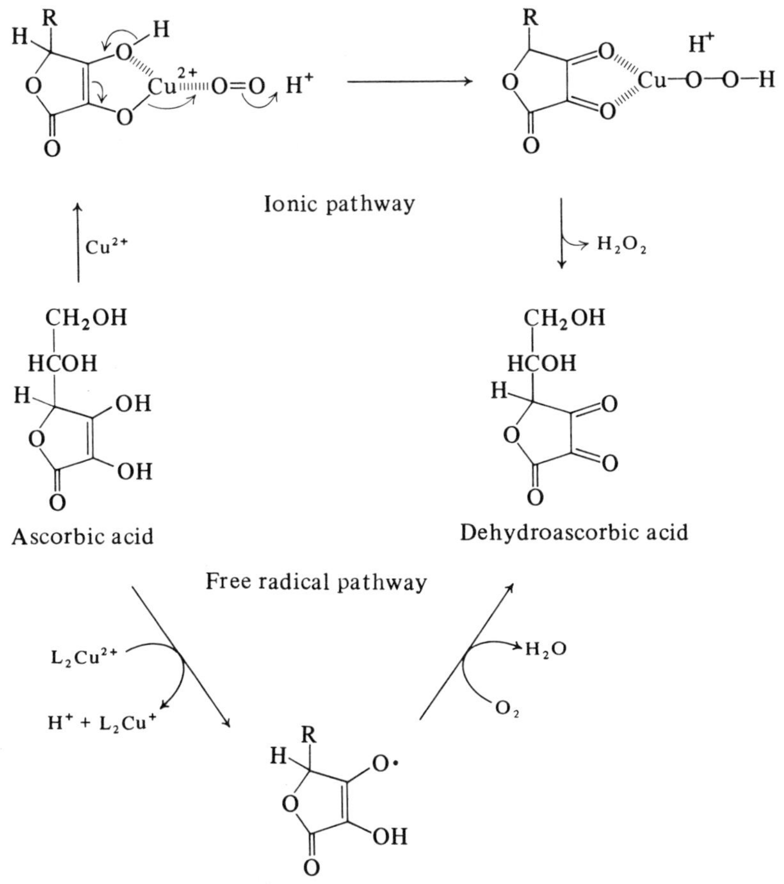

$$4\,H^+ + 4\,L_2Cu^+ + O_2 \longrightarrow 4\,L_2Cu^{2+} + 2\,H_2O$$

Fig. 7-11. Copper-catalyzed oxidation of ascorbic acid.[48]

which suggests a ternary copper complex, such as I in Fig. 7.11, as an intermediate. Two electrons pass directly from the enediol ligand through orbitals on copper to the oxygen molecule, which is reduced to hydrogen peroxide in this case. In the presence of extraneous ligands, as could be the case in a copper enzyme for example, the copper cannot simultaneously bind both ascorbate and oxygen; thus, the reaction must proceed stepwise. In this case the cupric complex oxidizes ascorbate to a resonance-stabilized semiquinonelike radical which is further oxidized by dissolved oxygen. The cuprous complexes are also very rapidly reoxidized by molecular oxygen, and in this case water rather than hydrogen peroxide is produced.

Similar studies have been done on the oxidation of catechols, and other "enediols" such as acetoin and benzoin.[49] In the iron-catalyzed autoxidation of acetoin, there is a metal-dependent pathway which produces hydrogen peroxide, but in this case there is a faster competing pathway in which iron catalyzes the oxidation of acetoin with the hydrogen peroxide produced in the first reaction. Ferric iron can also oxidize acetoin to biacetyl anaerobically and without hydrogen peroxide. Similar redox chemistry occurs in the body where various phenols, catechols, or mercaptans reduce the tightly bound ferric form of iron in ferritin to the loosely bound ferrous form. It appears that in no case is a "semiquinone" anion radical involved. In the autoxidation of catechols to produce o-quinones and hydrogen peroxide, the observed order of catalytic effectivity is Mn(II) > Co(II) > Fe(II) > Cu(II) ≫ Ni(II), which is quite different from the Irving-Williams sequence of reactivity. This same sequence has also been observed in the oxidative deamination of amino acids catalyzed by pyridoxal and a metal (reaction 7-13).[50] This order differentiates this particular pyridoxal reaction from the others discussed in Chapter 8, which clearly

$$R\diagdown C\diagup CO_2H \quad \xrightarrow[\text{H}_2\text{O, }100°\text{C}]{M^{2+},\text{ pyridoxal, O}_2} \quad NH_4^+ + R\!-\!\overset{\displaystyle O}{\underset{\displaystyle \|}{C}}\!-\!CO_2^- + H_2O_2 \qquad (7\text{-}13)$$

follow along lines of Irving–Williams and Lewis acidity behavior. Such sharp divergence from the Irving–Williams order can be considered *prima facie* evidence for the existence of redox processes.

Hydrogen peroxide is a metabolic by-product in living systems, and because of its toxicity nearly all aerobic organisms possess an enzyme known as catalase which scavenges hydrogen peroxide by disproportionating it to water and oxygen.[51] Catalase is a ferric hemoprotein and even hemoglobin and myoglobin have a weak catalaselike activity, as does free iron. The mechanism of the enzymatic reaction is not known, but it probably does not involve Fe(II)–Fe(III) interconversions since these would lead to highly reactive radical products derived from H_2O_2. A more likely mechanism would involve the intermediacy of a ferryl or oxoiron(IV) derivative of the heme. The mechanism shown in 7-14 is attractive, based on model studies, but unlikely since a histidine imidazole from the protein blocks one of the sides of the heme.

$$ \text{(7-14)} $$

The catalaselike activity of Cu(II) has been studied extensively by Sigel.[52] The reaction rate depends on the *number* of ligands in the coordination sphere of the copper,

$$Cu(en)^{2+} > Cu(dien)^{2+} \gg Cu(trien)^{2+}$$

as well as on the type of ligands present:

$$\text{bipy} > \text{en} > \text{glycinate} > P_2O_7{}^{2-}$$

From this information and the rate law for the reaction the mechanism shown in reaction 7-15 was deduced. Two open coordination sites are required for efficient catalysis, and the qualitative effects of the various ligands arise from

$$\text{Rate} = k[\text{bipy Cu}^{2+}][H_2O_2]^2[H^+]^{-1}$$

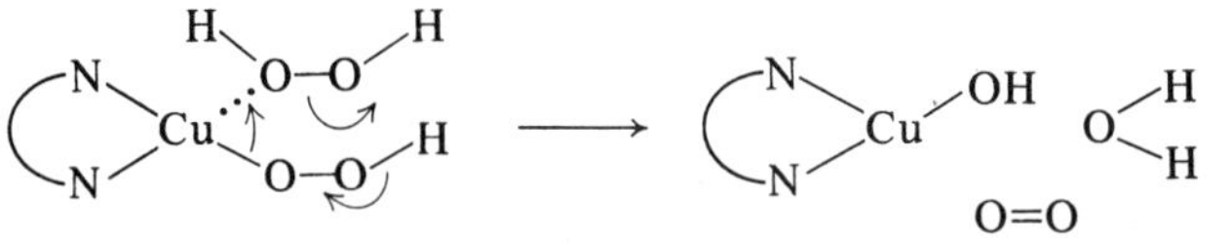

(7-15)

their stabilizing or destabilizing influence on oxyanion ligands like $HO_2{}^-$, as discussed in Chapter 6. The inverse pH dependence is consistent with the involvement of one molecule of H_2O_2 and one hydroperoxide anion. In a similar way bipyCu(II) catalyzes the reactions of hydrogen peroxide with hydroxylamine and hydrazine (reaction 7-16). Organic amines and hydrazones are

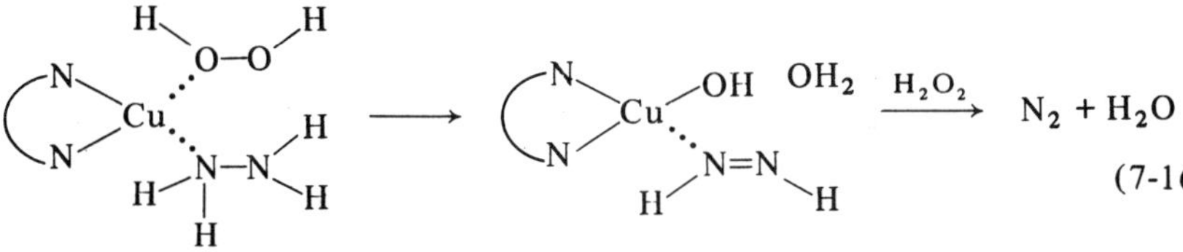

(7-16)

also oxidized by air in the presence of copper; for example, aniline and o-phenylenediamine are oxidized to azobenzene and diaminophenazine by catalytic amounts of copper (reactions 7-17 and 7-18, respectively). The oxidation of aromatic diamines takes an altogether different course if stoichiometric

$$C_6H_5NH_2 \xrightarrow{\text{Cu}^+/\text{air}} C_6H_5-N=N-C_6H_5 \qquad (7\text{-}17)$$

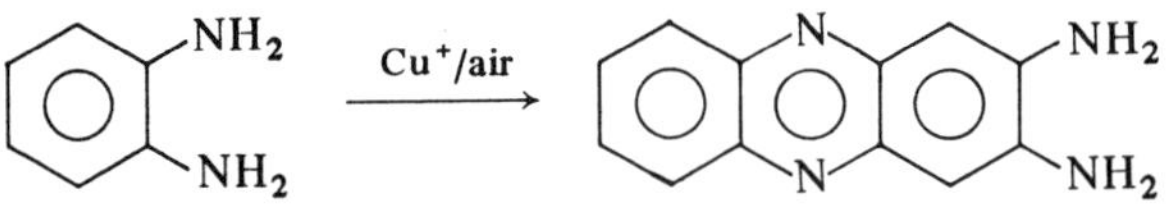

(7-18)

amounts of "oxygenated CuCl" in pyridine are used. In these cases the aromatic ring is cleaved to form a mucononitrile derivative (reaction 7-19).[53] The use of a large amount of metal species prevents the intramolecular dimerizations seen when the ligand is in large excess over the metal. Similar oxidations may be carried out on mono- and bishydrazone derivatives of α-

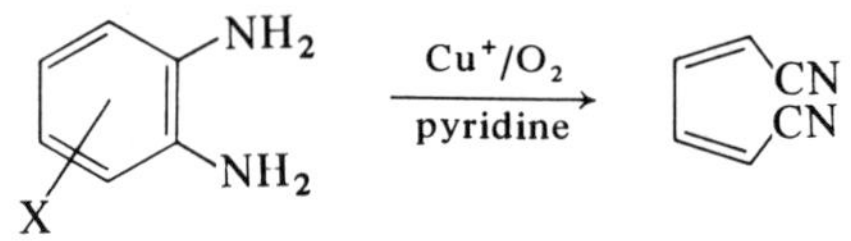

$$\text{(7-19)}$$

Diamine	cis,cis-mucononitrile (% yield)
X = H	95
X = 4-OCH$_3$	75
X = 4,5-diMe	95
X = benzo(c)	72

diketones. They constitute useful synthetic procedures for obtaining certain acetylenes and the otherwise hard to obtain α-diazoketones (reactions 7-19 and 7-20, respectively).[54]

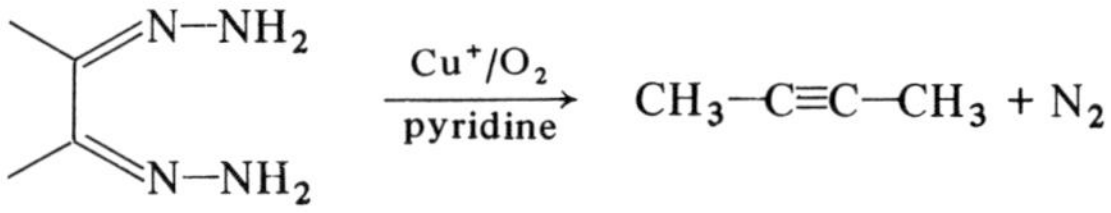

$$\text{(7-20)}$$

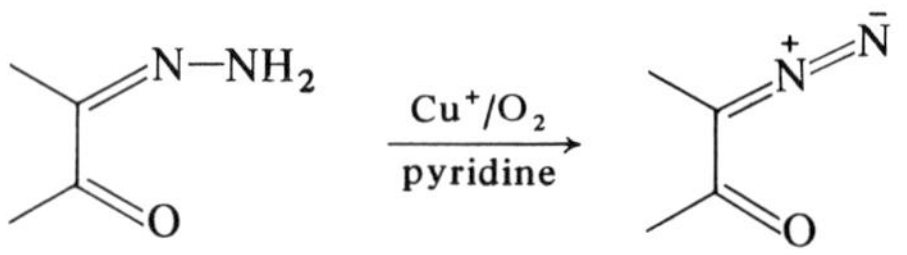

$$\text{(7-21)}$$

TABLE 7-4

Standard Oxidation Potentials at 25°C[a]

Couple		E_0 (V)
	Acid solutions	
Li	$\rightleftharpoons Li^+ + e^-$	3.01
Rb	$\rightleftharpoons Rb^+ + e^-$	2.92
K	$\rightleftharpoons K^+ + e^-$	2.92
Cs	$\rightleftharpoons Cs^+ + e^-$	2.92
Ba	$\rightleftharpoons Ba^{2+} + 2\,e^-$	2.90
Sr	$\rightleftharpoons Sr^{2+} + 2\,e^-$	2.89
Ca	$\rightleftharpoons Ca^{2+} + 2\,e^-$	2.87
Na	$\rightleftharpoons Na^+ + e^-$	2.71
Mg	$\rightleftharpoons Mg^{2+} + 2\,e^-$	2.37
Al	$\rightleftharpoons Al^{3+} + 3\,e^-$	1.66
Mn	$\rightleftharpoons Mn^{2+} + 2\,e^-$	1.18
Cr	$\rightleftharpoons Cr^{2+} + 2\,e^-$	0.91
Zn	$\rightleftharpoons Zn^{2+} + 2\,e^-$	0.76
AsH_3	$\rightleftharpoons As + 3\,H^+ + 3\,e^-$	0.60
SbH_3	$\rightleftharpoons Sb + 3\,H^+ + 3\,e^-$	0.51
Fe	$\rightleftharpoons Fe^{2+} + 2\,e^-$	0.44
Cr^{2+}	$\rightleftharpoons Cr^{3+} + e^-$	0.41
Cd	$\rightleftharpoons Cd^{2+} + 2\,e^-$	0.40
$Pb + 2\,I^-$	$\rightleftharpoons PbI_2 \downarrow + 2\,e^-$	0.365
$Pb + SO_4{}^{2-}$	$\rightleftharpoons PbSO_4 \downarrow + 2\,e^-$	0.356
Co	$\rightleftharpoons Co^{2+} + 2\,e^-$	0.277
$H_3PO_3 + H_2O$	$\rightleftharpoons H_3PO_4 + 2\,H^+ + 2\,e^-$	0.276
$Pb + 2\,Cl^-$	$\rightleftharpoons PbCl_2 \downarrow + 2\,e^-$	0.268
Ni	$\rightleftharpoons Ni^{2+} + 2\,e^-$	0.250
$Cu + I^-$	$\rightleftharpoons CuI \downarrow + e^-$	0.185
$Ag + I^-$	$\rightleftharpoons AgI \downarrow + e^-$	0.152
Sn	$\rightleftharpoons Sn^{2+} + 2\,e^-$	0.14
Pb	$\rightleftharpoons Pb^{2+} + 2\,e^-$	0.126
$Hg + 4\,I^-$	$\rightleftharpoons HgI_4{}^{2-} + 2\,e^-$	0.04
H_2	$\rightleftharpoons 2\,H^+ + 2\,e^-$	0.0000
$Ag + 2\,S_2O_3{}^{2-}$	$\rightleftharpoons Ag(S_2O_3)_2{}^{3-} + e^-$	−0.01
$Ag + Br^-$	$\rightleftharpoons AgBr \downarrow + e^-$	−0.095
$Cu + Cl^-$	$\rightleftharpoons CuCl \downarrow + e^-$	−0.137
H_2S	$\rightleftharpoons S^0 \downarrow + 2\,H^+ + 2\,e^-$	−0.141
$Sn^{2+} + 6\,Cl^-$	$\rightleftharpoons SnCl_6{}^{2-} + 2\,e^-$	−0.15
$2\,Sb + 3\,H_2O$	$\rightleftharpoons Sb_2O_3 \downarrow + 6\,H^+ + 6\,e^-$	−0.152
Cu^+	$\rightleftharpoons Cu^{2+} + e^-$	−0.153
$Bi + H_2O + Cl^-$	$\rightleftharpoons BiOCl \downarrow + 2\,H^+ + 3\,e^-$	−0.16
$SO_2 + 2\,H_2O$	$\rightleftharpoons SO_4{}^{2-} + 4\,H^+ + 2\,e^-$	−0.17
$Ag + Cl^-$	$\rightleftharpoons AgCl \downarrow + e^-$	−0.223
$As + 3\,H_2O$	$\rightleftharpoons H_3AsO_3 + 3\,H^+ + 3\,e^-$	−0.247
$Bi + H_2O$	$\rightleftharpoons BiO^+ + 2\,H^+ + 3\,e^-$	−0.32
Cu	$\rightleftharpoons Cu^{2+} + 2\,e^-$	−0.34
$Fe(CN)_6{}^{4-}$	$\rightleftharpoons Fe(CN)_6{}^{3-} + e^-$	−0.356

(*continued*)

TABLE 7-4—*continued*

Couple		E_0 (V)
2 HCN	$\rightleftharpoons C_2N_2 + 2\,H^+ + 2\,e^-$	-0.37
$S_2O_3^{2-} + 3\,H_2O$	$\rightleftharpoons 2\,H_2SO_3 + 2\,H^+ + 4\,e^-$	-0.40
$2\,Ag + CrO_4^{2-}$	$\rightleftharpoons Ag_2CrO_4\downarrow + 2\,e^-$	-0.446
$S^0 + 3\,H_2O$	$\rightleftharpoons H_2SO_3 + 4\,H^+ + 4\,e^-$	-0.45
$S_4O^{2-} + 6\,H_2O$	$\rightleftharpoons 4\,H_2SO_3 + 4\,H^+ + 6\,e^-$	-0.51
Cu	$\rightleftharpoons Cu^+ + e^-$	-0.52
$2\,I^-$	$\rightleftharpoons I_2 + 2\,e^-$	-0.535
$3\,I^-$	$\rightleftharpoons I_3^- + 2\,e^-$	-0.536
$CuCl\downarrow$	$\rightleftharpoons Cu^{2+} + Cl^- + e^-$	-0.538
$H_3AsO_3 + H_2O$	$\rightleftharpoons H_3AsO_4 + 2\,H^+ + 2\,e^-$	-0.559
MnO_4^{2-}	$\rightleftharpoons MnO_4^- + e^-$	-0.564
$2\,Sb(OH)_2^+ + H_2O$	$\rightleftharpoons Sb_2O_5\downarrow + 6\,H^+ + 4\,e^-$	-0.581
$2\,Ag + SO_4^{2-}$	$\rightleftharpoons Ag_2SO_4\downarrow + 2\,e^-$	-0.653
H_2O_2	$\rightleftharpoons O_2 + 2\,H^+ + 2\,e^-$	-0.682
$2\,SCN^-$	$\rightleftharpoons (SCN)_2 + 2\,e^-$	-0.77
Fe^{2+}	$\rightleftharpoons Fe^{3+} + e^-$	-0.771
$NO_2 + H_2O$	$\rightleftharpoons NO_3^- + 2\,H^+ + e^-$	-0.775
$2\,Hg$	$\rightleftharpoons Hg_2^{2+} + 2\,e^-$	-0.789
Ag	$\rightleftharpoons Ag^+ + e^-$	-0.799
$CuI\downarrow$	$\rightleftharpoons Cu^{2+} + I^- + e^-$	-0.86
Hg_2^{2+}	$\rightleftharpoons 2\,Hg^{2+} + 2\,e^-$	-0.92
$HNO_2 + H_2O$	$\rightleftharpoons NO_3^- + 3\,H^+ + 2\,e^-$	-0.94
$NO + 2\,H_2O$	$\rightleftharpoons NO_3^- + 4\,H^+ + 3\,e^-$	-0.96
$NO + H_2O$	$\rightleftharpoons HNO_2 + H^+ + e^-$	-1.00
$2\,Br^-$	$\rightleftharpoons Br_2\,(liq) + 2\,e^-$	-1.065
$ClO_2 + H_2O$	$\rightleftharpoons ClO_3^- + 2\,H^+ + e^-$	-1.15
$I_2 + 6\,H_2O$	$\rightleftharpoons 2\,IO_3^- + 12\,H^+ + 10\,e^-$	-1.195
$2\,H_2O$	$\rightleftharpoons O_2 + 4\,H^+ + 4\,e^-$	-1.229
$Mn^{2+} + 2\,H_2O$	$\rightleftharpoons MnO_2\downarrow + 4\,H^+ + 2\,e^-$	-1.23
$2\,Cr^{3+} + 7\,H_2O$	$\rightleftharpoons Cr_2O_7^{2-} + 14\,H^+ + 6\,e^-$	-1.33
$2\,Cl^-$	$\rightleftharpoons Cl_2 + 2\,e^-$	-1.36
$Pb^{2+} + 2\,H_2O$	$\rightleftharpoons PbO_2\downarrow + 4\,H^+ + 2\,e^-$	-1.456
Mn^{2+}	$\rightleftharpoons Mn^{3+} + e^-$	-1.51
$Mn^{2+} + 4\,H_2O$	$\rightleftharpoons MnO_4^- + 8\,H^+ + 5\,e^-$	-1.51
$Br_2 + 6\,H_2O$	$\rightleftharpoons 2\,BrO_3^- + 12\,H^+ + 10\,e^-$	-1.52
$Br_2 + 2\,H_2O$	$\rightleftharpoons 2\,HBrO + 2\,H^+ + 2\,e^-$	-1.59
$Cl_2 + 2\,H_2O$	$\rightleftharpoons 2\,HClO + 2\,H^+ + 2\,e^-$	-1.63
$PbSO_4\downarrow + 2\,H_2O$	$\rightleftharpoons PbO_2\downarrow + SO_4^{2-} + 4\,H^+ + 2\,e^-$	-1.685
$2\,H_2O$	$\rightleftharpoons H_2O_2 + 2\,H^+ + 2\,e^-$	-1.77
Co^{2+}	$\rightleftharpoons Co^{3+} + e^-$	-1.84
Ag^+	$\rightleftharpoons Ag^{2+} + e^-$	-1.98
$2F^-$	$\rightleftharpoons F_2 + 2\,e^-$	-2.65
	Basic solutions	
$Al + 4\,OH^-$	$\rightleftharpoons Al(OH)_4^- + 3\,e^-$	2.35
$Mn + 2\,OH^-$	$\rightleftharpoons Mn(OH)_2\downarrow + 2\,e^-$	1.55

TABLE 7-4—*continued*

Couple		E_0 (V)
$Zn + S^{2-}$	$\rightleftharpoons ZnS \downarrow + 2\,e^-$	1.44
$Zn + 4\,CN^-$	$\rightleftharpoons Zn(CN)_4^{2-} + 2\,e^-$	1.26
$Zn + 4\,OH^-$	$\rightleftharpoons Zn(OH)_4^{2-} + 2\,e^-$	1.22
$Cr + 4\,OH^-$	$\rightleftharpoons Cr(OH)_4^- + 3\,e^-$	1.2
$Cd + S^{2-}$	$\rightleftharpoons CdS \downarrow + 2\,e^-$	1.2
$Cd + 4\,CN^-$	$\rightleftharpoons Cd(CN)_4^{2-} + 2\,e^-$	1.03
$Zn + 4\,NH_3$	$\rightleftharpoons Zn(NH_3)_4^{2+} + 2\,e^-$	1.03
$SO_3^{2-} + 2\,OH^-$	$\rightleftharpoons SO_4^{2-} + H_2O + 2\,e^-$	0.93
$Sn + 3\,OH^-$	$\rightleftharpoons Sn(OH)_3^- + 2\,e^-$	0.91
$Sn(OH)_3^- + 3\,OH^-$	$\rightleftharpoons Sn(OH)_6^{2-} + 2\,e^-$	0.90
$H_2 + 2\,OH^-$	$\rightleftharpoons 2\,H_2O + 2\,e^-$	0.828
$Cd + 2\,OH^-$	$\rightleftharpoons Cd(OH)_2 \downarrow + 2\,e^-$	0.809
$Hg + S^{2-}$	$\rightleftharpoons HgS \downarrow + 2\,e^-$	0.72
$2\,Ag + S^{2-}$	$\rightleftharpoons Ag_2S \downarrow + 2\,e^-$	0.69
$AsO_3^{3-} + 2\,OH^-$	$\rightleftharpoons AsO_4^{3-} + H_2O + 2\,e^-$	0.67
$Cd + 4\,NH_3$	$\rightleftharpoons Cd(NH_3)_4^{2+} + 2\,e^-$	0.597
$S_2O_4^{2-} + 4\,NH_3 + 2\,H_2O$	$\rightleftharpoons 2\,SO_3^{2-} + 4\,NH_4^+ + 2\,e^-$	0.56
$Fe(OH)_2 \downarrow + OH^-$	$\rightleftharpoons Fe(OH)_3 \downarrow + e^-$	0.56
$Pb + 3\,OH^-$	$\rightleftharpoons Pb(OH)_3^- + e^-$	0.54
S^{2-}	$\rightleftharpoons S^0 + 2\,e^-$	0.48
$Ag + 2\,CN^-$	$\rightleftharpoons Ag(CN)_2^- + e^-$	0.31
$NH_3(g) + 7\,OH^-$	$\rightleftharpoons NO_2^- + 5\,H_2O + 6\,e^-$	0.18
$NH_3(g) + 9\,OH^-$	$\rightleftharpoons NO_3^- + 6\,H_2O + 8\,e^-$	0.13
$Cu + 4\,NH_3$	$\rightleftharpoons Cu(NH_3)_4^{2+} + 2\,e^-$	0.02
$Mn(OH)_2 \downarrow + 2\,OH^-$	$\rightleftharpoons MnO_2 \downarrow + 2\,H_2O + 2\,e^-$	0.05
$NO_2^- + 2\,OH^-$	$\rightleftharpoons NO_3^- + H_2O + 2\,e^-$	-0.01
$Cr(OH)_4^- + 4\,OH^-$	$\rightleftharpoons CrO_4^{2-} + 4\,H_2O + 3\,e^-$	-0.02
$Co(NH_3)_6^{2+}$	$\rightleftharpoons Co(NH_3)_6^{3+} + e^-$	-0.1
$Mn(OH)_2 \downarrow + OH^-$	$\rightleftharpoons MnOOH \downarrow + H_2O + e^-$	-0.1
$Co(OH)_2 \downarrow + OH^-$	$\rightleftharpoons Co(OH)_3 \downarrow + e^-$	-0.17
$2\,Ag + 2\,OH^-$	$\rightleftharpoons Ag_2O \downarrow + H_2O + 2\,e^-$	-0.342
$Ag + 2\,NH_3$	$\rightleftharpoons Ag(NH_3)_2^+ + e^-$	-0.373
$4\,OH^-$	$\rightleftharpoons O_2 + 2\,H_2O + 4\,e^-$	-0.401
$Ag_2O + 2\,OH^-$	$\rightleftharpoons 2\,AgO + 2\,H_2O + 3\,e^-$	-0.57
$MnO_2 + 4\,OH^-$	$\rightleftharpoons MnO_4^- + 2\,H_2O + 3\,e^-$	-0.59
$4\,OH^-$	$\rightleftharpoons O_2^{2-} + 2\,H_2O + 2\,e^-$	-0.88
$Cl^- + 2\,OH^-$	$\rightleftharpoons ClO^- + H_2O + 2\,e^-$	-0.89

[a] From "Qualitative Analysis and Electrolytic Solutions" by Edward J. King, Copyright 1959 by Harcourt Brace Jovanovich, Inc., and reprinted with their permission.

TABLE 7-5

Classification of Coenzymes, Carriers, and Substrates by Their E_0' Values (25°C, pH 7)[a]

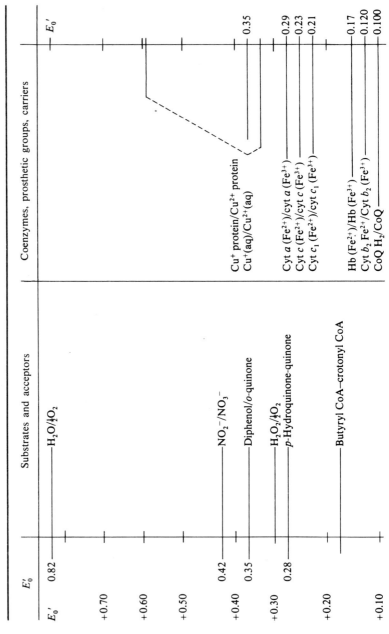

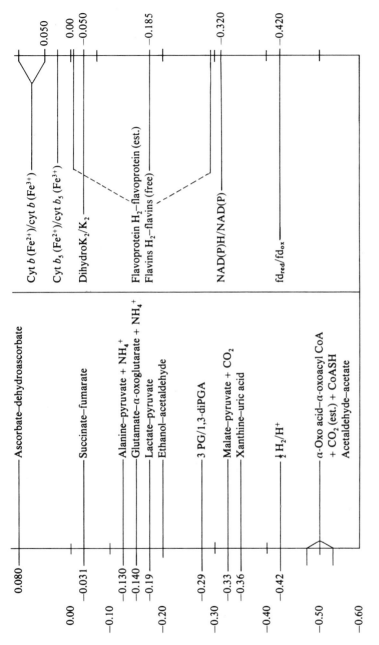

Cyt b (Fe^{2+})/cyt b (Fe^{3+}) — 0.050

Cyt b_5 (Fe^{2+})/cyt b_5 (Fe^{3+}) — 0.00

DihydroK$_2$/K$_2$ — −0.050

Flavoprotein H$_2$–flavoprotein (est.)
Flavins H$_2$–flavins (free) — −0.185

NAD(P)H/NAD(P) — −0.320

fd$_{red}$/fd$_{ox}$ — −0.420

Ascorbate–dehydroascorbate — 0.080

Succinate–fumarate — −0.031

Alanine–pyruvate + NH$_4^+$ — −0.130
Glutamate–α-oxoglutarate + NH$_4^+$ — −0.140
Lactate–pyruvate — −0.19
Ethanol–acetaldehyde — −0.20

3 PG/1,3-diPGA — −0.29

Malate–pyruvate + CO$_2$ — −0.33
Xanthine–uric acid — −0.36

$\frac{1}{2}$ H$_2$/H$^+$ — −0.42

α-Oxo acid–α-oxoacyl CoA + CO$_2$ (est.) + CoASH — −0.50
Acetaldehyde–acetate

0.00
−0.10
−0.20
−0.30
−0.40
−0.50
−0.60

$\Delta E_0' = 0.10$ equal to a $\Delta G'$ of 4600 cal for 2 e^-, 2300 cal for a 1 e^- change.

[a]Data from Mahler and Cordes.[2] [Copyright © 1966, 1971 by Henry R. Mahler and Eugene H. Cordes. By permission of Harper & Row, Publishers.]

TABLE 7-6

Standard Reduction Potentials for Systems of Biochemical Importance[a]

System	E_0' (pH 7), (V)
Oxygen–water	0.816
Ferric–ferrous	0.77
Nitrate–nitrite	0.42
Ferricyanide–ferrocyanide	0.36
Oxygen–hydrogen peroxide	0.30
Cytochrome a; ferric–ferrous	0.29
Cytochrome c; ferric–ferrous	0.25
Crotonyl-SCoA–butyryl-SCoA	0.19
Methemoglobin–hemoglobin	0.17
Adrenodoxin; ox–red	0.15
Cytochrome b_2; ferric–ferrous	0.12
Ubiquinone; ox–red	0.10
Dehydroascorbic acid–ascorbic acid	0.06
Metmyoglobin–myoglobin	0.046
Fumarate–succinate	0.03
Methylene blue, ox–red	0.01
Yellow enzyme, FMN/FMNH$_2$	−0.122
Pyruvate + ammonium–alanine	−0.13
α-Oxoglutarate + ammonium–glutamate	−0.14
Oxalacetate–Malate	−0.17
Pyruvate–lactate	−0.19
Acetaldehyde–ethanol	−0.20
Riboflavin, ox–red	−0.21
Glutathione, ox–red	−0.23
Acetoacetate–β-hydroxybutyrate	−0.27
Lipoic acid, ox–red	−0.29
NAD+–NADH	−0.32
Pyruvate–malate	−0.33
Cystine–cysteine	−0.34
Uric acid–xanthine	−0.36
Ferredoxin; ox–red	−0.41
Carbon dioxide–formate	−0.42
H+–H$_2$	−0.42
Acetate–acetaldehyde	−0.60
Succinate–α-oxoglutarate	−0.67
Acetate + carbon dioxide–pyruvate	−0.70

[a]Data from Mahler and Cordes.[2] [Copyright ©1966, 1971 by Henry R. Mahler and Eugene H. Cordes. By permission of Harper & Row, Publishers.]

REFERENCES

1. M. C. Day, Jr. and J. Selbin, "Theoretical Inorganic Chemistry," 2nd ed., pp. 335–360. Van Nostrand-Reinhold, Princeton, New Jersey, 1969.
2. H. R. Mahler and E. H. Cordes, "Biological Chemistry," 2nd ed., pp. 11–40 and 633–707. Harper, New York, 1971.
3. F. Haber and J. Weiss, *Proc. Roy. Soc., Ser. A* **147**, 332 (1934).
4. S. Fallab, *Angew. Chem., Int. Ed. Engl.* **6**, 496 (1967).
5. D. M. Jerina and J. W. Daly, *in* "Oxidases and Related Redox Systems" (T. S. King, H. S. Mason, and M. Morrison, eds.) 2nd ed. Vol. 1, p. 143, Univ. Park Press, Baltimore, Maryland, 1973.
6. P. George *in* "Oxidases and Related Redox Systems" (T. S. King, H. S. Mason, and M. Morrison, eds.) 1st ed. Vol. 1, p. 3, Wiley (Interscience), New York, 1965.
7. E. Hayon and M. Simic, *Accounts Chem. Res.* **7**, 114 (1974).
8. I. Fridovich, *Accounts Chem. Res.,* **5**, 321 (1973).
9. J. M. McCord, C. O. Beauchamp, S. Gascin, H. P. Mirsa, and I. Fridovich, *in* "Oxidases and Related Redox Systems" (T. S. King, H. S. Mason, and M. Morrison, eds.) 2nd ed., Vol. 1, p. 51, Univ. Park Press, Baltimore, Maryland, 1973.
10. H. Baltscheffsky and M. Baltscheffsky, *Annu. Rev. Biochem.* **43**, 868 (1974).
11. V. Ullrich, *Angew. Chem., Int. Ed. Engl.* **12**, 701 (1972).
12. G. A. Hamilton, *Progr. Bioorg. Chem.* **1**, 83 (1971).
13. R. Lemberg and J. Barrett, "The Cytochromes." Academic Press, New York, 1972.
14. H. A. Harbury and R. H. L. Marks, *in* "Inorganic Biochemistry," (G. Eichhorn, ed.) Vol. 2, pp. 902–954, Elsevier, Amsterdam, 1973.
15. For references, see Chapter 10.
16. R. E. Dickerson, T. Takano, D. Eisenberg, O. B. Kallai, A. Samson, A. Cooper, and E. Margolaish, *J. Biol. Chem.* **246**, 1511 (1971).
17. R. Malin, *in* "Inorganic Biochemistry" (G. Eichhorn, ed.) Vol. 1, p. 689. Elsevier, Amsterdam, 1973.
18. B. G. Malmstrom, *Pure Appl. Chem.* **24**, 393 (1970).
19. W. Lovenberg, ed., "Iron-Sulfur Proteins" Vol. 1. Academic Press, New York, 1973.
20. R. Mason and J. A. Zubieta, *Angew. Chem., Int. Ed. Engl.* **12**, 390 (1973).
21. S. J. Lippard, *Accounts Chem. Res.* **6**, 282 (1973).
22. W. H. Orme-Johnson, *in* "Inorganic Biochemistry" (G. Eichhorn, ed.), Vol. 1, p. 710, Elsevier, Amsterdam, 1973.
23. W. H. Orme-Johnson, *Annu. Rev. Biochem.* **42**, 159 (1973).
24. L. H. Jensen, *Annu. Rev. Biochem.* **43**, 461 (1974).
25. K. D. Watenpaugh, L. C. Sieker, J. R. Herriott, and L. H. Jensen, *Cold Spring Harbor Symp. Quant. Biol.* **36**, 359 (1971).
26. W. R. Dunham, A. Bearden, I. Salmeen, G. Palmer, R. H. Sands, W. H. Orme-Johnson, and H. Beinert, *Biochim. Biophys. Acta* **253**, 134 (1971).
27. E. T. Adman, L. C. Sieker, and L. H. Jensen, *J. Biol. Chem.* **248**, 3987 (1973).
28. C. W. Carter Jr., S. T. Freer, Ng. H. Xuong, R. A. Alden, and J. Kraut, *Cold Spring Harbor Symp. Quant. Biol.* **36**, 381 (1971).
29. M. R. Churchill and J. Wormold, *Inorg. Chem.* **10**, 1778 (1971).
30. A. Davison and D. L. Reager, *Inorg. Chem.* **10**, 1967 (1971).
31. A. Ali, F. Fahrenholz, J. C. Garing, and B. Weinstein, *J. Amer. Chem. Soc.* **94**, 2556 (1972).
32. T. Herskovitz, B. A. Averill, R. H. Holm, J. A. Ibers, W. D. Phillips, and J. F. Weiher, *Proc. Nat. Acad. Sci. U.S.* **69**, 2437 (1972).

33. B. A. Averill, T. Herskovitz, R. H. Holm, and J. A. Ibers, *J. Amer. Chem. Soc.* **95**, 3523 (1973).
34. H. Taube, "Electron Transfer Reactions of Complex Ions in Solution." Academic Press, New York, 1970.
35. N. Sutin, *Chem. Brit.* **8**, 148 (1972).
36. N. Sutin, *in* "Inorganic Biochemistry" (G. Eichhorn, ed.) Vol. 1, p. 611, Elsevier, Amsterdam, 1973.
37. R. G. Linck, *in* "Transition Metals in Homogeneous Catalysis" (G. N. Schrauzer, ed.) p. 297, Dekker, New York, 1971.
38. G. Eichhorn and R. Latif, *J. Amer. Chem. Soc.* **76**, 5180 (1959).
38a. H. Taube, H. Meyers, and R. L. Rich, *J. Amer. Chem. Soc.* **75**, 4118 (1953).
39. J. E. Lyons and J. O. Turner, *J. Org. Chem.* **37**, 2881 (1972).
40. C.-C. Su, J. W. Reed, and E. S. Gould, *Inorg. Chem.* **12**, 337 (1973).
41. W. F. Brill, *J. Amer. Chem. Soc.* **85**, 141 (1963).
42. C. S. Foote, *Pure Appl. Chem.* **27**, 635 (1971).
43. A. P. Schapp, A. L. Thayer, G. R. Faler, K. Goda, and T. Kimura, *J. Amer. Chem. Soc.* **96**, 4025 (1974).
44. E. A. Mayeda and A. J. Bard, *J. Amer. Chem. Soc.* **96**, 4023 (1974).
45. M. R. Egmond, J. F. G. Vliegenthart, and J. Boldingh, *Biochem. Biophys. Res. Commun.* **48**, 1055 (1972).
46. J. I. Teng and L. L. Smith, *J. Amer. Chem. Soc.* **95**, 4060 (1974).
47. J. W. Peters, J. N. Pitts Jr., I. Rosenthal, and H. Fuhr, *J. Amer. Chem. Soc.* **94**, 4348 (1972).
48. M. M. Taqui Khan and A. E. Martell, *J. Amer. Chem. Soc.* **89**, 4167 and 7104 (1967); **90**, 6011 (1968).
49. P. K. Adolf and G. A. Hamilton, *J. Amer. Chem. Soc.* **93**, 3420 (1971).
50. A. E. Martell, *Pure Appl. Chem.* **17**, 129 (1968).
51. B. C. Saunders, *in* "Inorganic Biochemistry" (G. Eichhorn, ed.) Vol. 2, p. 988, Elsevier, Amsterdam, 1973.
52. H. Sigel, *Angew. Chem., Int. Ed. Engl.* **8**, 167 (1969).
53. H. Takahashi, T. Kajimoto, and J. Tsuji, *Syn. Commun.* **2**, 181 (1972).
54. J. Tsuji, H. Takahashi, and T. Kajimoto, *Tetrahedron Lett.* p. 4573 (1973).

VIII

Influencing Equilibria with Metal Ions: Synthesis via Chelation

When a metal ion associates with a ligand in solution the stability of the complex formed is a measure of the stabilizing influence of the metal and the ligand *on each other*. The complexation of a ligand by a metal ion can have great consequences for any other concurrent equilibrium processes in which the ligand molecule may be involved. Such processes might include protonation–deprotonation equilibria, equilibration of conformers, tautomers, anomers, or isomers of the ligand, or equilibria in which the ligand is a hydrolysis fragment or condensation product of other molecules in the solution. The selective complexation of one of the molecular species participating in an equilibrium will differentially stabilize that species and displace the equilibrium by an amount proportional to the increment in stability. Since the above equilibria often constitute useful synthetic reactions, or may be part of a multistep synthetic sequence, metal complexation can influence the outcome of the reaction both qualitatively and quantitatively. The deliberate use of metals to manipulate equilibria in this way is the basis not only for some exotic and esoteric chemical curiosities but also for some valuable and specific synthetic reactions. The number of examples of truly useful "metal-directed" synthetic reactions has been increasing at an accelerating pace in recent years.

185

It will undoubtedly continue to increase and attain a high level as the factors which govern the selectivity that may be obtained become better understood.

In the discussion of ligand exchange reactions in Chapter 6 it became important to distinguish between kinetic and thermodynamic control of products and processes, i.e., inert vs stable complexes. In discussing metal-catalyzed and metal-directed reactions of ligands, a parallel distinction must be made between thermodynamic and kinetic effects before a meaningful analysis of a reaction can be undertaken.

It is a simple matter to define metal-catalyzed reactions as those in which the metal species is regenerated in each reaction cycle. In these cases it is very probably the *transition state* of the catalyzed reaction, rather than a reaction product, which is most strongly complexed by the metal. Thus it is the *rate of establishment of equilibrium*, rather than the *position* of the equilibrium that is altered, although the latter may appear to have occurred if only one of several competing reaction pathways is catalyzed by the metal. In contrast, the product of a metal-directed reaction is formed *as a metal complex*, and stoichiometric amounts of metal are consumed in the process. In many cases when the desired product is the free ligand rather than the complex, the initially formed complex is decomposed by the reaction workup conditions and only the free ligand appears as a final "reaction product." In virtually all cases the driving force for metal-directed reactions is the stabilization associated with the formation of a chelating ligand from monodentate ligands or the conversion of a weakly chelating ligand to a strong chelator.

Metal-Directed Reactions[1-6]

Often the course of metal-directed reactions is such that the major ligand product formed is not even detectable from the same reactants in the absence of the metal. It then becomes difficult to decide whether the metal has merely caused an extreme displacement of an equilibrium, or if it has in fact catalyzed a new and rapid reaction pathway, by complexation and therefore stabilization of an otherwise inaccessible transition state, which only coincidentally leads to a product that is a good metal-binding agent. Reactions which yield product ligands that are unstable and decompose once freed of the metal appear to be examples of the latter case. These possibilities must be borne in mind when attempting to analyze a reaction proceeding in the presence of metal ions capable of complexing ligands in solution.

The control of chemical reactions by the judicious application of metal ions is both a fascinating and profitable endeavor which has attracted the attention of a large number of chemists, and more recently, biochemists. Consequently, this area has also been the subject of numerous reviews and attempts to categorize according to mechanistic criteria. Since there have been very few detailed mechanistic studies of metal-directed reactions, much of the mechanistic interpretation of these reactions is pure speculation, only some of which is justifiable through analogy to simpler better known systems. Therefore, the emphasis in this chapter is neither completeness of coverage nor detailed categorization of reaction types, but rather an attempt to provide a broad spectrum of examples which illustrate the several factors now clearly recognized as being important in governing metal-directed reactions of organic ligands. To begin by stating the conclusions of this chapter, the following principles are of prime importance in governing metal-directed reactions, although it is entirely possible that other factors may eventually be recognized. It is hoped that some of the abundant speculation about the mechanisms of these reactions will stimulate research leading to their elucidation.

Chelation

This is probably the most recurrent theme if not the most important factor in the analysis of metal-dependent reactions of organic ligands. In nearly all cases it is the formation of a stable metal chelate as the primary reaction product which drives the equilibrium to favor that product, although this fact is sometimes obscured by the decomposition of the chelate, e.g., by protonation during workup of the reaction to allow isolation of the metal-free ligand. Chelation is also an important aspect of both metal-dependent and metal-catalyzed reactions at the kinetic level. This derives from the increased likelihood of a ligand reacting while under the stereochemical or electronic influence of a metal, if the lifetime of the complex in solution is increased by chelation.

Ligand Polarization

Nucleophilic and electrophilic reactions of ligands, such as condensation, hydrolysis, alkylation, or solvolysis reactions, can be greatly enhanced by coordination to metal ions. The metal ions then act variously as Lewis acids, π acids, or π donors to alter electron density or distribution on the ligand, and thereby alter the character of the ligand as a nucleophile or electrophile. In some respects this is similar to altering ligand reactivity by protonation, but metal complexation often produces larger effects than does protonation, and sometimes they are in the opposite direction. It is expected, but *not always observed* that anionic ligands may attenuate the activity of a metal ion by lowering the charge on the overall complex and/or reducing the Lewis acidity of the metal. Metal coordination may also be used as a device to *block* reactions at a particular site in a molecule, i.e., as a protective device.

TEMPLATE EFFECTS

A metal ion can act as a collector of ligands as well as promoter of ligand reactivity by means of chelation and polarization effects. This combination has led to some spectacular syntheses of macrocyclic and cage ligands via polycondensation reactions (see Figs. 8-1 and 8-2). Template reactions are also discussed in Chapter 2.

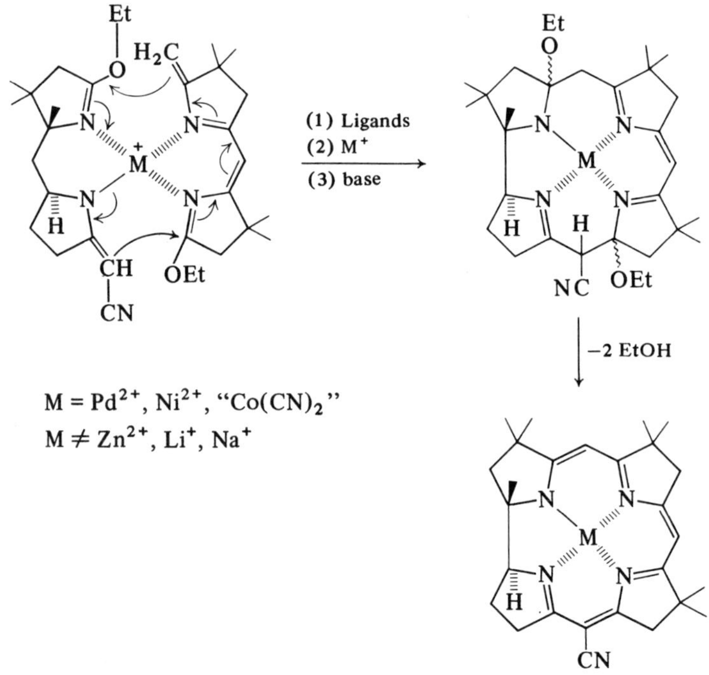

$$M = Pd^{2+}, Ni^{2+}, \text{``Co(CN)}_2\text{''}$$
$$M \neq Zn^{2+}, Li^+, Na^+$$

Fig. 8-1. Corrin synthesis via template effects.[10] The reaction is not promoted by zinc, which normally requires tetrahedral geometry, nor by lithium and sodium which are small ions preferring hard oxygen donors.

REDOX EFFECTS

Metals in high oxidation states are useful as stoichiometric oxidizing agents, and because of their variable valences, metals may also facilitate electron transfer reactions by undergoing alternate oxidation and reduction. In some cases both the oxidant and reductant may be ligands in the coordination sphere of a ternary metal complex. Redox reactions were discussed in Chapter 7.

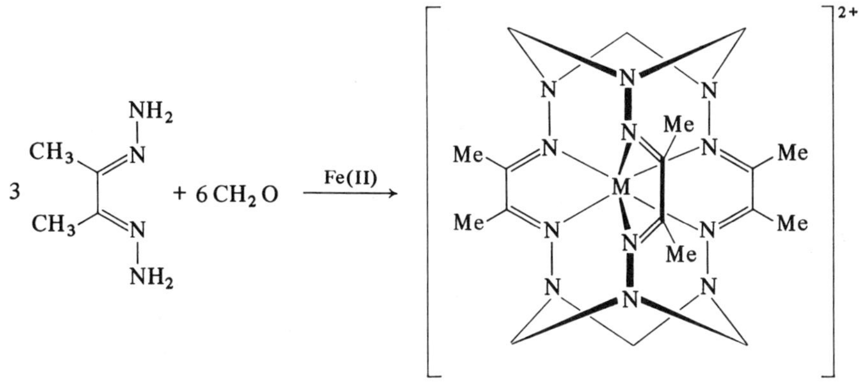

Fig. 8-2. Template synthesis of a crown-capped clathrochelate (birdcage) complex.[11]

ENANTIOMER DISCRIMINATION

The binding of a ligand to a metal, and hence its reactivity while coordinated, can depend on the other ligands already present in the coordination sphere of the metal. If these ligands are optically active and bulky enough near the metal, they can effect a differential binding of another chiral *or prochiral* ligand. Similar effects can be observed for chemical reactions occurring at metal centers rendered chiral by a dissymmetric rigid ligand. One interesting class of metal complexes of rigid dissymmetric ligands may be metalloenzymes.

METAL ION LABILITY

The ability of a metal ion to exchange at least some of its ligands very rapidly is obviously of prime importance if it is to be used as a template or promoter of ligand reactivity. Lability is also important if the metal is to function catalytically as a Lewis acid or in inner sphere redox processes.

Reactions of Schiff Base Ligand Systems

The equilibrium involving a carbonyl compound and an amine with their corresponding Schiff base is one of the oldest and best systems for demonstrating and studying the manipulation of an equilibrium with metal ions (reaction 8-1). In addition, Schiff bases are important intermediates in many

$$\underset{R_2}{\overset{R_1}{\diagdown}}C=O + H_2N-R_3 \rightleftharpoons \underset{R_2}{\overset{R_1}{\diagdown}}C=N\overset{R_3}{\diagdown} + H_2O \qquad (8-1)$$

chemical and enzymatic reactions, and for this reason the control of their formation and subsequent reactions by the use of metal ions is of particular interest. The extent of Schiff base formation at equilibrium is a function of the
nature of the carbonyl and amine components. If both are aromatic compounds the equilibrium will favor the Schiff base since this brings two aromatic
systems into conjugation. Conversely, aliphatic Schiff bases hydrolyze very
easily and otherwise react to give dark polymeric products. This equilibrium
can be shifted dramatically by the presence of transition metal ions if chelating
products can be produced. The following reactions illustrate the effects that a
metal ion can exert on some simple Schiff base systems. The facile formation
of six-membered organic rings prevents extensive polymer formation in the
ethylenediamine reaction (8-2) as compared with the propylenediamine reaction, but when Cu(II) or Ni(II) is present, the course of the latter reaction is

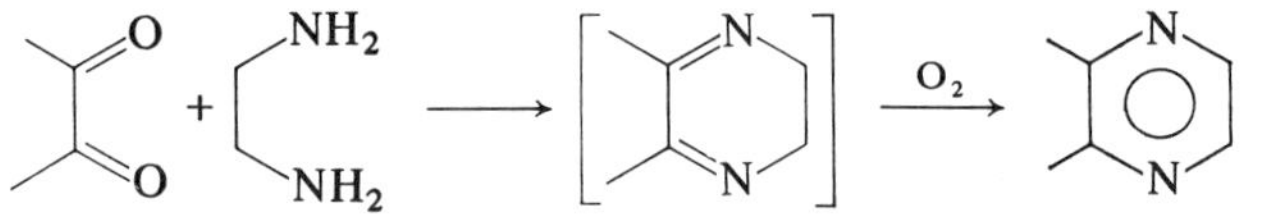

(8-2)

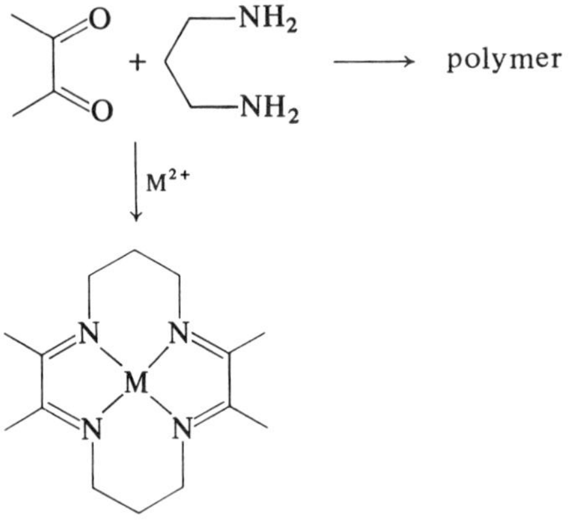

(8-3) 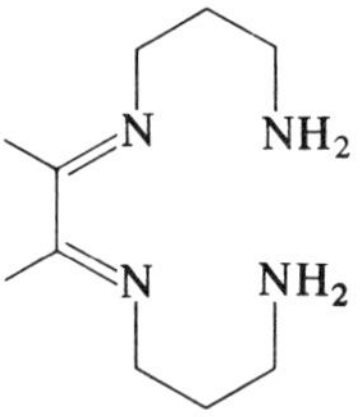   (8-4)

altered to produce a macrocyclic complex which is stable even in boiling acid
or base. The formation of complex (8-3) is probably a stepwise process involving an intermediate such as (8-4). α-Diimines are known to be exceptionally
good ligand systems, much as bipyridyl or 1,10-phenanthroline are, and a
ligand such as (8-4) is also a likely intermediate in the polymerization which occurs in the absence of the metal.

    In the presence of amines and diamines, acetone and other simple aliphatic
carbonyl compounds give extensively condensed and polymerized products

(reaction 8-5), probably derived from base-catalyzed aldol and Michael reactions. The efficient formation of (8-6) *suggests* that the same type of condensations can occur within the coordination sphere of a transition metal ion to

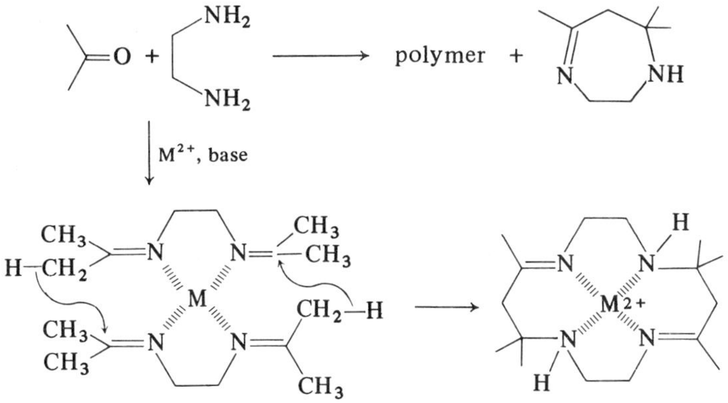

produce a specific *macrocyclic* oligomer instead of a random polymer. The inductive effect of the metal ion facilitates both the deprotonation of one ligand *and* its nucleophilic attack at a second, the electrophilic character of which is increased by the Lewis acidity of the metal ion.

Two of the earliest[1] examples of macrocylic syntheses begin with strictly monodentate reactants. They are the metal-directed synthesis of the meso-tetrasubstituted porphyrins (8-7)[7] and the phthalocyanines (8-8).[8] It is interesting that the porphyrin synthesis proceeds nicely with Zn(II), since that

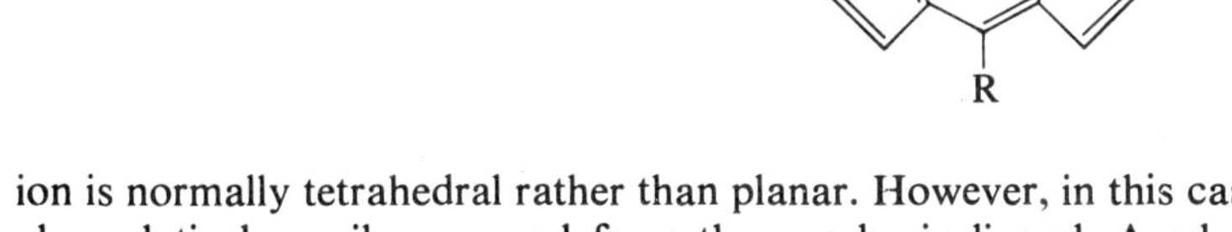

ion is normally tetrahedral rather than planar. However, in this case the zinc is also relatively easily removed from the porphyrin ligand. A related template synthesis directed by a lithium cation was given in Table 1-5. Many permutations of the template synthesis of macrocyclic ligands have been explored,[6,9–11] and have produced such interesting complexes as (8-9) and (8-10).

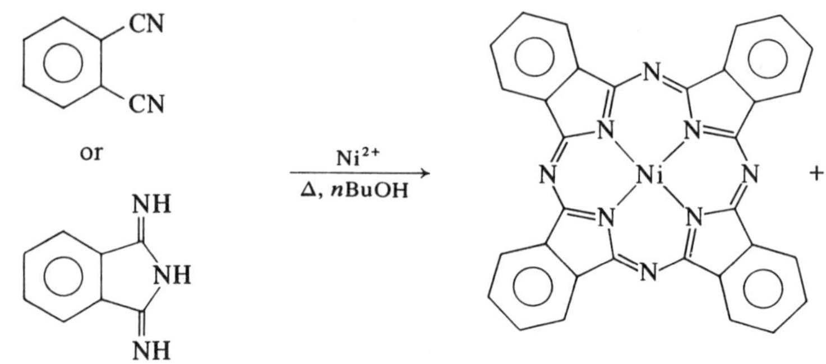

$$+ C_3H_7CHO \qquad (8\text{-}8)$$

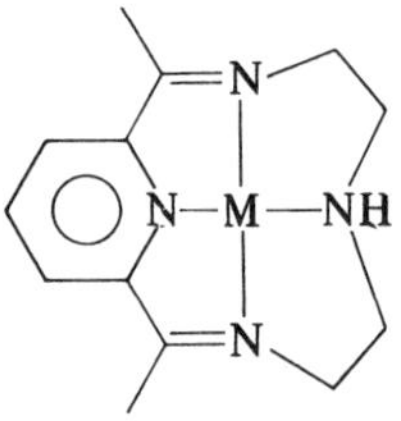

(8-9)

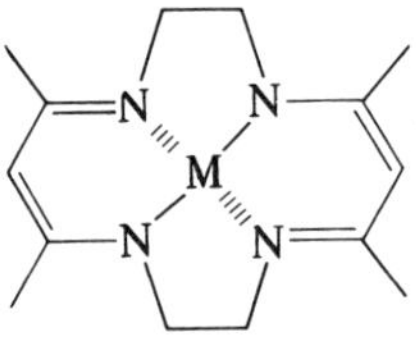

(8-10)

The formation of (8-11) shows that coordinated anions can still be good nucleophiles. In this case the coordinated mercaptide (8-12) is actually more reactive than the metal-free mercaptan.

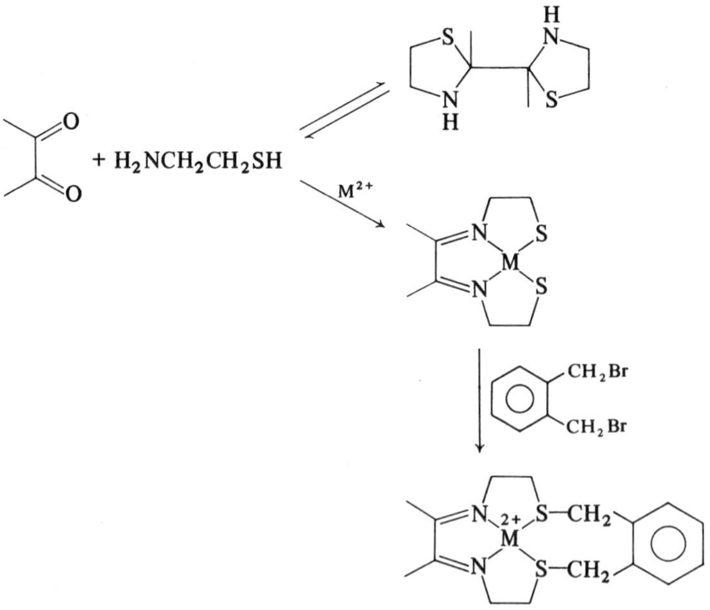

$+ H_2NCH_2CH_2SH$

(8-12)

(8-11)

While complexation of a Schiff base normally sensitive to hydrolysis may reverse that equilibrium to favor condensation, other reactions of the Schiff base may not be affected; for example, salicylaldehyde-derived Schiff base complexes resist hydrolysis yet easily undergo amine exchange under equilibrium conditions. Ammonia is displaced by methylamine to form (8-13)

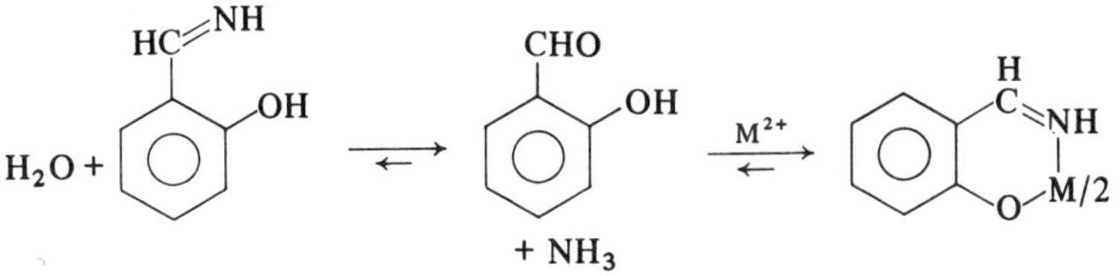

which further reacts with ethylenediamine (en) to form the Salen ligand system (8-14). The latter is the most stable complex since the ligand is tetradentate, planar, unstrained, has a net charge of 2−, and has good back-bonding characteristics. Since exchange of amine groups in these coordinated Schiff bases is so facile, their stability toward hydrolysis must lie in the metal ion's greater equilibrium preference for nitrogen over oxygen donors (e.g., see Fig. 6-7 and Table 6-3). Similar reactivity might be expected for acetylacetonate

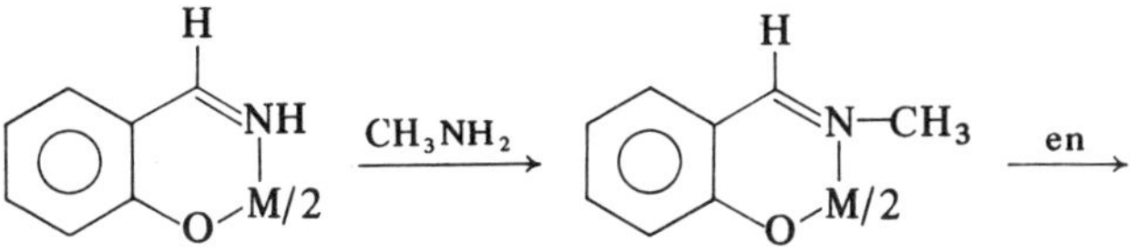

(8-13)

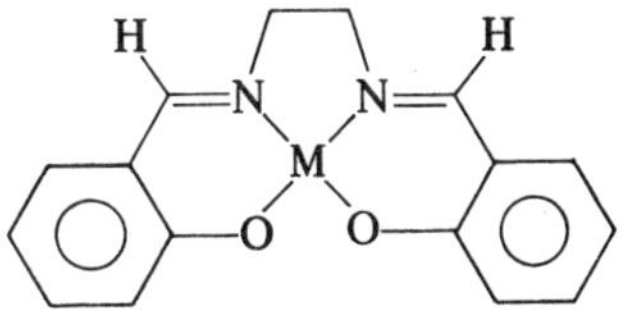

(8-14)

(acac) complexes (e.g., 8-15), but surprisingly it is generally *not* observed. One reason for the lack of nucleophilic attack at a "carbonyl carbon" in an acac chelate may be that the extensive $p\pi-d\pi$ cycloconjugation in (8-15), approaching that of a Hückel $(4n + 2)$ aromatic system, would be lost in the carbinolamine intermediate (8-16). However, it is possible to synthesize the Schiff base ligand (8-17) and then form a complex which is stable toward hydrolysis. The aromatic character of many acac complexes is supported by the fact that they undergo many of the same type of electrophilic substitution reactions characteristic of benzenoid compounds, such as bromination,

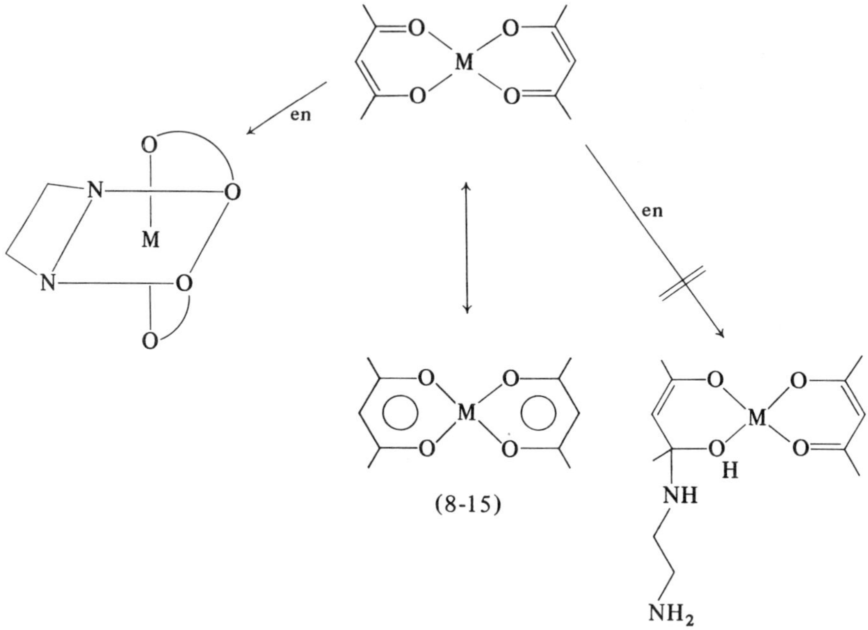

(8-15)

(8-16)

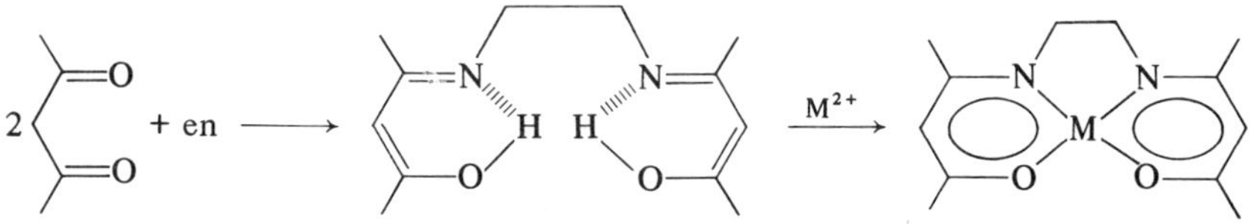

acacH   acacenH$_2$ Macacen

(8-17)

nitration, and acylation (Scheme 8-1). Metal coordination of the diketone ligand facilitates its deprotonation, i.e., stabilizes the corresponding enolate and shifts an acid-base equilibrium. As a result the carbonyl carbons lose their electrophilic character and the ligand becomes a good nucleophile, undergoing electrophilic attack at the methine carbon. Even relatively weak electrophiles such as isocyanates will attack to give an amide product (8-18) from which the ligand may be liberated by treatment with H$_2$S in benzene. Optically active bromination products were obtained from optically active Cr(acac)$_3$, suggesting that all three chelate rings remained intact during the bromination.[12] Such behavior is not restricted to pseudobenzenoid β-diketone complexes. Glyoxal condenses with ammonia or methylamine in the presence of Fe(II) to form a *tris-α*-diimine chelate (9-19). When treated with bromine this Fe(II)

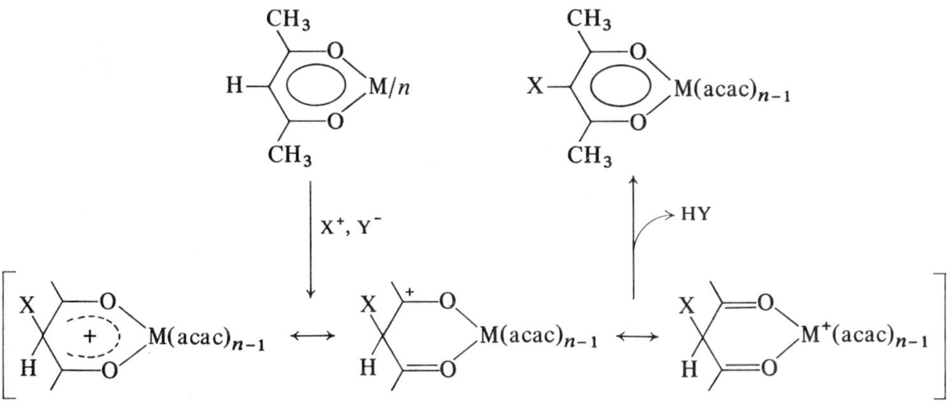

Scheme 8-1

$$XY = Br_2, HONO_2, AcCl, etc.$$

Stoichiometric if M = Cr(III) > Co(III)

Catalytic if M = Cu(II) > Ni(II) > Zn(II) > Mn(II) $\gg$ H$^+$

(8-15) $\xrightarrow{\text{RNCO}}$ (acac)M ... $\longrightarrow$ M/2 (8-18)

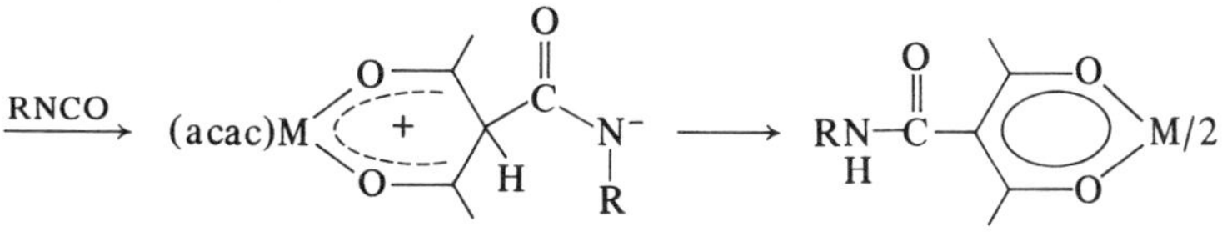

R = aryl > alkyl; M = Cu(II), Ni(II)[13].

complex undergoes ligand bromination, much like thiophene or furan, rather than oxidation to a Fe(III) complex.[14]

Schiff bases are not always stabilized toward hydrolysis by metal complexation, as shown by (8-20) and (8-21). The former is rapidly hydrolyzed to give the corresponding salicylaldehyde complex, which probably reflects the ligand–ligand complementarity of o-phen and hard oxyanions in mixed complexes (e.g., see Table 6-8). Complex (8-21) also hydrolyzes to give Ni(en)$_2$$^{2+}$ and the free aldehyde. Evidently an isolated amino group is a better (softer?) ligand than an isolated imino group, but the opposite is true for the conjugated α-diimines. The fact that complexes (8-20) and (8-21) have a net positive charge may also be a significant factor in promoting the attack of water on the

$$\text{(8-19)} \quad \text{Fe}^{2+}/3 \xrightarrow[\text{FeBr}_3]{\text{Br}_2} \text{Fe}^{2+}/3 + \text{HBr}$$

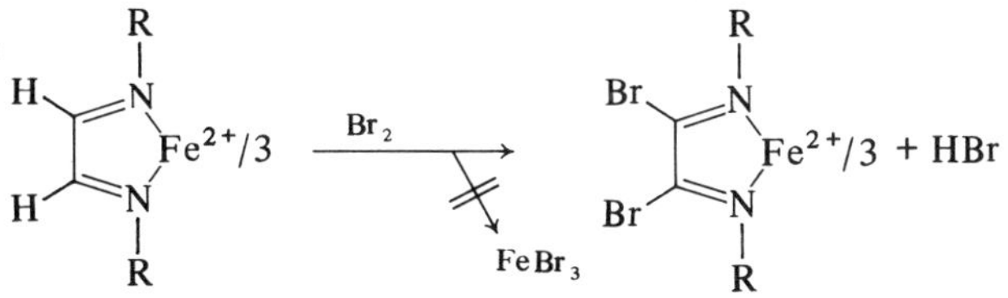

coordinated ligand. The importance of this factor is more clearly shown by the reactions of (8-22) and (8-23) with the electrophile methyl iodide (cf. Ru complexes in Chapter 5).[15]

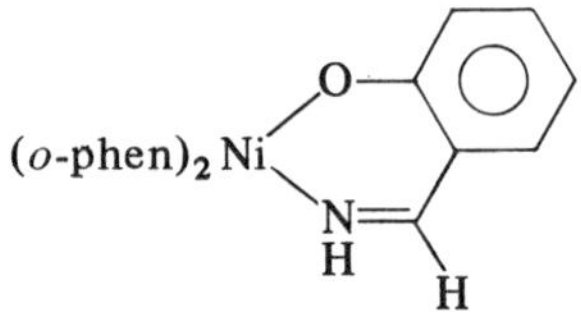

(8-20)

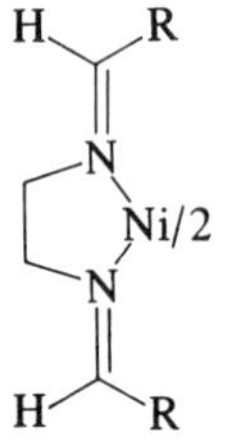

R = 2-thienyl

(8-21)

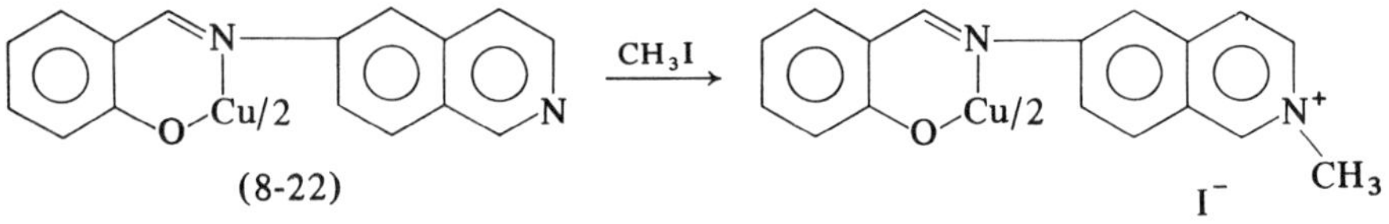

(8-22)

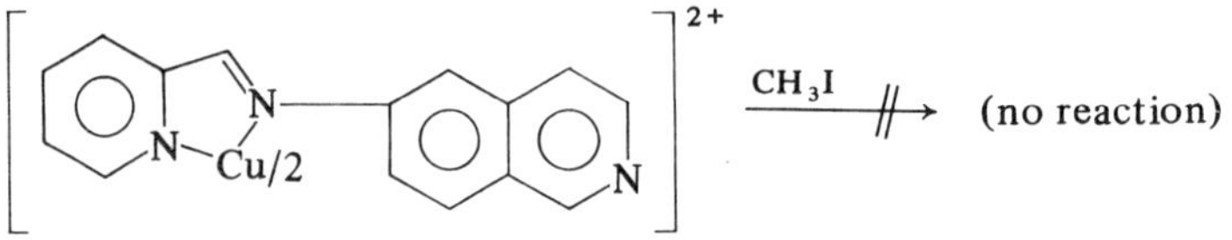

(8-23)

Metal-Directed Formation and Reactions of Enols

A large number of reactions may be grouped under the heading of metal-directed formation and reaction of enols. Some are useful as synthetic reactions, and others are interesting as models for the catalytic step of important enzymatic reactions, which as often as not are catalyzed by metalloenzymes. They depend heavily on the formation of a chelate as a means of directing the equilibrium and on the inductive effects of metal ions ($M^{2+} < M^{3+} < M^{4+}$) for promoting enolization and deprotonation to generate a nucleophilic enolate. In turn it is possible to use these reactions to generate particular product stereochemistries by capitalizing on the properties of the chelated complex intermediate.

THE ALDOL CONDENSATION AND RELATED EQUILIBRIA

The aldol condensation is well known as a simple method for the formation of new carbon–carbon bonds, but its practical utility is often hampered by the formation of mixtures of isomeric products, especially when two different carbonyl compounds are being used. Furthermore, the condensation is easily reversible, provided dehydration does not occur, and the equilibrium does not favor the aldol product. This problem is overcome by the Wittig "directed aldol condensation" shown in Fig. 8-3, which depends on metal chelation to shift the

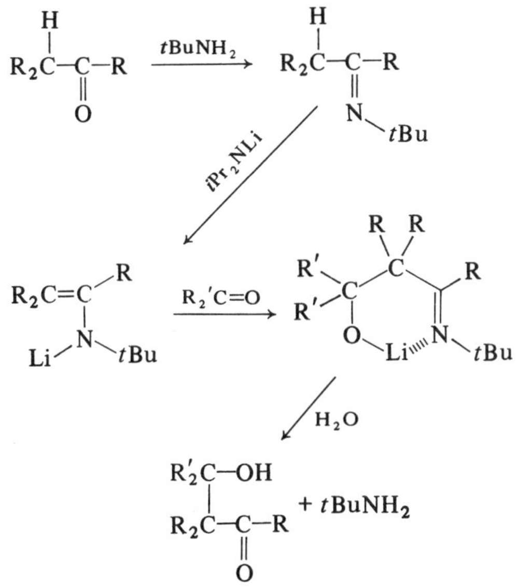

Fig. 8-3. Wittig procedure for directed aldol condensations.[16]

equilibrium in a useful direction.[16] House et al.[17] have shown that other attributes of a chelated intermediate, i.e., conformational preferences, can also be used to achieve great selectivity among possible isomeric products from an aldol condensation. This method involves the condensation of a preformed lithium enolate with an aldehyde in the presence of anhydrous zinc chloride.

The formation of a single aldol in high yield has been rationalized by the scheme shown in Fig. 8-4, in which zinc preferentially complexes one of two equilibrating aldoxide anions. An alternative explanation would involve equilibration of reactants between the two pseudo-chair intermediate complexes (8-24) and (8-25) before C—C bond formation. The complex with the least number of axial substituents reacts to form the product, but in the other form the approach of the two reacting carbons is impeded by the developing steric interactions. This explanation is supported by the observations that it is necessary to add the $ZnCl_2$ *before* the addition of the carbonyl compound, that the best results are obtained in nonpolar solvents (e.g., ether-benzene mixtures)

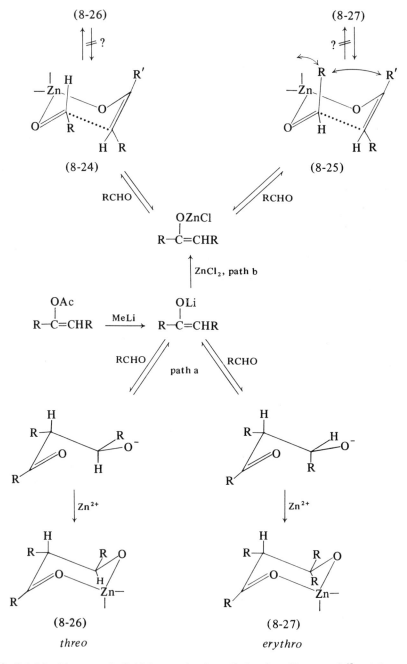

Fig. 8-4. Metal ion control of aldol stereochemistry. Path a from House *et al.*;[17] path b see text.

in which a large amount of LiCl precipitates when the ZnCl$_2$ is added, and that in those cases which give mixtures the ratio of products does not change from 10 sec after addition of the carbonyl compound until workup 5 min later. It is also possible that once formed, steric interactions in (8-27) would tend to force ring opening so that an equilibrium favoring (8-26) would be established rapidly.

The aldolase reaction of the glycolytic pathway is but one example of the importance of aldol reactions in biochemistry. As a counterpoint to the above discussion it should be noted that aldolases from animal sources, like the Wittig directed aldol synthesis, depend on a Schiff base intermediate to promote enolization, and that yeast aldolase, like the House procedure, depends on a zinc ion at the reaction center.[18]

Related to these aldol condensations are the reactions of aldehydes with enolates derived from chelates of α-amino acids. Enolization is greatly facilitated by the inductive effects of the metal, and possibly by the rigidity of the planar five-membered chelate ring, much as cyclopentanone is more acidic than cyclohexanone. Treatment of Cu(II) α-amino acid chelates with base and an aldehyde, followed by removal of copper with H$_2$S, gives the hydroxyalkyl amino acids serine and threonine. Complex (8-28) was originally thought to be an intermediate in these reactions, but recent work has shown that complexes

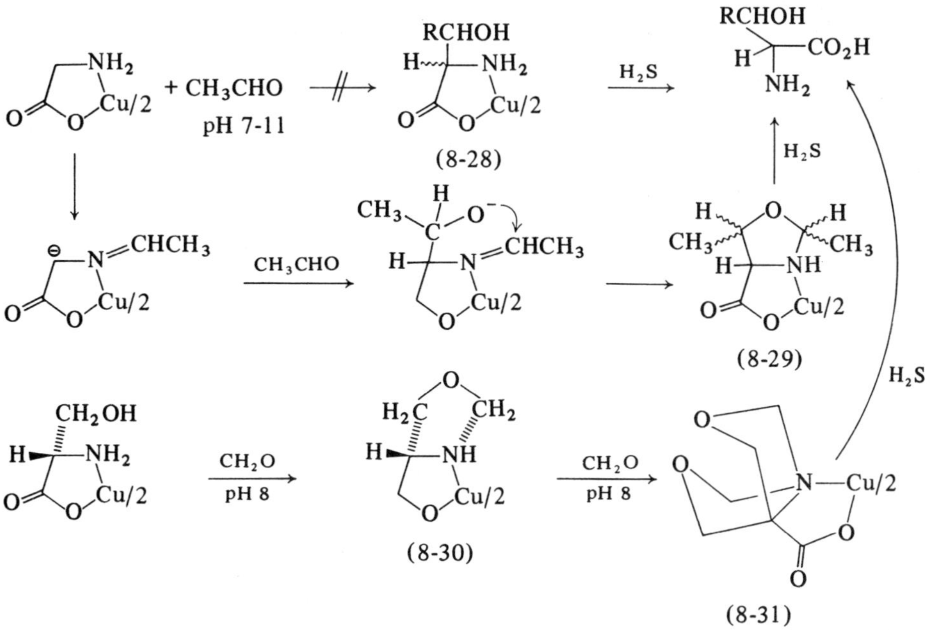

Fig.8-5. Hydroxyalkylation of chelated α-amino acids.

like (8-29–8-31) are in fact the actual intermediates.[19] These oxazoline amino acids can be isolated, but they are hydrolyzed to serine and threonine by the hydrogen sulfide workup. Figure 8-5 shows how these adducts might be formed in the solution. It might be expected that if a glycine molecule were activated by an optically active metal center, the new amino acid produced might be optically active as well. This is indeed the case, as shown by the reaction of (8-32) with acetaldehyde to produce in 80% overall yield a mixture of glycine (10%), *allo*-threonine (8-33, 20%), and threonine (8-34, 70%, 8% optically active).[20] It is not known whether this reaction involves oxazoline intermediates, but enolization at the Co(III) center is much faster than observed for Cu(gly)$_2$ because of the influence of net charge on the metal's inductive effects. The low optical yield is to be expected since the steric interactions of the ligands are small and since Co(III) chelates of alanine and valine undergo base catalyzed α-carbon H/D exchange and racemization at identical rates.[21] In contrast,

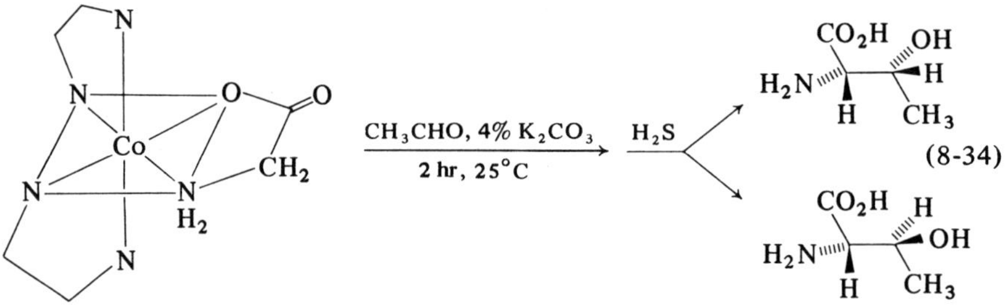

(8-32)

(8-34)

(8-33)

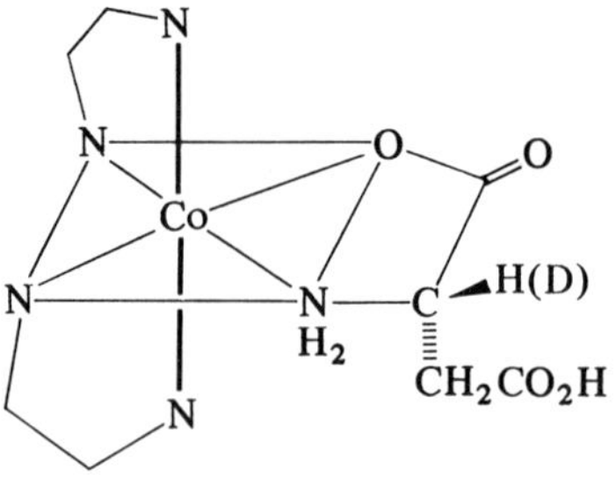

(8-35)

base-catalyzed α-carbon H/D exchange of the optically active aspartic acid chelate (8-35) proceeds with 75% *retention* of configuration at the α-carbon.[22] Other related reactions of metal enolates are given in Fig. 8-6.

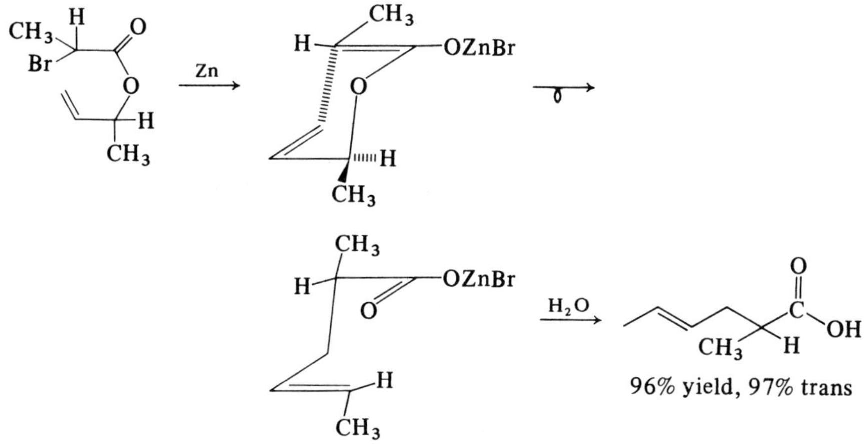

(a)

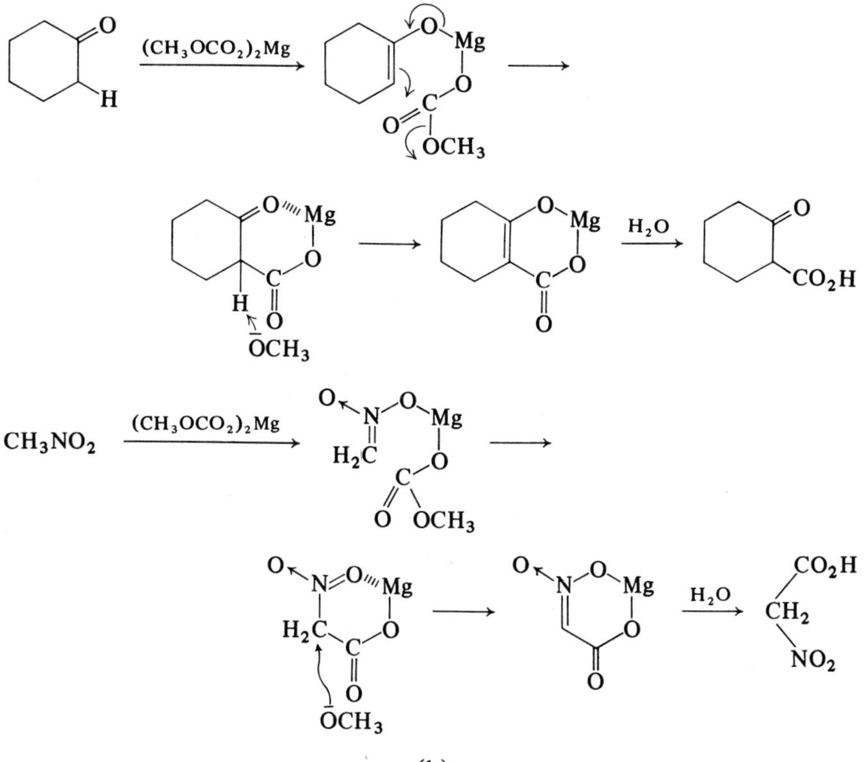

(b)

Fig. 8-6 (for caption see over).

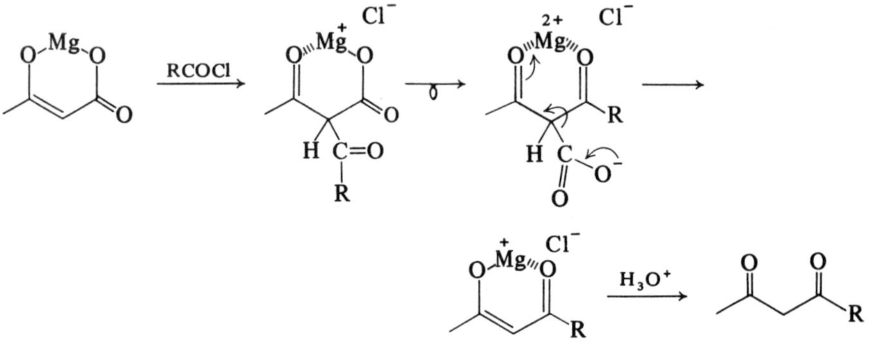

(c)

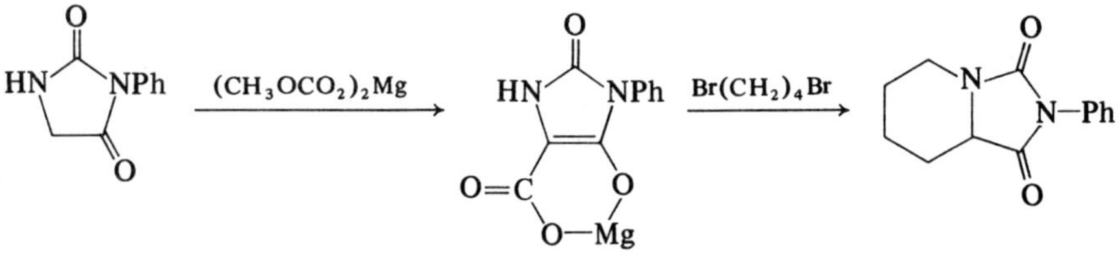

(d)

Fig. 8-6. Synthetically useful reactions of metal enolates. (a) Sigmatropic rearrangement,[25] (b) carboxylation,[26,27] (c) acylation,[28] and (d) alkylation.[29]

CARBOXYLATION AND DECARBOXYLATION

As seen in Fig. 8-6, both of these processes can be directed by metal ions. Under most circumstances the equilibrium favors decarboxylation since the carbon dioxide produced is a gas and is either lost from the reaction or converted to carbonate ions. Consequently, the carboxylation of enolates will necessarily involve the formation of a new chelated product to reverse this equilibrium. Similarly, the only ligands that will undergo metal-directed decarboxylation are those in which at least one chelate ring is preserved in the final product. One example of this is the decarboxylation step of the acylation sequence given in Fig. 8-6. This reaction reaches its normal equilibrium but at an accelerated rate because of the influence of the inductive effect of the metal in promoting the decarboxylation. The stability of the chelated product prevents the metal from functioning catalytically.

Decarboxylations which take place at asymmetric metal centers can proceed with a high degree of stereospecificity. This was first shown by Asperger and Liu using the complex (8-36, Fig. 8-7) which underwent thermal decarboxylation to give a product 25–30% enriched in the L-(S)-isomer of alanine. A

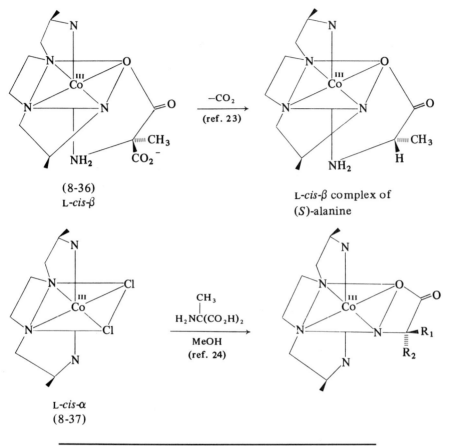

(8-36)
L-*cis*-β

L-*cis*-β complex of
(*S*)-alanine

L-*cis*-α
(8-37)

L-*cis*-α-complex	R_1	R_2	Yield (%)
(8-38) Malonate	CO_2	CH_3	100
(8-39) (*S*)-Alanine	H	CH_3	65
(8-40) (*R*)-Alanine	CH_3	H	35

Fig. 8-7. Asymmetric reactions of nondissymmetric ligands at chiral metal centers.[23,24]

similar system was studied by Job and Bruice[24], as shown in Fig. 8-7. They found that the asymmetric metal center (8-37) exhibited 100% chiral differentiation between the two prochiral carboxyl groups of the meso ligand. This process is very similar to the "three-point attachment" of citric acid at a Fe(II) center in the enzyme aconitase (q.v.) which enables the enzyme to distinguish between the two $-CH_2CO_2H$ groups of citric acid. When complex (8-38) is refluxed in methanol, decarboxylation proceeds to give a 65:35 mixture of the

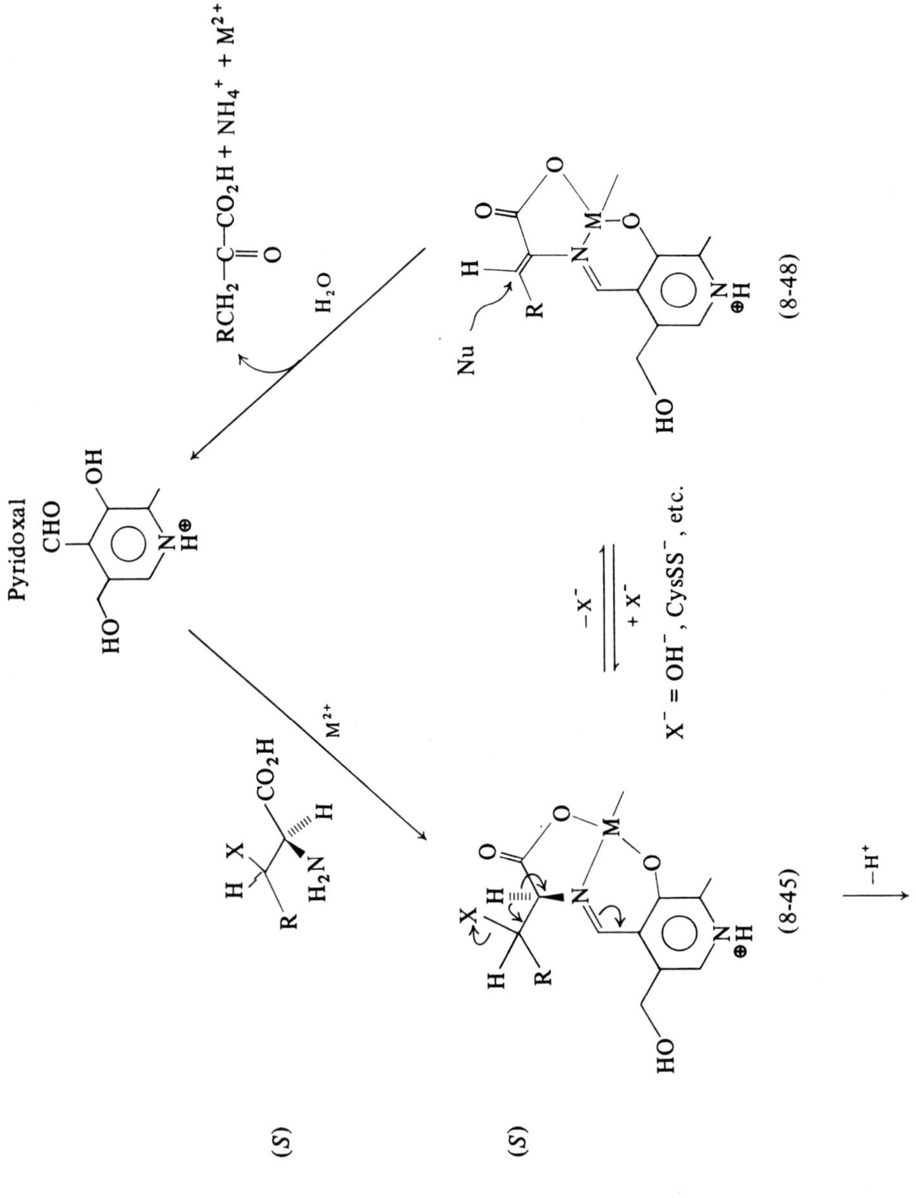
Pyridoxal
CHO
OH
HO
N
H⊕
RCH2—C—CO2H + NH4+ + M2+
O
H2O
M2+
H X
R
CO2H
H2N
H
(S)
Nu
O
M
N
R
H
X
(8-45)
(8-48)
−X−
+X−
X− = OH−, CysSS−, etc.
−H+

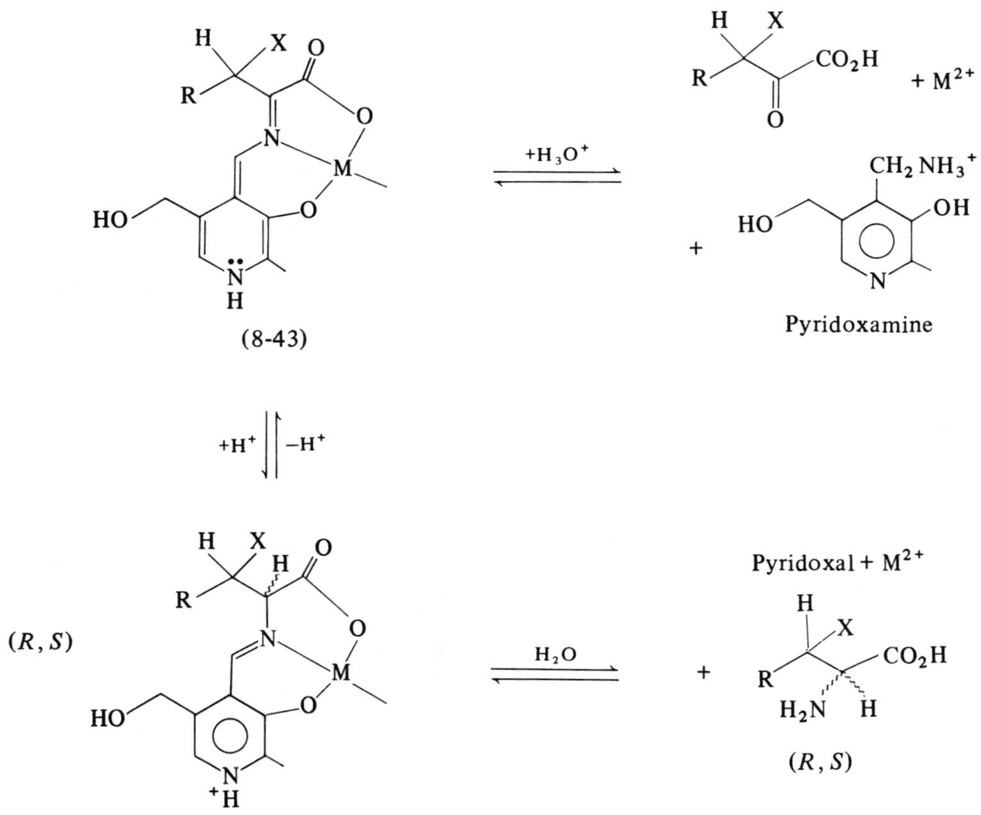

Fig. 8-8. Pyridoxal-metal ion catalysis of transamination, and β elimination, and epimerization of α-amino acids.

(*S*)- and (*R*)-alanine complexes (8-39) and (8-40), respectively. Since the central malonate carbon becomes sp^2 hybridized during decarboxylation, asymmetric induction must be occurring during protonation of the intermediate enolate similar to that in (8-35).

Coordination Chemistry of Vitamin B₆

Vitamin B_6 is a name for a group of biologically interconvertible pyridine derivatives (e.g. 8-41 and 8-42) which are essential as coenzymes for a multitude of enzymatic reactions of amino acids, including transamination, oxidative deamination, decarboxylation, racemization, elimination, and certain aldollike reactions. As yet there is little evidence that metal ions are involved in more than a token number of these reactions, but the fact that they all have been simulated in "model systems" involving chelated transition metal ions has kept interest in this area high.[30] Another factor in this continuing interest is the diversity of reaction types which can be studied within a single family of compounds. Underlying this diversity, however, are two processes of central importance, i.e., the formation and subsequent reaction of coordinated Schiff bases and enolic intermediates.

(8-41)   (8-42)

Pyridoxal   Pyridoxamine

R = H, vitamin form; $R = PO_3^{2-}$, coenzyme form

The mechanisms of pyridoxal-mediated transamination, racemization, and β elimination of amino acids are given in Fig. 8-8. It can be seen that the methyl and hydroxymethyl groups do not participate in these reactions, although they are very important for *enzymatic* recognition and binding of the cofactor. The importance of the pyridinium nitrogen is indicated by the failure of salicylaldehyde to react beyond the formation of the Schiff base complex, probably because of the lack of a low energy resonance form such as (8-43) in Fig. 8-8 with which to stabilize an enolized amino acid-Schiff base.

SCHIFF BASE FORMATION

The kinetics of condensation of α-amino acids with salicylaldehyde (as a model for pyridoxal) were shown by Hopgood and Leussing[31] to involve both metal-dependent and metal-independent pathways. The efficiency of various metal ions in catalyzing the condensation was found to be $Pb^{2+} \gg Cd^{2+} > Mn^{2+}, Mg^{2+} > Zn^{2+} \gg Co^{2+} > Ni^{2+} > Cu^{2+}$. The rates for Co^{2+}, Ni^{2+}, and Cu^{2+} were considerably lower than that of the metal-independent pathway. This order was interpreted as arising from the relatively relaxed geometrical requirements of the d^0, d^5, and d^{10} ions which could more readily allow the two coordinated ligands to form the carbinolamine intermediate (8-44). This has

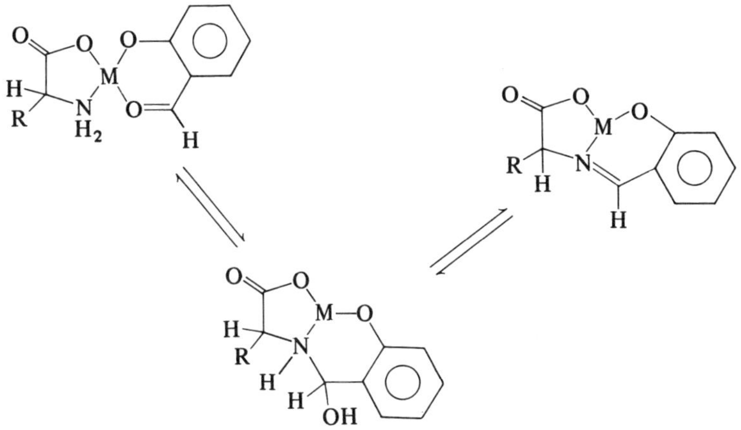

(8-44)

been termed[31] the "promnastic effect" to describe the effects of the metal as "a catalytic agent which forms a ternary complex with the two reactants but which imposes a minimum of steric requirements upon them." The substantial crystal field effects on ions with incompletely filled d subshells impose rigid coordination geometries which are not conducive to carbinolamine formation or other ligand exchange processes (Fig. 5-3). Hence, *catalysis* by these metals is inefficient despite the fact that they form the most stable Schiff base complexes.

α-CARBON ENOLIZATION AND SUBSEQUENT REACTIONS

The inductive effects of the metal are important in facilitating loss of a proton from the α-carbon, but the protonated pyridinium ring also makes a considerable contribution in this case, i.e., salicylaldehyde analogs of (8-45, Fig. 8-8) enolize with great difficulty. Proton return to the opposite side of the enolate ligand plane constitutes an epimerization of the amino acid. Protonation of the aldehyde carbon in (8-43, Fig. 8-8) gives a complex which can hydrolyze to form pyridoxamine and an α-keto acid, although the tautomeric equilibrium favors the aldimine tautomers over the ketimine tautomers. The tautomerization of (8-46) to (8-47) is accelerated around 10^3-fold by Cu^{2+} > Zn^{2+} > Ni^{2+}, and is further enhanced by added base.[32] Catalytic transamination between an α-keto acid and an α-amino acid can be accomplished with

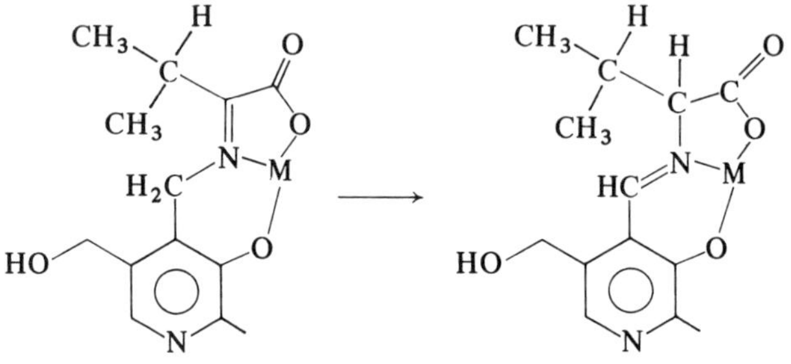

(8-46)   (8-47)

pyridoxal, pyridoxamine, or 4-nitrosalicylaldehyde and a variety of divalent and trivalent metal ions.[33] The order of metal efficiencies is difficult to interpret, probably because the rate limiting step changes from aldimine Schiff base formation to tautomerization to ketimine hydrolysis as the metal is varied. Amino acids such as serine or cystathionine (Homocys–S–S–Cys) undergo β elimination more easily than transamination under these conditions, yielding α-aminoacrylate intermediates (8-48, Fig. 8-8), which are susceptible to Michael attack by nucleophiles.[33]

OTHER REACTIONS INVOLVING PYRIDOXAL

As might be expected, the decarboxylation of pyridoxal-amino acid Schiff bases is *inhibited* rather than catalyzed by metal ions. However, it is facilitated by protonation of the pyridine nitrogen. The reverse reaction, carboxylation of a coordinated pyridoxal-amine Schiff base has not been reported, but such a process might be anticipated since the product could be stabilized by chelation. Pyridoxal is also important in the enzymatic interconversion of serine and glycine via an aldol- or Mannich-like condensation process (Fig. 8-9). The "formaldehyde" comes from a variety of sources included in the so-called one-carbon pool of metabolites. Free formaldehyde is extremely toxic so it is transported as a derivative of another vitamin, tetrahydrofolic acid (THFA).

Acid-Base Equilibria: Acidity of Coordinated Water

In numerous places in this as well as preceding chapters we have made reference to the (Brønsted) acidity of a ligand being increased upon complex-

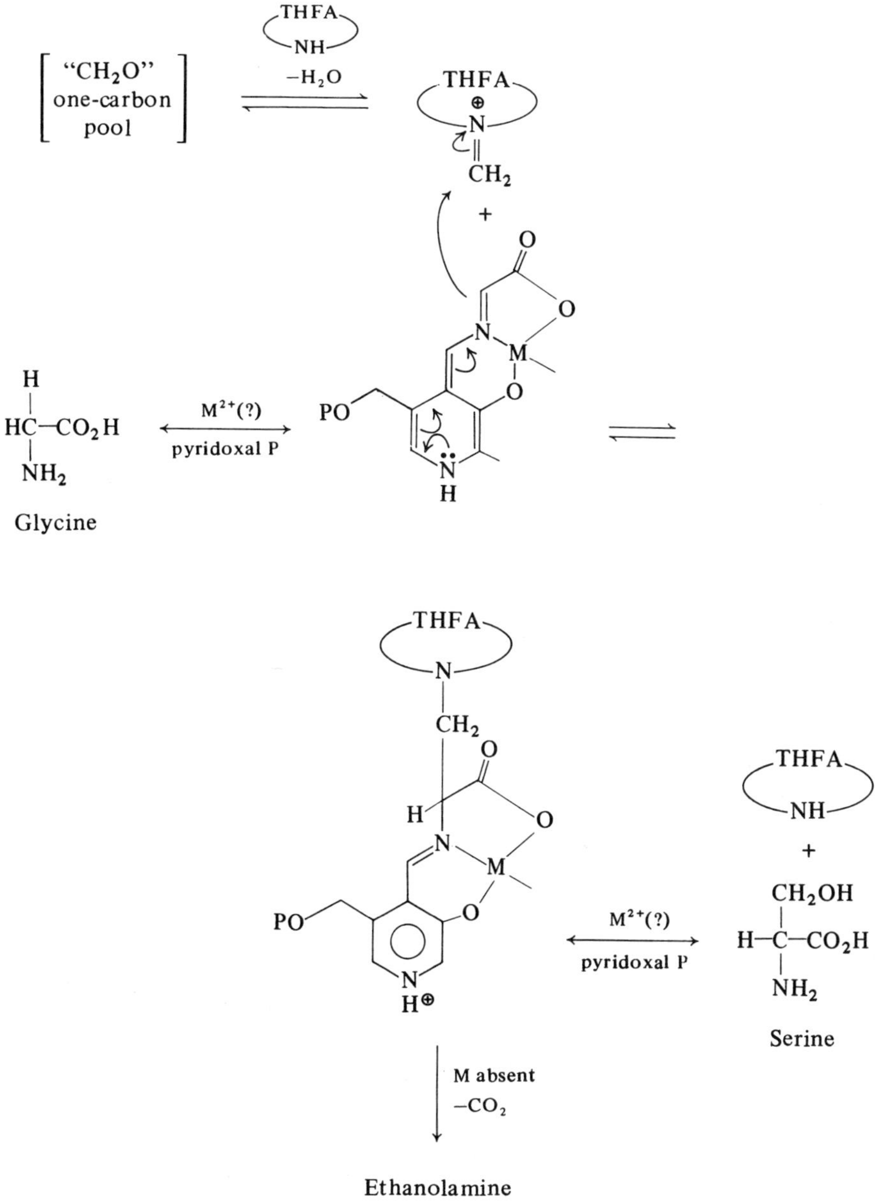

Fig. 8-9 Enzymatic hydroxymethylation of pyridoxylideneglycine. The involvement of a metal ion is speculative; see Fig. 8-8 for intermediates. In human metabolism the serine → glycine + "CH$_2$O" conversion is the most important.

TABLE 8-1

Acidity of Coordinated Water and Alcohol Ligands

Ion or complex	pK_a	Ref.[a]
Hg^{2+}	3.7	1
Fe^{2+}	8.3	1
Cu^{2+}	8.3	1
Ni^{2+}	9.3	1
Mn^{2+}	>9	2
Zn^{2+}	9.6	1
	<9	2
Carbonic anhydrase $ZnOH_2$	6.9–7.1	3
	(7.3–7.8)	4

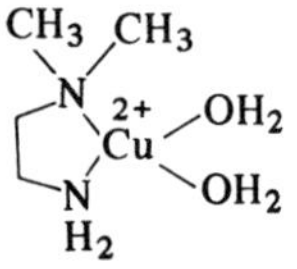

	$pK_1 = 8$	5
	$pK_2 = 9.8$	

bipyCu(OH$_2$)$_2$$^{2+}$	8	5

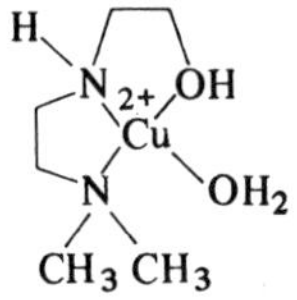

	$pK_1 = 7.3$	5
	$pK_2 = 9.3$	

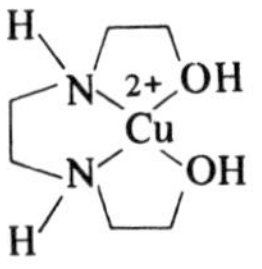	7.3	5

ation to a metal ion. Since the acidity of a molecule is a measure of a proton dissociation equilibrium, the increase in acidity can be understood as a preferential complexation of the *conjugate base* form of the ligand. When the ligand in question is a water molecule, its deprotonation equilibrium is called "hydrolysis." This process is thought to be important in facilitating water and ligand exchange for small hard ions such as Be^{2+} and Al^{3+} and is central to the Sn1cb mechanism for ligand substitution in inert octahedral complexes. A related issue of considerable importance in many reactions of coordinated ligands, and of some importance to the mechanisms of hydrolytic metallo-

TABLE 8-1-continued

Ion or complex	pK_a	Ref.[a]
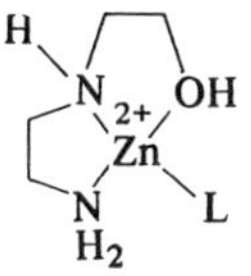 L = p-nitrophenylpicolinate	8.4	6
$Fe(OH_2)_6^{3+}$	2.21	7
$Fe(OH)(OH_2)_5^{2+}$	3.05	8
$Hematin(OH_2)_2$	7.5	9
$Co(NH_3)_5(OH_2)^{3+}$	5.69	7
$Co(NH_3)_4(OH_2)_2^{3+}$	5.21	7
$Co(NH_3)_3(OH_2)_3^{3+}$	4.70	7
$Co(NH_3)_2(OH_2)_4^{3+}$	3.40	7
$Cr(OH_2)_6^{3+}$	3.90	7
$Al(OH_2)_6^{3+}$	4.89	7

[a] Key to references:
1. J. E. King, "Qualitative Analysis and Electrolytic Solutions," pp. 434–442. Harcourt, New York, 1959.
2. E. T. Kaiser and B. L. Kaiser, *Accounts Chem. Res.* **5**, 219 (1972).
3. J. E. Coleman, *Progr. Bioorg. Chem.* **1**, 292 (1971).
4. S. Lindskog, L. E. Henderson, K. K. Kannan, A. Liljas, P. O. Nyman, and B. Strondberg, *in* "The Enzymes" (P. D. Boyer, ed.), 3rd ed., Vol. 5, p. 587. Academic Press, New York, 1971.
5. A. E. Martell, *Pure Appl. Chem.* **17**, 129 (1968).
6. D. S. Sigman and C. T. Jorgensen, *J. Amer. Chem. Soc.* **94**, 1724 (1972).
7. L. Pakras, *J. Chem. Educ.* **33**, 152 (1955).
8. B. Mahan, "University Chemistry," p. 565. Addison-Wesley, Reading, Massachusetts, 1965.
9. H. R. Mahler and E. H. Cordes, "Biological Chemistry," 2nd ed., p. 665. Harper, New York, 1971.

enzymes, is whether coordinated conjugate bases such as hydroxide or alkoxide ions still retain appreciable nucleophilic character. There is some good evidence and at least one conclusive demonstration that a M—OH group is still nucleophilic enough to hydrolyze an ester; these will be discussed further in chapter 9. Our concern here is only with the equilibria which involve coordinated hydroxide and alkoxide groups. For the divalent cations, the pK_a of coordinated water or alcohols is usually in the range of 8–10. With trivalent ions, stronger and harder Lewis acids, the pK_a of coordinated water is generally around 3–4 but may be lower in some cases. However, the apparent pK_a's of metal ions are very sensitive to temperature and to the ionic strength of the measuring medium. Examples of specific complexes whose pK_a's have been determined are given in Table 8-1. For further information Special Publications No. 17 and No. 25 of The Chemical Society (London) should be consulted.[33a]

Optical Resolution of Amino Acids

As discussed in Chapter 6 the nature of one ligand bound to a metal influences the binding of a second ligand. Steric as well as electronic inter-actions are important, and if the first bound ligand has an asymmetric center near its metal-binding site, the two enantiomers of a second asymmetric ligand may be differentially bound. While the equilibrium constants for the reactions

$$Cu(\text{L-A})^+ + \text{L-A}^- \rightleftharpoons Cu(\text{L-A})_2$$
$$Cu(\text{L-A})^+ + \text{D-A}^- \rightleftharpoons Cu(\text{L-A})(\text{D-A})$$

are identical for A = alanine, phenylalanine, valine, and proline, the copper complex of L-valine-N-monoacetate (LVMA) binds the L-enantiomer of leucine, phenylalanine, alanine, and serine, 3.3–6.5 times more strongly than the D-enantiomer, but curiously for valine the binding of the D-enantiomer is 2.5 times greater than that of the L-enantiomer.[34] This selectivity doubtless originates in the steric interactions of the amino acid ligands, particularly in the area of their bulky side-chain groups. Figure 8-10 shows a possible complexa-tion arrangement consistent with the preferential binding of L-isomers, but another arrangement must be involved in the Cu(LVMA)(D-Val) complex since opposite results are obtained.

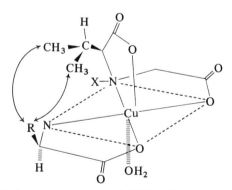

Fig. 8-10. Possible solution structure for the Cu(LVMA)(aa) complex. In this case the less stable D-amino acid is shown to illustrate the side-chain interactions; X = H, or a polystyrene-derived support.[34,35]

Such differences in binding form the basis for a method for resolving racemic amino acid mixtures[35]. For example, the LVMA complex can be attached via the nitrogen to an inert polystyrene support (Fig. 8-10) to form a solid adsorbent which can be packed into a column. Solutions of amino acids are passed through the column and both L and D forms are adsorbed. Since the D form is less strongly bound, it can be more easily displaced by running a solution of ethylenediamine through the column. The early effluent from the column is considerably enriched (80%) in D-isomer. In another case the complete resolution of DL-proline on a column with L-proline:copper bound to polystyrene has been achieved.

REFERENCES

1. A. E. Martell, *Pure Appl. Chem.* **17**, 129 (1968).
2. D. H. Busch, *Science* **171**, 241 (1971).
3. D. H. Busch, ed., "Reactions of Coordinated Ligands," Advan. Chem. Ser. No. 37. Advan. Chem. Ser., Washington, D.C., 1963.
4. A. E. Martell, "Metal Ions in Biological Systems" (H. Sigel, ed.), Vol. 2, p. 208. Dekker, New York, 1973.
5. M. M. Jones, "Ligand reactivity and Catalysis." Academic Press, New York, 1968.
6. N. F. Curtis, *Coord. Chem. Rev.* **3**, 3 (1968).
7. E. B. Fleischer, *Inorg. Chem.* **1**, 493 (1962).
8. T. J. Hurley, M. A. Robinson, and S. I. Trotz, *Inorg. Chem.* **6**, 389 (1967).
9. D. H. Busch, K. Farmery, V. Goedken, V. Katovic, A. C. Melnyk, C. R. S. Sperati, and N. Tokel, *Advan. Chem. Ser.* **100**, 44 (1971).
10. A. Eschenmoser, *Quart. Rev., Chem. Soc.* **24**, 366 (1970).
11. V. L. Goedken and S.-M. Peng, *Chem. Commun.* 62 (1973).
12. J. P. Collman, *Advan. Chem. Ser.* **37**, 78 (1963).
13. J. W. Kenney, J. H. Nelson, and R. A. Henry, *Chem. Commun.* 690 (1973).
14. E. Bayer, *Angew. Chem.* **73**, 533 (1961).
15. D. H. Busch, *Advan. Chem. Ser.* **37**, 1 (1963).
16. G. Wittig and A. Hesse, *Org. Syn.* **50**, 66 (1970).
17. H. O. House, D. S. Crumrine, A. Y. Teranishi, and H. D. Olmstead, *J. Amer. Chem. Soc.* **95**, 3310 (1973).
18. B. L. Horecker, O. Tsolas, and C. Y. Lau, *in* "The Enzymes" (P. D. Boyer, ed.), 3rd ed., Vol. 7, p. 213. Academic Press, New York, 1972.
19. J. R. Brush, R. J. Magee, M. J. O'Connor, S. B. Teo, R. J. Gene, and M. R. Snow, *J. Amer. Chem. Soc.* **95**, 2034 (1973).
20. M. Murakami and K. Takahashi, *Bull. Chem. Soc. Jap.* **32**, 308 (1959).
21. D. A. Buckingham, L. G. Marzilli, and A. M. Sargeson, *J. Amer. Chem. Soc.* **89**, 5133 (1967).
22. W. E. Keyes and J. I. Legg, *J. Amer. Chem. Soc.* **95**, 3431 (1973).
23. R. G. Asperger and C. F. Liu, *Inorg. Chem.* **6**, 796 (1967).
24. R. Job and T. C. Bruice, *Chem. Commun.* p. 332 (1973); *J. Amer. Chem. Soc.* **96**, 809, 5523, 5533, and 5741 (1974).
25. J. E. Baldwin and J. A. Walker, *Chem. Commun.* p. 117 (1973).
26. M. Stiles, *J. Amer. Chem. Soc.* **81**, 2598 (1959).
27. M. Stiles and H. L. Finkbeiner, *J. Amer. Chem. Soc.* **81**, 505 (1959); **85**, 616 (1963).

28. R. E. Ireland and J. E. Marshall, *J. Amer. Chem. Soc.* **81**, 2907 (1959).
29. H. Finkbeiner, *J. Org. Chem.* **30**, 3414 (1965).
30. R. H. Holm, in "Inorganic Biochemistry," (G. Eichhorn, ed.) Vol. 2, pp. 1137–1165. Elsevier, Amsterdam, 1973.
31. D. Hopgood and D. L. Leussing, *J. Amer. Chem. Soc.* **91**, 3740 (1969).
32. Y. Matsushima and A. E. Martell, *J. Amer. Chem. Soc.* **89**, 1331 (1967).
33. D. E. Metzler, M. Ikawa, and E. E. Snell, *J. Amer. Chem. Soc.* **76**, 648 (1954).
33a. L. G. Sillen and A. E. Martell, *Chem. Soc.* (London) *Spec. Pub. No.* 17 (1964); *No.* 25 (1970).
34. B. E. Leach and R. J. Angelici, *J. Amer. Chem. Soc.* **91**, 6296 (1969).
35. R. V. Snyder, R. J. Angelici, and R. B. Meck, *J. Amer. Chem. Soc.* **94**, 2660 (1972).

IX

Catalysis by Metal Ions, Metal Complexes, and Metalloenzymes

The Origin of Catalytic Effects

According to the transition state theory of reaction mechanisms and rates, a reactant or set of reactants must pass from its normal ground state to an energetically activated *transition state* as part of its journey to products. The rate of the overall process is determined by the ease with which the ground state reactants can reach the transition state energy, i.e., on the energy difference $\Delta G\ddagger$ such as shown in Fig. 9-1. Catalysts increase the rates of reactions by providing a mechanism for reducing $\Delta G\ddagger$ so that there will be an increased probability of populating the transition state from the ground state of the reactants. Metal ion catalysts are no exception, acting to stabilize transition states by complexing them more strongly than the reactants, or in the case of redox reactions, by providing an alternate pathway for electron transfer which involves a transition state of lower energy than in the uncatalyzed reaction (see Chapter VII). Figure 9-1 shows a free energy diagram representing the catalyzed and uncatalyzed decarboxylation of oxalacetic acid. In this case, because both the reactant and the products are relatively poor chelating agents (i.e., $\Delta G_1 \approx \Delta G_3$, and both are small), only a catalytic amount of metal ion is

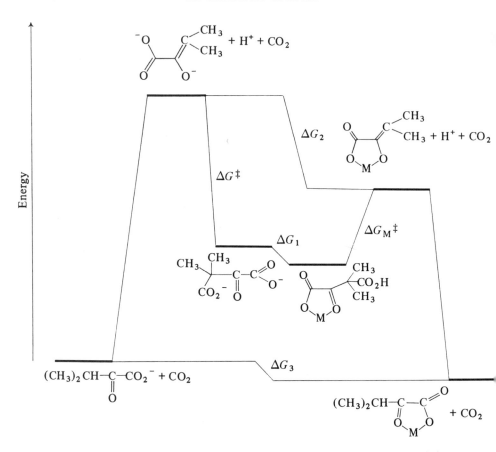

Fig. 9-1. Approximate free energy diagram for α,α-dimethyloxalacetate decarboxylation.

required. The source of the catalysis lies in the fact that the transition state is stabilized by complexation to a greater extent than either reactants or products ($\Delta G_2 \gg \Delta G_1 \approx \Delta G_3$). Since the product does not bind the metal very strongly, the metal can recycle and catalyze the decarboxylation of many molecules of oxalacetate.

In many cases stoichiometric amounts of metal ion are requied to "catalyze" a given reaction because the metal is bound by the reaction products more strongly than by any other species present, thus preventing its recycling or turn-over. Nevertheless, it is the greater complexation of the transition state, relative to complexation of the reactant, that gives rise to the catalysis. This is shown for the hydrolysis of ethyl glycinate in Fig. 9-2. In this case the reactant is a reasonably good chelating agent, but the product, glycine, chelates strongly and competes for metals much more effectively than the reactant ($\Delta G_3 \gg$

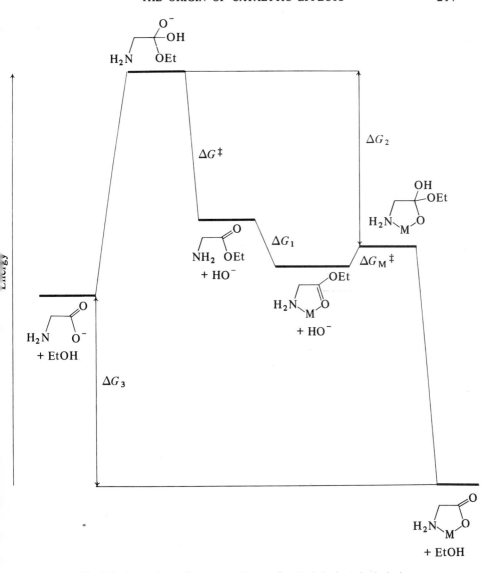

Fig. 9-2. Approximate free energy diagram for ethyl glycinate hydrolysis.

ΔG_1). The observation of rate accelerations of 10^4–10^6 with ethyl glycinate and metals such as Co(III) and Cu(II) suggests that the tetrahedral intermediate is also a good chelating ligand, its complex being easily formed from the ground state complex ($\Delta G_M^{\ddagger}$ small). The tight binding of the reaction product actually has nothing to do with the catalysis, it merely restricts catalyst turnover to one complete cycle. The catalysis still arises from the stabilization of the transition state ($\Delta G_2 > \Delta G_1$).

Even in the case of metal-assisted solvolysis reactions, where an insoluble metal halide such as AgCl or HgBr$_2$ forms, the catalysis still arises from stabilization of the transition state as shown in Fig. 9-3. In Chapter VI it was

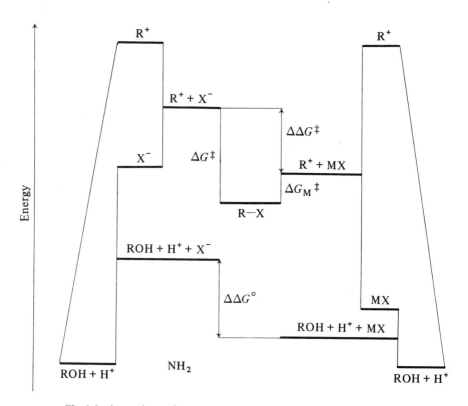

Fig. 9-3. Approximate free energy diagram for the solvolysis of R–X.

noted that changes in the R–X bond strength during solvolysis, where R may be a carbon or a metal center, parallel changes in the solvation of X$^-$. This is often paraphrased by saying that the transition state of the reaction resembles the products more closely than the reactants, although in this case it would be more accurate to say that one of the reaction products, X$^-$, is actually formed as an integral part of the transition state. Therefore, shifting the reaction equilibrium by metal complexation and stabilization of this reaction product also stabilizes the transition state by a comparable amount ($\Delta\Delta G° \approx \Delta\Delta G^{\ddagger}$). Thus, metals which complex the product, X^-, accelerate the solvolysis of R–X, and conversely the addition of a large amount of X^- to the system retards the solvolysis. The latter phenomenon is known as the "common ion effect" and represents the decrease in efficiency of trapping R^+ by solvent when the concentration of X^- becomes high. Another interpretation is that the free energy of X^- in solution, and therefore the free energy of the transition state, increases with the concentration of X^-. A final requirement in this system is that there be a small activation energy for the trapping of R^+ by solvent; only then will the transition state be accurately represented by R^+ + X^-.

Metal catalysis of a reaction depends (1) on the reactants being able to enter the coordination sphere of the metal ion and remain there long enough to be influenced or "activated" by the metal, (2) on the properties of the metal and its inherent ability to modify the chemical properties of ligands by means of complexation, and (3) on the ability of the reaction products to separate from the metal. Although the latter requirement is often overlooked for the sake of a broader definition of catalysis, it is of considerable importance in the application of the art and science of catalysis to real-life situations. Many of the factors governing the association of a reactant ligand and a metal catalyst and the subsequent chemical transformation of the ligand under the metal's influence have been mentioned in Chapters VI and VIII. However, they are important enough to bear brief but collective restatement before any individual reactions are discussed.

Clearly the lability of metal-coordinated water, solvent, or other ligands will affect the rate at which a reactant can enter the coordination sphere *and* the rate at which unreacted starting materials as well as reaction products can leave. Chelation stabilizes the metal–ligand association and thus increases the chance of the ligand reacting while under the influence of the metal. Extraneous ligands reduce the number of coordination sites available on the metal, as well as modify the properties of the metal ion by $p\pi$–$d\pi$ interactions and other "mixed ligand complex" effects. Because metal ions are Lewis acids, they polarize neutral ligands and neutralize some of the charge on anionic ligands, rendering them more susceptible to attack by nucleophiles or anionic reagents. Metal ions with inflexible coordination geometries may bind reactants or products very strongly but may not allow facile rearrangement to a particular transition state, as, for example, in the "promnastic effect" of some metal ions on pyridoxal Schiff base formation (Chapter VIII). Finally, just as complexation of a reaction product lowers its energy and displaces the equilibrium of that reaction, the complexation of a transition state, or a portion of a transition state, lowers its net free energy and results in an increase in the rate of the reaction.

Survey of Metal-Catalyzed Reactions

In attempting to analyze a metal-catalyzed reaction in terms of the aforementioned factors, it is often not possible to attribute the rate enhancement exclusively or even predominately to one particular factor, although one can approach this simplicity in the case of some "solvolysis-like" reactions. Despite this limitation, and for organizational purposes only, we will group our examples into categories according to the factors they best serve to illustrate. In compiling this survey the intent was to be selective and illustrative rather than comprehensive. Additional examples may be found in the other reviews listed at the end of the chapter.[1-5]

Metal Enhancement of Leaving Group Reactivity

SOLVOLYTIC REACTIONS

The hydrolysis of simple phosphate monoesters and anhydrides is a reaction of considerable biological importance, and one which is often catalyzed by metalloenzymes. Since nucleophilic attack at phosphorus is unlikely in the doubly ionized monoesters $ROPO_3^{2-}$ ($pK_a \sim 1.5$ and ~ 6.5), the most important pathway is usually a dissociative reaction to form metaphosphate and an alkoxide ion (Scheme 9-1). This rate limiting step is then followed by rapid

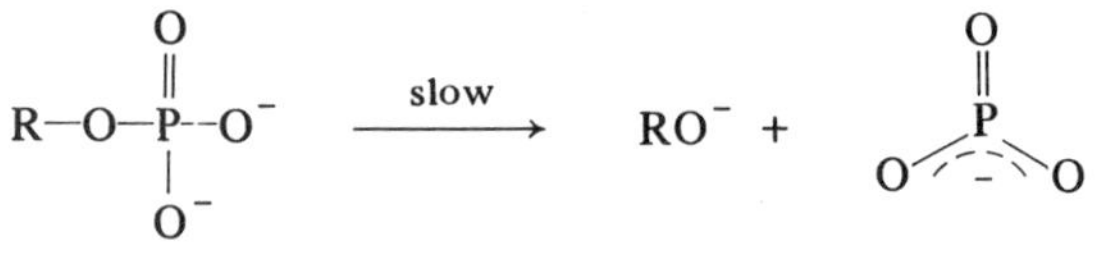

$$R-O-\overset{\overset{\textstyle O}{\|}}{\underset{\underset{\textstyle O^-}{|}}{P}}-O^- \xrightarrow{\text{slow}} RO^- + \overset{\overset{\textstyle O}{\|}}{\underset{O^{-\cdots}O}{P}}$$

Scheme 9-1

protonation of the alkoxide and rapid reaction of the metaphosphate species (a monomeric anhydride of phosphoric acid) with water or hydroxide. As expected for such a mechanism the rate of hydrolysis of $ROPO_3^{2-}$ increases as the acidity of ROH and the stability of RO^- increase. The RO^- species can be stabilized by making R more electron withdrawing or by complexing it to a suitable metal ion. The general case for the latter type of catalysis is shown in the free energy diagram of Fig. 9-3 and a specific example is given in Scheme 9-2. The hydrolysis of 2-(imidazol-4-yl)phenyl phosphate is accelerated 10^3- to 10^4-fold by Cu(II) in the range of pH 4–7.[6] Stabilization of the phenoxide via chelation thus lowers the energy of the transition state, which otherwise would resemble $RO^-\cdots PO_3^-$.

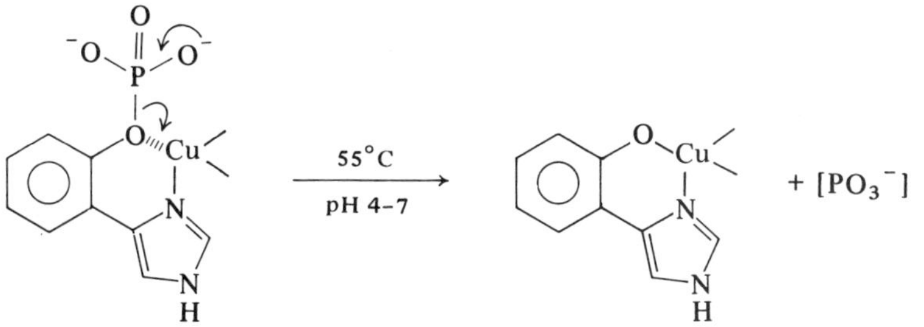

(9-1)   (9-2)

Scheme 9-2

A similar effect may be operating in the hydrolysis of phenyl phosphosulfate catalyzed by Mg^{2+} in mixtures of water and acetonitrile (Scheme 9-3). The inverse dependence of the rate acceleration on the water

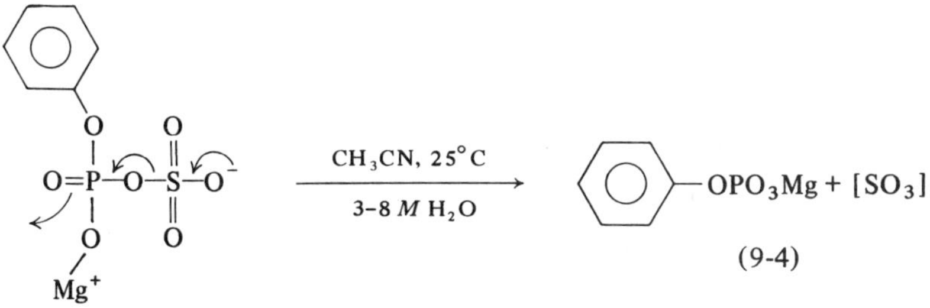

(9-3)

Scheme 9-3

concentration suggests that the magnesium–phosphate interaction, weakened by increasing hydration of the ions, is critical to the catalytic effect.[7] The observed "medium effect" could be relevant to enzymatic catalysis for an enzyme whose active site is relatively anhydrous. An example of a related enzymatic reaction would be the formation of phenolic sulfates from phenols and phosphoadenosyl phosphosulfate (PAPS) catalyzed by phenol-O-sulfotransferase, which requires magnesium ions.

Insofar as CO_2 is also an acid anhydride, the metal-catalyzed decarboxylation of certain β-keto acids also falls into this category of reaction.[1,3,4,8,9] Here the "leaving group" is an enolate ion which is stabilized by complexation. Catalysis is restricted to divalent and trivalent metal ions, and the pH dependence indicates that the dianion form of diacid reactant is the best substrate. The catalyzed decarboxylation of dimethyloxalacetic acid is shown in Fig. 9-1, and rate data for various metal ions are given in Table 9-1. During the

TABLE 9-1

Effect of Metal Ions on α,α-Dimethyloxalacetate Decarboxylation[a]

Metal ion	(Conc.)	pH	$10^3 k$ (min^{-1})
—		4.6	2.4
Cu^{2+}	(0.001)	4.6	143
Ni^{2+}	(0.01)	4.6	21.6
Mn^{2+}	(0.01)	4.6	5.8
—		2.4	3.2
Fe^{2+}	(0.002)	2.4	10.2
Fe^{3+}	(0.002)	2.3	301

[a] Data from Speck.[9]

reaction there is a transient yellow color attributed to a steady-state concentration of the pyruvate–enol complex, although some very recent work[9a] disputes this interpretation.

Bender has pointed out that "those reactions of carboxylic and phosphoric acid derivatives which are susceptible to metal ion catalysis in non-enzymatic systems are almost without exception catalyzed by enzymes which contain metals. This strongly implies that the metals are involved in *catalysis* and not in merely binding the substrate."[10] However, the success of attempts to correlate the catalytic effectivity of a series of metals with stability constants for metal–substrate complexes depends on how well the complexes resemble the transition state for the reaction; that is, the metal must stabilize the transition state and *not* the ground state of the reactant in order to have catalytic activity.

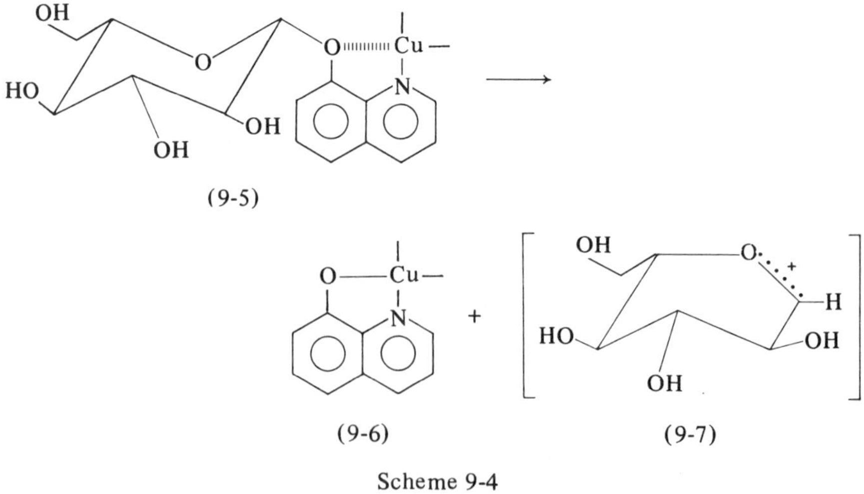

(9-5)

(9-6)   (9-7)

Scheme 9-4

Thus the activity of various metals in catalyzing the decarboxylation of oxalacetic and 3-ketoglutaric[11] acids correlates with the stability of the metal–oxalate and metal–malonate complexes, respectively; acetoacetic acid is not decarboxylated by metal ions. The *enhancement* of a metal's catalytic effect by bipyridyl,[12] resulting from mixed ligand symbiotic effects, was noted at the end of Chapter 6. This contrasts sharply the *inhibitory* effects of coligands such as citrate in this system.

The hydrolysis of 8-quinolyl-β-D-glucopyranoside (oxine glucoside, 9-5) shown in Scheme 9-4 is an example of a metal-catalyzed solvolysis reaction at a carbon center.[13] In this system the relative catalytic effectivities of Cu(II), Ni(II), and Co(II) are 1350:5:1, respectively, with copper giving a rate acceleration approaching 10^6 that of the metal-free reaction. In contrast the protonated quinolinyl glucoside solvolyzes only 16 times faster than phenyl glucoside.[14] A good correlation of log K_f for the metal–oxine complex and the log of the relative rate of hydrolysis also suggests that the rate limiting step is a dissociative process.[13] (For further perspective on such correlations the reader is referred to Fig. 9-3 and the discussion of leaving and entering group effects in the early part of Chapter VI.) In a similar fashion copper also catalyzes the hydrolysis of the sulfate monoester of 8-hydroxyquinoline.[15] However, the metal-catalyzed hydrolysis of corresponding carboxylate esters is somewhat different and will be discussed later in this section.

The metal-catalyzed hydration (Scheme 9-5) of 2-pyridyloxirane (9-8), also involves enhancement of leaving group reactivity, but this reaction is probably

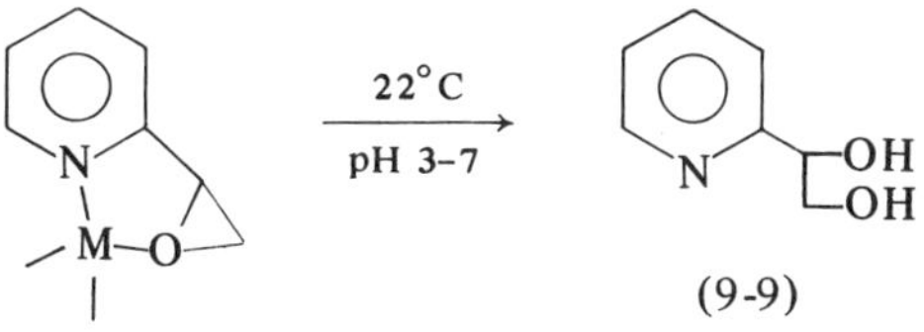

Scheme 9-5

more nucleophilic than solvolytic in nature.[16] The reaction is catalyzed by Cu(II) > Co(II) > Zn(II), with copper giving a rate acceleration of 1.8×10^4. The reaction is first order in copper and has a bell-shaped pH-rate profile with a sharp maximum at pH 5 and inflection points at pH 6 and 3.7. The latter inflection point corresponds to the pK_a of the pyridine-epoxide and reflects proton/Cu(II) competition for the pyridine nitrogen. The origin of the other inflection point is not known, but it probably does not result from the formation of a hydroxocopper complex, which tends to exclude intramolecular attack of a coordinated hydroxide nucleophile as a reaction pathway. This type of pathway has also been essentially excluded for Cu^{2+}-catalyzed hydrolysis of amino acid esters.

One of the more interesting features of this hydration is its regiospecificity: All of the copper-catalyzed reactions of (9-8) were found to involve C_β–O bond breaking. This was shown using methanol, chloride, and bromide as nucleophiles, and also by the hydration of ^{18}O-labeled epoxide with copper ion. Since the solvent deuterium isotope effect for the copper-catalyzed hydration is the same as that for the uncatalyzed neutral pH hydration of this and other simple epoxides ($k_{H_2O}/k_{D_2O} = 1.08$), this reaction may be pictured as involving nucleophilic attack on the coordinated epoxide as shown in (9–10). The β-carbon is the least sterically hindered, and the C_β–O bond is probably quite

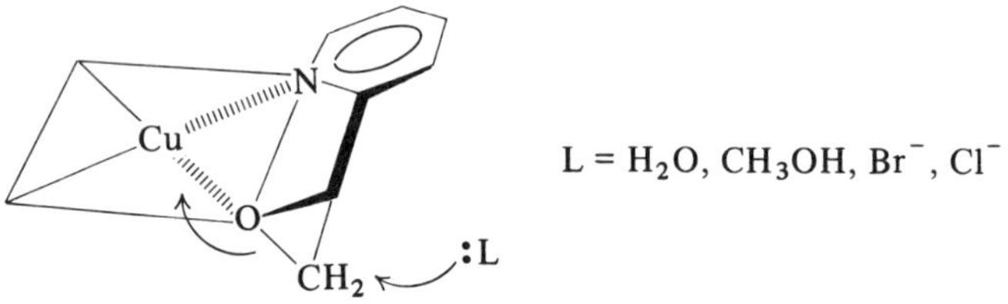

$$L = H_2O, CH_3OH, Br^-, Cl^-$$

(9-10)

strained compared to the stable five-membered chelate ring. Hence, the reactions occur at the β-carbon both to a greater extent (95% with H_2O, ~ 100% with others) and at much faster rates in the presence of the metal ion. It is probably only a coincidence, but an intriguing one, that the liver microsomal enzyme epoxide hydrase hydrates 2-pyridyloxirane, styrene oxide, and other simple epoxides with regiospecificity identical to the above model system.[16a]

Esters and amides of carboxylic acids, rather than undergoing "solvolysis" to form an acylium ion, usually hydrolyze by way of an associative mechanism involving a tetrahedral intermediate (Scheme 9-6). Metal coordination of the

$$X = OR, NR_2, SR$$

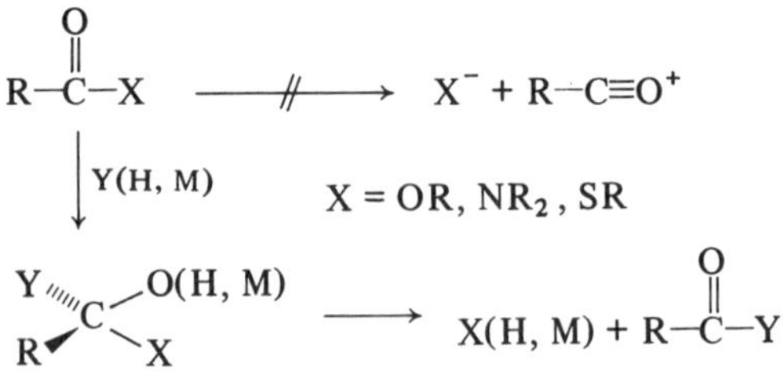

Scheme 9-6

carbonyl oxygen can accelerate the process by stabilizing the tetrahedral intermediate; this type of catalysis will be discussed in a later section of this chapter. Metal coordination of the leaving group X can accelerate the process in two ways. First, it can reduce the delocalization of lone pairs of electrons from X into the carbonyl group in the ground state. This will render the carbonyl more susceptible to nucleophilic attack, just as a thioester is more reactive than an oxyester or an amide, even though S is less electronegative than O or N. Second, coordination of X can stabilize it during its departure from the tetrahedral intermediate. Examples of both effects may be seen in the hydrolysis of carboxylate esters of 8-hydroxyquinoline,[17,18] and the methanolysis of amides of N,N-di(2-picolyl)amine (9-11) (Scheme 9-7).[19] For

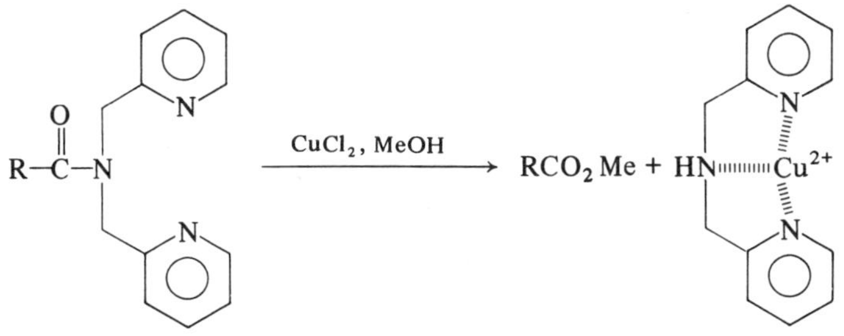

(9-11a), R = p-NO$_2$C$_6$H$_4$—
(9-11b), R = Me$_3$C—
(9-11c), R = CH$_3$CH=CH—

Scheme 9-7

example, addition of (9-11a) to a hot solution of CuCl$_2$ in methanol resulted in the "almost instantaneous" formation of methyl p-nitrobenzoate and the deep blue color of (9-12). In contrast no ester formed upon heating N,N-dibenzyl-p-nitrobenzamide with CuCl$_2$ in methanol for 24 hr, or when (9-11a) was heated in methanol without copper. Thus, chelation of the metal by the "leaving group" resulted in an enormous increase in the methanolysis rate. Hydrolysis rates were also increased, although the effect of copper was less dramatic

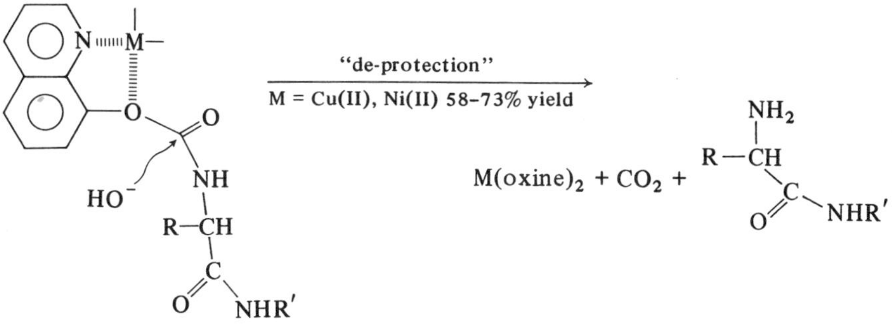

Scheme 9-8

because the tendency of the amide–copper complex to dissociate is much greater in water than in methanol.

Often in the course of an organic synthetic sequence it is necessary to mask the reactivity of a particular functional group by forming a temporary derivative of it. This problem is particularly acute in the synthesis of polynucleotides and polypeptides. The usual strategy is to form a derivative which will be very stable toward certain reaction conditions (e.g., acidic, basic, or oxidizing) yet very labile and easily removed under others. The observation that metal ions greatly accelerate the hydrolysis of carboxylate esters of 8-hydroxyquinoline (oxine) led to the development of the carbo(8-quinoloxy) substituent as an amino-protecting group for peptide synthesis (Scheme 9-8).[17] Kinetic studies on the Cu^{2+}-catalyzed hydrolysis of 8-acetoxyquinoline have shown that although the reaction is not a simple one, it is nevertheless extremely effective.[18]

ACONITASE ACTION

As a final example of metal enhancement of leaving group reactivity we may consider the enzyme, aconitase. This enzyme is interesting from several points of view. It was discovered that aconitase can discriminate between the two *prochiral* $-CH_2CO_2H$ groups in its substrate, citric acid. Long before aconitase was known to be a metalloenzyme, Ogston formulated his famous "three-point attachment" theory to explain how an asymmetric reagent (enzyme) could discriminate between two identical groups attached to a *prochiral* center. It is now known that the part of the enzyme responsible for the chiral differentiation is a ferrous ion asymmetrically bound at the active site. Recently, a similar differentiation has been achieved in a simple model system (Fig. 8-7). A second point of interest concerning aconitase is that it is a *nonredox* iron protein, although *in vitro* it is necessary to supply reducing agents such as mercaptoethanol to keep iron in the ferrous form.

Aconitase interconverts citric acid, isocitric acid, and the dehydration product, *trans*-aconitic acid, producing an equilibrium mixture in the ratio of 90:7:3. Its mechanism of action was clarified over a period of more than 20 years[20,21] by several groups using stable isotope tracers, X-ray crystallography of metal–citrate model complexes, and NMR studies of dynamic enzyme–substrate interactions with both the native Fe(II) enzyme and a catalytically inactive Mn(II)-reconstituted enzyme. The dehydration–rehydration sequence has been established as shown in Scheme 9-9. Both reactions occur in a trans fashion, but note that although the tritium is conserved in the citrate $\rightleftharpoons$ isocitrate interconversion, the hydroxyl group is exchanged with bulk water. This information was accommodated in a mechanism of action for aconitase proposed by Glusker and dubbed the "ferrous wheel" mechanism.[20] Later, Villafranca and Mildvan estimated the average distances of citrate and

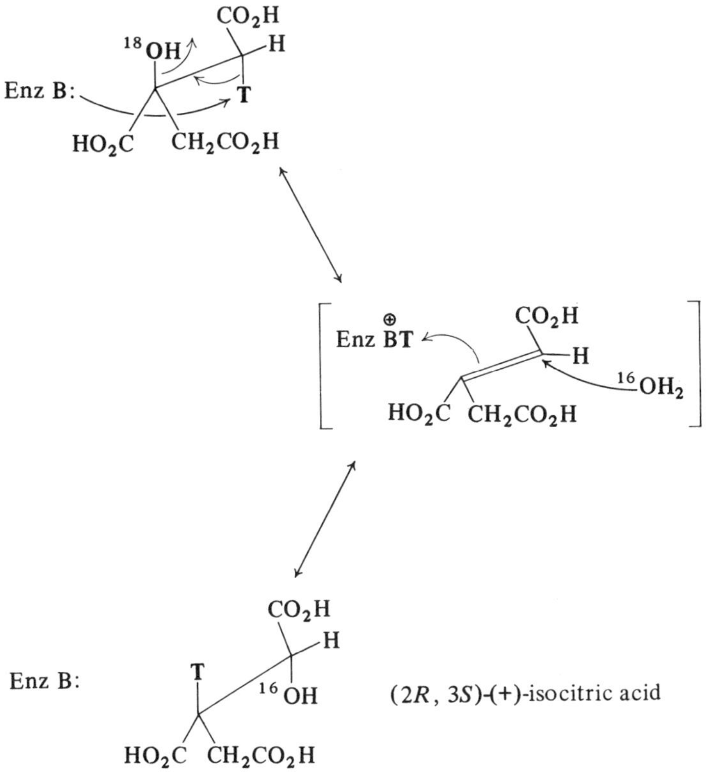

CO_2H

^{18}OH H

Enz B: T

HO_2C CH_2CO_2H

Enz $\overset{\oplus}{BT}$ CO_2H H $^{16}OH_2$

HO_2C CH_2CO_2H

CO_2H H

Enz B: T ^{16}OH (2R, 3S)-(+)-isocitric acid

HO_2C CH_2CO_2H

Scheme 9-9

isocitrate protons from the metal center by NMR techniques and suggested an alternative mechanism for aconitase.[21] However, in both mechanisms the iron is important for (1) chiral recognition of the prochiral functional group (via three-point attachment as shown in Fig. 9-4A) and (2) assistance rendered in the elimination of the coordinated hydroxyl group (Fig. 9-4B). The flip-over of the enzyme bound aconitate needed to give the proper stereochemical outcome may be accomplished by one of three mechanisms. Figure 9-4C shows the ferrous wheel mechanism, in which water from the citrate hydroxyl leaves and solvent water enters from the opposite side of the iron center. Figure 9-4D shows an ion pair variation for the ferrous wheel which may be less attractive as an intermediate because of the unfavorable energetics of Fe(II)-carboxylate bond breaking. Figure 9-4E shows a third possibility for the flip-over of aconitate. It is based on a "Bailar twist" mechanism, in which three cis ligands that form one triangular face of the octahedral iron complex (in this case they

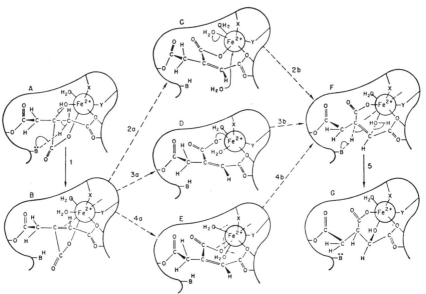

Fig. 9-4. Structures for complexes of substrates for aconitase determined by NMR by Villafranca and Mildvan. In structures A to B, water is eliminated from citrate by the aid of the Fe(II) and a base on the enzyme to form *cis*-aconitate. The flip of the double bond in B to F can take place by three possible transitions. Pathway 2a to 2b is the ferrous wheel mechanism involving a trigonal bipyramidal intermediate C; pathway 3a to 3b is the reverse ferrous wheel mechanism involving dissociation of a carboxyl group in intermediate D; and pathway 4a to 4b is a Bailar twist mechanism with a rotation axis in intermediate E about the dashed line. Water adds to *cis*-aconitate to form isocitrate in F to G. (Reprinted from Villafranca and Mildvan[21] with permission of the *J. Biol. Chem.*)

are the two waters and the aconitate carboxyl) rotate about an axis perpendicular to their plane. The iron center thus passes from trigonal antiprismatic (octahedral) to trigonal prismatic to a new trigonal antiprismatic geometry, as shown in Fig. 9-5. Having accomplished the dehydration, and the "flip" which allows the loss of ^{18}O label, reversal of these steps places the hydrogen

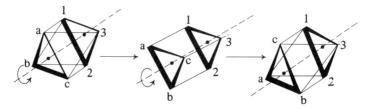

Fig. 9-5. Isomerization of octahedral complexes via a Bailar twist.

originally taken from C-2 back onto C-3 while a new hydroxyl is added to C2 giving isocitric acid.

Metal Catalysis via Ligand Polarization

A significant proportion of all chemical reactions proceed via intermediates or transition states which are considerably more polar or ionic than the ground state reactants. As a rule these reactions are facilitated by polar solvents because they stabilize polar *solutes* (i.e., transition states) through favorable dipolar interactions. Metal ions, by virtue of their Lewis acidity or π acidity, are able to induce or to increase the polarization of molecules to which they become coordinated. If the polarization of a coordinated ligand is in the direction which leads from the ground state toward the transition state of a particular reaction, the reaction of the coordinated ligand will be faster than that of an uncomplexed ligand molecule. Conversely, if the ligand polarization by a metal is not in the direction of the transition state polarization, the coordinated ligand will be less reactive than an uncomplexed ligand with respect to that particular reaction pathway, although the change in polarization could conceivably give rise to a totally new mode of reaction.

Metal-ligand polarization effects often involve metals in upper oxidation states ($2+$ to $4+$), and electron withdrawal from the ligand toward the metal. However, for low valent transition metals ($2-$ to $1+$) rich in d electrons, back-bonding to some ligands can cause a net *increase* in electron density on the ligand. One example of this effect is provided by the complex $(NH_3)_5Ru(piperazine)^{2+}$ discussed at the end of Chapter 5, and other examples may be found in Chapters 10, 11, and 12. In this section we will focus chiefly on divalent and trivalent first-row transition metal ions. Among these the cupric ion has been extensively studied because of its great Lewis acidity and complexing strength, and because it undergoes facile ligand exchange reactions. The Co(III) ion is also a strong Lewis acid, but in contrast to Cu(II) its inertness toward ligand exchange via heterolytic dissociation often permits the synthesis and/or isolation of intermediates along a reaction pathway.

HYDROLYSIS OF AMINO ACID DERIVATIVES

In 1952, Kroll reported that divalent metal ions greatly accelerated the hydrolysis of amino acid esters under neutral conditions.[22] He reasoned that since amino acid esters ordinarily hydrolyze very slowly at neutral pH, the species undergoing rapid hydrolysis must be the metal-complexed ester, and he rationalized this by assuming that the carbonyl group of the coordinated ester would be a stronger Lewis acid than an ordinary ester carbonyl (Scheme 9-10).

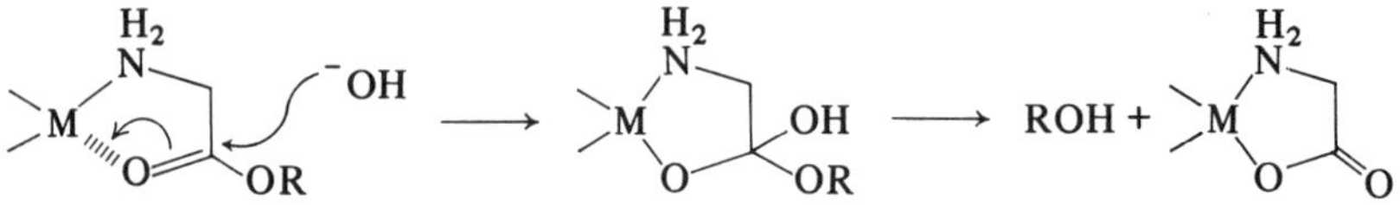

$$M = Cu^{2+} > Co^{2+} > Mn^{2+} > Ca^{2+} \sim Mg^{2+}$$

Scheme 9-10

A priori, Kroll might well have written a structure involving metal coordination of the "ether" or leaving group oxygen of the ester. However, 20 years of study, including several elegant isotopic labeling and kinetic investigations, have not significantly changed his original conception of the reaction mechanism.

Bender and co-workers studied the kinetics of the copper-catalyzed hydrolysis of methyl glycinate.[23,24] They used equimolar amounts of cupric ion and ester, and a glycine buffer was employed so the concentration of the 1:1 copper–glycinate complex (the actual catalytic species in this case) would remain essentially constant thereby allowing for a first-order kinetic analysis. The rate acceleration they observed with copper monoglycinate was considerably less than that with free-cupric ion (10^4 vs 10^8 times the uncatalyzed rate). Their data were consistent with a mechanism involving a rapid equilibrium complexation of the ester followed by a rate limiting attack by water (see reactions below). They also studied the copper monoglycinate-catalyzed hydrolysis of phenylalanine ethyl ester which was labelled with ^{18}O

$$glyCu^+ + ester \underset{fast}{\overset{K}{\rightleftharpoons}} glyCu(ester)^+ \xrightarrow[H_2O]{k} products$$

$$-d[ester]/dt = k[glyCu(ester)^+] = kK[glyCu^+][ester]$$

in the carbonyl oxygen. Analysis of ester recovered at intermediate times during the reaction showed that its ^{18}O content decreased at a rate approximately one-fourth its rate of hydrolysis, and similar results were obtained for the slower *uncatalyzed* hydrolysis. These results indicate the formation of an addition intermediate which is *symmetrical* with respect to the isotopic oxygen atom, as shown in Scheme 9-11.

The lability of Cu(II) complexes, as illustrated by the facile ^{18}O exchange of phenylalanine ester, stands in sharp contrast to the rigid inertness of some analogous Co(III) systems. Interest in the reactions of amino acid derivatives coordinated to Co(III) stemmed from a remarkable observation made by Collman and Buckingham, namely, that *cis*-trienCo(OH)(OH$_2$)$^{2+}$ (9-13) effects a selective N-terminal hydrolysis of simple polypeptides and forms an inert complex of the cleaved N-terminal amino acid.[25]

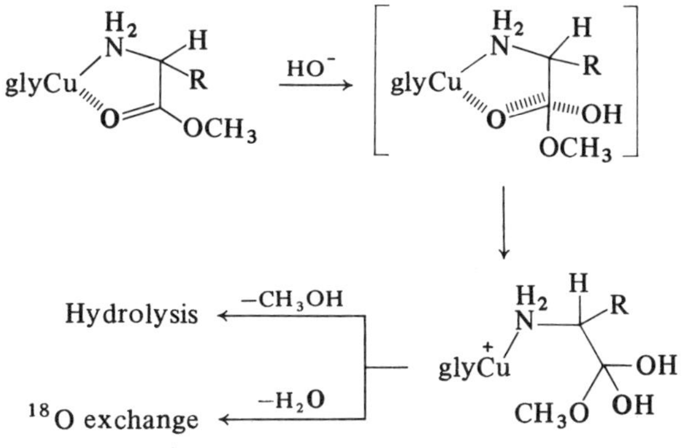

$$\text{glyCu} \underset{O}{\overset{H_2\,N\diagdown}{\big|}} \overset{H}{\underset{OCH_3}{\big|}} R \quad \xrightarrow{HO^-} \quad \left[\text{glyCu} \underset{O}{\overset{H_2\,N\diagdown}{\big|}} \overset{H}{\underset{OCH_3}{\big|}} \overset{R}{\underset{OH}{\big|}} \right]$$

$$\downarrow$$

$$\text{Hydrolysis} \xleftarrow{-CH_3OH} \quad \overset{}{\underset{}{\big|}} \quad \text{glyCu} \overset{+}{\underset{CH_3O}{\overset{H_2\,N}{\diagup}}} \overset{H}{\underset{OH}{\big|}} R \quad -OH$$

$$^{18}O \text{ exchange} \xleftarrow{-H_2O}$$

Scheme 9-11

The amino acid chelates may be separated and identified by paper chromatography, and the reaction is rapid and selective enough to be used for sequencing peptides from the amino terminus. The reaction is similar to one catalyzed by leucine aminopeptidase, an N-terminal exopeptidase which requires Mg^{2+} or Mn^{2+} for activity, but this similarity is probably superficial and does not necessarily imply similar mechanisms of action. The cis-hydroxyaquo

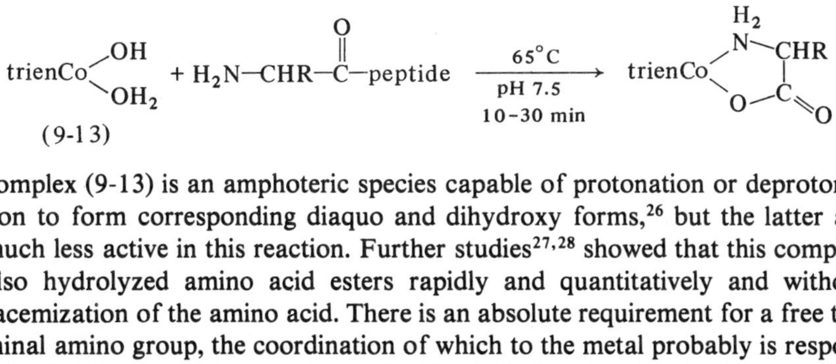

$$\text{trienCo} \overset{OH}{\underset{OH_2}{\diagdown}} + H_2N-CHR-\overset{O}{\overset{\|}{C}}-\text{peptide} \xrightarrow[\substack{pH\ 7.5 \\ 10-30\ min}]{65°C} \text{trienCo} \overset{H_2\,N\diagdown}{\underset{O}{\diagdown}} \overset{CHR}{\underset{C}{\big|}} {\diagup}_{O}$$

(9-13)

complex (9-13) is an amphoteric species capable of protonation or deprotonation to form corresponding diaquo and dihydroxy forms,[26] but the latter are much less active in this reaction. Further studies[27,28] showed that this complex also hydrolyzed amino acid esters rapidly and quantitatively and without racemization of the amino acid. There is an absolute requirement for a free terminal amino group, the coordination of which to the metal probably is responsible for the specificity for N-terminal hydrolysis. Most interesting was their observation of identical rates of hydrolysis for glycinamide and several glycine esters, which suggested that RCO–X bond breaking was not involved in the rate limiting step. To account for these observations two limiting cases of reaction mechanisms were proposed: "After the amino group combines with the cobalt ion by displacement of a water molecule (a) either the adjacent coordinated hydroxyl group attacks the peptide carbonyl group through a five-ring intermediate or (b) the carbonyl becomes activated to attack by external hydroxide through prior coordination of the carbonyl oxygen with the cobalt atom. In the former mechanism the complex ion acts both as a template and a buffered source of hydroxide" (see Fig. 9-6).

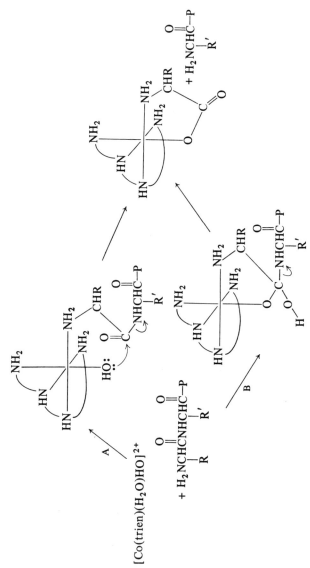

Fig. 9-6. Proposed mechanisms of peptide hydrolysis.

Continuing the study of the mechanisms of these and related reactions, Buckingham, Sargeson, and their collaborators have shown that in fact both of these limiting mechanisms contribute significantly to the hydrolysis reactions, although the exact extent of their contribution may depend on the reaction conditions employed.[29] The design of the critical experiments hinged on the inertness of Co(III) toward ligand exchange, which allowed the synthesis and resolution of enantiomers of cis-β_2-Co(trien)Cl(glyOEt) (9-14, Scheme 9-12).

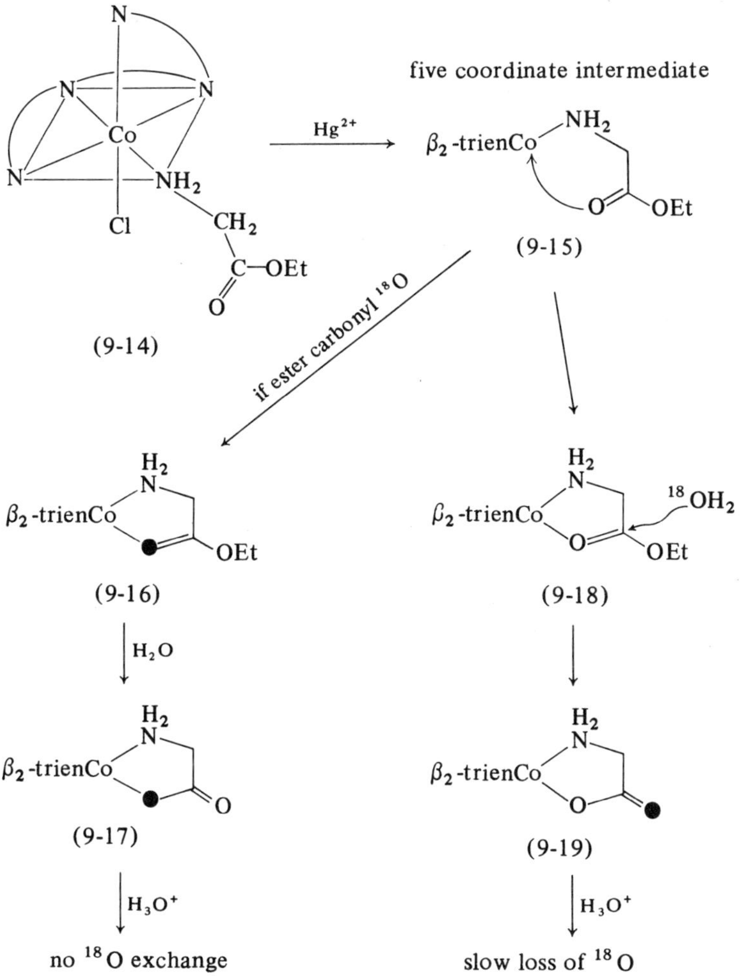

Overall rate = k [(9-14)] [Hg^{2+}]

Scheme 9-12

From this intermediate it is possible to approach the two limiting transition states (carbonyl coordination vs carbonyl attack by coordinated hydroxide) in several ways. Treatment of (9-14) labeled with ^{18}O in the *carbonyl* oxygen with Hg^{2+} to remove the chloride ion results in the extremely rapid formation of β_2-Co(trien)gly^{2+} containing 88% of the expected amount of ^{18}O label, all of which is located in the Co–O bonded position of the product. A parallel hydrolysis of unlabeled ((9-14) in ^{18}O-enriched water resulted in the formation of a glycine chelate containing 70% of the expected amount of ^{18}O label. Control experiments showed that this product gradually lost ^{18}O by exchange with solvent upon standing, whereas the product from the ^{18}O-labeled ester lost no ^{18}-O upon standing in acidic solution. These results are summarized in Scheme 9-12. The asymmetric β_2 configuration about the cobalt center is maintained in the Hg^{2+}-induced hydrolysis, and the process is not impeded by the presence of NO_3^- or HSO_4^-, even though these ions normally compete more effectively than solvent water for five coordinate intermediates similar to (9-15). Taken with the ^{18}O-labeling results, this strongly suggests that the only

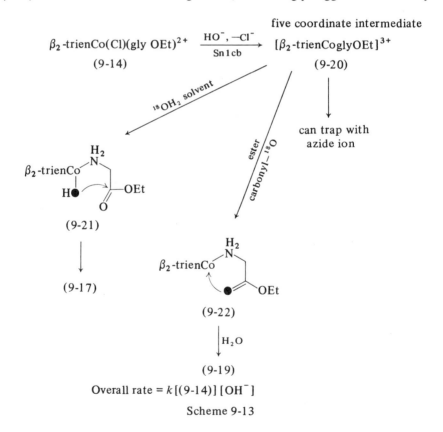

Scheme 9-13

significant hydrolysis pathway involves internal carbonyl trapping of the five-coordinate intermediate followed by rapid hydrolysis of the chelated ester.

The base hydrolysis of (9-14) follows a rather different course. Elimination of chloride via a Snlcb mechanism also produces a five-coordinate intermediate, which is probably different from that formed in the Hg^{2+} reaction because it can be trapped by solvent water and azide ion if present, although retention of the asymmetric configuration about the cobalt center is still observed. Sudies with ^{18}O tracers showed that 16% of the oxygen in the Co–O

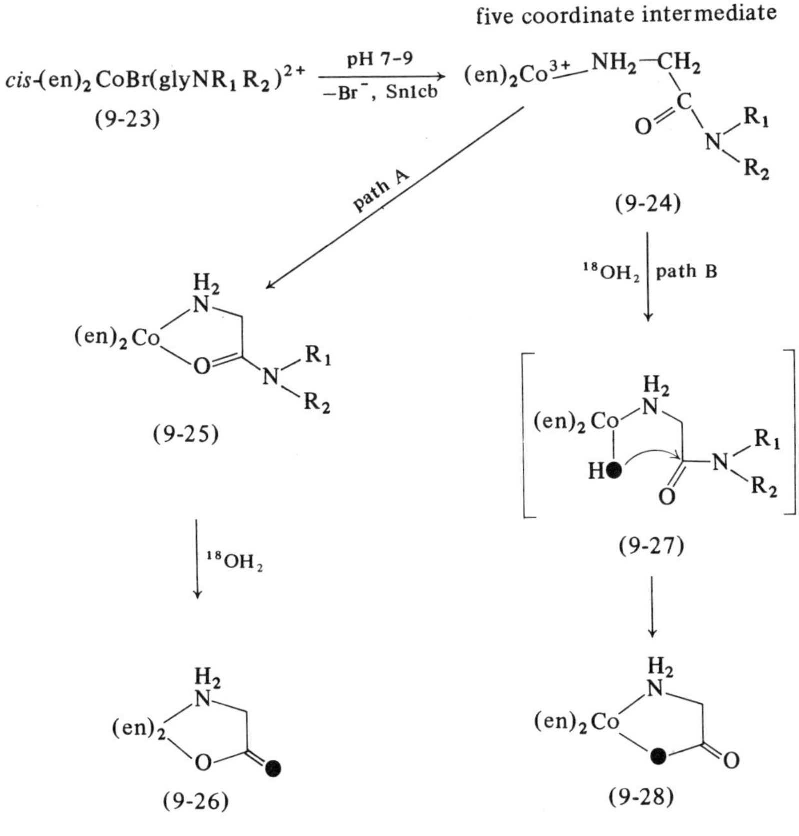

Scheme 9-14

	R_1	R_2	Path A (%)	Path B (%)
(a)	H	H	46	54
(b)	H	CH_3	66	34
(c)	CH_3	CH_3	82	18

position arose from solvent water while 84% was originally the ester carbonyl oxygen. No oxygen exchange was observed in a related complex, $(en)_2Cogly^+$, at the pH of these experiments (pH 8.6). Thus the results of base hydrolysis of (9-14) can be expressed as shown in Scheme 9-13).

The general conclusion of the above studies, i.e., that both intramolecular attack of coordinated hydroxide and polarization of the ester carbonyl group by metal coordination are viable reaction pathways, is supported by a related ^{18}O tracer and kinetic study of the hydrolysis of glycine amides. Buckingham *et al.*[30] have shown that glycine amides hydrolyze more slowly than esters by a factor of $\sim 10^5$ when coordinated to a $(en)_2Co^{3+}$ center. Thus from the hydrolysis of (9-23, Scheme 9-14) at pH 7–9 it has been possible to isolate the intermediates corresponding to the limiting mechanisms of hydrolysis proposed for the analogous esters.

The hydrolysis of (9-25a) is much slower than the loss of bromide from (9-

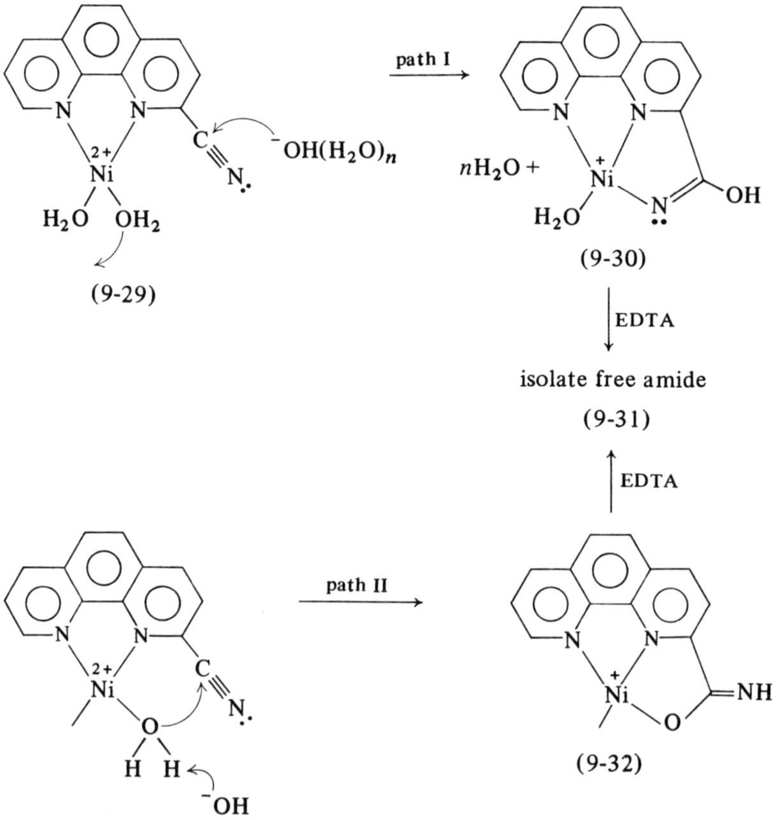

Scheme 9-15

23), and hence it could be isolated from reaction mixtures. The internal attack by hydroxide in (9-27a) was so rapid ($\geqslant$ 10 times the rate of bromide loss) that its isolation was not possible although an analog of (9-27) with $R_1 = H$ and $R_2 = CH_2CO_2C_3H_7$ has recently been isolated and shown to form (9-28) very rapidly above pH 9.[31] Thus, despite the 10^8 reduction of the basicity of the coordinated OH^- (pK_a of coordinated $H_2O \sim 6$), amide hydrolysis by coordinated OH^- is more efficient than hydrolysis by free OH^- by a factor of 10^9.

These results naturally raise the question of metal-coordinated hydroxide being involved in the hydrolysis of amino acid esters by labile metal ions such as Ni^{2+} or Cu^{2+}. Because of the lability of these ions it is not possible to distinguish the two mechanisms by ^{18}O tracer studies. Furthermore, the kinetic rate laws for hydrolysis of amino acid esters by internal or external attack of hydroxide are identical, i.e., rate = k[complex][OH^-]. However, in at least one system involving a labile metal ion Breslow and coworkers have obtained evidence suggesting that *external* attack on a coordinated substrate is involved.[32] They found that the hydration of 2-cyano-1,10-phenanthroline (9-29) to the corresponding amide is strongly promoted by Cu^{2+}, Ni^{2+}, and Zn^{2+} (Scheme 9-15). The reaction of the Ni^{2+} complex is first order in complex and first order in hydroxide, and proceeds 10^7 times faster than the bimolecular hydration of the free ligand with hydroxide. The coordinated amide product hydrolyzes further to the coordinated carboxylic acid, but this reaction is only accelerated 400-fold. Experiments in aqueous ethanol showed that ethanol was more reactive than water, as generally observed for simple nucleophilic reactions. Since ethanol is much poorer than water as a ligand, this argues against a mechanism involving attack by coordinated nucleophiles. Similarly, polydentate coordinating nucleophiles which are good substrates for intercomplex reactions (see next section) could not substitute for hydroxide.

The conclusion that attack by external hydroxide (path I) is involved is supported by comparison of the activation parameters for the Ni^{2+}-catalyzed and base-catalyzed reactions. For the hydroxide reaction $\Delta H^{\ddagger} = 15.1$ kcal/mole and $\Delta S^{\ddagger} = -20$ eu. These values are comparable to those for other *bimolecular* nitrile hydration reactions. For the Ni^{2+}-catalyzed reaction $\Delta H^{\ddagger} = 15.7$ kcal/mole, but $\Delta S^{\ddagger} = +14$ eu. It is rather unusual to find *positive* activation entropies for bimolecular reactions. This probably represents a substantial liberation of solvent water molecules in the transition state, in addition to loss of one water molecule from the coordination sphere of the metal. A substantial loosening of the solvation of the transition state would be consistent with the decrease of net charge on the complex from 2+ to 1+ in the transition state. Thus, although the reactant, transition state, and product of this reaction are each coordinated to the metal ion, the rate acceleration is accounted for almost entirely by the difference in $\Delta S^{\ddagger}$ for the base and Ni^{2+} reactions. A large positive activation entropy has also been observed in the base hydrolysis of $(en)_2Co(glyOC_3H_7)^{3+}$ in dimethylsulfoxide $(DMSO)$[33] and of $(NH_3)_5$ $Co(OCHNMe_2)^{3+}$ in water.[34]

Angelici and co-workers have obtained additional evidence suggesting that catalysis of amino acid ester hydrolysis by labile metal ions involves external attack by hydroxide upon the carbonyl of the chelated ester. They studied a series of mixed ligand chelates of the type $[LCu(ester)]^n$, where the net charge on the complex depends on the charge on the ancillary ligand, L (see reactions below). For the hydrolysis of a series of amino acid esters catalyzed by

$$Cu^{2+} + L \quad \overset{K_L}{\rightleftharpoons} \quad CuL$$

$$CuL + ester \quad \overset{K_f}{\rightleftharpoons} \quad CuL(ester)$$

$$CuL(ester) + HO^- \quad \underset{k_{OH}}{\longrightarrow} \quad CuL(amino\ acid)$$

$Cu(NTA)^-$, they observed a linear free energy relationship between $\log k_{OH}$ and $\log K_f$, which strongly implies that $chelation\ of\ the\ ester$ is important in its hydrolysis.[35] This is supported by the observation that the ratio of catalyzed to uncatalyzed hydrolysis rates for HisOMe, β-AlaOEt, and AlaOEt are 7, 30, and 285, respectively, which reflects the relative ability of the ester carbonyl in these substrates to interact with (i.e., chelate) the copper ion once the amine group has bound.

Another factor which was investigated was the effect of the ancillary ligand, L, on the activity of the CuL complex. The net charge on CuL depends on the charge L, but as Table 9-2C shows, the hydrolytic activity of the complex does not seem to depend on the net charge of the complex. Rather, it was found that both K_f and k_{OH} depended on the magnitude of K_L.[36] Although amine ligands do not reduce the net positive charge on the complex, they are bound much more strongly than carboxylate ligands and thus are more effective than the latter in reducing the Lewis acidity of the CuL complex. This may not be true for Co(III) complexes however. Note the decrease in rate and shift in pH optimum for dipeptide hydrolysis by $[trienCo(OH)(OH_2)]^{2+}$ vs the neutral complex $[eddaCo(OH)(OH_2)]$ (Table 9-2B).[37] Temperature dependence studies of the base hydrolysis of $[Cu(NTA)(glyOEt)]^-$ indicate that the catalytic effect arises exclusively from an $enthalpy$ effect; $\Delta H^{\ddagger} = 4.9$ kcal/mole and $\Delta S^{\ddagger} = -33$ eu, whereas for GlyOEt + HO^-, $\Delta H^{\ddagger} = 10.3$ kcal/mole and $\Delta S^{\ddagger} = -22$ eu.

CARBOXYPEPTIDASE ACTION

The enzymatic hydrolysis of proteins and peptides is a process of enormous biological importance, and a number of important proteolytic enzymes are

TABLE 9-2

Effects of Ancillary Ligands and Net Charge on Hydrolytic Activity of Metal Chelates[a]

A. Peptide hydrolysis, 65°C[b]

$[trienCo(OH)(OH_2)]^{2+} \gg Cu^{2+} \gg Cu(gly)^+$

B. L-Ala-L-Phe hydrolysis, 65°C[c]

	k_2, M^{-1} sec^{-1} (at pH optimum)	
$[trienCo(OH)(OH_2)]^{2+}$	5.00	(7.5–8)
$[eddaCo(OH)(OH_2)]$	3.02	(10.0)

C. GlyOCH$_3$ hydrolysis, 25°C[d]

	k_{OH}, M^{-1} sec^{-1}
Cu^{2+}	1.5×10^5
Cu(IMDA)	3.2×10^4
Cu(NTA)$^-$	4.6×10^2
Cu(DPA)$^{2+}$	1.7×10^2
Cu(dien)$^{2+}$	1.4×10^2
H$^+$	5.8×10

[a] Abbreviations: edda, $(CH_2NHCH_2CO_2^-)_2$; NTA, $N(CH_2CO_2^-)_3$; IMDA, $HN(CH_2CO_2^-)_2$; Cu(DPA)$^{2+}$, (9–12); and DPA, N,N-di-2-picolylamine.
[b] Data from Bender and Turnquest[23] and Bender and Thomas.[25]
[c] Data from Oh and Storm.[37]
[d] Data from Nakon et al.[36]

metalloenzymes. Among these carboxypeptidase A has probably received the greatest amount of study.[38–40] Carboxypeptidase is a pancreatic enzyme composed of 307 amino acids and one zinc ion, with a molecular weight of 34,600 daltons and an overall ellipsoidal shape about 50 × 42 × 38 Å. There is a cleft or depression on one side of the ellipsoid. At the bottom of this cleft lies the zinc ion, bound by two histidine imidazoles and one glutamate carboxyl group. The fourth coordination site on the zinc is occupied by a water molecule. The overall structure of carboxypeptidase A is rather similar to that of carbonic anhydrase, another zinc enzyme to be discussed later.

As with carbonic anhydrase, the zinc ion of carboxypeptidase can be removed by dialysis against chelating agents, and other metals can be put back into the apoenzyme. Nickel and manganese ions partially restore the original peptidase activity, but with Co(II) a *more active* enzyme results. The native enzyme is inhibited by the dipeptide glycyltyrosine. This inhibitor also binds to the apoenzyme and prevents its re-uptake of zinc ion, suggesting that the zinc ion is in fact located at the active site of the enzyme. Various chemical modifications of specific amino acid residues in the enzyme revealed the importance of one tyrosine, one histidine, one arginine, and one free glutamate carboxyl group. X-ray structure determinations by Lipscomb and co-workers[39,40] have shown that substantial conformational changes occur upon

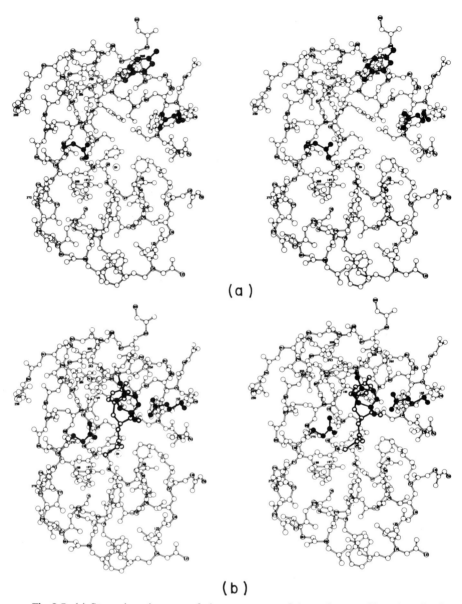

(a)

(b)

Fig. 9-7. (a) Stereoview along $-y$ of about a quarter of the carboxypeptidase A molecule, showing the cavity, the Zn atom, and the functional groups Arg-145 (right), Tyr-248 (above), and Glu-270 (left). (b) Stereoview of the same region, after the addition of glycyl-L-tyrosine (heavy open circles), showing the new positions of Arg-145, Tyr-248, and Glu-270. The guanidinium movement is 2 Å, the OH of Tyr-248 moves 12 Å, and the carboxylate of Glu-270 moves 2 Å when Gly-Tyr binds to the enzyme.[39,40]

CARBOXYPEPTIDASE A

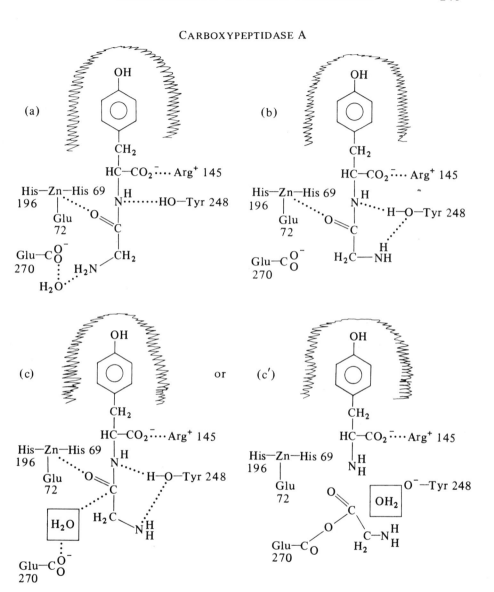

Fig. 9-8. (a) Specific binding interactions of Gly-Tyr as deduced from the difference electron density. (b) Probable productive mode of binding of Gly-Tyr at the start of catalysis. (c) General base path in which Glu-270 promotes attack of a lone pair from H_2O at the carbonyl carbon, probably preceded by or concurrent with proton transfer from Tyr-248 to the NH group. (c') Anhydride intermediate pathway in which H_2O later attacks the acylenzyme intermediate. Further studies are required in order to resolve the ambiguity of mechanism.[39,40] (Reprinted with permission from *Accounts of Chemical Research*. Copyright by the American Chemical Society.)

binding of the inhibitor glycyltyrosine to the α isozyme of carboxypeptidase A. This can be seen by comparing the two stereoviews in Fig. 9-7.

Based on structure–activity relationships among substrates and inhibitors, and on the results of the crystal structure determinations by Lipscomb's group, the following mechanism has been pieced together for peptide hydrolysis. Substrate binding to the active site involves (1) the displacement of some structured water from a "hydrophobic pocket" by the large aromatic substituent of the C-terminal amino acid, (2) the ionic interaction of the terminal carboxylate of the substrate with a cationic arginine group of the enzyme, and (3) the displacement of a coordinated water from zinc by the carbonyl oxygen of the siscile peptide bond. This much is shown diagrammatically in Fig. 9-8. Following these events a conformational change in the enzyme occurs: the arginine side chain moves about 2 Å closer to the substrate carboxylate, and the phenolic group of Tyr-248 swings 12 Å to come within hydrogen bonding distance of the —NH— of the C-terminal amino acid. The peptide bond cleavage depends on the action of the carboxylate group of Glu-270 as either a general base promoting attack of water at the coordinated carbonyl or as a nucleophile which would attack the coordinated carbonyl directly to produce an anhydride intermediate. In either case the original C-terminal amino acid would be released in this step. In light of the demonstration that metal ions may catalyze the hydrolysis of amides and esters by coordinating either the substrate carbonyl or the attacking hydroxide nucleophile, carboxypeptidase A presents a clear example of the former. Whether other metalloproteases will also be shown to utilize this mechanism, or some other, will have to await further study.

Metal Catalysis via Template and Charge-Neutralization Effects

A metal cation can catalyze an *inter*molecular reaction of two potential ligands by serving as a template for collection of the two ligands. The reaction may thus be facilitated by the entropy advantages of intramolecular vs intermolecular formation of transition states, as well as by any metal-induced polarization of the ligand in the direction of the transition state. If both reactants are negatively charged the effect is particularly prominent since their electrostatic repulsion is more easily overcome in the coordination sphere of a metal ion. This can be clearly seen in the effects of metal ions on the rates of base hydrolysis of the half-esters of oxalic, malonic, and adipic acids as shown in Table 9-3.[41] The relative rates can be explained on the basis of cyclic transition states (9-33) and (9-34) involving the metal, the half-ester, and a

TABLE 9-3

Effects of Metal Ions on the Rate of Base Hydrolysis of Monoesters of Dicarboxylic Acids[a]

	Monoethyl ester		
Metal hydroxide	Oxalate	Malonate	Adipate
K^+, OH^-	29.2	0.778	2.01
TlOH	1330	5.65	4.37
$BaOH^+$	9200	38.0	
$CaOH^+$		39.0	7.94
$Co(NH_3)_6(OH)_2{}^+$	1740	16.0	17.1

[a] Data from Hoppe and Prue.[41]

hydroxide nucleophile, with the order of preference for ring size being $5 > 6 \gg$ larger rings. The fact that the substitutionally inert $Co(NH_3)_6{}^{3+}$ ion is also a

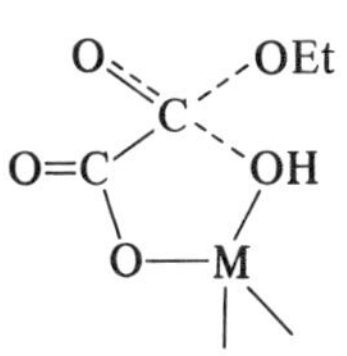

(9-33) 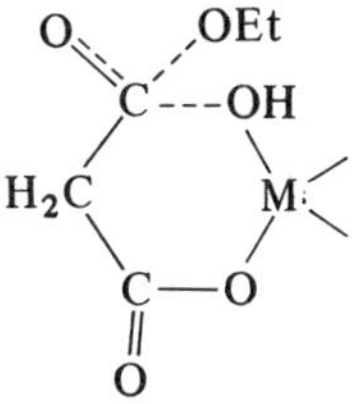   (9-34)

good catalyst must be attributed solely to its ability to gather the two negatively charged reactants, HO^- and $EtO_2CCO_2{}^-$, in its outer coordination sphere. A similar ordering of ring formation rates with ring size may be seen in the rates of conversion of the following half-esters to cyclic anhydrides by release of p-bromophenolate ion (X^-).[42]

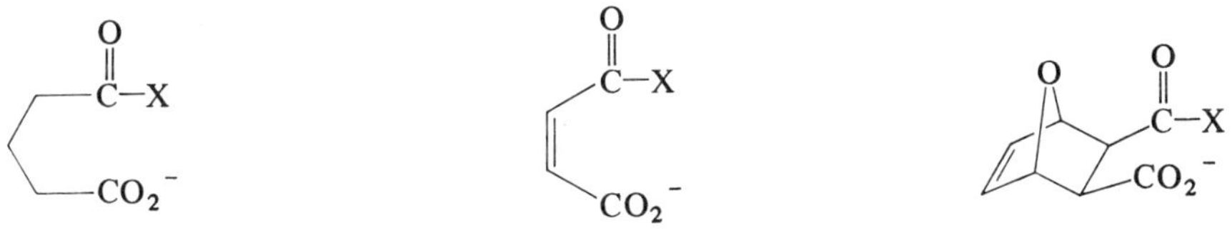

Relative rates of ring closure

1   230 5.3×10^4

Ligand gathering effects are not confined to small nucleophiles like hydroxide. Indeed, some of the more complicated template syntheses discussed in earlier chapters are examples of this effect. In some enzymatic reactions a metal ion serves as a bridge between the enzyme and its substrate, and in this way may promote a reaction between them to form a covalent intermediate. Imitation of this form of metal catalysis has been achieved in several chemical systems. The first of these was reported by Breslow and Chipman[43] and serves to illustrate the general characteristics of intracomplex or template catalysis. In their system zinc ion catalyzed the transfer of an acetyl group from 8-acetoxyquinoline-5-sulfonate (9-35) to the bidentate anion of 2-pyridinecarboxaldoxime (9-36). The *O*-acetate of (9-36), which could be

(9-35)   (9-36)

isolated after quenching with EDTA, also underwent a subsequent zinc-catalyzed hydrolysis at a rate about one-tenth the rate of the acetyl transfer. Because the anion (9-36) is extremely nucleophilic its reaction with the zinc complex of (9-35) was only 11 times faster than its reaction with *p*-nitrophenylacetate, a noncomplexing substrate analog.

The zinc-catalyzed transesterification of *N*-(*β*-hydroxyethyl)-ethylenediamine (9-37) and *p*-nitrophenyl picolinate (9-38) has been studied by Sigman and Jorgensen.[44] They found a rate law for the appearance of *p*-nitrophenolate (measured by UV) which had the form

$$\text{Rate} = k_{\text{obsd}}[(9\text{-}38)]$$

$$k_{\text{obsd}} = k_{\text{o}} + k_{\text{m}}[\text{Zn}] + k_{\text{a}}[(9\text{-}37)] + k_{\text{c}}[\text{Zn}\cdot(9\text{-}37)]$$

Kinetic analysis showed that the catalysis (k_{c}) indeed involved a ternary zinc complex of the two reactants, and indicated that the k_{c} term was linearly dependent on hydroxide concentration in the pH range 6.5 to 7.5. The fact that pyridine, when added in equal concentration to (9-38), reduced k_{c} by 50% suggested that *only the pyridine nitrogen* of (9-38) was binding the zinc in the active complex, which would also be consistent with the inertness of ethyl picolinate as a substrate for the transacylation of (9-37). Therefore, the authors proposed the intermediate (9-39), with an estimated pK_{a} of 8.4, for the coordinated hydroxyl group.

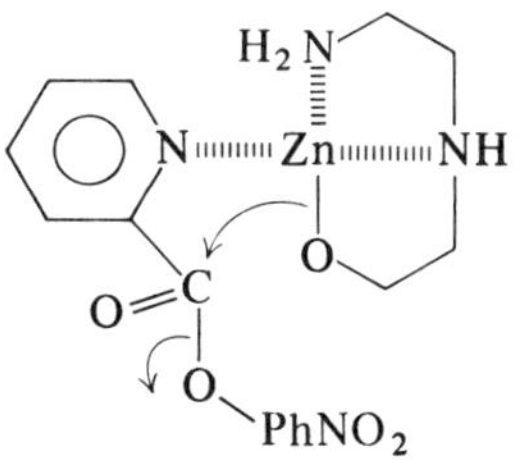

(9-39)

Most, if not all, enzymatic reactions involving adenosine triphosphate (ATP) or other nucleoside phosphates require metal ions for full activity. For example, the enzyme, nucleoside diphosphotransferase, requires Mg^{2+} in order to catalyze the net reaction

$$ATP + GDP \rightleftharpoons ADP + GTP$$

The reaction involves a phosphorylated enzyme intermediate in which the phosphate is thought to be bound to an imidazole of histidine.

$$ATP + EHis \xrightarrow{Mg^{2+}} ADP + EHisP$$

$$EHisP + GDP \xrightarrow[Mg^{2+}]{} GTP + EHis$$

The enzymatic dephosphorylation step has been "modeled" in a chemical system studied by Lloyd and Cooperman,[45] using the anion (9-36) as a nucleophile, zinc ion, and imidazole-N-phosphoric acid (9-40). In the absence of zinc only the slow hydrolysis of (9-40) is observed, but in the presence of zinc, phosphate is transferred to the nucleophile 10^3 times faster than it is lost by hydrolysis. There is a pH-rate plateau from pH 4.8 to 6.4, above which the rate drops off. However, if the N-3-imidazole nitrogen is methylated (9-41), the rate does not decrease above pH 6.4, suggesting that the protonated imidazole is the leaving group from phosphorus. The mechanism is thus pictured as (9-42).

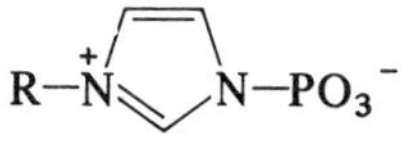

(9-40) R = H
(9-41) R = CH$_3$

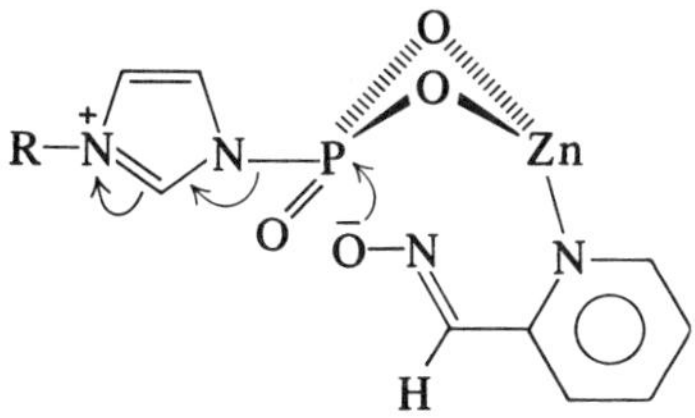

(9-42)

REACTIONS OF COORDINATED NUCLEOPHILES

We have already encountered one extensively documented example of the reactivity of coordinated hydroxide, i.e., the hydrolysis of amino acid derivatives. The majority of the reactions in this category involve the attack of a coordinated nucleophile on the carbonyl group of an ester or amide, which leads ultimately to a transacylation or hydrolysis reaction. One of the earliest reactions (Scheme 9-16) demonstrating the reactivity of a coordinated nucleophile was reported without comment by Kroll in his original paper on

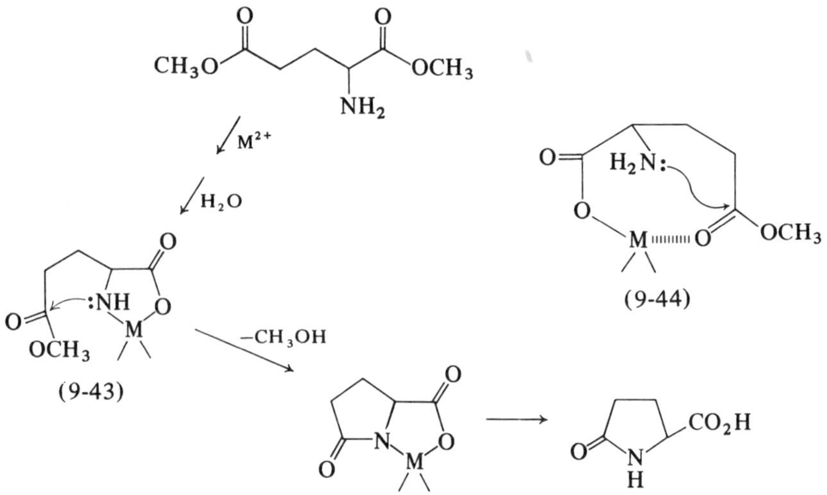

Scheme 9-16

the metal-catalyzed hydrolysis of amino acid esters.[22] He observed that dimethyl glutamate was converted to pyroglutamic acid by Co^{2+} ions. In retrospect this reaction might result from intramolecular attack of the coordinated amino group on the carbonyl group of the terminal ester. Although the pH of the reaction was only around 8–9, the deprotonated amido form of the ligand (9-43) could have been involved. An alternative mechanism involving attack of a free amino group on a coordinated ester group would involve a rather unlikely eight-membered bridge transition state (9-44).

Coordinated hydroxide and amido groups can be very reactive participants in nucleophilic displacements at saturated carbon centers as well as carbonyl groups. Buckingham *et al.*[46] studied the "hydrolysis" of *cis*-$Co(en)_2(NH_2CH_2CH_2Br)Br^{2+}$ (9-45) with hydroxide in the pH range 8–14. As shown in Scheme (9-17) the reaction proceeds in two steps, the first being the replacement of the *coordinated* bromine by a hydroxide by means of an Sn1cb mechanism. In the second step the coordinated hydroxide is alkylated by the

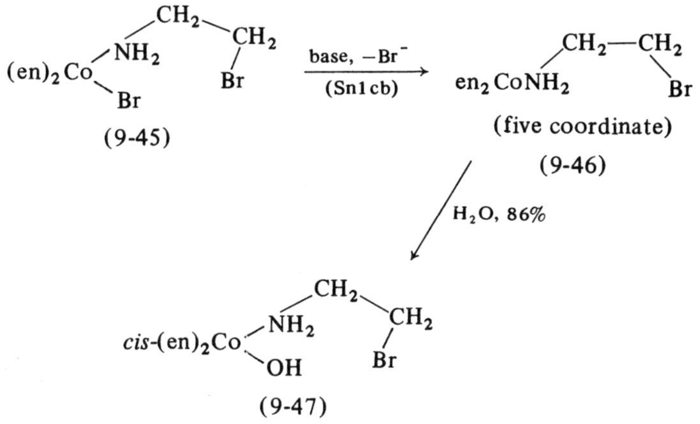

Scheme 9-17

bromoalkyl side chain to give a coordinated ethanolamine ligand. Below pH 11 this is by far the major product. However, at pH 14 where the nitrogen of the coordinated bromethylamine is readily ionized, the major reaction product (77%) is the hydroxoaziridino complex (9-50, Scheme 9-18). The stability of the coordinated aziridine is surprising in comparison to the great reactivity of coordinated epoxides such as (9-8). Perhaps one factor is the additional ring strain wrought upon the epoxide via chelate formation.

Breslow and co-workers have reported the zinc-catalyzed hydrolysis of anhydrides (9-52) and (9-53).[47] Their study was designed as a chemical model

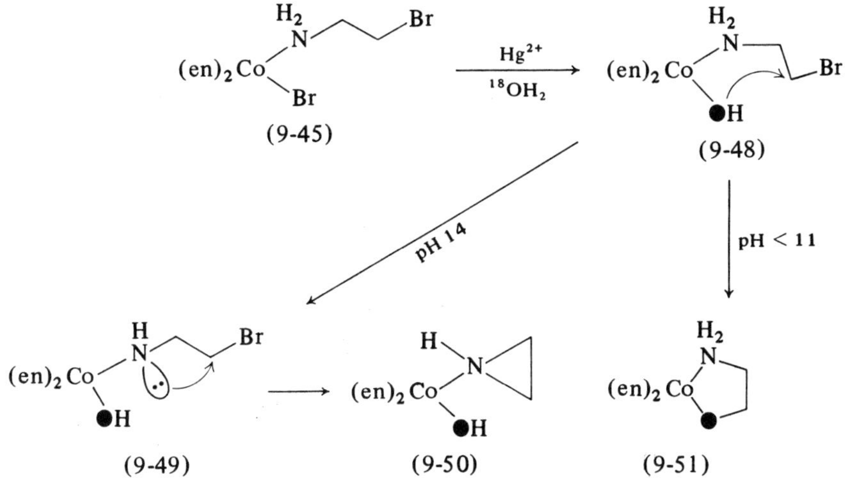

Scheme 9-18

to test the feasibility that the zinc ion in carboxypeptidase could promote the hydrolysis of an intermediate acyl enzyme such as Fig. 9-8C′. From the pH dependence and effects of added nucleophiles on the hydrolysis of (9-53), they determined that zinc ion catalyzed the attack of *hydroxide* upon the anhydride group but *did not catalyze water or hydroxylamine attack*, i.e.,

$$-d[(9\text{-}53)]/dt = k_1[(9\text{-}53)\cdot Zn^{2+}][OH^-] + k_2[(9\text{-}53)][H_2NOH]$$

The mechanism proposed to explain these results involved attack on a *non-coordinated* anhydride group by a hydroxide ion coordinated to the zinc ion in complex (9-54). As long as the anhydride group is not activated by coordination to zinc, its reaction with H_2O or H_2NOH would not be facilitated. Similarly, coordination of H_2O or H_2NOH to zinc, without their deprotonation, would decrease rather than increase their nucleophilicity. Since coordinated water deprotonates readily in the range of pH 5–8, whereas much more highly basic coordinations are required for N-deprotonation, the zinc ion catalysis is specific for hydroxide attack.

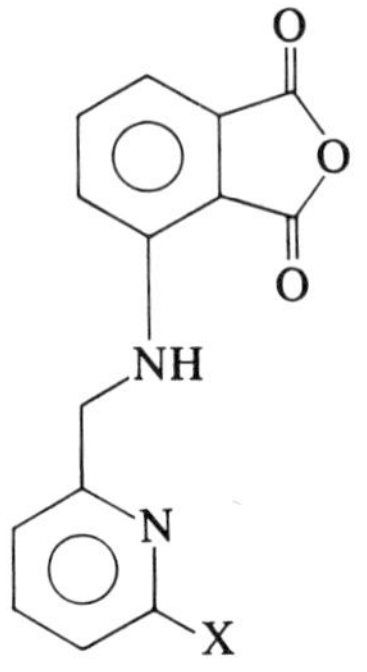

(9-52) X = H
(9-53) X = CO_2H

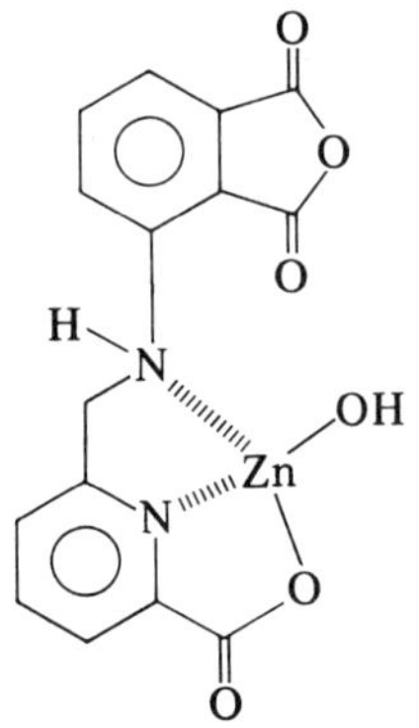

(9-54)

CARBONIC ANHYDRASE ACTION

The hydration of *cis*-aconitric acid by aconitase is one example of an enzymatic reaction of a coordinated hydroxide. Another important enzymatic reaction of a coordinated hydroxide is the hydration of carbon dioxide to bicarbonate by carbonic anhydrase.[38,48,49] This enzyme, present virtually wherever CO_2 and bicarbonate are involved in physiological buffering or ion transport, has one of the largest turnover numbers known for an enzyme. The rates of hydration of CO_2 by water and hydroxide are 3×10^{-2} sec^{-1} and 8×10^3 M^{-1} sec^{-1}, respectively, while the enzyme-catalyzed rate is 6×10^5 sec^{-1}. Similarly, the rate of spontaneous dehydration of HCO_3^- is 2×10^{-4} sec^{-1}

while the enzyme-catalyzed rate is 15 sec^{-1}. Thus, carbonic anhydrase is a tremendously efficient enzyme.

Several isozymes of carbonic anhydrase are known. X-ray structural determinations have shown that the human C isozyme is a globular protein about $41 \times 41 \times 47$ Å with a zinc ion near its center, at the bottom of a deep cleft filled with water molecules. In this respect it is similar to carboxypeptidase A. The zinc ion is bound by three imidazole nitrogens from histidine side chains, and around pH 7–9 the fourth coordination site can be occupied by water or hydroxide. The facile hydrolysis of several nonphysiological substrates by carbonic anhydrase (Fig. 9-9), suggests that its zinc hydroxyl can be very nucleophilic. This is also evident in the postulated mechanism for its hydration

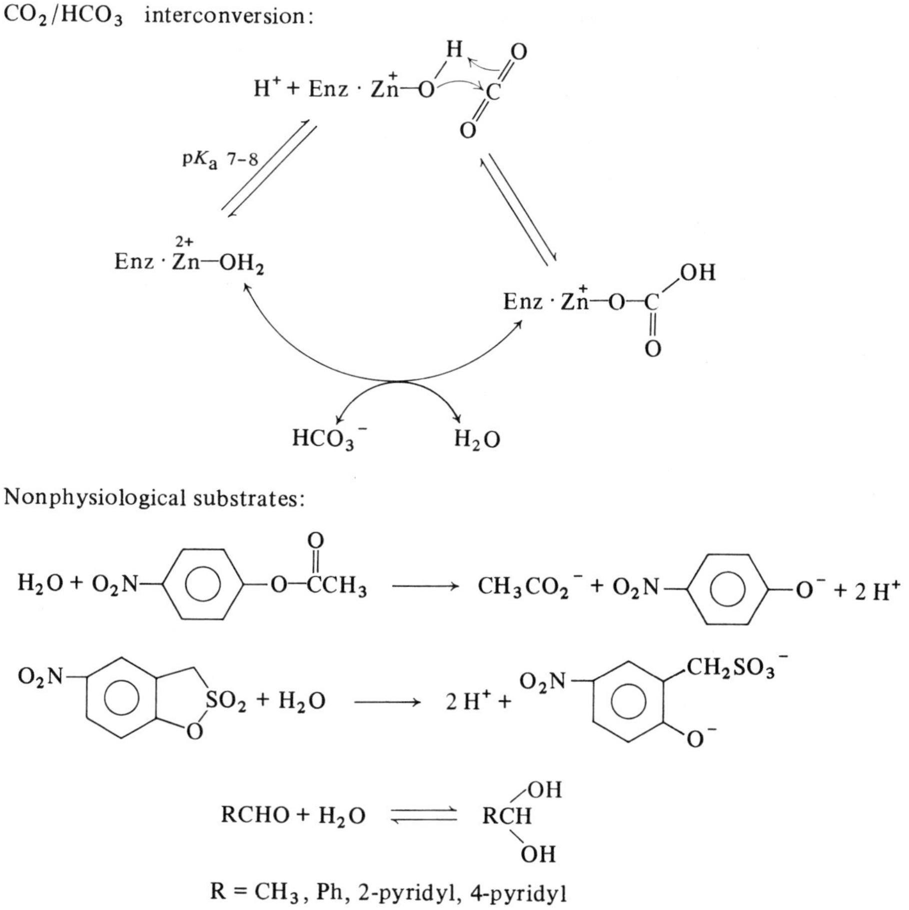

CO_2/HCO_3 interconversion:

Nonphysiological substrates:

$R = CH_3$, Ph, 2-pyridyl, 4-pyridyl

Fig. 9-9. Actions of carbonic anhydrase.[38,48]

of carbon dioxide. Thus, in facilitating the deprotonation of water at physiological pH, *the zinc ion performs the same role as a general base, despite the fact that it is actually a Lewis acid.*

REFERENCES

1. A. E. Martell, *Pure Appl. Chem.* **17**, 129 (1968).
2. D. H. Busch, *Science* **171**, 241 (1971).
3. D. H. Busch, ed., "Reactions of Coordinated Ligands," Advan. Chem. Ser. No. 37. Advan. Chem. Ser., Washington, D.C., 1963.
4. A. E. Martell, *in* "Metal Ions in Biological Systems" (H. Sigel, ed.), Vol. 2, p. 208. Dekker, New York, 1973.
5. M. M. Jones, "Ligand Reactivity and Catalysis." Academic Press, New York, 1968.
6. S. J. Benkovic and L. K. Danikoski, *J. Amer. Chem. Soc.* **93**, 1526 (1971).
7. W. Tagaki, Y. Asai, and T. Eiki, *J. Amer. Chem. Soc.* **93**, 3037 (1973).
8. R. Sternberger and F. H. Westheimer, *J. Amer. Chem. Soc.* **73**, 429 (1951).
9. J. F. Speck, *J. Biol. Chem.* **178**, 315 (1948).
9a. W. D. Covey and D. L. Leussing, *J. Amer. Chem. Soc.* **96**, 3860 (1974).
10. M. L. Bender, *Advan. Chem. Ser.* **37**, 19 (1963).
11. J. Prue, *J. Chem. Soc., London* p. 2331 (1952).
12. R. W. Hay and K. N. Leong, *Chem. Commun.* p. 800 (1967); *J. Chem. Soc., A* p. 3639 (1971).
13. C. R. Clark and R. W. Hay, *J. Chem. Soc., Perkin Trans 2* p. 1943 (1973).
14. B. Capon, M. C. Smith, E. Anderson, R. H. Dahm, and G. H. Sankey, *J. Chem. Soc., B* p. 1038 (1969).
15. R. W. Hay and J. A. G. Edwards, *Chem. Commun.,* p. 969 (1967).
16. R. P. Hanzlik and W. J. Michaely, *Chem. Commun.* p. 113 (1975).
16a. R. P. Hanzlik, W. J. Michaely, M. Edelman, and G. Scott, *J. Amer. Chem. Soc.* **98**, 1952 (1976).
17. E. J. Corey and R. L. Dawson, *J. Amer. Chem. Soc.* **84**, 4899 (1962).
18. C. R. Wasmuth and H. Freiser, *Talanta* **9**, 1059 (1962).
19. R. P. Houghton and R. R. Puttner, *Chem. Commun.* p. 1270 (1970).
20. J. P. Glusker, *in* "The Enzymes" (P. D. Boyer, ed.), 3rd ed., Vol. 5, p. 434, Academic Press, New York, 1971.
21. J. J. Villafranca and A. N. Mildvan, *J. Biol. Chem.* **247**, 3454 (1972).
22. H. Kroll, *J. Amer. Chem. Soc.* **74**, 2036 (1952).
23. M. L. Bender and B. W. Turnquest, *J. Amer. Chem. Soc.* **79**, 1889 (1957).
24. M. L. Bender and R. J. Thomas, *J. Amer. Chem. Soc.* **83**, 4189 (1961).
25. J. P. Collman and D. A. Buckingham, *J. Amer. Chem. Soc.* **85**, 3039 (1963).
26. J. Bjerrum and S. E. Rasmussen, *Acta Chem. Scand.* **6**, 1265 (1952).
27. D. A. Buckingham, J. P. Collman, D. A. R. Hopper, and L. G. Marzilli, *J. Amer. Chem. Soc.* **89**, 1082 (1967).
28. M. D. Alexander and D. H. Busch, *J. Amer. Chem. Soc.* **88**, 1130 (1966).
29. D. A. Buckingham, D. M. Foster, L. G. Marzilli, and A. M. Sargeson, *Inorg. Chem.* **9**, 11 (1970).
30. D. A. Buckingham, D. M. Foster, and A. M. Sargeson, *J. Amer. Chem. Soc.* **92**, 6151 (1970).
31. D. A. Buckingham, F. R. Keene, and A. M. Sargeson, *J. Amer. Chem. Soc.* **96**, 4981 (1974).

32. R. Breslow, R. Fairweather, and J. Keana, *J. Amer. Chem. Soc.* **89**, 2135 (1967).
33. D. A. Buckingham, J. Dekkers, A. M. Sargeson, and M. Wein, *J. Amer. Chem. Soc.* **94**, 4032 (1972).
34. D. A. Buckingham, J. M. Harrowfield, and A. M. Sargeson, *J. Amer. Chem. Soc.* **96**, 1726 (1974).
35. R. J. Angelici and D. Hopgood, *J. Amer. Chem. Soc.* **90**, 2514 (1968).
36. R. Nakon, P. R. Rochani, and R. J. Angelici, *J. Amer. Chem. Soc.* **96**, 2117 (1974).
37. S. K. Oh and C. B. Storm, *Bioinorg. Chem.* **3**, 89 (1973).
38. J. E. Coleman, *Prog. Bioorg. Chem.* **1**, 159 (1971).
39. W. N. Lipscomb, *Chem. Soc. Rev.* **1**, 319 (1972); *Tetrahedron* **30**, 1725 (1974).
40. M. L. Ludwig and W. N. Lipscomb, *in* "Inorganic Biochemistry" (G. Eichhorn, ed.), Vol. 1, p, 438. Elsevier, Amsterdam, 1973.
41. J. I. Hoppe and J. E. Prue, *J. Chem. Soc., London* p. 1775 (1957).
42. T. C. Bruice and W. C. Bradbury, *J. Amer. Chem. Soc.* **87**, 4846 and 4851 (1965).
43. R. Breslow and D. Chipman, *J. Amer. Chem. Soc.* **87**, 4195 (1965).
44. D. S. Sigman and C. T. Jorgensen, *J. Amer. Chem. Soc.* **94**, 1724 (1972).
45. G. J. Lloyd and B. S. Cooperman, *J. Amer. Chem. Soc.* **93**, 4883 (1971).
46. D. A. Buckingham, C. E. Davis, and A. M. Sargeson, *J. Amer. Chem. Soc.* **92**, 6159 (1970).
47. R. Breslow, D. E. McClure, R. S. Brown, and J. Eisenach, *J. Amer. Chem. Soc.* **97**, 194 (1975).
48. J. E. Coleman, *in* "Inorganic Biochemistry" (G. Eichhorn, ed.), Vol. 1, p. 488. Elsevier, Amsterdam, 1973.
49. S. O. Lindskog, *in* "The Enzymes" (P. D. Boyer, ed.), 3rd ed., Vol. 5, p. 587. Academic Press, New York, 1971.

X

Oxygen and Nitrogen

With the exception of water vapor, and in certain unfortunate localities CO, SO_2, and NO_x, oxygen and nitrogen comprise more than 99% of our gaseous environment. The oxygen molecule, dioxygen, is highly reactive and its relationship to both combustion and respiration has been appreciated for a very long time. In contrast the nitrogen molecule, dinitrogen, is extremely inert and reacts chemically only under the most severe of conditions. Despite the widely disparate chemical properties of these two molecules, biological organisms have developed enzymatic equipment for catalyzing their complete reduction smoothly and efficiently. Both biological and abiological redox reactions of dioxygen were discussed in Chapter VII. In this chapter we explore the reversible formation of metal complexes of dioxygen and dinitrogen, and the ways in which complexation affects the chemical properties of these molecules.

Binuclear μ-Peroxide and μ-Superoxide Complexes

The reversible absorption of atmospheric oxygen by inorganic solids ("ammonia-cobalt salts") was first noted by Fremy in 1852. Toward the end of the nineteenth century, Werner studied the oxygenation of ammoniacal solutions of Co(II) and obtained a brown diamagnetic complex which easily released oxygen when treated with excess ammonia (Scheme 10-1). Similar binuclear cobalt peroxides are formed with other ligand systems, including cyanide ion, chelating polyamines, peptide amides having dissociable amide hydrogens, the amino acid histidine, and certain tetradentate Schiff bases.[1-11a]

$$2\,Co(NH_3)_6{}^{2+} + O_2 \rightleftharpoons [(NH_3)_5Co{-}O{-}O{-}Co(NH_3)_5]^{4+} + 2\,NH_3$$

(10-1)

Ce(IV) or $S_2O_8{}^{2-}$

$$[(NH_3)_5Co{-}O{-}O{-}Co(NH_3)_5]^{5+}$$

(10-2)

Scheme 10-1

TABLE 10-1

Bond Geometries in Various Oxygen Compounds

Compound[a]	Structure	O–O (Å)	Ref.
$O_2{}^+$	Gaseous	1.12	12
O_2	Gaseous	1.21[b]	12
$K^+O_2{}^-$	Solid	1.28	12
H_2O_2	Solid	1.47[c]	13
BaO_2	Solid	1.49	10
pyCo(bzacen)·O_2	Fig. 10-4	1.26	14
pyCo(acacen)·O_2	Bent Co–O–O	[d]	15
$\left[(NH_3)_5Co{-}O\diagdown_{O{-}Co(NH_3)_5}\right]^{+5}$	[e]	1.31	16
$\left[(NH_3)_5Co{-}O\diagdown_{O{-}Co(NH_3)_5}\right]^{+4}$	[f]	1.47	17
(DMF·CoSalen)$_2O_2$	Fig. 10-1	1.34	18
$\left[(NH_3)_4Co\diagup^{O{-}O}\diagdown_{NH_2}\diagdown Co(NH_3)_4\right]^{+5}$	[g]	1.32	19
$[H_2O\cdot Co(3\,FSalen)\cdot O_2\cdot Co(3\,FSalen)]_2$	[h]	1.30	20

[a] Abbreviations: py, pyridine; bzacen^{2-}, [PhCOCHC(CH$_3$)NCH$_2$]$_2-$; acacen^{2-}, [CH$_3$COCHC(CH$_3$)NCH$_2$]$_2-$; CoSalen, N,N'-ethylenebis(solicylidene)iminatoCobalt(II); DMF, dimethylformamide.

[b] O–O stretch 1555 cm^{-1} in Raman.

[c] Dihedral angle ∼90°, 1380 cm^{-1} in Raman.

[d] O–O stretch 1123 cm^{-1} in ir.

[e] Cobalts octahedral, Co–O–O–Co nearly planar (175°C).

[f] Cobalts octahedral, Co–O–O–Co dihedral angle 146°C.

[g] Di-μ-ring nearly planar, Co–O–O angles 120°.

[h] Asymmetrical O–O bridge, not known if complex is paramagnetic.

X-ray structural determinations of several such compounds (Table 10-1[10,12–20] and Fig. 10-1[18]) generally reveal O—O bond lengths and dihedral angles expected of a peroxidelike group. Not all of these systems can be cycled through oxygenation and deoxygenation very many times before irreversible degradation occurs. The cyanide and peptide systems are particularly sensitive in this way, but solid CoSalen (8-14) retains 50% of its original oxygen binding capacity after 3000 uptake/regeneration cycles. The ease of reversibility of oxygenation

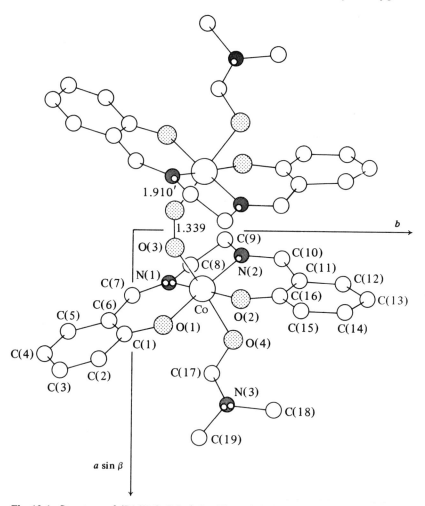

Fig. 10-1. Structure of (DMF·CoSalen)$_2$O$_2$. The cobalt is slightly out of the N$_2$O$_2$ ligand plane toward the oxygen and away from the DMF ligand. The Co—O$_2$ bonds are 1.91 Å, the O—O bond is 1.34 Å, the Co—O—O angles are 118°, and there is a CoOOCo dihedral angle of 110°.[18]

as well as the turnover lifetime of the complex seems to be related to the nature of the ligand system around cobalt. In the absence of strong axial σ-donor ligands the oxygen affinity is low but the lifetime is long and reversibility is good. The importance of this factor is more clearly demonstrated by studies of related monomeric superoxide adducts as discussed in the next section.

Typical of many binuclear cobalt peroxides,[11] complex (10-1) can be oxidized easily in solution to the green paramagnetic ion (10-2). Electron paramagnetic resonance studies show that the unpaired electron interacts equally with both cobalt nuclear spins ($I = \frac{7}{2}$ for ^{59}Co), although 80% of the spin density is centered on the oxygen atoms. X-ray determinations reveal that the Co–O–O–Co group now has a planar s-trans geometry with an O–O bond length very similar to that observed in the superoxide anion radical.

On the basis of stoichiometry and diamagnetic character, the formation of a binuclear cobalt peroxide can be represented as (10-3). Additional support for

$$2\,Co^{2+} + O_2 \; \rightleftharpoons \; [Co^{3+}(O_2{}^{2-})Co^{3+}] \qquad\qquad (10\text{-}3)$$

this formulation comes from potentiometric studies of the formation of a binuclear cobalt peroxide with trien as well as other ligands. Nakon and

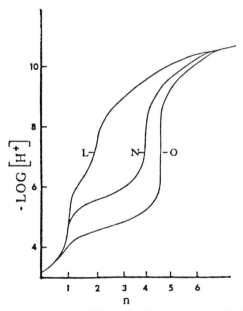

Fig. 10-2. Potentiometric titrations of (L) trien, (N) 1:1 trien to Co(II) solution under a nitrogen atmosphere, (O) 1:1 solution under an oxygen atmosphere.[21] (Reprinted with permission from the *Journal of the American Chemical Society.* Copyright by the American Chemical Society.)

Martell[21] titrated the tetrahydrochloride of trien with hydroxide and obtained the pH curve labeled L in Fig. 10-2. When the titration was repeated in the presence of Co(II) under nitrogen (curve N), the same amount of hydroxide was consumed but the buffer region had shifted to a considerably lower pH range reflecting the competition for ligand by Co(II) and protons. When this titration was repeated under oxygen (curve O), a further shift of the buffer region to lower pH was apparent. This may be attributed to the even greater competition of Co(III) for trien ligand. The consumption of an additional one-half mole of base per mole of (trien)Co introduces a new facet of the chemistry of μ-peroxides, the formation of a second bridging ligand, the μ-hydroxo group. Kinetic studies of the oxygenation of this system by Wilkins and co-workers[22] indicated two parallel pathways for the formation of the μ-peroxo-μ-hydroxo complex as shown in Fig. 10-3. Although Co(III) complexes are

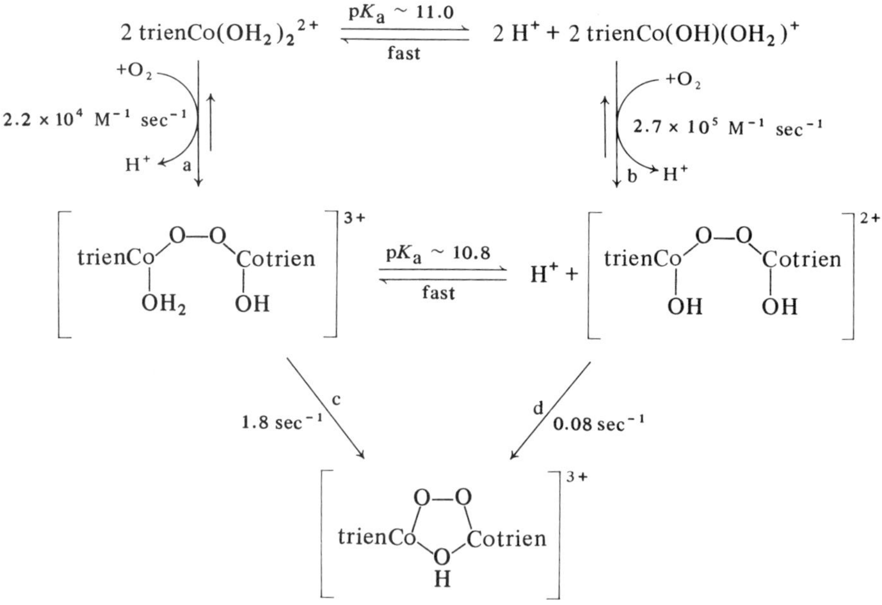

Fig. 10-3. Equilibria and rate constants in the Co(II)–trien–O_2 system (25°C, $I = 0.2\ M$).[9] Dissociative processes are rate limiting, hence $k_b > k_a$ and $k_c > k_d$.

relatively inert toward (heterolytic) ligand dissociation or exchange the oxygen uptake is freely reversible via reversal of the redox process at lower pH. At higher pH (11) the cyclic dibridged complex forms, and this species is deoxygenated only by treatment with acid. Bridging (NH_2^-) groups are also known to occur, and an example is given in Table 10-1.

Mononuclear Superoxide Complexes

If it is reasonable to formulate the 2:1 metal:oxygen adducts of Co(II) ammines as μ-peroxocobalt(III) complexes, it should be as reasonable to postulate the existence of mononuclear superoxide complexes as transient intermediates in the formation of the binuclear peroxides. Wilkins and co-workers have

$$M \cdot + O_2 \quad \rightleftharpoons \quad M{-}O{-}O \cdot$$

$$M{-}O{-}O \cdot + M \rightleftharpoons M{-}O{-}O{-}M$$

recently obtained convincing kinetic evidence for the existence of such a two-step mechanism in the formation of several amine and amino acid cobalt peroxides.[9] The oxygenation reactions are extremely fast even at low temperatures. Stopped-flow mixing techniques were used for the kinetic studies, and it was found that most systems gave very similar oxygenation rates. This was interpreted as a rate limiting dissociation of coordinated water followed by a very fast combination with dissolved oxygen.

The study of 1:1 oxygen adducts of cobalt complexes took a giant step forward when Floriani and Calderazzo[23] reported the isolation of (pyridine)Co(3-methoxysalen)·O_2 (10-4). This adduct had a magnetic moment

(10-4)   (10-5)

indicative of one unpaired electron, weak IR absorptions at 1140 and 1060 cm^{-1} not present in the parent complex, and easily gave up oxygen by heating *in vacuo*. Very shortly thereafter Hoffman, Basolo, and co-workers reported the isolation and characterization of monomeric oxygen adducts of a series of related BCo(acacen) complexes (10-5), where B is dimethylformamide (DMF) or a 4-substituted pyridine.[24-26] These complexes rapidly and reversibly absorb oxygen in toluene solution below room temperature. However, at room temperature or above only a slow uptake of oxygen is observed, corresponding to the slow oxidation of the ligand. The isolated oxygen adducts displayed ir absorptions at 1120–1140 cm^{-1} (O–O stretch), the intensities of which were suggestive of low symmetry and a strong dipole for the Co–O_2 unit, e.g., structures B or C but not A (Diagram 10-1). The oxygen adducts gave an 8-line electron spin resonance (ESR) spectrum, as a result of the splitting caused by the ^{59}Co nucleus ($I = \frac{7}{2}$), which also showed superhyperfine splitting from the quadrupole of the pyridine nitrogen but not from the Schiff base nitrogens.[25] From this it was concluded (1) that the Co–O$_2$ unit was not linear as in C, and (2) that the unpaired electron was in an orbital which included the cobalt atom although most of the spin density was on the oxygens. Thus a bent Co–O–O geometry such as B was suspected for the BCo(acacen)·O$_2$ complexes.

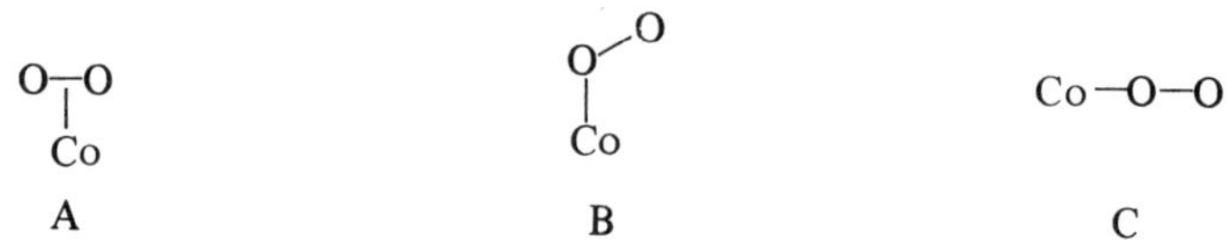

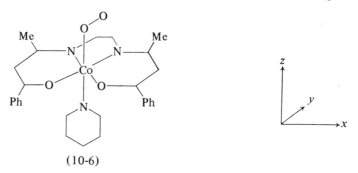

Diagram 10–1

Determinations of the crystal structures of pyCo(acacen)·O$_2$ (10-5)[15] and a related complex pyCo(bzacen)·O$_2$ (10-6)[14] confirmed the bent geometry of the Co–O–O unit. Disorder in the crystal of pyCo(acacen)·O$_2$, as a result of the existence of rotamers about the Co–O bond, prevented accurate determination of the O–O bond length. The structure of (10-6) is shown in Fig. 10-4. The

(10-6)

Fig. 10-4. Structure of the monomeric oxygen complex pyCo(bzacen)·O$_2$. The Co–O–O unit lies in the *yz* plane with a bond angle of 126° and Co–O and O–O bond lengths of 1.86 and 1.26 Å, respectively. The phenyl groups are coplanar with the Schiff base ligand, and the pyridine ring lies in the *xz* plane.[14]

O–O bond length of 1.26 Å is very similar to that of the superoxide ion. Additional studies of (10-6) by X-ray photoelectron spectroscopy indicated Co2p$_{3/2}$ binding energies for Co(bzacen), pyCo(bzacen)NO$_2$, and (10-6) of 780.0, 781.4, and 781.4 eV, respectively.[27] The identical increase in binding energies for the nitro and oxygen complexes clearly indicates that oxygen binding involves the essentially complete oxidation of Co(II) to Co(III) with concomitant reduction of dioxygen to the superoxide ion O$_2^{-}$.

At this point it is important to mention that *in solution*, the monomeric Co(II)O$_2$ complexes may not have the rigid "bent superoxide" structure observed in the crystalline adducts. Electron spin resonance studies of $^{17}O={}^{16}O$ and $^{17}O={}^{17}O$ adducts of (10-6) have shown that both oxygen atoms are magnetically equivalent.[28] This was interpreted as evidence favoring a symmetrical type A structure rather than a rapid flipping of the O—O group between two bent positions. The analog of CoSalen (10-7) has been claimed to form adducts with O_2, CO, and MeNC in solution. On the basis of ESR and NMR spectra it was suggested[29] that the oxygen adduct of (10-7) should be regarded as a singlet oxygen complex of Co(II), since CO and MeNC are unlikely to be reduced to species analogous to superoxide anion radicals. However a reinvestigation of this report has shown that the "CO and MeNC adducts" were artifacts caused by traces of oxygen in the system.[29a] Finally, the recently isolated oxygen complex $[Bu_4N][O_2Co(pfp)(Hpfp)]\cdot EtOH$, where H_2pfp is perfluoropinacol, has the ESR spectrum and magnetic susceptibility expected of a Co(II) complex rather than a superoxide complex. It forms reversibly in solution under oxygen or nitrogen, but is stable to 100°C in the solid state.[30]

(10-7) R = CH$_3$
(10-8) R = H

A number of factors influence the equilibrium constant for oxygen binding by Co(II) Schiff base complexes. The effect of the axial base on oxygen affinity and turnover lifetime was mentioned earlier. In the case of the 1:1 cobalt superoxide adducts the axial base has a profound effect on the tendency of the complex to bind oxygen. In the series BCo(acacen), oxygen affinity increases with the pK_a of BH$^+$: DMF $\ll$ 4-cyanopyridine $\ll$ pyridine $<$ 4-methylpyridine $<$ 4-aminopyridine.[24] Evidence was obtained for the existence of $H_2O\cdot Co(acacen)\cdot O_2$ in aqueous solution, but the coordinated water was rapidly replaced by added pyridine.[25] Similar results were obtained for the oxygenation of Co(II) protoporphyrin IX-dimethyl ester (10-9): B = 1-methylimidazole (CH$_3$-Im) $>$ 4-t-butylpyridine $>$ pyridine.[31] Stynes and Ibers determined the oxygen pressures required for 50% oxygenation of the cobalt porphyrin complexes (10-9).[31] In toluene at −23°C CH$_3$-Im · (10-9) required 417 Torr O$_2$ for half-oxygenation. In DMF only 12.6 Torr O$_2$ was required for the same oxygenation, but the corresponding value for half-oxygenation of

DMF·(10-9) in DMF was 40 Torr. Thus, the nature of solvent can be *more* important than the axial base in determining the equilibrium constant. The "medium" effect probably results from stabilization of the polar Co(III)–O$_2$·⁻ species in more polar aprotic solvents such as CH$_3$CN and DMF. Such medium effects are no doubt of profound importance in governing the oxygenation of heme in hemoglobin and myoglobin.

Another parameter which influences oxygenation equilibria is the redox potential of the complex. Carter *et al.*[23] found a linear correlation between the equilibrium constant for oxygenation and the redox potential for a series of neutral Co(II) complexes of tetradentate ligands (Fig. 10-5), although a related

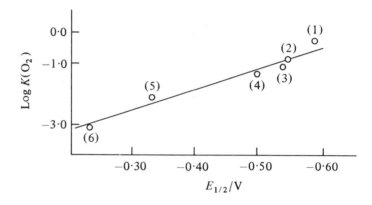

Compound	R$_1$	R$_2$	X
1	Me	H	O
2	Me	Ph	O
3	Me	Me	O
4	Ph	H	O
5	Me	H	S
6	CoTMPP		

Fig. 10-5. Correlation of oxygen affinity and redox of some Co(II) complexes. CoTMPP = α,β,γ,δ-tetra(*p*-methoxyphenyl)porphinatocobalt(II). *K* determined in toluene at −31°C, $E_{1/2}$ in pyridine vs SCE.

complex with a net positive charge exhibited an oxygen affinity much lower than its redox potential would predict. Evidently, a high redox potential or a preexisting positive charge makes the

$$Co(II) + O_2 \longrightarrow Co(III) \cdot O_2^-$$

change less favorable and reduces the oxygen affinity of the complex. No doubt some of the effects of axial bases on oxygen affinity can be attributed to their effects on the Co(II) → Co(III) redox potential according to their balance of σ-donor and π-acceptor properties as ligands.

Under ordinary circumstances the oxygen molecule is slow to react with organic molecules despite the fact that such reactions are very exothermic. There are three reasons for this: (1) the direct concerted combination of triplet dioxygen with singlet organic molecules is spin-forbidden, (2) C–H bonds are generally much stronger than the ·OO–H bond, and (3) in the stepwise reduction of dioxygen the addition of the first electron is endothermic. The formation of a superoxide complex might therefore be regarded as a form of "oxygen activation." Metal complexation of the superoxide ion could overcome the endothermal nature of the reduction process and the oxygen molecule would thus be poised for accepting a second electron in a very exothermal process. Such activation has been demonstrated, although in reality the activation is not very great. Only two types of substrates are known to be "oxidized" by M–O–O· superoxide complexes. One of these is the low valent metal "oxygen carrying" complex itself, in which case the reaction product is merely the μ-peroxo-bridged dimer. The other substrates oxidized by M–O–O· complexes are phenolic compounds, which are inherently moderately sensitive to oxidation.[33,34] One case in point is the oxidation of 4-methyl-2,6-di(t-butyl)-phenol in methanol by the "oxygen carrier" (10-8) or by CoSalen.[35]

Below 10°C in methanol (10-8) catalyzed the formation of hydroperoxide (10-10) in 80% yield with a turnover number of about 4 or 5 moles (10-10) per mole (10-8). Above 10°C in methanol the major product was (10-11), the formation of which was accompanied by the oxidation of solvent CH_3OH (or $PhCH_2OH$ if added). Oxidation of 2,6-di(t-butyl)phenol gave only (10-12) in high yield. A good clue to the mechanism of these oxygenations came from the

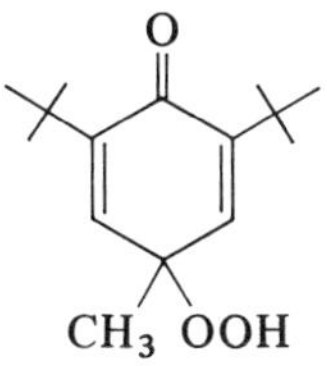

CH_3 OOH
(10–10)

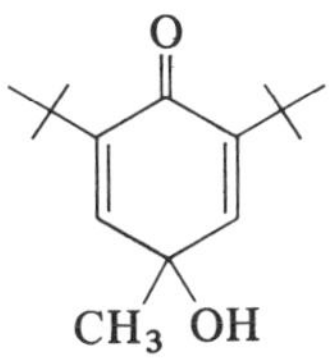

CH_3 OH
(10–11)

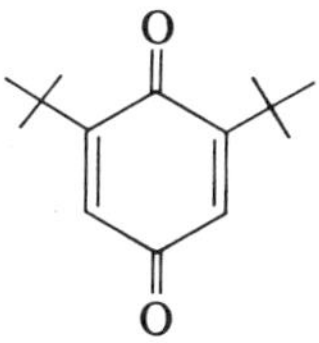

(10–12)

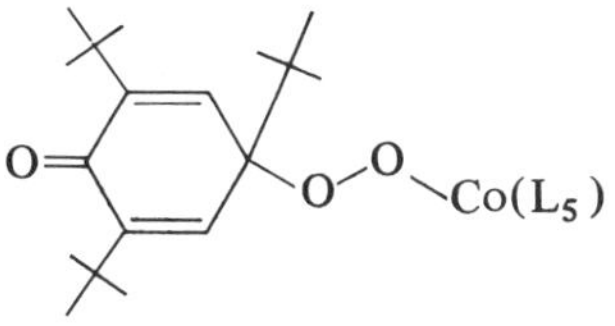

(10–13)

isolation of (10-13) from the stoichiometric oxidation of 2,4,6-tri(t-butyl)phenol with (10-8).

Metal–Oxygen π Complexes

In 1963, Vaska reported the reversible oxygenation of an iridium(I) complex, $IrCl(CO)(Ph_3P)_2$, (7-4), in benzene solution.[36] This complex, now usually referred to as "Vaska's complex," thus became the forerunner of a large and still growing class of metal compounds forming 1:1 π complexes with molecular oxygen.[7,8] Members of this group of metal compounds have a number of characteristic features which set them clearly apart from the oxygen carriers which form superoxide or μ-peroxide adducts.

1. Both the metal compound and its derived oxygen adduct usually are diamagnetic, although exceptions are known.[30]

2. Most of the metal compounds contain d^8 or d^{10} metals in their lower oxygenation states. Dioxygen complexes of the metals shown in the following tabulation are known.

d^6		d^8		d^{10}
			Co(I)	Ni(0)
Ru(II)	Ru(0)		Rh(I)	Pd(0)
	Os(0)		Ir(I)	Pt(0)

3. Very often the other ligands in the complex are phosphines, although arsines, isocyanides, halides, and carbonyl groups are not uncommon.

4. The metal–oxygen unit of the complex is symmetrical, i.e., both O atoms are essentially equidistant from the metal.

5. Those complexes which can be oxygenated are "coordinatively unsaturated"; that is, they have 14 or 16 but not 18 electrons, and often form complexes with other π acids such as tetracyanoethylene, tetrafluoroethylene, or carbon monoxide.

As in the case of the superoxide- and peroxide-type dioxygen complexes, two of the most important parameters which provide information on the nature of coordinated oxygen are the O–O bond length and the ir stretching frequency for the $M \cdot O_2$ group. Unfortunately, it appears that the latter is characteristic of the entire group and is not a simple O–O stretch. In general, however, there is a good correlation between reversibility and O–O bond length. The most easily reversible adducts have the shortest O–O bonds (and the longest M–O bonds) and their M·O_2 group absorptions generally come at lower energies in the ir.

Several theories have been proposed to explain the diamagnetic nature of the oxygen adducts. At first it was proposed that the actual ligand was the excited singlet state of oxygen.[37] A later theory postulated a "three-center" bonding orbital formed from a (filled) metal d orbital and a π^* molecular orbital on oxygen.[38,39] Two subtypes within this scheme would be valence forms involving superoxide or cyclic peroxide complexes, e.g.,

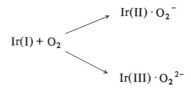

For the Ir(II) · O_2^- case it would be necessary to postulate that the spins on the d^7 metal and superoxide canceled each other in order to explain the diamagnetic behavior.

Probably the most generally accepted view of bonding in these oxygen adducts is similar to that originally proposed by Dewar[40] and Chatt and Duncanson[41] for Pt(II)–ethylene complexes. Both oxygen and ethylene are π acids, and electron-rich metals in their low oxidation states are Lewis bases. The bonding thus involves weak σ donation from a ligand π-MO to an empty metal d or d hybrid orbital, and a stronger more significant "back-bond" from a metal d orbital to an empty π^*MO on the ligand. Examples of this type of

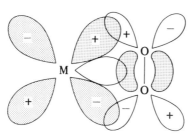

Fig. 10-6. Bonding in a metal–oxygen π complex. Based on the Dewar–Chatt–Duncanson proposal for Pt(II)-ethylene complexes.[40,41]

bonding may be found in Figs. 5-3, 5-4, and 10-6. The strength of the π complex formed depends on three main factors:

1. The "π acidity" of the ligand (e.g., TCNE > $CF_2=CF_2$ > SO_2 > O_2 > $CH_2=CH_2$) which is a function of the electronegativity of the ligand atoms and the π^* energy level

2. The electron density on the metal, which depends on the net charge of the complex, and the balance between σ donor and π acceptor properties of the ancillary ligands in the complex

3. The extent of overlap of metal and ligand orbitals, which depends on their complementarity in terms of orbital energy levels and symmetry

For oxygen complexes of transition metals in a given triad, it would be expected that the strongest complexes would be formed with the first-row metal, and the weakest with the third-row metal, e.g., Co(I) > Rh(I) > Ir(I). In this particular case, however, the observed order (Co > Ir > Rh) correlates with the energies observed for spectroscopic $d–d$ transitions in the parent complex

TABLE 10-2

Structural Features of Metal–Oxygen π Complexes

Complex, structure[a]		O–O (Å)	ir[b] (cm^{-1})	R[c]	Ref.
[Co(2=phos)$_2$O$_2$]BF$_4$	I	1.420	909	–	46
[Rh(diphos)$_2$O$_2$]PF$_6$	I	1.418	880	+	47
[Rh(2=phos)$_2$O$_2$]BF$_4$	I		880	+	42
[Ir(diphos)$_2$O$_2$]PF$_6$	I	1.52[d]	845	–	49a
[Ir(2=phos)$_2$O$_2$]BF$_4$	I		840	–	42
O$_2$IrCl(CO)L$_2$	II	1.30	860	+	36, 50, 51
O$_2$IrCl(CO)(Ph$_2$PEt)$_2$	II	1.461	856	+	52
O$_2$IrBr(CO)L$_2$	II	1.36	862	±	36, 50, 51
O$_2$IrI(CO)L$_2$	II	1.525	862	–	36, 50, 51
O$_2$Ru(NO)(NCS)L$_2$	III		860	–	53
O$_2$PtL$_2$	III	1.45	830	–	54, 55
O$_2$PdL$_2$	III		875	–	11, 56
O$_2$Ni(tBuNC)$_2$	III		898	–	56

[a] Structures I and II are basically trigonal pyramids while III is essentially planar. Abbreviations: L, Ph$_3$P; diphos, Ph$_2$PCH$_2$CH$_2$PPh$_2$; and (2=phos), cis-Ph$_2$PCH=CHPPh$_2$.

[b] M · O$_2$ group absorption, not a simple O–O stretch.

[c] + Indicates readily reversible oxygenation, and – indicates not readily reversible.

[d] Originally reported[48,49] as 1.625 Å; later work showed this overestimate was caused by deterioration of the crystal during the collection of the X-ray data.[49a]

[M(bisphosphine)$_2$]$^+$ rather than with the size of the central metal.[42] In contrast to Vaska's iridium complex (7-4), the rhodium analog RhCl(CO)(Ph$_3$P)$_2$ does not interact detectably with oxygen in solution. Analysis of the oxygenation equilibria for these complexes in terms of the rates and activation parameters for oxygenation and deoxygenation shows that the activation *enthalpy* ($\Delta H_2^\ddagger$) for the oxygenation process correlates with the energy of the *d–d* spectral transition.[43]

A large number of square planar analogs of Vaska's complex have been prepared by substituting other ligands for the chloride, carbonyl, and phosphines in (7-4).[43–45] Oxygen complexes of most of these have been characterized by ir, X-ray determinations of O–O bond lengths, and/or determinations of various kinetic and activation parameters for their oxygenation equilibria. Some of this information is collected in Tables 10-2[11,36,42,46–56] and 10-3.[43–45] Inspection of these tables using IrCl(CO)(Ph$_3$P)$_2$ or its oxygen adduct as a reference point reveals several pronounced trends, all of which indicate dependence of the oxygenation process on the electron density on the

TABLE 10-3

Equilibrium, Kinetic, and Activation Parameter Data for Oxygenation of IrX(CO)L$_2$ in Chlorobenzene at 40°C

$$O_2 + IrX(CO)L_2 \underset{k_{-1}}{\overset{k_2}{\rightleftharpoons}} O_2IrX(CO)L_2, \quad K = k_2/k_{-1}$$

Complex X	Complex L	$10^2 k_2$ (M^{-1} sec^{-1})	$10^{-3} K$ (M^{-1})	$\Delta H_2^\ddagger$ (kcal)	$\Delta H_{-1}^\ddagger$ (kcal)	$\Delta S_2^\ddagger$ (eu)	Ref.
F	Ph$_3$P	1.48	0.29	13.6	23.7	−24	43
Cl	Ph$_3$P	10.1	7.32	9.50	26.5	−33	43
Br	Ph$_3$P	20.6	62.0	8.42	28.8	−35	43
I	Ph$_3$P	72.3	857	5.67	29.0	−41	43
Cl	Ph$_3$As	14.2	17.2	8.1		−37	44
Cl	Ph$_2$PEt	14.2	21.0	8.4		−36	44, 45
Cl	(p-ClC$_6$H$_4$)$_3$P	3.10	3.58	10.8		−31	44, 45
Cl	(p-CH$_3$C$_6$H$_4$)$_3$P	21.6	17.7	9.3		−32	44, 45
Cl	(p-CH$_3$OC$_6$H$_4$)$_3$P	48.4	38.7	8.5		−33	44, 45

metal. Thus, as the halide is varied from highest to lowest electronegativity (F > Cl > Br > I) several effects are observed.

1. In the absorption spectra of the complexes the energy of the lowest *d–d* transition shifts to lower energies, and a linear free energy correlation is observed for this value and $\Delta G°$ for oxygenation of the complex.

2. Both the rate of oxygenation and the oxygen affinity increase, while the rate of deoxygenation *decreases* along this series. Thus, the iodo complex requires boiling in DMF at 150°C for deoxygenation.

3. The length of the O–O bond increases, indicating that more electron density from the metal is moving into the π^* antibonding orbital on oxygen.

In all cases the activation entropy for oxygenation, $\Delta S_2^{\ddagger}$, is very close to the standard entropy change $\Delta S°$. This suggests that the transition state for oxygenation closely resembles the oxygen adduct. The complexes with the less electronegative ligands are more basic and therefore form π complexes with dioxygen more rapidly as well as more strongly. The data for the complexes in which the phosphine ligands are varied also support this hypothesis. Alkylphosphine complexes oxygenate faster and more strongly than (7-4) unless the phosphines are too bulky. A Hammett correlation (with $\rho < 0$) has been found for the rate of oxygenation of a series of $IrCl(CO)[(p\text{-}XC_6H_4)_3P]_2$ complexes (Table 10-3). This clearly establishes the importance of electron donation from the ligands, through their π system, to the metal and ultimately to the oxygen molecule. Replacing the carbonyl group (itself a strong π acid) and the halide ion in (7-4) by a chelating phosphine produces a complex whose oxygen adduct is extremely stable, having one of the longest O–O bonds yet measured, 1.52 Å for $O_2Ir(diphos)_2PF_6$. This distance is somewhat longer than the O–O distances usually observed for various peroxides, both organic and inorganic.

The chemical properties of the more stable oxygen π complexes are often suggestive of peroxide chemistry. For example, $O_2IrI(CO)(Ph_3P)_2$ will liberate I_2 from acidic KI solution via initial formation of H_2O_2 and Ir(III) compounds.[57] The oxygen in π complexes can also show the nucleophilic behavior characteristic of peroxy anions,[58] as in the examples of Fig. 10-7. Finally, oxygen π complexes can oxidize small molecules such as CO, SO_2, NO, and NO_2.[59] In these cases the metal center is also oxidized; thus, the reactions are stoichiometric rather than catalytic. In some cases, however, catalytic oxygenation of organic compounds has been achieved. Characteristically, the organic substrates are good reducing agents and strong soft ligands, while the products are poor ligands for soft metals. Platinum(0), palladium(0), and nickel(0) are powerful catalysts for the oxidation of phosphines and isocyanides.[55,56] Kinetic Scheme 10-2 has been proposed by Halpern and Pickard for the $(Ph_3P)_3Pt$-catalyzed oxidation of triphenylphosphine.[60]

$$(Ph_3P)_3Pt + O_2 \xrightarrow{k_1} (Ph_3P)_2PtO_2 + Ph_3P$$

$$(Ph_3P)_2PtO_2 + Ph_3P \xrightarrow{k_2} (Ph_3PO)_2Pt(Ph_3P)$$

$$(Ph_3PO)_2Pt(Ph_3P) + 2\,Ph_3P \xrightarrow{fast} 2\,Ph_3PO + (Ph_3P)_3Pt$$

Scheme 10-2

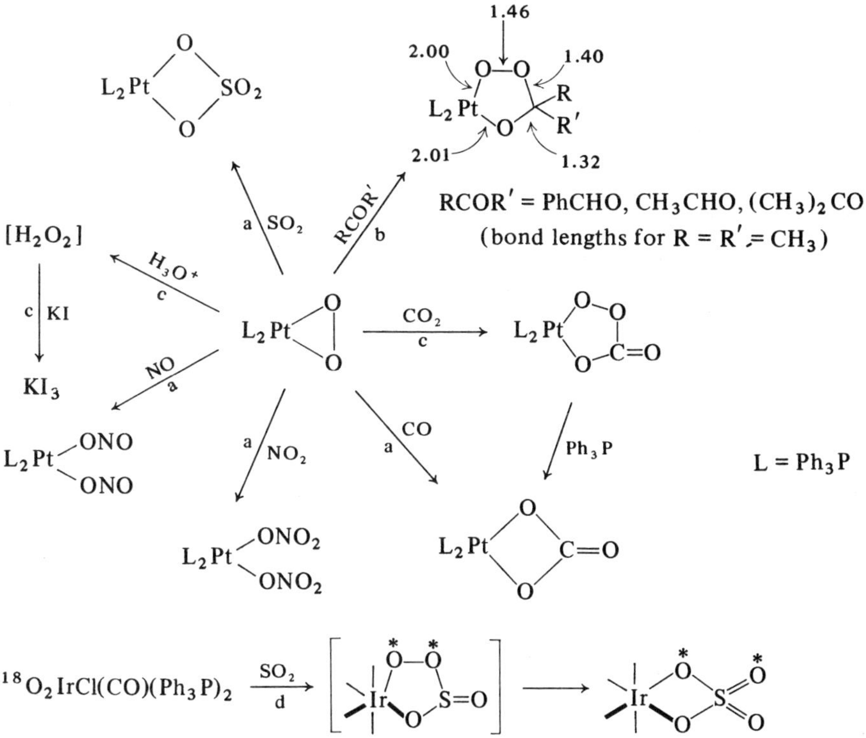

Fig. 10-7. Reactions of metal–oxygen π complexes. a, Collman[59]; b, Ugo *et al.*[58]; c, Hayward *et al.*[57]; d, Horn *et al.*[47]

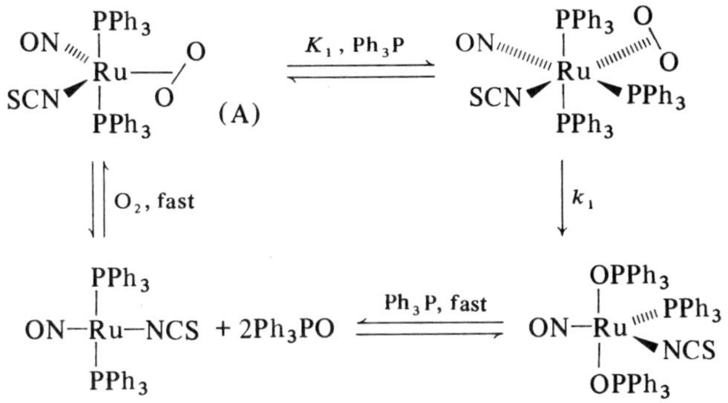

Rate = $k_1 K_1 [P][A]/(1 + K_1[P])$

At 80°C in xylene $k_1 = 1.1 \times 10^{-2}$ sec^{-1} and $K_1 = 1.62$ M^{-1}

Scheme 10-3

At 25°C in benzene k_1 and k_2 are 2.6 and 0.15 M^{-1} sec^{-1}. The rate limiting step k_2 presumably involves phosphine attack at Pt together with migration of two phosphines from Pt to coordinated oxygen, thus forming the phosphine oxide ligands.

In the oxidation of triphenylphosphine catalyzed by $O_2Ru(NO)(NCS)(Ph_3P)_2$,

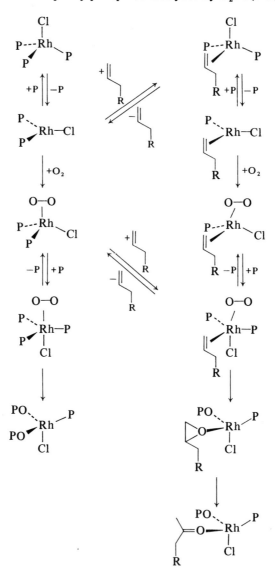

Fig. 10-8. Rhodium(I)-catalyzed oxidation of 1-hexene to 2-hexanone.[64]

a similar mechanism (Scheme 10-3) is followed, except that breakdown of the adduct rather than its formation is rate limiting.[53]

The oxidation of other organic substrates, e.g., olefins, by similar catalytic processes is currently a formidable challenge to chemists. There are two major obstacles to reckon with in olefin oxidation. One is the fact that dioxygen requires four electrons for full reduction, while a $C=C$ bond is a two-electron reducing agent (unless cleavage to carbonyl products occurs). The other problem is the inherent susceptibility of most olefins to metal-promoted free radical chain autoxidation, which can obscure any specific metal-catalyzed oxygenation processes that might be occurring. One attractive mechanism by which olefins could perhaps be oxygenated would be to bring the olefin and oxygen together as cis ligands in the coordination sphere of a suitable metal complex. Despite the attractiveness of such a mechanism "on paper," the complex $O_2IrCl(CH_2=CH_2)(Ph_3P)_2$ is quite stable with respect to forming ethylene oxide, formaldehyde, or acetaldehyde.[61] However, coordinated oxygen apparently can react at the allylic positions of coordinated olefins. One possible example of this is the formation of hydroperoxide (7-8) catalyzed by Vaska's complex.[62] Recently, an oxygen-cyclooctene mixed complex of Rh(I) has been shown to rearrange intramolecularly to give a complexed allylic hydroperoxide.[63] Perhaps the most interesting olefin oxidation is the *cooxidation* of 1-hexene and triphenylphosphine catalyzed by $(Ph_3P)_3RhCl$ (Fig. 10-8).[64] Free radical intermediates have been excluded and the mechanism appears similar to the oxidation of Ph_3P by Pt(0) or Ru(II) except that the olefin substitutes for one molecule of phosphine. The overall process is *formally* (i.e., stoichiometrically) similar to the action of mixed function oxidase enzymes which use nicotinamide-adenine dinucleotide (NADPH) as a two-electron cosubstrate in olefin oxidation (Chapter VII). In order to become synthetically useful, however, it will be necessary to find a cheaper, more convenient cosubstrate to replace the triphenylphosphine. The most desirable cosubstrate, of course, would be a second molecule of olefin.

Oxygen Complexes in Biology

Having surveyed the various types of oxygen complexes known to occur, we may now turn to the question of oxygen binding under physiological conditions. All organisms whose size places critical limitations on the distribution of oxygen by simple diffusion face a problem in connection with oxygen *transport*. Organisms whose oxygen supply is interruptible, such as the diving mammals, may also face a problem in connection with oxygen *storage*. Consequently, many of these organisms have developed special oxygen-binding metalloproteins to serve their specialized needs. Probably the most familiar of these are hemoglobin (Hb) and myoglobin (Mb), which respectively transport bound oxygen in erythrocytes and store surplus oxygen in skeletal muscle tissue. Hemerythrin is an oxygen storage protein found in several groups of marine invertebrates including the seashore sandworm *Golfingia gouldii*, and hemocyanin is the oxygen transport pigment found in many arthropods and mollusks. The properties of these "respiratory pigments" are compared in Table 10-4.

TABLE 10-4

Comparison of Some Properties of Oxygen-Carrying Proteins[a]

	Hemoglobin	Hemerythrin	Hemocyanin
Function	Transport	Storage	Transport
Location	Erythrocytes	Extracellular	Extracellular
Molecular weight	64,450	108,000	$4–200 \times 10^5$
Number of subunits	4; $2\alpha, 2\beta$	8	Many
Metal content, deoxy	4 Fe(II), heme	16 Fe(II), Nonheme	Cu(I)
Metal: O_2 ratio	1 : 1	2 : 1	2 : 1
Color:			
deoxy	Red-purple	Colorless	Colorless
oxy	Red	Violet-pink	Blue-bluegreen
Magnetic behavior			
deoxy	High spin Fe(II)	High spin Fe(II)	Diamagnetic Cu(I)
oxygenated	Diamagnetic	Fe(III), diamagnetic[b]	Cu(II), diamagnetic[b]
oxidized	High spin Fe(III)	Fe(III), diamagnetic[b]	Cu(II), diamagnetic[b]
Oxygen binding	Cooperative, $n = 3$	Sites independent	Cooperative, $n = 3.5–4$
Bohr (pH) effect	Yes	No	No

[a] Data from refs. 5, 10, 65–68.
[b] Apparently diamagnetic as a result of antiferromagnetic coupling of the two metals.

HEMERYTHRIN AND HEMOCYANIN[5,10,65–67]

In the case of hemerythrin the 2 Fe : O_2 stoichiometry of oxygen binding, together with the absence of oxygen binding by the ferric form methemerythrin, is immediately suggestive of a binuclear μ-peroxide structure for the oxygenated protein. This otherwise hasty conclusion is also supported by Mössbauer studies which indicate that all the iron in oxyhemerythrin is in the ferric form. The diamagnetism of the oxygenated form apparently results from antiferromagnetic coupling of two high spin Fe(III) centers through the peroxo bridge. In the oxidized protein methemerythrin the ferric irons are also coupled, which may indicate that they both share a permanently bridging ligand such as a carboxylate or hydroxo group.

A plot of oxygen uptake vs oxygen pressure gives a simple hyperbolic curve for both myoglobin and hemerythrin, indicating that the binding sites act independently rather than cooperatively as in hemoglobin and hemocyanin. In the latter cases binding of the first oxygen causes conformational changes in the protein which *increase* the oxygen affinity of the remaining empty sites. Thus oxygenation curves for these proteins are sigmoidal, indicating the binding of the first three oxygens almost simultaneously. In addition, hemoglobin liberates protons when oxygen is bound (Bohr effect) but the others do not. This acid-base reaction is very important physiologically: Since oxygen deprived tissues will be rich in CO_2 and lactic acid, oxyhemoglobin will unload its oxygen preferentially where it is needed most. Myoglobin, which has only one oxygen storage site, gives a regular hyperbolic oxygenation curve and lacks a Bohr effect.

Despite its name, hemocyanin contains neither heme nor iron; it is a copper protein. Although it is a larger protein and not as well understood as hemerythrin, most evidence (Table 10-4) suggests that it too forms a μ-peroxide structure when oxygenated. Again the diamagnetism of both the oxygenated and the oxidized forms is attributed to antiferromagnetic coupling via bridging ligands. The oxygenation of hemocyanin is cooperative but is not tied to an acid-base equilibrium as in the case of hemoglobin.

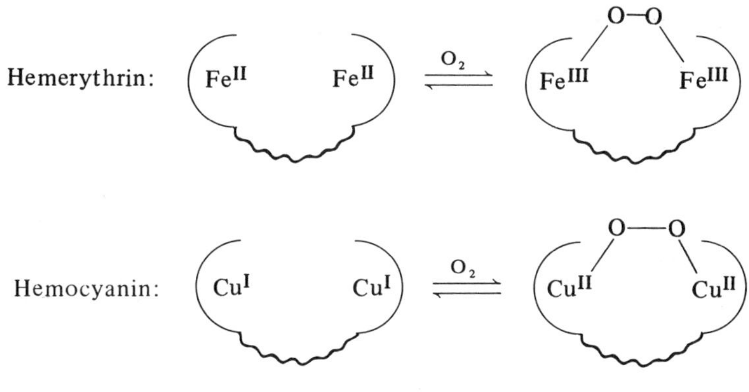

THE HEMOGLOBIN PROBLEM

The many fascinating aspects of the reversible oxygenation of hemoglobin can be conveniently divided into two areas. One area is concerned with the complex protein chemistry associated with the oxygenation equilibrium such as site–site cooperativity and the Bohr (pH) effect, features which are not displayed by the monomeric subunits of hemoglobin nor by myoglobin. The other area is concerned with the chemistry of the oxygen-binding site and the detailed manner in which the oxygen is bound. It is the latter area that is of immediate concern to us, although as we probe the oxygenation mechanism we shall see that the two areas are quite closely interrelated.

Hemoglobin and myoglobin have been subjected to study by almost every conceivable chemical and spectroscopic method applicable to proteins,[68] and their crystal structures have been determined by Perutz (hemoglobin)[69] and Kendrew (Myoglobin, Fig. 10-9).[70] The overall shapes of myoglobin and the α

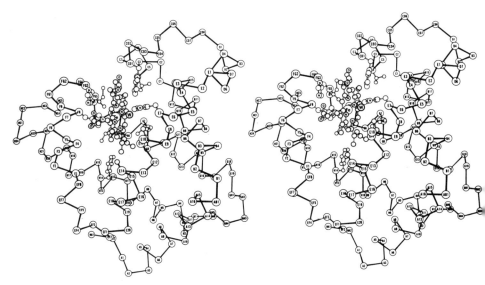

Fig. 10-9. A stereo drawing of sperm whale metmyoglobin based on the X-ray diffraction studies of Kendrew *et al.*[70] The water molecule and the imidazole side chain of histidine F8 coordinated to the iron are shown. Besides several amino acid residues in the region of the heme for which the side chains are included, only the polypeptide backbone is indicated. (From the stereo supplement to the *Structure and Action of Proteins* by permission of Harper and Row. (Reprinted from *The Structure and Action of Proteins* by Richard E. Dickerson and Irving Geis, W. A. Benjamin, Inc., Publishers, Menlo Park, California. Copyright 1969 by Dickerson and Geis.)

and β subunits of hemoglobin are quite similar, and there is even a good correspondence of their primary amino acid sequences. Differences in the oxygen-binding sites are minor, and will therefore generally be ignored in the following discussions.

The most important feature of the oxygen-binding site is, of course, the heme and its iron atom. Hemoglobin and myoglobin contain a type-*b* heme group embedded in a sea of hydrophobic amino acid side chains. No covalent bonds hold it to the globin, although the iron is coordinated by an imidazole from a histidine residue referred to as the proximal histidine. Thus iron is five-coordinate with a sixth site open for a small mobile ligand trans to the imidazole base. This axial imidazole actually makes an important contribution to the oxygen-binding site, just as the axial bases in the simple 1 : 1 cobalt–oxygen complexes discussed earlier. Several genetically inherited types of abnormal hemoglobins are known in which this histidine is missing or replaced by another amino acid. Some of these hemoglobins have a low oxygen affinity or a decreased cooperativity, while others are more readily oxidized to the ferric forms. In one case heme is not retained by the β chain which carries the defect.

X-ray crystallographic studies of many heme complexes indicate that the iron is quite often located *out of the plane of the porphyrin* by 0.3–0.5 Å.[71] In methemoglobin (Hb$^+$) the iron is 0.3 Å out of the porphyrin plane toward the proximal histidine, while in deoxyhemoglobin this distance is 0.75 Å, corresponding to the larger size of Fe(II) than Fe(III). The ferrous iron in Hb is in the high spin form, but the presence of certain ligands in the sixth position can cause it to change to the low spin form. This is true only for strong field ligands such as NO, CO, O_2, or isocyanides, and these are the *only* ligands which bind to deoxyHb. Even cyanide (which is isoelectronic with CO) is not detectably bound, although it and other anions are tenaciously bound by Hb$^+$. The change in spin state on the ferrous ion is accompanied by a change in the *size* of the ion, which allows *it to move approximately into planarity with the porphyrin system* away from the proximal histidine and toward the sixth ligand. It is postulated that this motion is translated via the protein chains in connection with site–site cooperativity and the Bohr effect.[72,73]

Any proposal for the structure of the Fe–O_2 unit in oxygenated hemoglobin must be able to account for the fact that oxyhemoglobin is diamagnetic. In 1956, Griffith proposed a structure with oxygen parallel to the heme.[74] In 1964, Pauling proposed that a bent Fe–O–O unit was formed.[75] Much earlier, Pauling had suggested a linear Fe–O–O unit, but this was abandoned because oxygen bound in this manner could not cause the d orbital splitting needed to achieve a diamagnetic configuration for iron.

Pauling geometry   Griffith geometry

The bonding in oxyHb has been variously viewed as a π-oxygen complex of low spin Fe(II), as an antiferromagnetically coupled Fe(III)-superoxide complex, and as a low spin Fe(IV)-peroxide complex.[65] Mössbauer studies do indicate the transfer of electron density from iron to oxygen, but no ir absorptions attributable to an unsymmetrically bound O_2 or O_2^- can be detected. Recent studies with a "hemoglobin model" and a "myoglobin model" to be described below have shown that oxygen is bound with Pauling geometry. However, the actual mode of oxygen binding in hemoglobin and myoglobin must still be regarded as an open question.

COBOGLOBINS

Hoffmann and Petering[76] found that it was possible to replace the heme in Mb and Hb with the Co(II) derivative of protoporphyrin IX (10-14). These reconstituted "coboglobins" were found to complex oxygen reversibly, and the hemoglobin analog even exhibited cooperativity in oxygen binding. The ESR spectra of oxyCoHb showed no significant differences from the spectra of oxygenated (10-9), indicating that oxygen in coboglobin is also bound as superoxide.

The globin protein is thought to serve three principal functions in heme oxygenation. (1) It furnishes a ligand to the iron, i.e., the imidazole of the proximal histidine. (2) It provides the heme with a "medium of low dielectric strength," which should tend to modulate the oxygen affinity by disfavoring extensive charge transfer from iron and reduction of oxygen to superoxide. (3) By restricting the motion of the heme it prevents the formation of the μ-oxo dimeric oxidation products which always form when heme is exposed to oxygen in solution at room temperature. (Interestingly, Wang[77] has shown that heme suspended in a polystyrene matrix can be oxygenated reversibly.) Evidently the globin also makes other important contributions to the oxygenation process, because the oxygen affinities of CoHb and CoMb are 600 and 300 times greater than the oxygen affinity of the cobalt porphyrin (10-9) in solution.[76] The origin of this unpredicted effect is not yet understood.

IRON COMPLEXES OF HINDERED PORPHYRINS

Several attempts have been made to construct Fe(II) complexes containing a built-in hydrophobic pocket in the hopes of creating a myoglobin-like hydrogen carrier.[78] The most successful of these is the "picket-fence porphyrin" complex prepared by Collman's group.[78a] This ligand is a porphyrin having four o-pivalamidophenyl groups attached to the alpha positions. These substituents are all rotated such that the pivalamide side chains point in the same direction and form a hydrophobic pocket on one side of the porphyrin. In benzene the Fe(II) complex of this ligand binds two molecules of nitrogen bases (L) such as 1-methylimidazole to form a diamagnetic complex, $L_2Fe(\alpha,\alpha,\alpha,\alpha$-TpivPP) (10-15, Fig. 10-10). Because of nonbonded interactions with the pivalamide side chains one of the bases dissociates relatively easily in solution to give a five-coordinate complex (10-16, Fig. 10-10). If oxygen is added to a solution of

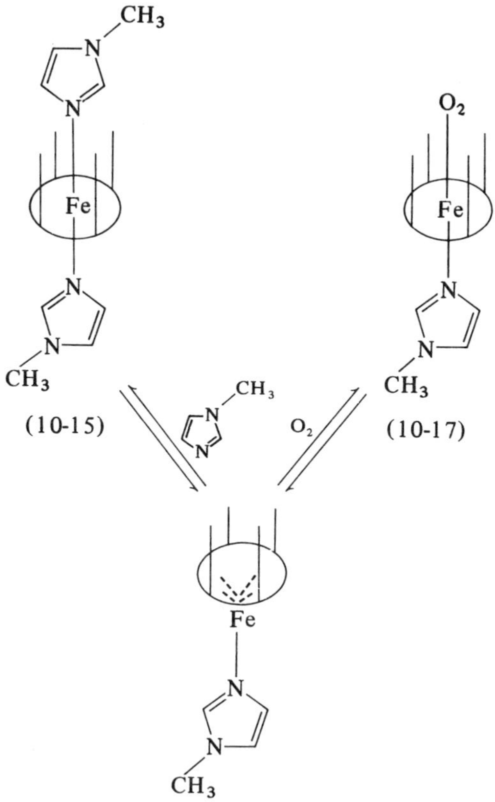

Fig. 10-10. Probable equilibria involved in the oxygenation and deoxygenation of Fe-1-Me(imid)$_2$($\alpha,\alpha,\alpha,\alpha$-TpivPP).[78]

(10-15), the equilibrium of Fig. 10-10 develops, and purple-red crystals of the oxygen adduct form upon addition of heptane. Treatment of the oxygen complex with excess pyridine or 1-methylimidazole gives off exactly one equivalent of oxygen and returns a quantitative yield of the starting complex. The oxygen adduct (10-17, Fig. 10-10) is diamagnetic and has a Mössbauer spectrum similar to oxyHb. The ir and Raman fail to show absorptions ascribable to coordinated oxygen, even when the $^{18}O_2$ complex is made. However, the absorptions may be obscured by porphyrin absorption bands. An X-ray structural determination for this oxygen complex has recently been completed, and shows that oxygen is bound with Pauling geometry (Fig. 10-11).[79]

There have also been several interesting extensions of this work. For example, if the methylimidazole in (10-17) is replaced by tetrahydrofuran (THF),

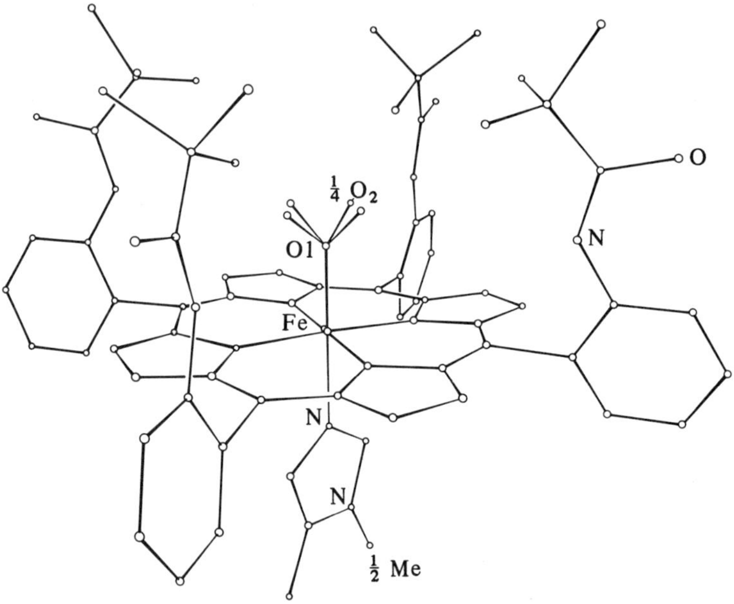

Fig. 10-11. Perspective view of one molecule of the iron dioxygen complex (10-17) showing the crystallographic twofold axis of symmetry and the four-way disorder of the terminal O(2) oxygen atom of dioxygen.[79]

the resulting dioxygen complex is *paramagnetic* and its oxygenation is reversible even in the solid state.[80] This may be relevant to studies of oxygen binding by other hemoproteins, notably cytochrome P_{450}. The cobalt analog of (10-16) has also been prepared.[81] It binds oxygen reversibly in solution, and this process generates ESR spectral changes which parallel the oxygenation of simple cobalt porphyrin complexes. However, as with CoHb and CoMb, *the oxygen affinity of the "Co-picket fence porphyrin" is greater than that of the simple porphyrin complexes.* Collman attributed this to the oxygen-binding site of the hemoproteins having an environment more like a solid than a solution. Finally, if the pivaloyl groups in (10-15) are replaced by *p*-toluenesulfonyl groups, the complex is irreversibly oxidized rather than reversibly oxygenated.[78a] This has been attributed to the greater acidity of the sulfonamide hydrogens, leading to the formation of hydrogen peroxide and oxidation of the iron.

Although a great deal is now known about oxygen complexes, their formation, and their reactions, we still do not know how oxygen is bound by hemoglobin. At present we know even less about the details of biological oxygen activation, for example, by mixed function oxidase enzymes which depend on cytochrome P_{450}. However, considering the current interest in these areas we can predict that the answers to these questions may not be far off. Somewhat further removed, but not out of reach, lies the attractive possibility of imitating the action of mixed function oxidases in the laboratory and on an industrial scale.

Preparation and Characterization of Dinitrogen Complexes[5,82-88]

As is often the case with important discoveries in science, the first metal complex containing a dinitrogen ligand was prepared unintentionally.[89] In 1965, Allen and Senoff were attempting to prepare Ru(II) amine complexes by reduction of commercial "RuCl$_3$," with hydrazine hydrate at room temperature. What they isolated was a yellow diamagnetic solid which eventually was proved to contain the complex ion [Ru(NH$_3$)$_5$N$_2$]$^{2+}$. This complex has now been prepared by several other methods, which also serve to exemplify the various types of general methods by which dinitrogen complexes of other metals are prepared.

1. Direct coordination of dinitrogen by ligand replacement

$$Ru(NH_3)_5OH_2{}^{2+} + N_2 \rightleftharpoons Ru(NH_3)_5N_2{}^{2+} + H_2O \qquad \text{(ref. 90)}$$

2. Formation of dinitrogen from ligands containing catenated nitrogen atoms

$$Ru(NH_3)_5NNO^{2+} \xrightarrow[\text{or Cr}^{2+}]{\text{Zn/Hg}} Ru(NH_3)_5N_2{}^{2+} \qquad \text{(ref. 91)}$$

$$2[Ru(NH_3)_5N_3]^{2+} \xrightarrow{\text{acid}} N_2 + 2\,Ru(NH_3)_5N_2{}^{2+} \qquad \text{(ref. 92)}$$

3. Formation of dinitrogen by catenation of nitrogen atoms

$$Ru(NH_3)_6{}^{3+} + NO + OH^- \longrightarrow Ru(NH_3)_5\,N_2{}^{2+} + 2\,H_2O \qquad \text{(ref. 93)}$$

The Ru(NH$_3$)$_5$N$_2$$^{2+}$ ion is not typical of dinitrogen complexes in that it is an ionic water-soluble species, while probably the majority of dinitrogen complexes are neutral species soluble in organic solvents. Most of these complexes are prepared directly from dinitrogen, a metal complex, and a powerful reducing agent, as shown by the examples below.

1. Cobalt and cobalt hydride complexes of dinitrogen

$$Co(acac)_3 + Ph_3P + AlEt_3 + N_2 \xrightarrow[-10°C]{Et_2O} CoH(N_2)(Ph_3P)_3 \qquad \text{(ref. 94)}$$

$$Co(acac)_3 + Ph_3P + iBu_3Al + N_2 \xrightarrow[10°C]{\text{toluene}} Co(N_2)(Ph_3P)_3 \qquad \text{(ref. 95)}$$

2. Molybdenum dinitrogen complexes

$$Mo(acac)_3 + diphos + AlEt_3 + N_2 \xrightarrow{\text{toluene}} (diphos)_2 Mo(N_2)_2 \quad \text{(ref. 96)}$$

$$MoCl_3(THF)_3 + diphos + N_2 \xrightarrow[\text{THF, 15°C}]{\text{Na·Hg}} (diphos)_2 Mo(N_2)_2 \quad \text{(ref. 97)}$$

In all, about one-hundred dinitrogen complexes are now known, involving the following metals:

Ti	–	–	Mn	Fe	Co	Ni
–	–	Mo	–	Ru	Rh	–
–	–	W	Re	Os	Ir	Pt

Despite the range of metals involved (d^1 to d^{10}), these complexes share several characteristic features. In all of the *isolable* dinitrogen complexes, the M–N–N unit is *linear*, and in many cases the coordinated dinitrogen is reactive enough to serve as a bridging ligand between two metal centers. With one exception, it has not been possible to reduce an isolable dinitrogen complex to ammonia. In fact, the organic soluble nitrogen complexes are all very sensitive to oxidation and must be handled under an inert atmosphere.

The free nitrogen molecule is characterized by a bond length of 1.098 Å[98] and an absorption in the Raman spectrum of 2330 cm^{-1}. Because of its symmetry dinitrogen, like dioxygen, does not absorb in the ir. However, coordinated dinitrogen usually gives a sharp strong absorption in the ir around 2200–1900 cm^{-1}, indicating a high degree of asymmetry in the M·N$_2$ unit and some lowering of the N–N bond order. When a dinitrogen complex is isotopically substituted with ^{15}N$_2$, the ir absorption shifts by a fairly constant value of –70 cm^{-1}. Metal-nitrogen stretching frequencies occur around 450–550 cm^{-1} and are usually reciprocally related to the N–N frequencies, i.e., as the N–N bond weakens the M–N bond strengthens. X-ray studies have shown that the M–N–N unit in both 1:1 and 2:1 μ-dinitrogen complexes is *linear* (Figs. 10-12, 13, 14).[99,104] The linearity of the M–N–N–M unit contrasts to the trigonal geometry observed in organic compounds such as azobenzene, PhNNPh, and indicates that there has been little reduction in N–N bond order as a result of metal complexation. The X-ray studies have also shown that there is very little lengthening of the N–N bond upon coordination to a metal. Even less change

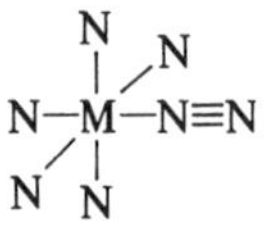

Fig. 10-12. Structure of [(NH$_3$)$_5$OsN$_2$]$^{2+}$ and [(NH$_3$)$_5$RuN$_2$]$^{2+}$. The geometry in both cases is octahedral, with bond lengths (Å) as follows: Ru–NH$_3$ (2.12), Ru–N$_2$ (2.12), N–N (1.12),[99] Os–NH$_3$ (2.14), Os–N$_2$ (1.84), N–N (1.12).[103]

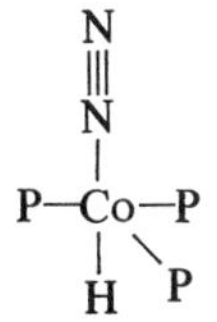

$$
\begin{array}{c}
N \\
\|\| \\
N \\
| \\
P-Co-P \\
| \quad \backslash \\
H \quad P
\end{array}
$$

Fig. 10-13. Structure of $(Ph_3P)_3CoH(N_2)$. X-ray studies of crystals of this complex grown from di-*n*-butyl ether revealed two discrete but very close structures in the lattice. The structure presented here represents the averaged bond angle and bond length (Å) parameters. The Co is ca. 0.3 Å above the plane of the three phosphorus atoms, the CoNN unit is essentially linear, and the PCoP angles are close to 120°; Co–N (1.80), N–N(1.123), Co–P (2.16–2.20). The Co–H distance was estimated at 1.6 Å[104].

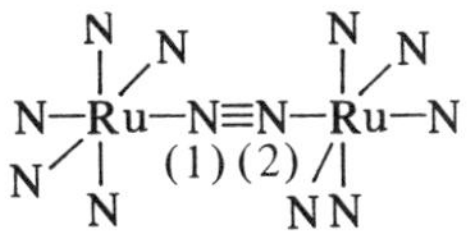

$$
\begin{array}{c}
N \quad N \qquad N \quad N \\
| \diagup N \qquad | \diagup N \\
N-Ru-N\equiv N-Ru-N \\
N \diagup | \quad (1)(2) \diagup | \\
N \qquad\quad N N
\end{array}
$$

Fig. 10-14. Structure of $[Ru_2(NH_3)_{10}N_2]^{4+}$. The geometry around the metals is octahedral and the RuN$_2$Ru unit is essentially linear with bond lengths (Å) as follows: Ru–Ru (4.979), Ru–N(1) (1.928), N(1)–N(2) (1.124), Ru–NH$_3$ (2.12–2.14).[100]

is produced upon coordination of a second metal to form a μ-dinitrogen complex. For example, the bond lengths and N–N frequencies for N$_2$, $[(NH_3)_5RuN_2]Cl_2$, and $[(NH_3)_5Ru-N_2-Ru(NH_3)_5]^{4+}$ are 1.098 Å[98] (2330 cm^{-1}), 1.120 Å[99] (2105 cm^{-1}), and 1.124 Å[100] (2100 cm^{-1}),[101] respectively. This is consistent with the relative ease with which coordinated dinitrogen dissociates or is displaced from its complexes, and also with the general lack of chemical reactivity of coordinated dinitrogen.

Stable complexes in which dinitrogen is bound "sideways" are unknown, although such complexes may exist as intermediates in the interconversion

$$[(NH_3)_5 Ru-^{15}N\equiv^{14}N]^{2+} \rightleftharpoons [(NH_3)_5 Ru-^{14}N\equiv^{15}N]^{2+}$$

The energy of activation for the interconversion is 21 kcal/mole, while for loss of N$_2$ it is 28 kcal/mole.[102] It is also possible that sideways or "edge-on" bonding of dinitrogen is involved in the highly labile titanium complexes in which the nitrogen can be reduced to the oxidation state of hydrazine or ammonia (see (10-18) below).

A molecular orbital energy diagram for dinitrogen is given in Fig. 5-15. The energies of the highest filled and lowest unfilled orbitals on dinitrogen are given by the ionization potential and electron affinity of N$_2$, +15.6 and +3.6 eV, respectively. The corresponding values for dioxygen are +12.1 and −0.9 eV, which immediately gives a clue to the striking difference in chemical reactivity between nitrogen and oxygen. Nitrogen is very much more difficult both to oxidize and reduce than is oxygen.

The bonding in nitrogen complexes is basically like that in carbonyl complexes, a combination of σ donation and π back-bonding. One reason that dinitrogen is not encountered as a ligand as commonly as CO is that its donor and acceptor orbitals are at energies that require a difficult-to-achieve combination of metal orbitals for overlap, i.e., an empty σ orbital of very low energy and a filled π orbital of high enough energy to overlap with a π^* orbital on nitrogen. These conditions are most easily met by low spin, low valent metals with large Δ splittings such as those found with very strong soft ligands (phosphines) and with second and third-row metals. Ligands with good π-acceptor properties strongly disfavor coordination of dinitrogen, which suggests that π-back-bonding contributes more than σ bonding to the stability of the complexes. Thus, the coligands are critically important to the stability of dinitrogen complexes.

Reactions of Dinitrogen Complexes

The scope of reactions which dinitrogen complexes undergo is limited at least by the medium in which the complex is soluble. Thus, the chemical behavior of the ionic ruthenium and osmium complexes *appears* to differ greatly from the behavior of, for example, $CoH(N_2)(Ph_3P)_3$.

RUTHENIUM AND OSMIUM AMINE COMPLEXES

$Ru(NH_3)_5N_2^{2+}$ and its Os(II) analog (Fig. 10-12)[99,103] are reasonably stable toward oxygen in aqueous solution. However, these species can be easily oxidized electrochemically. The oxidized species lose nitrogen very easily, and have N–N stretching frequencies 80–100 cm^{-1} higher than the parent complexes (Scheme 10-4).

$$(NH_3)_5OsN_2^{2+} \xrightleftharpoons{+0.3\ V^*} (NH_3)_5OsN_2^{3+} \xrightarrow{0.02\ sec^{-1}} (NH_3)_5OsOH_2^{3+} + N_2$$

(ref. 105)

$$(NH_3)_5RuN_2^{2+} \xrightarrow{+0.72\ V^*\ fast} (NH_3)_5RuOH_2^{3+}$$

(ref. 87)

$$(NH_3)_5RuOH_2^{2+} \xrightleftharpoons{-0.25\ V^*} (NH_3)_5RuOH_2^{3+}$$

(ref. 106)

* vs SCE

Scheme 10-4

In the ruthenium series dinitrogen can be displaced by other ligands such as halide, DMSO, pyridine, or ammonia, but not by CO. Remarkably stable binuclear complexes with a bridging dinitrogen group are formed in aqueous solution, usually by simply combining solutions of $(NH_3)_5MOH_2^{2+}$ and $(NH_3)_5MN_2^{2+}$. These complexes can be oxidized electrochemically to form mixed valence complexes which are rather prone to dissociation and hydration (Scheme 10-5). Electron spin resonance studies of the oxidized binuclear complexes have not been reported, but they would be expected to show that the odd

$$(NH_3)_5\,RuOH_2^{2+} + N_2 \xrightarrow{\ 7.3 \times 10^{-2}\ M^{-1}\ sec^{-1}\ } (NH_3)_5\,RuN_2^{2+} \text{ (ref. 106)}$$

$$(NH_3)_5\,RuOH_2^{2+} + (NH_3)_5\,RuN_2^{2+} \xrightarrow{\ 4.2 \times 10^{-2}\ M^{-1}\ sec^{-1}\ } \text{ (ref. 106)}$$
$$(NH_3)_5\,RuNNRu(NH_3)_5^{4+}$$

$$(NH_3)_5\,RuNNOs(NH_3)_5^{4+} \underset{}{\overset{+0.1\ V^*}{\rightleftharpoons}} (NH_3)_5\,RuNNOs(NH_3)_5^{5+} \qquad \text{(ref. 107)}$$

$$(NH_3)_5\,RuNNRu(NH_3)_5^{4+} \underset{}{\overset{+0.5\ V^*}{\rightleftharpoons}} (NH_3)_5\,RuNNRu(NH_3)_5^{5+} \qquad \text{(ref. 105)}$$

$$\Big\downarrow {\scriptstyle 0.1\ sec^{-1}}$$

$$(NH_3)_5\,RuN_2^{2+} + (NH_3)_5\,RuOH^{2+} + H^+$$

* vs SCE

Scheme 10-5

electron was in a delocalized four-center molecular orbital as in the case of the μ-superoxide complexes discussed earlier.

COBALT AND IRON COMPLEXES

Nitrogen in these complexes is weakly bound and very easily removed by purging a solution of the complex with argon. The result is a coordinatively-unsaturated metal complex which will tend to add the most convenient ligand available. Species such as H_2, CH_3CN, or NH_3 can displace nitrogen *reversibly* from $(Ph_3P)_3CoH(N_2)$ (Fig. 10-15), but CO displaces N_2 irreversibly. Interestingly, if D_2 is used instead of H_2, HD is formed, and the phenyl rings on the phosphine ligands become labeled with deuterium in the ortho positions specifically.[108] The mechanism postulated to explain the phenyl labeling involves the *oxidative insertion* of the metal into a C–H bond at the ortho position. Similar labeling of ortho positions of arylphosphine ligands is observed with the iron analog $(PhPEt)_3FeH_2(N_2)$.[109] The facile *exchange* of N_2 and NH_3 in the cobalt complexes is not seen in the ruthenium or osmium complexes; with the latter ammonia displaces nitrogen irreversibly. Since the enzymatic reduction of nitrogen to ammonia is not subject to product inhibition,

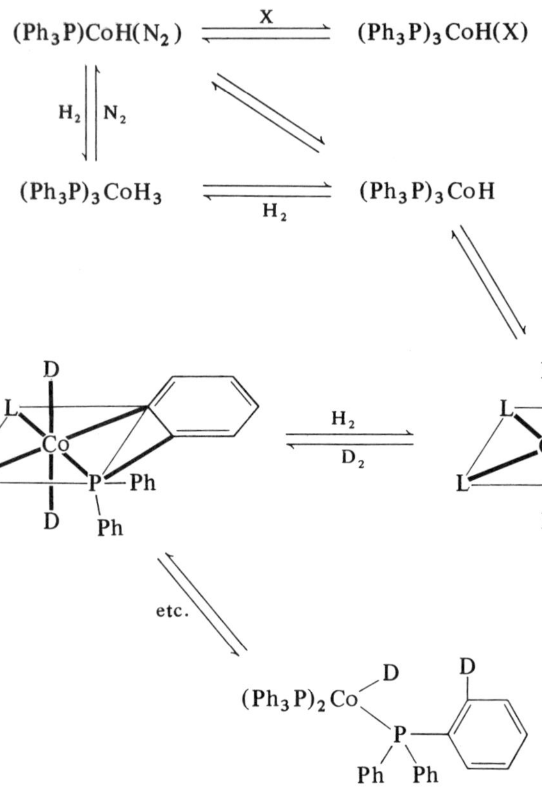

Fig. 10-15. Reactions of $(Ph_3P)_3CoH(N_2)$. N_2 and X exchange reversibly if X = CH_3CN, C_2H_4, or NH_3; CO displaces N_2 irreversibly.

the behavior of the organic soluble phosphine-hydride complexes may be regarded as a better model for the nitrogen complexation site of the nitrogenase enzyme system discussed in a later section of this chapter.

MOLYBDENUM DINITROGEN COMPLEXES

The complex trans-$(diphos)_2Mo(N_2)_2$ is interesting in several respects. First, molybdenum is a key constituent of nitrogenase, the enzyme which reduces nitrogen to ammonia. It is also unique in that it contains two dinitrogen ligands,[110] which have a N–N stretching frequency of 1975 cm^{-1}. Finally, it has the distinction of being the first well-defined dinitrogen complex in which the nitrogen could be reduced to form ammonia.[111] Perhaps the key to the success of this reduction was the use of a ferredoxin-like [4 Fe–4S]$^{4-}$ cluster complex as the reducing agent (Table 10-6). Nitrogenase contains similar Fe:S clusters, and utilizes reduced ferredoxin as an electron source.

TITANIUM DINITROGEN COMPLEXES

In hydrocarbon solvents at room temperature and reduced pressure, permethyltitanocene dihydride releases H_2 and forms a rapidly equilibrating mixture of $(C_5Me_5)_2Ti$ (10-18) and $(C_5Me_5)TiH(C_5Me_4CH_2)$ (10-19).[112] Addition of deuterium gas to this mixture results in the complete exchange of all the ligand hydrogens via the equilibria shown in Scheme 10-6. When solutions of

$$(C_5Me_5)_2TiD_2 \rightleftharpoons (C_5Me_5)_2Ti + D_2$$

$$(C_5Me_5)_2TiD_2 + (C_5Me_5)_2Ti \rightleftharpoons 2(C_5Me_5)TiD$$

$$(C_5Me_5)_2TiD + (C_5Me_5)TiH(C_5Me_4CH_2) \rightleftharpoons$$
$$(C_5Me_5)_2Ti + (C_5Me_5)TiH(C_5Me_4CH_2D)$$

Scheme 10-6

(10-18) and (10-19) are exposed to nitrogen, any or all of three different dinitrogen complexes are formed. The least soluble complex, (10-20) is a diamagnetic blue-black solid having the composition $[(C_5Me_5)_2Ti]_2N_2$. It is moderately soluble in hydrocarbon and ether solvents and slowly releases its nitrogen under reduced pressure. The lack of any ir absorption in the 1600–2300 cm^{-1} region suggests a symmetrical structure for the Ti_2N_2 unit.

On cooling toluene solutions of (10-20) below $-10°C$, there is a color change to intense purple-blue, the stoichiometry of which indicates the formation of a monomeric complex $(C_5Me_5)_2TiN_2$ (10-21). Proton and ^{13}C nmr studies indicate the existence of two rapidly interconverting forms of (10-21). The ^{15}N nmr spectrum of a $^{15}N_2$-enriched sample shows two sets of resonances, a pair of doublets, and a sharp singlet further downfield. The doublets were attributed to an "end-on" complex in which the nitrogens are nonequivalent and coupled ($J = 7$), while the singlet was attributed to a symmetrical "edge-on" complex. The ir spectrum of (10-21) shows a strong band at 2023 cm^{-1} and a band of medium intensity at 2056 cm^{-1}, which provides additional evidence for the existence of two isomers in solution. The small difference in wavelengths suggests that there is very little difference in energy between the end-on and edge-on modes of dinitrogen binding, if the edge-on mode is in fact involved in this case.

Protonation of (10-21) in toluene with excess HCl at $-80°C$ produces a quantitative yield of $(C_5Me_5)_2TiCl_2$ and in addition gives a mixture of hydrogen, nitrogen, *and hydrazine*, isolated as the dihydrochloride. Complex 10-21 reacts with HCl as shown here

$$(10\text{-}21) + 2\,HCl \quad
\begin{cases}
\xrightarrow{45\%} (C_5Me_5)_2\,TiCl_2 + N_2 + H_2 \\[2ex]
\xrightarrow{55\%} (C_5Me_5)_2\,TiCl_2 + \tfrac{1}{2}\,N_2 + \tfrac{1}{2}\,N_2H_4
\end{cases}$$

Since protonation of other end-on dinitrogen complexes leads only to quantitative evolution of dinitrogen, it may be that an edge-on isomer of (10-21) is the species which gives rise to the hydrazine. This reaction may bear a highly significant relationship to certain titanium-based systems which reduce nitrogen to ammonia.

Chemical Nitrogen Fixation

Dinitrogen is a very stable species which only reluctantly undergoes reduction ammonia, even though the reduction is thermodynamically favored. Considering that enzymatic catalysis is ordinarily much more facile than chemical catalysis, the enormous requirement of nitrogenase for ATP and for powerful reducing agents foretell the difficulty encountered in the chemical reduction of nitrogen. From one point of view the chemical systems for nitrogen fixation may be grouped into two types. In the commercial Haber–Bosch process which has been operated for more than 60 years, nitrogen and hydrogen are combined directly over an iron catalyst at high temperatures and pressures (450°C and 200–300 atm). Although extreme conditions are required, the process is relatively efficient in a thermodynamic sense as well as from a commercial point of view. In contrast to this stand the very recent systems for reduction of dinitrogen in homogeneous solution under mild conditions (1 atm, room temperature). These processes, like nitrogenase catalysis, are not very efficient thermodynamically because they consume large amounts of very powerful reducing agents. They are useful systems for elucidating the fundamental chemistry of nitrogenase and coordinated dinitrogen, but in their present state they offer little prospect for commercial competition with the Haber–Bosch process.

Some of the reasons for the difficulty in reducing dinitrogen, compared to the reduction of acetylene or oxygen, can be found in the bond strength data of Table 10-5. Perhaps the most important factors are the enormous strength of the $N{\equiv}N$ triple bond ($226 - 98 = 128$ kcal/mole, compared to 46 kcal/mole for $C{\equiv}C$) and the fact that the formation of partially reduced species such as diimide or hydrazine is highly endothermic.

TABLE 10-5

Bond Strengths and Heats of Formation for Some Carbon, Nitrogen, and Oxygen Compounds

		Bond dissociation energies (kcal/mole)			
$C\equiv C$	194	$N\equiv N$	226		
$C=C$	148	$N=N$	98	$O=O$	117
$C-C$	83	$N-N$	38	$O-O$	52
		Heat of formation (gas phase, kcal/mole)			
C_2H_2	+54				
C_2H_4	+12	N_2H_2	+44		
C_2H_6	−20	N_2H_4	+23	H_2O_2	−32
CH_4	−17	NH_3	−11	H_2O	−68

Some characteristics of several chemical systems which reduce dinitrogen are given in Table 10-6. Except for systems 8 and 9, all reactions involve aprotic solvents and extremely powerful reducing agents.

Probably the earliest observation of nitrogen fixation at mild temperatures and pressures was made by workers using the Ziegler–Natta type of olefin polymerization catalysts (e.g., $TiCl_4/Et_3Al$).[82] Dinitrogen often reduces the efficiency of these systems, and some ammonia can be detected upon hydrolysis of the reaction mixture. Similar systems were independently observed by Vol'pin (Table 10-6, system 2)[113] in Russia and van Tamelen (Table 10-6, systems 3–5)[114–116] in the United States. The heart of all these systems seems to be a soluble form of titanium in a low oxidation state (II or lower?) which absorbs and, in the presence of excess reducing agent, reduces dinitrogen to the equivalent of N_2^{4-} or $2 N^{3-}$. When a proton source is added, ammonia and/or hydrazine are formed in various ratios dependent upon the amount of excess reducing agent. The reduction can also be performed electrolytically, but nitrogen fixation is negligible until the Ti is reduced below the Ti(III) state. Ammonia yields of several hundred percent are obtained when $Al(OR)_3$ is added, presumably because it complexes the reduced nitrogen species such as nitride and keeps the Ti species free for further N_2 complexation and reduction. Titanium–dinitrogen complexes have not been *isolated* from systems 2–5, although an ir absorption at 1960 cm^{-1}, attributed to $(C_5H_5)_2TiN_2$, has been reported for system 5. Thus, systems such as 6 and 7,[112,117] in which unstable nitrogen complexes are both isolable *and* reducible, bridge an important gap between the nitrogen-fixing systems and the stable, structurally characterized dinitrogen complexes. The completion of X-ray, ESR, and Raman spectroscopic characterization of complexes such as (10-20) and (10-21) is thus of considerable importance and is urgently awaited.

Of the two systems for nitrogen fixation in aqueous solution (i.e., 8 and 9) relatively little information is available for system 9 involving the

TABLE 10-6

Characteristics of Chemical Systems Which Fix Nitrogen

Catalysts	Reducing agents[a]	Substrates	Products	Remarks	Reference
1. Porous iron	H_2	N_2	NH_3	Haber–Bosch process, 450°C, 200 atm	82, 83
2. Ti(IV), also Cr(III), Mo(VI), W(VI)	$LiAlH_4$, $EtMgBr$, or iBu_3Al	N_2, 1–100 atm	NH_3 (5–50%)	Via hydrolysis of metal nitrides	113
3. $(RO)_2TiCl_2$ or $Ti(OR)_4$	NaNp	N_2, air	NH_3	Cyclic generation of NH_3 by alternate addition of NaNp and ROH; via metal nitrides(?)	114
4. $Ti(OR)_4 + Al(OR)_3$	Nichrome cathode, Al anode	N_2	NH_3	Reduction by Ti species, $Al(OR)_3$ complexes nitrides formed until hydrolysis	115
5. $(C_5H_5)_2TiCl_2$	NaNp	N_2	NH_3	$(C_5H_5)_2TiN_2$ forms and is quantitatively reduced to nitride or ammonia	116
6. $(C_5Me_5)_2TiH_2$	(Ti species)	N_2, H^+	N_2H_4	Low temperature protonation of $(C_5Me_5)_2TiN_2$ gives N_2H_4 along with H_2 and $(C_5Me_5)_2TiCl_2$	112
7. $(C_5H_5)_2TiR$; R = Ph, alkyl	NaNp	N_2	NH_3, N_2H_4	Probably similar to 5 and 6; reduction of $[(C_5H_5)_2TiR]_2N_2$, which can be isolated	117
8. meso-tetra(p-sulfophenyl)-porphinato Co(III), also Mo analog	aq. $NaBH_4$	H_2C_2, N_3^-, CN^-, MeCN	C_2H_4, NH_3, others NH_3	Forms 0.8–1.0 mole NH_3 per mole Co, others reduced catalytically	118
9. Mo(VI) + thiols, $Na_2MO_2O_4(cys)_2$	aq. $NaBH_4$ or $Na_2S_2O_4$	N_2, N_2H_4, CN^-, N_3^-, N_2O, C_2H_2	$[N_2H_2]$, NH_3 CH_4, NH_3 $N_2 + NH_3$ $N_2 + H_2O$ C_2H_4	Activated by ATP at pH 10, catalyzes $D_2 + H^+ \rightarrow HD + D^+$	119, 119a
10. $(EtS)_4Fe_4S_4^{4-}$	NaNp	$(diphos)_2Mo(N_2)_2$ RCN RNC $RC\equiv CH$	NH_3 — RNH_2 $RCHCH_2$	Up to 0.3 mole NH_3 per mole Mo complex upon hydrolysis; Fe/Mo pair also reduces RNC and RCCH	111

cobalt–porphyrin complex.[118] The "molybdothiol" system (8) has been extensively studied by Schrauzer and co-workers[119] and found to mimic many of the features of the molybdenum containing enzyme nitrogenase including D_2/H^+ exchange, reduction of many of the nonphysiological substrates reduced by nitrogenase (C_2H_2, N_2O, N_3^-, N_2), and utilization of $Na_2S_2O_4$ or Ferredoxin-like reducing agents. Most remarkable is the fact that this system is *stimulated by the addition of ATP*, and to a lesser extent ADP. Although the ATP is hydrolyzed, this step is not per se coupled to nitrogen reduction.

Biological Nitrogen Fixation

The reduction of atmospheric nitrogen to ammonia, a process of tremendous ecological importance, is carried out by several diverse groups of organisms. Certain free-living soil bacteria, both aerobic (e.g., *Azotobacter vinelandii*) and anaerobic (e.g., *Clostridium pasteurianum*) carry out nitrogen fixation. These species are not particularly important in terms of enriching the content of fixed nitrogen in the soils, but they have contributed greatly to our knowledge of the biochemistry of nitrogen fixation. Agriculturally more important are the various species of *Rhizobium*, the symbiotic bacteria found in root nodules of various leguminous plants. These bacteria provide the host plant with NH_3 and in return receive carbohydrates which they are incapable of manufacturing. Certain photosynthetic blue-green algae, as well as lichens containing these algae, also reduce nitrogen to ammonia.

Several factors operate to control the nitrogen fixing activity of these organisms. Nitrogen fixation requires large quantities of ATP, usually obtained by the oxidative metabolism of carbohydrates. The enzymes of nitrogen fixation are highly sensitive to oxygen and must function in an anaerobic compartment of the cell, even in the aerobic organisms like *Azotobacter* or the blue-green algae. Such conditions might be maintained locally by the exaggerated respiratory rate of these organisms. Finally, the end product ammonia controls nitrogenase activity, not by inhibiting the enzyme itself but by repressing, directly or indirectly, the expression of the gene which codes for synthesis of the nitrogenase proteins.

Cell-free preparations active in nitrogen fixation, chiefly from *Azotobacter* and *Clostridium*, have been known since 1960 when it was discovered that the protein components of the system were irreversibly sensitive to oxygen.[120] Biochemical fractionation of these extracts yielded a Mo–Fe protein and a Fe protein which recombined[121] to give an active nitrogenase enzyme:

$$\text{Mo–Fe protein} + 2 \text{ Fe protein} \rightleftharpoons \text{nitrogenase}$$

The Mo–Fe protein is a very large complex, having a molecular weight up to

270,000 daltons and containing Mo, Fe, S^{2-}, and Cys in the ratios (roughly, and depending on the source) of 1–2: 15–30: 10–20: 15–30 per 200,000–270,000 daltons.[83,84] Electron spin resonance and Mössbauer studies indicate that the majority of the iron is in the high spin ferric form, easily reducible with $Na_2S_2O_4$. The Fe protein is smaller (55,000) and contains 4 Fe and 4 S^{2-} per molecule. Unless it is denatured by storage at low temperatures, only two of the four Fe are complexed by dipyridyl, suggesting that there are two types of iron-binding sites. The low temperature esr spectrum of the Fe protein indicates that it resembles the ferredoxin-type nonheme iron proteins.

In addition, to the Mo–Fe protein and the Fe protein, reduction of nitrogen requires about 12–14 molecules of ATP per mole of dinitrogen reduced, and a source of reducing power. The reducing equivalents come from oxidation of carbohydrates such as pyruvate and are carried to the nitrogenase complex by ferredoxin or flavodoxin proteins.[122] Ferredoxins are among the most powerful reducing agents known in biochemical systems (Chapter VII). *In vitro* $Na_2S_2O_4$ and KBH_4 can substitute for pyruvate as source of reducing power. Various functions have been suggested for the ATP, including electron activation, maintenance of an anhydrous active site through its hydrolysis, and the induction of essential changes in protein conformation.

All the reactions catalyzed by nitrogenase are reductions. In the absence of dinitrogen or other artificial substrates active nitrogenase preparations reduce H_3O^+ and evolve H_2 at a constant rate. As dinitrogen is added to the system the rate of hydrogen evolution decreases and ammonia production ensues, the rate of electron utilization remaining constant. Nitrogenase also catalyzes the N_2-dependent formation of HD from D_2 and H_3O^+, although the exchange reaction

$$^{15}N_2 + {}^{14}N_2 \rightleftharpoons 2\,^{15}N{\equiv}N^{14}$$

has not been detected. Hydrogen evolution is not inhibited by CO or H_2, while the reduction of dinitrogen and other substrates is. This implies that electron activation and nitrogen reduction occur at two different sites in the enzyme complex. Nitrogenase also reduces many other small unsaturated molecules, consuming variable numbers of electrons in the process (Table 10-7). The reduction of acetylene is stereospecific, *cis*-HDC=CHD being formed cleanly from both H_2C_2 in D_2O and D_2C_2 in H_2O. This is thought to imply that a metal acetylide (i.e., M—C≡CH, which would be isoelectronic and isostructural with known dinitrogen complexes) is *not* an intermediate in acetylene reduction. However, EtC≡CH is reduced to 1-butene, while $CH_3C{\equiv}CCH_3$, which has no acetylenic proton and cannot form an acetylide complex, is not reduced. A possible chemical parallel is Wilkinson's catalyst, $RhCl(Ph_3P)_3$, which also catalyzes the cis reduction of acetylenes, apparently without going through an acetylide intermediate (Chapter XI). A convenient assay of

TABLE 10-7

Nitrogenase Substrates and Products[a]

Substrates	Products	No. of electrons
H_3O^+	H_2	2
N_2	$2 NH_3$	6
N_2O	N_2, H_2O	2
N_3	N_2, H_2O	2
C_2H_2	C_2H_4	2
HCN	CH_3NH_2	4
	$CH_4 + NH_3$	6
CH_3CN	$C_2H_6 + NH_3$	6
$CH_2=CHCN$	$C_3H_6 + NH_3$	6
	C_3H_8	8
CH_3NC	$CH_3NH_2 + CH_4$	6
	C_2H_4, C_2H_6	8, 10
H_2	Competitive inhibition of N_2 reduction	
CO, NO, N_2H_4	Strong inhibition of N_2 reduction	

[a] Data from Hardy et al.[83,88]

nitrogenase, free from interference by endogenous NH_3, and not requiring the use of $^{15}N_2$, is based upon gas chromatographic monitoring of the acetylene to ethylene reduction.[121]

The reduction of CH_3NC to ethylene by nitrogenase is remarkable in that a two-carbon chain is formed from a precursor having two unlinked carbons. Since ammonia is apparently not a product of this reduction, while methylamine is, an insertion mechanism has been proposed for the formation of C_2 fragments (Scheme 10-7). Such a mechanism would not be inconsistent with the observed lack of $^{14}N_2-^{15}N_2$ isotopic scrambling by nitrogenase.

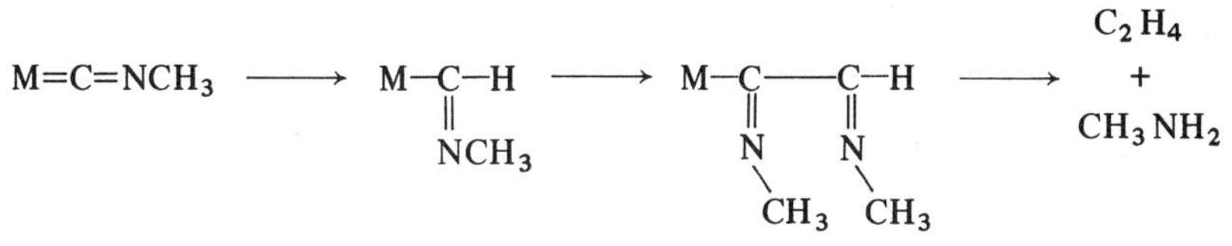

$$M=C=NCH_3 \longrightarrow \underset{NCH_3}{M-\overset{\|}{C}-H} \longrightarrow \underset{\underset{CH_3}{\overset{\backslash}{N}}\quad\underset{CH_3}{\overset{\backslash}{N}}}{M-\overset{\|}{\underset{}{C}}\overline{\quad\quad}\overset{\|}{\underset{}{C}}-H} \longrightarrow \begin{array}{c} C_2H_4 \\ + \\ CH_3NH_2 \end{array}$$

Scheme 10-7

Neither hydrazine nor diimide (HN=NH) can be detected as intermediates in the nitrogenase reaction, nor does nitrogenase reduce exogenous hydrazine of diimide. However, it is possible that equivalent "enzyme bound" species are involved. There is no direct evidence to indicate that an enzyme–metal–dinitrogen complex is the active species for reduction, but a large body of indirect evidence would suggest that this is the case, e.g., the occurrence of metals which are known to form dinitrogen complexes in both nitrogenase proteins, and the fact that the substrates and inhibitors of nitrogenase are good metal binding agents but interact poorly if at all with simple proteins. Probably the most difficult aspect of nitrogenase action to understand is the coupling of ATP hydrolysis to the redox processes. Of course, the same situation holds for the *synthesis* of ATP via oxidative phosphorylation and the electron transport cytochrome chain which "fixes" dioxygen to water. It is entirely possible that the fundamental chemistry of these two systems is related, even though one involves ATP synthesis while the other involves its hydrolysis.

There have been a number of proposals for the way in which dinitrogen is actually reduced and cleaved to give ammonia. The stepwise reduction of nitrogen atoms via nitride intermediates, e.g.,

$$M \equiv N + 3\,H^+ \longrightarrow NH_3 + M^{3+}$$

is generally disfavored because of the large energy requirement for this type of reduction, and because inorganic systems which form nitrides (e.g., $TiCl_4/Et_3Al$) also interact strongly with ethylene. Several other proposals involve bimetallic reduction sites and the intermediacy of metal hydrides as reducing agents. One example is shown in Fig. 10-16.[124] There are several

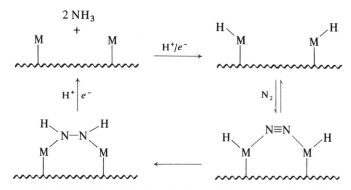

Fig. 10-16. Hypothetical mechanism of nitrogenase action.[83,124]

possible chemical precedents for these proposals. Two involve the reduction of the benzene diazonium cation (an organic analog of a $M-N \equiv N$ complex) with hydrogen and platinum or palladium catalysts (Fig. 10-17).[125-126] A third is the formation of hydrazine by low temperature protonation of a nitrogen complex of "permethyltitanocene" discussed earlier.[112] However, much more work with both the enzyme and model systems will be required to clarify the mechanism of nitrogenase action.

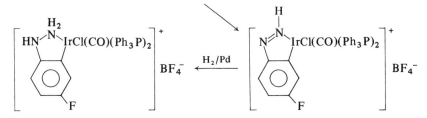

$[p\text{-}FC_6H_4N\equiv N]BF_4 + H\text{-}\underset{\underset{PEt_3}{|}}{\overset{\overset{PEt_3}{|}}{Pt}}\text{-}Cl \longrightarrow \left[p\text{-}FC_6H_4\text{-}N=N\text{-}\underset{\underset{PEt_3}{|}}{\overset{\overset{H\ \ PEt_3}{|}}{Pt}}\text{-}Cl \right]^+ BF_4^-$

Pt/H₂ or
Na₂S₂O₄ or
electrolysis
at −0.4 V

$p\text{-}FC_6H_4NH_3 + NH_3$

$[p\text{-}FC_6H_4N\equiv N]BF_4 + IrCl(CO)(Ph_3P)_2$

H₂/Pd

Fig. 10-17. Mixed organic–inorganic models for nitrogenase action.[125,126]

REFERENCES

1. G. Henrici-Olive and S. Olive, *Angew. Chem., Int. Ed. Engl.* **13**, 29 (1974).
2. L. Klevan, J. Peone, and S. K. Madan, *J. Chem. Educ.* **50**, 670 (1973).
3. S. Fallab, *Angew. Chem., Int. Ed. Engl.* **6**, 496 (1967).
4. O. Hayaishi, ed., "Molecular Mechanisms of Oxygen Activation." Academic Press, New York, 1974.
5. M. M. Taqui Khan and A. E. Martell, "Homogeneous Catalysis by Metal Complexes," Vol. I. Academic Press, New York, 1974.
6. A. E. Martell and M. M. Taqui Khan, *in* "Inorganic Biochemistry" (G. Eichhorn, ed.) Vol. 2, p. 654. Elsevier, Amsterdam, 1973.
7. J. S. Valentine, *Chem. Rev.* **73**, 235 (1973).
8. V. J. Choy and C. J. O'Connor, *Coord. Chem. Rev.* **9**, 145 (1972/1973).
9. R. G. Wilkins, *Advan. Chem. Ser.* **100**, 111 (1971).
10. E. Bayer, P. Krauss, A. Röder, and P. Schretzmann, *in* "Oxidases and Related Redox Systems" (T. E. King, H. S. Mason, and M. Morrison, eds.), 2nd ed., Vol. 1, p. 227. Univ. Park Press, Baltimore, Maryland, 1973.
11. A. G. Sykes and J. A. Weil, *Progr. Inorg. Chem.* **13**, 1 (1971).
11a. F. Basolo, B. M. Hoffman, and J. A. Ibers, *Accounts Chem. Res.* **8**, 384 (1975).

12. S. C. Abrahams, *Quart. Rev., Chem. Soc.* **10**, 407 (1956).
13. S. C. Abrahams, R. L. Collin, and W. N. Lipscomb, *Acta Crystallogr.* **4**, 15 (1951.
14. G. A. Rodley and W. T. Robinson, *Nature (London)* **235**, 438 (1972).
15. M. Calligaris, G. Nardin, L. Randaccio, and G. Tauzher, *Inorg. Nucl. Chem. Lett.* **9**, 419 (1973).
16. W. P. Schaefer and R. E. Marsh, *Acta Crystallogr.* **16**, 247 (1963).
17. W. P. Schaefer, *Inorg. Chem.* **7**, 729 (1968).
18. M. Calligario, G. Nardin, L. Randaccio, and A. Ripamonti, *J. Chem. Soc. A.* p. 1069 (1970).
19. C. G. Cristoph, R. E. Marsh, and W. P. Schaefer, *Inorg. Chem.* **8**, 291 (1969).
20. B.-C. Wang and W. P. Schaefer, *Science* **166**, 1404 (1969).
21. R. Nakon and A. E. Martell, *J. Amer. Chem. Soc.* **94**, 3026 (1972).
22. J. Simplicio and R. G. Wilkins, *J. Amer. Chem. Soc.* **91**, 1325 (1969).
23. C. Floriani and F. Calderazzo, *J. Chem. Soc., A* p. 946 (1969).
24. A. L. Crumbliss and F. Basolo, *J. Amer. Chem. Soc.* **92**, 55 (1970).
25. B. M. Hoffman, D. L. Diemente, and F. Basolo, *J. Amer. Chem. Soc.* **92**, 61 (1970).
26. D. Diemente, B. M. Hoffman, and F. Basolo, *Chem. Commun.* p. 467 (1970).
27. J. W. Lauher and J. E. Lester, *Inorg. Chem.* **12**, 244 (1973).
28. E. Melamud, B. L. Silver, and Z. Dori, *J. Amer. Chem. Soc.* **96**, 4689 (1974).
29. B. S. Tovrog and R. S. Drago, *J. Amer. Chem. Soc.* **96**, 6765 (1974).
29a. B. M. Hoffman, T. Szymanski, and F. Basolo, *J. Amer. Chem. Soc.* **97**, 673 (1975).
30. C. J. Willis, *Chem. Commun.* p. 117 (1974).
31. H. C. Stynes and J. A. Ibers, *J. Amer. Chem. Soc.* **94**, 15 and 2125 (1972).
32. M. J. Carter, P. D. Rillema, and F. Basolo, *J. Amer. Chem. Soc.* **96**, 392 (1974).
33. E. W. Abel, J. M. Pratt, R. Whelan, and P. J. Wilkinson, *J. Amer. Chem. Soc.* **96**, 7119 (1974).
34. A. Nishinaga, T. Tojo, and T. Matsumra, *Chem. Commun.* p. 896 (1974).
35. A. Nishinaga, K. Watanabe, and T. Matsuura, *Tetrahedron Lett.* p. 1291 (1974).
36. L. Vaska, *Science* **140**, 809 (1963).
37. R. Mason, *Nature (London)* **217**, 543 (1968).
38. J. A. McGinnety, R. J. Doedens, and J. A. Ibers, *Inorg. Chem.* **6**, 2243 (1967).
39. J. A. McGinnety and J. A. Ibers, *Chem. Commun.* p. 235 (1968).
40. M. J. S. Dewar, *Bull. Soc. Chim. Fr.* **18**, C71 (1951).
41. J. Chatt and L. A. Duncanson, *J. Chem. Soc., London* p. 2939 (1953).
42. L. Vaska, L. S. Chen, and W. V. Miller, *J. Amer. Chem. Soc.* **93**, 6671 (1971).
43. L. Vaska, L. S. Chen, and C. V. Senoff, *Science* **174**, 587 (1971).
44. L. Vaska and L. S. Chen, *Chem. Commun.* p. 1080 (1971).
45. E. E. Mercer, W. M. Peterson, and B. F. Jordan, *J. Inorg. Nuc. Chem.* **34**, 3290 (1972).
46. N. W. Terry, E. L. Amma, and L. Vaska, *J. Amer. Chem. Soc.* **94**, 652 (1972).
47. R. W. Horn, E. Weissberger, and J. P. Collman, *Inorg. Chem.* **10**, 219 (1971).
48. L. Vaska and D. L. Catone, *J. Amer. Chem. Soc.* **88**, 532 (1966).
49. J. A. McGinnety, N. C. Payne, and J. A. Ibers, *J. Amer. Chem. Soc.* **91**, 6301 (1969).
49a. M. J. Nolte, E. Singleton and M. Laing, *J. Amer. Chem. Soc.* **97**, 6396 (1975).
50. S. J. Laplaca and J. A. Ibers, *J. Amer. Chem. Soc.* **87**, 2581 (1965).
51. J. P. Collman and W. R. Roper, *Advan. Organometal. Chem.* **7**, 63 (1968).
52. M. S. Weininger, I. F. Taylor, and E. L. Amma, *Chem. Commun.* p. 1172 (1971).
53. B. W. Graham, K. R. Laing, C. J. O'Connor, and W. R. Roper, *Chem. Commun.* p. 1272 (1970).
54. P.-T. Cheng, C. D. Cook, S. C. Nyburg, and W. Y. Wan, *Can. J. Chem.* **49**, 3772 (1971).
55. G. Wilke, H. Schott, and P. Heimbach, *Angew. Chem. Int. Ed. Engl.* **6**, 92 (1967).
56. S. Otsuka, A. Nakamura, and Y. Tatsuno, *J. Amer. Chem. Soc.* **91**, 6994 (1969).

57. P. J. Hayward, D. M. Blake, C. J. Nyman, and G. Wilkinson, *Chem. Commun.* p. 987 (1969).

58. R. Ugo, F. Conti, S. Cenini, R. Mason, and G. B. Robertson, *Chem. Commun.* p. 1498 (1968).

59. J. P. Collman, *Accounts Chem. Res.* **1**, 136 (1968).

60. J. Halpern and A. L. Pickard, *Inorg. Chem.* **9**, 2798 (1970).

61. H. vanGaal, H. Cuppers, and A. vander Ent, *Chem. Commun.* p. 1694 (1970).

62. J. E. Lyons and J. O. Turner, *J. Org. Chem.* **37**, 2881 (1972).

63. B. R. James and E. O. Chaiai, *Can. J. Chem.* **49**, 975 (1971).

64. C. Dudley and G. Read, *Tetrahedron Lett.* p. 5273 (1972).

65. H. B. Gray, *Advan. Chem. Ser.* **100**, 365 (1971); also Tagui Kahn and Martell[5] and Bayer *et al.*[10]

66. M. Y. Okamura and I. M. Klotz, *in* "Inorganic Biochemistry" (G. Eichhorn, ed.), Vol. 1, p. 320. Elsevier, Amsterdam, 1973.

67. R. Lontic and R. Witters, *in* "Inorganic Biochemistry" (G. Eichhorn, ed.), Vol. 1, p. 344. Elsevier, Amsterdam, 1973.

68. J. M. Rifkind, *in* "Inorganic Biochemistry" (G. Eichhorn, ed.), Vol. 2, p. 832. Elsevier, Amsterdam, 1973.

69. M. F. Perutz, H. Muirhead, J. M. Cox, and L. C. G. Goaman, *Nature (London)* **219**, 131 (1968).

70. J. C. Kendrew, R. E. Dickerson, B. E. Strandberg, R. G. Hart, D. R. Davies, D. C. Phillips, and V. C. Shore, *Nature (London)* **185**, 422 (1960).

71. E. B. Fleischer, *Accounts Chem. Res.* **3**, 105 (1970).

72. J. L. Hoard, *Science* **174**, 1295 (1971).

73. M. F. Perutz, *Nature (London)* **228**, 726 (1970).

74. J. S. Griffith, *Proc. Roy. Soc., Ser. A* **235**, 23 (1956).

75. L. Pauling, *Nature (London)* **203**, 182 (1964).

76. B. M. Hoffman and D. H. Petering, *Proc. Nat. Acad. Sci. U.S.* **67**, 637 (1970); **69**, 2122 (1972).

77. J. H. Wang, *Accounts Chem. Res.* **3**, 90 (1970).

78. J. Almog, J. E. Baldwin, and J. Huff, *J. Amer. Chem. Soc.* **97**, 227 (1975); J. Geibel, C. K. Chang, and T. G. Traylor, *J. Amer. Chem. Soc.* **97**, 5924 (1975).

78a. J. P. Collman, R. R. Gagne, T. R. Halbert, J.-C. Marchon, and C. A. Reed, *J. Amer. Chem. Soc.*, **95**, 7868 (1973).

79. J. P. Collman, R. R. Gagne, C. A. Reed, W. T. Robinson, and G. A. Rodley, *Proc. Nat. Acad. Sci. U.S.* **71**, 1326 (1974).

80. J. P. Collman, R. R. Gagne, and C. A. Reed, *J. Amer. Chem. Soc.* **96**, 2629 (1974).

81. J. P. Collman, R. R. Gagne, J. Kouba, and H. Ljusberg-Wahren, *J. Amer. Chem. Soc.* **96**, 6800 (1974).

82. J. Chatt, *Chem. Soc. Rev.* **1**, 121 (1972).

83. R. W. F. Hardy, R. C. Burns, and G. W. Parshall, *in* "Inorganic Biochemistry" (G. Eichhorn, ed.), Vol. 2, p. 745. Elsevier, Amsterdam, 1973.

84. H. Dalton and L. E. Mortenson, *Bacteriol. Rev.* **36**, 231 (1972).

85. A. D. Allen and F. Bottomley, *Accounts Chem. Res.* **1**, 360 (1968).

86. E. E. van Tamelen, *Accounts Chem. Res.* **3**, 361 (1970).

87. A. D. Allen, R. O. Harris, B. R. Loescher, J. R. Stevens, and R. N. Whiteley, *Chem. Rev.* **73**, 11 (1973).

88. R. W. F. Hardy, R. C. Burns, and G. W. Parshall, *Advan. Chem. Ser.* **100**, 219 (1971).

89. A. D. Allen and C. V. Senoff, *Chem. Commun.* p. 621 (1965).

90. D. E. Harrison and H. Taube, *J. Amer. Chem. Soc.* **89**, 5706 (1967).

91. J. N. Armour and H. Taube, *J. Amer. Chem. Soc.* **93**, 6476 (1971).

92. L. A. P. Kane-Maguire, P. S. Sheridan, F. Basolo, and R. G. Pearson, *J. Amer. Chem. Soc.* **92**, 5865 (1970).
93. S. Pell and J. N. Armor, *J. Amer. Chem. Soc.* **94**, 686 (1972).
94. A. Misono, Y. Uchida, and T. Saito, *Bull. Chem. Soc. Japan* **40**, 799 (1967).
95. A. Yamamoto, M. Ookawa, and S. Skeda, *Chem. Commun.* p. 841 (1969).
96. M. Hidai, K. Tominari, Y. Uchida, and A. Misono, *Chem. Commun.* p. 814 (1969).
97. B. Bell, J. Chatt, and G. J. Leigh, *Chem. Commun.* p. 842 (1969).
98. B. P. Stoichef, *Can. J. Phys.* **82**, 630 (1954).
99. F. Bottomley and S. C. Nyburg, *Acta Crystallogr., Sect. B* **24**, 1289 (1968).
100. I. M. Treitel, M. T. Flood, R. E. March, and H. B. Gray, *J. Amer. Chem. Soc.* **91**, 6512 (1969).
101. J. Chatt, G. J. Leigh, and R. Richards, *Chem. Commun.* p. 515 (1969).
102. J. N. Armor and H. Taube, *J. Amer. Chem. Soc.* **92**, 2560 (1970).
103. J. E. Fergusson, J. L. Love, and W. T. Robinson, *Inorg. Chem.* **11**, 1662 (1972).
104. B. R. Davis, N. C. Payne, and J. A. Ibers, *J. Amer. Chem. Soc.* **91**, 1240 (1969).
105. C. M. Elson, J. Gulens, I. J. Itzkovitch, and J. A. Page, *Chem. Commun.* p. 875 (1970).
106. I. J. Itzkovitch and J. A. Page, *Can. J. Chem.* **46**, 2743 (1968).
107. C. M. Elson, J. Gulens, and J. A. Page, *Can. J. Chem.* **49**, 207 (1971).
108. G. W. Parshall, *J. Amer. Chem. Soc.* **91**, 1669 (1968).
109. A. Sacco and M. Aresta, *Chem. Commun.* p. 1223 (1968).
110. M. Hidai, K. Tominari, and Y. Uchida, *J. Amer. Chem. Soc.* **94**, 110 (1972).
111. E. E. van Tamelen, J. A. Gladysz, and C. Brulet, *J. Amer. Chem. Soc.* **96**, 3020 (1974).
112. J. E. Bercaw, *J. Amer. Chem. Soc.* **96**, 5087 (1974).
113. M. E. Vol'pin and V. B. Shur, *Nature (London)* **290**, 1236 (1966).
114. E. E. van Tamelen, G. Boche, and R. H. Greeley, *J. Amer. Chem. Soc.* **90**, 1674 (1968).
115. E. E. van Tamelen and D. G. Seeley, *J. Amer. Chem. Soc.* **91**, 5194 (1969).
116. E. E. van Tamelen, R. B. Fechter, S. W. Schneller, G. Boche, R. H. Greeley, and B. Åkermark, *J. Amer. Chem. Soc.* **91**, 1551 (1969).
117. J. H. Teuben and H. J. de Liefde Meijer, *Rec. Trav. Chim. Pays-Bas* **90**, 360 (1971).
118. E. B. Fleischer and M. Krishnamurthy, *J. Amer. Chem. Soc.* **94**, 1382 (1972).
119. G. N. Schrauzer, G. W. Kiefer, K. Tano, and P. A. Doemeny, *J. Amer. Chem. Soc.* **96**, 641 (1974).
119a. G. N. Schrauzer, G. W. Kiefer, P. A. Doemeny, and H. Kisch, *J. Amer. Chem. Soc.* **95**, 5582 (1973).
120. J. E. Carnahan, L. E. Mortensen, H. F. Mower, and J. E. Castle, *Biochim. Biophys. Acta* **44**, 520 (1960).
121. R. W. F. Hardy, R. D. Holsten, E. K. Jackson, and R. C. Burns, *Plant Physiol.* **43**, 1185 (1968).
122. L. E. Mortensen, *Biochim. Biophys. Acta* **81**, 473 (1964).
123. E. K. Jackson, G. W. Parshall, and R. W. F. Hardy, *J. Biol. Chem.* **243**, 4952 (1968).
124. A. Shilov, *Kinet. Katal.* **11**, 256 (1970).
125. G. W. Parshall, *J. Amer. Chem. Soc.* **89**, 1822 (1967).
126. A. B. Gilchrist, G. W. Rayner-Canham, and D. Sutton, *Nature (London)* **235**, 44 (1972).

XI

Organometallic Complexes: Structure, Bonding, and Reaction Mechanisms

The term "organometallic" has unfortunately come to acquire several different connotations. It is sometimes used to refer to organic derivatives of the "a" type metals such as lithium, beryllium, magnesium, or aluminum. In other cases, it refers to organic derivatives of metalloids or even nonmetals such as arsenic, germanium, silicon, or phosphorus. For our purposes we will adhere to a third and perhaps the most common connotation for the term "organometallic," i.e., that of derivatives or complexes of the transition metal elements in which at least some of the ligands are carbon-donor species, either as neutral organic molecules (e.g., C_2H_4) or fragments thereof (e.g., CH_3, CH_2). In general, "organometallic complexes" may be recognized by the low (in some cases zero or negative) formal oxidation states of the central metal, as well as by the occurrence of hydride, phosphorus and carbon donors, i.e., soft ligands. In contrast, the "coordination complexes" discussed in the earlier chapters generally involve metals in their higher oxidation states (2+ to 4+) surrounded by halide, oxygen or nitrogen donors, i.e., hard ligands.

In this discussion of organometallic chemistry we will consider first the nature of some types of carbon-donor ligands and their bonding to transition metals, the nature of basic organometallic reaction mechanisms (which should be compared to discussions in Chapter VI), and, finally, some reactions of organometallic reagents or catalysts of practical importance (Chapter XII).

Ligands in Organometallic Chemistry

Although ligands with O, N, S, or P donor atoms are important in organometallic complexes, carbon donor ligands and hydrogen are, in fact, the *raison d'être* for much of organometallic chemistry. For this reason carbon donor ligands and hydrogen are singled out for discussion here; the others have been discussed in Chapter VI.

Displacement

$$M-X + LiAlH_4 \longrightarrow M-H + X^-$$

Oxidative addition

$$M + H_2 \longrightarrow H-M-H$$
$$M + HX \longrightarrow X-M-H$$

Protonation

$$M + HX \longrightarrow H-M^+ + X^-$$

Deinsertion

$$M-CH_2CH_2-H \longrightarrow M-H + C_2H_4$$
$$M-O_2CH \longrightarrow M-H + CO_2$$
$$M-O-CH_3 \longrightarrow M-H + CH_2O$$

$$\overset{\displaystyle \overset{L}{|}}{\underset{\displaystyle \underset{L}{|}}{R-Pt-Cl}} + Py \underset{}{\overset{EtOH}{\rightleftharpoons}} \overset{\displaystyle \overset{L}{|}}{\underset{\displaystyle \underset{L}{|}}{R-Pt^+-Py}} + Cl^-$$

$$k_{obs} = k_1 + k_2[Py]$$ Py = pyridine, L = PEt$_3$

R	Temp (°C)	k_1 (sec^{-1})	k_2 (M^{-1} sec^{-1})[a]
-H	0	1.8×10^{-2}	4.2
-CH$_3$	25	1.7×10^{-4}	6.7×10^{-2}
-C$_6$H$_5$	25	3.3×10^{-5}	1.6×10^{-2}
-Cl	25	1.0×10^{-6}	4.0×10^{-4}

[a] Data from Basolo et al.[3]

Scheme 11-1

TRANSITION METAL HYDRIDES

As a substituent or ligand nothing could be simpler than a hydrogen atom.[1,2] When bonded directly to a transition metal hydrogen is usually regarded as hydride (H^-) rather than H or H^+, although transition metal hydrides can be prepared by protonation of an electron-rich metal center as well as by hydride displacement of another ligand such as halide.

Hydride is a very strong ligand with nearly pure σ donor properties, and the presence of a hydride ligand weakens the bonding of other ligands in the complex, as can be seen from the kinetic data in Scheme 11-1.[3] The protonation of metal carbonyl anions (pK_a's range from < 0 to 14 or higher) leads to enormous increases in the rate of CO exchange via a rate limiting dissociative process, and a shift of ca. $+150$ cm^{-1} for ν_{CO} in the ir spectrum. The latter results from the tightening of the C—O bond with weakening of the M—C bond.

TRANSITION METAL ALKYLS

The synthesis of stable or isolable transition metal alkyls frustrated chemists' attempts for nearly a century before the late 1960's when the factors contributing to the stability of the M—C bond became sufficiently understood to allow for the now routine synthesis of this class of compounds. An interesting historical account has been given by Wilkinson,[4] who in 1973 shared the Nobel Prize in chemistry with E. O. Fischer for contributions to organometallic chemistry.

Relatively little is known of the *thermodynamic* stability of the metal-alkyl bond, but several factors are now clearly recognized to affect its *kinetic* stability.[5,6] According to one view, homolytic rupture of the metal-alkyl bond may depend on promotion of an electron from a σ-bonding orbital to an e_g type antibonding orbital [e.g., Co(III) alkyls are notoriously sensitive to visible light]. Thus it is observed that the metal-alkyl unit is stabilized by the presence of strong σ donors and/or back-bonding ligands which increase the energy separation Δ between the bonding and antibonding orbitals (cf. Figs. 5-16 and 5-17); transition metal alkyls also become more stable, and Δ increases, with increasing atomic number of the metal, e.g., Pt $\gg$ Pd $>$ Ni.

One of the first approaches to the synthesis of transition metal alkyls involved displacing a coordinated halide with a nucleophilic alkyllithium or Grignard reagent (reaction 11-1). In many cases reduction of the metal occurred at the expense of oxidative coupling of the alkyl units. However, in

$$M—X + RLi \underset{\searrow}{\overset{\nearrow}{\bigg\langle}} \begin{array}{l} \tfrac{1}{2}\,R—R + M + LiX \\ \\ R—M + LiX \end{array} \qquad (11\text{-}1)$$

many of those cases in which alkyl derivatives were formed, it was quickly noticed that methyl derivatives were more stable than the higher n-alkyls and that secondary or tertiary alkyl complexes were extremely susceptible to decomposition to an α-olefin and a metal hydride. This "β-elimination" phenomenon, sometimes now used as a preparative method for metal hydrides, depends on the number of β-hydrogens available for elimination. Complexes with alkyl ligands having no β-hydrogens are kinetically much more stable than those which have β-hydrogens. Furthermore, alkyls having β-hydrogens and an electron-withdrawing β substituent are even more prone to eliminate M—H and form an olefin. Perfluoroalkyl ligands, on the other hand, are very stable because of their lack of β-hydrogens, and possibly because of the greater electronegativity of perfluoroalkyl vs alkyl groups.

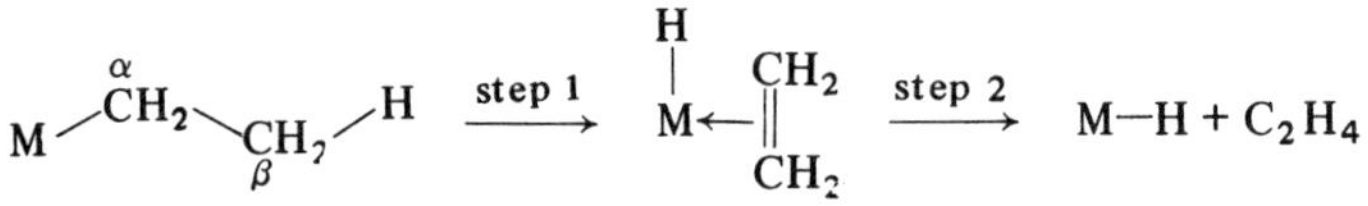

$$M \overset{\alpha}{\underset{\beta}{\diagup}} CH_2 \diagdown CH_2 \diagup H \quad \xrightarrow{\text{step 1}} \quad M \leftarrow \| \overset{H}{\underset{CH_2}{|}} CH_2 \quad \xrightarrow{\text{step 2}} \quad M-H + C_2H_4$$

M—R "stability"	Alkyl group R	
│	$-(CF_2)_nF$	$n \geqslant 1$
│	$-CH_2-X$	No β-hydrogens; X = H, Ph, OR, NR_2, CR_3, SiR_3
Decreasing	$-CH_2CH_3, -OCH_3$	
│	$-CH(CH_3)_2$	
↓	$-C(CH_3)_3, -CH_2CH_2CO_2R, -CH_2CH_2CN$	

Scheme 11-2

The nature of the metal may influence the tendency toward β-elimination in one very important way. If the transition metal alkyl has an 18-electron configuration with no easily dissociable ligands, the deinsertion (q.v.) of the olefin (step 1, Scheme 11-2) will not be facile, indeed may not occur at all, since the hydridometal-olefin complex would have 20 electrons and would therefore be a high energy intermediate. The recent preparation of a stable t-butyl complex

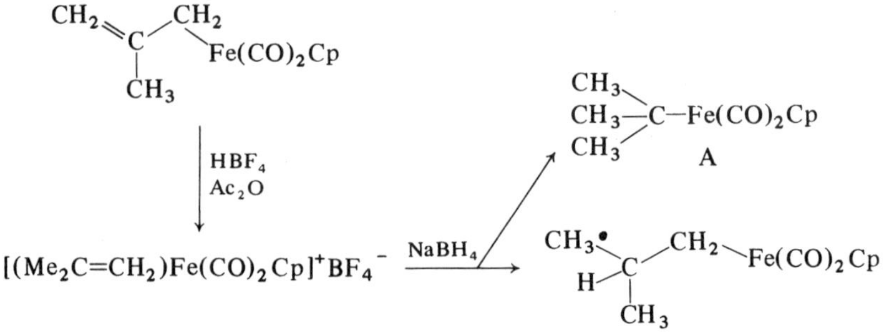

$$CH_2 \overset{}{\diagdown} \underset{CH_3}{\overset{|}{C}} \diagdown CH_2 \diagdown Fe(CO)_2Cp$$

HBF$_4$
Ac$_2$O

$[(Me_2C=CH_2)Fe(CO)_2Cp]^+BF_4^-$ $\xrightarrow{\text{NaBH}_4}$

$$CH_3 \overset{}{\underset{CH_3}{\overset{|}{C}}} \diagdown CH_2 \diagdown Fe(CO)_2Cp \quad A$$

$$CH_3 \overset{\bullet}{\diagdown} \underset{H}{\overset{|}{\underset{CH_3}{C}}} \diagdown CH_2 \diagdown Fe(CO)_2Cp$$

Scheme 11-3

(Scheme 11-3) illustrates this point.[7] If the metal ion is substitutionally inert, stable alkyl derivatives may be formed,[5] e.g., $PhCH_2Cr(OH_2)_5^{2+}$ and $C_2H_5Ru(NH_3)_5^{2+}$. Cobalt(III) is inert toward *heterolytic* ligand dissociation, but its high oxidation potential favors facile *homolytic* loss of easily oxidizable ligands. Visible light accelerates this process, probably by promoting the charge-transfer step

$$R-Co(III) \longrightarrow R\cdot + Co(II)$$

Extremely high formal oxidation states also can stabilize metal alkyl units via extensive covalent bonding, as in Et_3Al or $(PhCH_2)_4Ti$.

Other general routes to alkyl and perfluoroalkyl complexes include decarbonylation and oxidative addition (Scheme 11-4).

$$\begin{array}{c} RCOCl \\ R_fCOCl \end{array} + M^- \longrightarrow Cl^- + \begin{array}{c} M-COR \rightarrow CO + M-R \\ M-COR_f \rightarrow CO + M-R_f \end{array}$$

$$R-X + M \longrightarrow RM + MX \text{ or } R-M-X$$

Scheme 11-4

sp²-CARBON-DONOR LIGANDS

The stability of a M–C unit may be further enhanced if π bonding in addition to σ bonding is possible. This effect has already been mentioned for CO and CN^- ligands (cf. Fig. 5-3). Other carbon ligands in which back-bonding effects are probably important include the following:

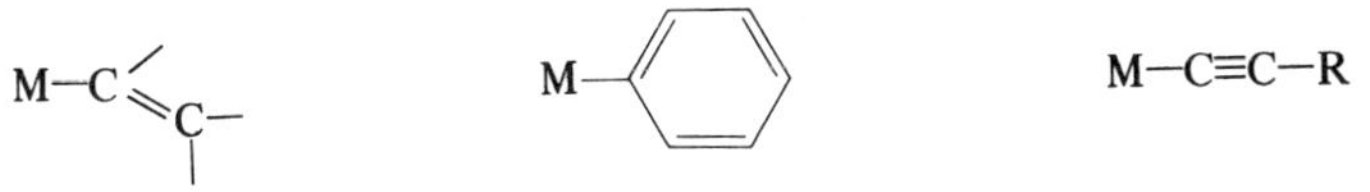

Alkenyl   Aryl   Acetylide (Alkynyl)

Acyl   Carbene

Alkenyl and aryl complexes are often prepared by decarbonylation of the corresponding acyl complexes (reaction 11-2), although in certain cases alkenyl derivatives may be formed by metal hydride addition to an acetylene (reaction 11-3).[8]

Acyl complexes may be formed by direct acylation of an electron-rich nucleophilic metal center (as in reactions 11-2 and 11-3) or by attack of a carbanionic reagent on coordinated CO (Scheme 11-5).

$$RCOCl \xrightarrow{Mn(CO)_5^-} RCOMn(CO)_5 \xrightarrow{\Delta} R-Mn(CO)_5 + CO$$

$$R = C_2H_3, Ph \qquad (11\text{-}2)$$

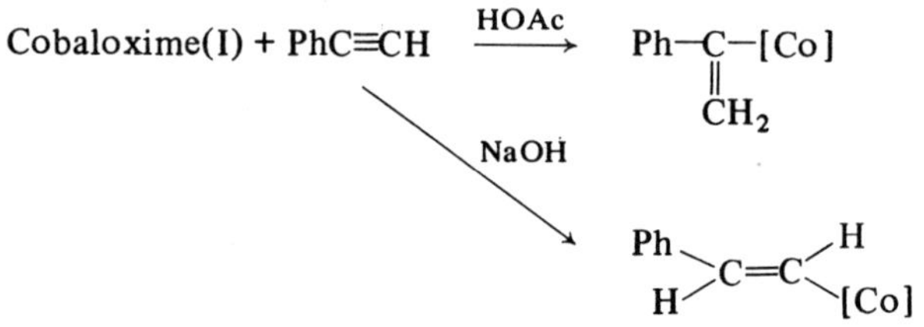

Cobaloxime(I) + PhC≡CH $\xrightarrow{HOAc}$ $\underset{\overset{\|}{CH_2}}{Ph-C-[Co]}$

$\xrightarrow{NaOH}$

$$\underset{H}{\overset{Ph}{>}}C=C\underset{[Co]}{\overset{H}{<}} \qquad (11\text{-}3)$$

$$RLi + M(Co) \longrightarrow \left[M-C\underset{R}{\overset{\displaystyle O}{<}} \right]^- Li^+$$

$$\Big\Updownarrow$$

$$M \leftarrow C\underset{R}{\overset{OCH_3}{<}} \xleftarrow{Me_3O^+BF_4^-} M \leftarrow C\underset{R}{\overset{OLi}{<}}$$

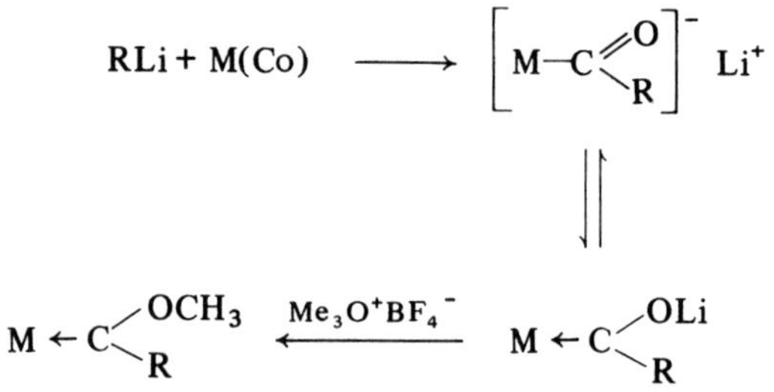

Scheme 11-5

Stable carbene complexes are formed in a number of ways including electrophilic attack on the oxygen or an acyl ligand (reaction 11-4), gem-dehalogenations (reaction 11-5), substitution reactions of coordinated carbenes (reaction 11-6), and cleavage of reactive electron-rich olefins (reaction 11-7).[9]

$$Cp(CO)_2FeCOCH_3 + HCl \longrightarrow [Cp(CO)_2 Fe \leftarrow C(OH)CH_3]^+Cl^- \qquad (11\text{-}4)$$

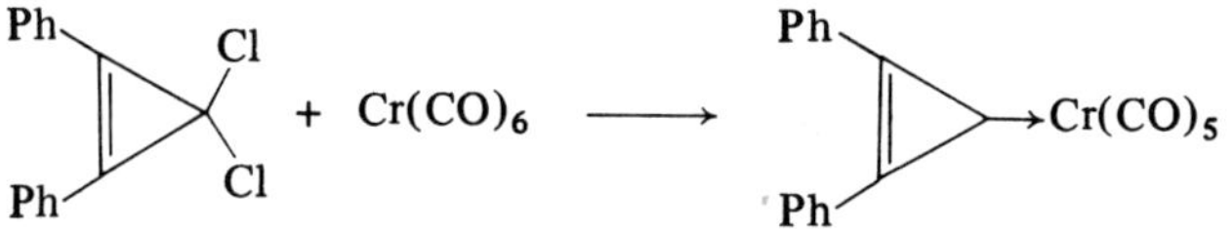

+ Cr(CO)_6 $\longrightarrow$ →Cr(CO)_5   (11-5)

$$(CO)_5Cr \leftarrow C\underset{CH_3}{\overset{OCH_3}{<}} + RSH \longrightarrow CH_3OH + (CO)_5Cr \leftarrow C\underset{CH_3}{\overset{SR}{<}} \qquad (11\text{-}6)$$

+ Fe(CO)_5 $\longrightarrow$ (11-7)

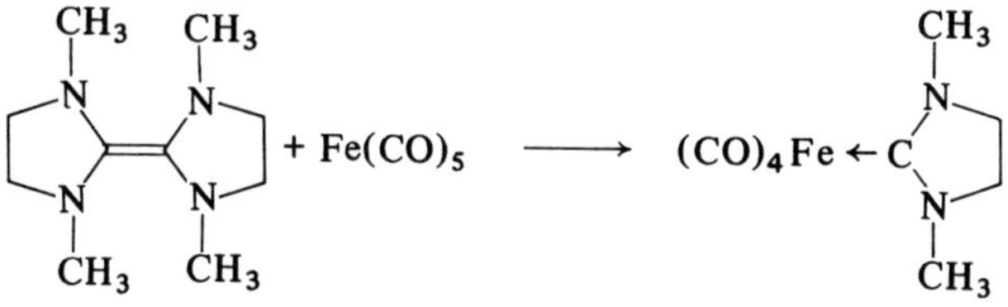

The extent of metal carbon π back-bonding with the above types of ligands, as manifest by the chemical and spectroscopic properties of the complexes *vis-à-vis* simple alkyl complexes or even totally organic compounds, varies considerably. Aryl and vinyl complexes are generally less reactive at the M–C bond than the corresponding methyl derivatives (compare the relative trans effects of H, C_6H_5, and CH_3 discussed earlier). In the case of transition metal aryls, X-ray structure determinations have shown the metal–carbon distance to be shorter than expected for a single bond, possibly because of partial double bond character. However, ^{19}F NMR studies[10] of mono- and pentafluorophenyl complexes suggest that resonance effects were less important than inductive effects in determining the chemical shift of o-, m-, and p-fluorines, which suggested that π bonding is less important than σ bonding in these complexes.

Acyl derivatives of transition metals exhibit C=O stretching frequencies lower than most organic ketones (1650 cm^{-1} vs 1725 cm^{-1}) and considerably lower than coordinated CO (1800–2000 cm^{-1}). Metal to carbonyl back-bonding renders the carbonyl oxygen reactive toward electrophiles, as seen in their ready conversion to carbene complexes by O-alkylation. The electron donating effects of a transition metal α substituent can be seen in the facile cleavage of the ether linkage in the methoxymethyl complex, which is thought to proceed via a carbene intermediate[11] (Scheme 11-6). Complexes of CH$_2$ or

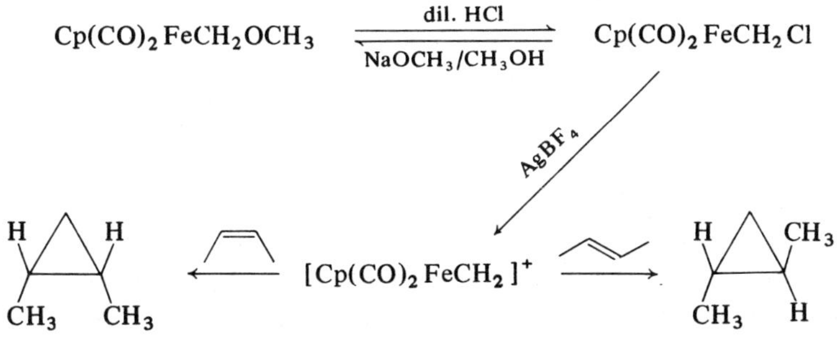

Scheme 11-6

its monosubstituted derivatives have not yet been isolated, although a very large number of complexes of carbene with one or two heteroatom substituents are known. In these complexes the central carbon is rather electrophilic, but stabilization by lone pair donation from the heteroatoms is more important than back donation from the metal. Consequently, these complexes readily undergo nucleophilic substitution at carbon (possibly via a tetrahedral carbon intermediate, just as, for example, in a transesterification reaction). Base-catalyzed H/D exchange at the β-carbon is another reflection of the electrophilic character of the central carbon (Scheme 11-7). X-ray structural

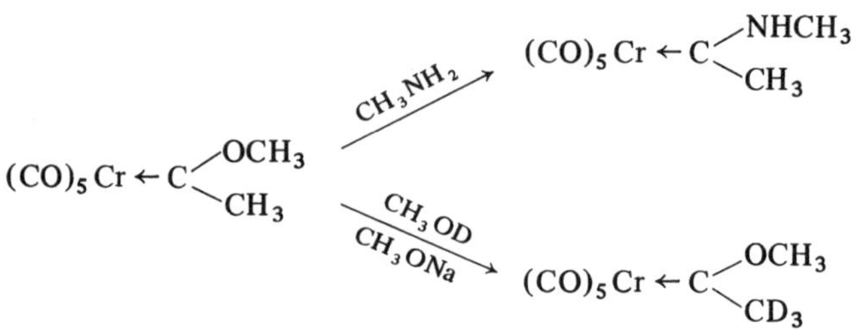

Scheme 11-7

data on a number of carbene complexes also suggest the absence of any significant back-bonding inasmuch as the metal–C* carbon bond is not shortened (e.g., 2.04 Å vs 1.87 Å for M–CO) while the carbon-heteroatom bonds are shortened (e.g., 1.33 Å vs 1.46 Å for O–CH$_3$).[12] In addition, nmr studies show a large energy of activation for rotation about the C*–OCH$_3$ bond.

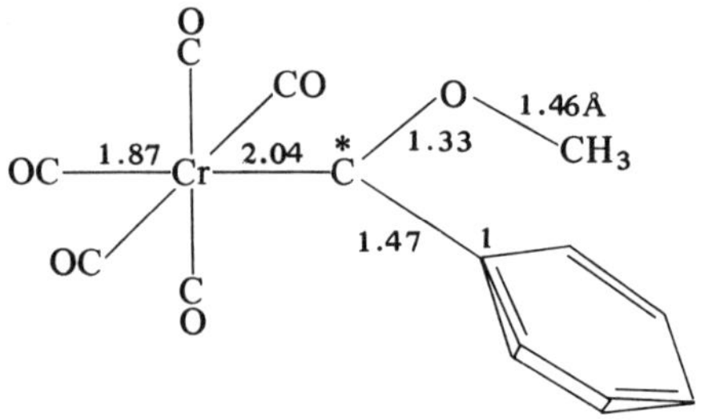

Phenyl plane 90° to Cr—C*—C$_1$ plane[12]

Finally, it should be mentioned that carbene complexes are *diamagnetic*, whereas the free carbene ligands would be paramagnetic triplets (and consequently quite reactive) in their ground states. Thus, the metal apparently complexes the first excited state of the carbene. We have seen this situation before with diamagnetic dioxygen complexes, and Mason has pointed out[13] that in many complexes the geometry of the ligands, as well as the magnetic and other properties of the complex, approach those of the first excited state of the ligand. This is usually attributed to partial population of ligand antibonding orbitals by means of back-bonding.

OLEFIN AND ACETYLENE COMPLEXES

The first organometallic complex ever made was an ethylene complex, prepared in 1827 by W. C. Zeise, a Danish pharmacist, through the interaction of ethanol and chloroplatinic acid. The structure of the product

TABLE 11-1

Structural Data for Representative Olefin Complexes

Complex	C=C bond, Å		Remarks	Ref.
	Complex	Free		
(a)	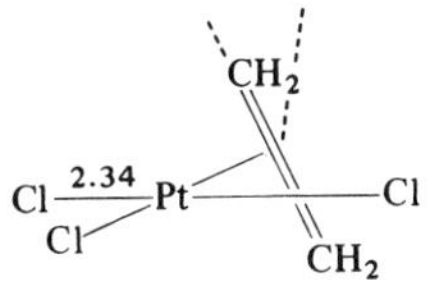 1.37	1.34	Olefin rotated 6° from normal, Pt–C = 2.16 Å	16, 17
(b)	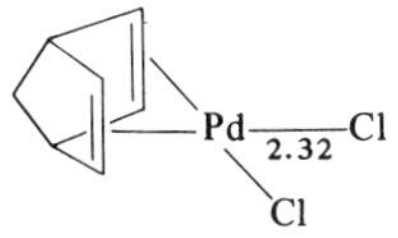 1.37	1.37		18
(c)	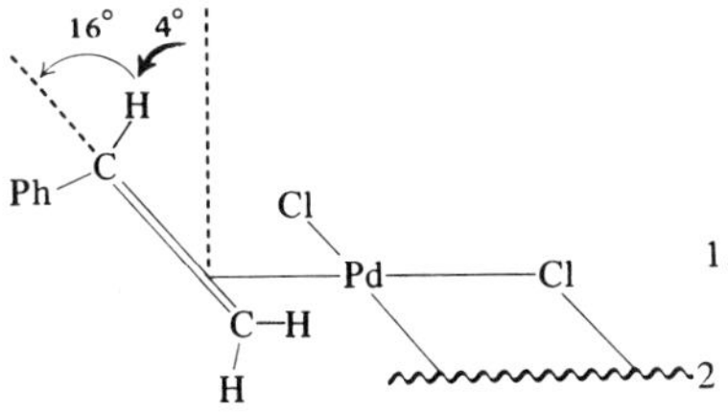 1.32	—	Olefin rotated 4° and tipped back 16° from normal, carbons are not equidistant to Pd	19
(d)	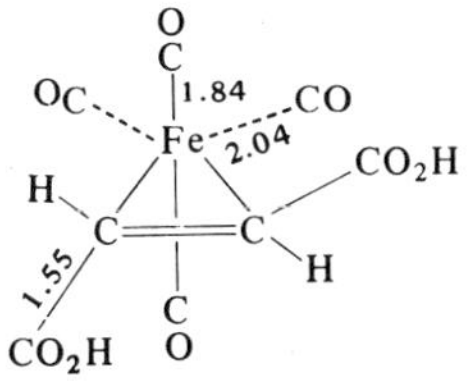 1.40	1.35	(One of three optical isomers) Ligand folded in 128° dihedral about central C—C	20
(e)	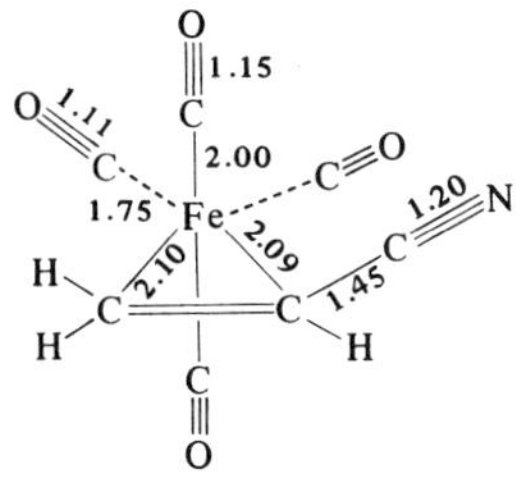 1.40	1.34	∠ C–C–CN = 116°	21

$K[(C_2H_4)PtCl_3]$, now known as Zeise's salt, was not fully elucidated until 1954 when its crystal structure was first determined (Table 11-1).[16-21] Since then numerous other olefin and acetylene complexes have been isolated or identified as intermediates in reaction sequences.[14,15]

Discussion of the bonding in olefin and acetylene complexes usually centers around the Dewar–Chatt–Duncanson theory which considers σ bonding from a filled ligand π MO to an empty d or d hybrid metal orbital, together with π back-bonding from a filled metal orbital to an empty π^* ligand orbital. This same scheme was used earlier in connection with π-dioxygen complexes (cf. Fig. 10-6). By varying the relative contribution of the π and σ components of the net bonding one can nicely explain many of the characteristics of olefin and acetylene complexes. Occasionally it is convenient to use a valence bond approach to describe olefin and acetylene complexes. In this model the complexation is regarded as an oxidative addition process wherein the metal furnishes two electrons to formally reduce the ligand to a dianion which then chelates the metal ion. In contrast the Dewar–Chatt–Duncanson model involves π and σ Lewis acid-base complexation with no formal redox change (Fig. 11-1).

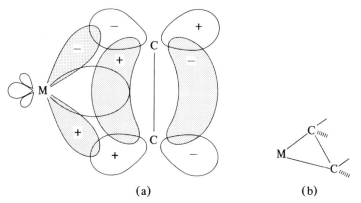

(a)   (b)

Fig. 11-1. Metal-olefin bonding schemes. (a) Dewar–Chatt–Duncanson model based on σ and π Lewis acid-base complexation. (b) Valence bond model based on formal oxidative addition similar to formation of a cyclopropane from a carbene and an olefin.

Based on the changes experienced by an olefin or acetylene upon coordination to a metal, and on certain properties of the complexes themselves, it is possible to discern two broad categories of olefin and acetylene complexes.[14] The origins of the features which distinguish the two are best interpreted in terms of the relative importance of σ vs π contributions to the net bonding of the ligand to the metal. However, the reader is reminded that the more extensive the generalization, the more likely it is that exceptions will be found.

Furthermore, *three* considerations are of prime importance here: the metal center and its electron configuration, the nature of the olefin or acetylene under consideration, and the coligands as they modify the electronic environment of the metal center.

Back-bonding from metal to olefin or acetylene is clearly important for electron-rich metals such as Ni(0), Pd(0), Pt(0), Ir(I), Rh(I), and Fe(0). These metal centers form olefin or acetylene complexes which are trigonal or trigonal bipyramidal. In the solid state both carbons of the ligand are in or very near the trigonal plane, with small distortions resulting from crystal packing forces. In this configuration the back-bonding overlap is maximized, as evidenced by substantial lengthening of C–C distances and lowering of ν_{C-C} by as much as 300 cm^{-1} for olefins and 500 cm^{-1} for acetylenes. Olefin and acetylene substituents are bent away from the metal substantially, indicating the importance of a cyclopropane-like bonding scheme (i.e., oxidative addition, cf. Table 11-2).[14,22–25] This is also suggested by measurements of various metal, ^{1}H, ^{13}C, and ^{19}F coupling constants, as well as by the chemistry of coordinated acetylenes.[14] (See reactions 11-8 and 11-9.)

$$(Ph_3P)_2Pt \overset{C-CN}{\underset{C-CN}{\big\|}} \xrightarrow{HCl} \underset{PPh_3}{\overset{PPh_3}{Cl-Pt-C}} \overset{CN}{\underset{C-CN}{=C}} \; H \qquad (11\text{-}8)$$

$$(CH_3)_2C \overset{C-CN}{\underset{C-CN}{\big\|}} \xrightarrow{HCl} \underset{CH_3}{\overset{CH_3}{Cl-C-C}} \overset{CN}{\underset{C-CN}{=C}} \; H \qquad (11\text{-}9)$$

Nuclear magnetic resonance studies of Pt(0) and Pd(0) complexes show that coordinated olefins and acetylenes do not rotate about the metal–ligand bond, as is observed in Pt(II) and Pd(II) complexes. This may be a consequence of the importance of π back-bonding, which would cause the ligand to tend to maintain a fixed geometry favoring maximum orbital overlap. Alternatively, the valence bond model in which the ligand adds oxidatively to the metal can also explain the lack of rotation in terms of ring formation. In the Rh(I) complexes $(C_5H_5)Rh(C_2H_4)_2$, $(C_5H_5)Rh(C_2H_4)(SO_2)$, and $(C_5H_5)Rh(C_2H_4)(C_2F_4)$, the ethylene unit does rotate about an axis perpendicular to the C–C bond and passing through the metal, and the activation energies for rotation are 15.0, 12.2, and 13.6 kcal/mole, respectively.[26] The activation energies for rotation thus *decrease* as other ligands (e.g., SO$_2$ and C$_2$F$_4$) *compete for back-bonding*. Furthermore, rotation of the C$_2$F$_4$ ligand, a better back-bonder than C$_2$H$_4$, was not observed.

Olefins and acetylenes coordinated to electron-rich metal centers are highly

TABLE 11-2

Structural Data for Representative Acetylene Complexes

Complex	C≡C bond, Å		Remarks	Ref.
	Complex	Free		
(a)	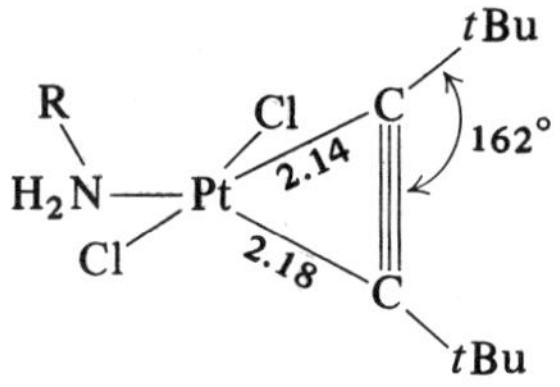 1.24	1.20	R = p-CH$_3$C$_6$H$_4$—	22
(b)	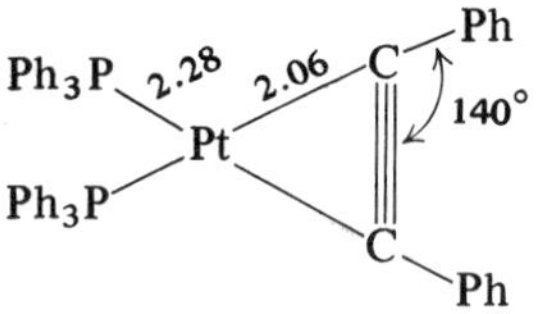 1.32	1.19	C—C at 14° to PtP$_2$ plane	23
(c)	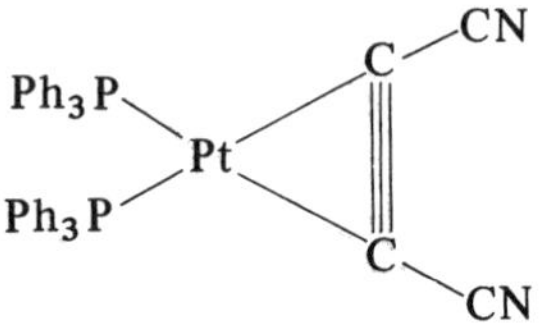 1.40	1.19	C—C at 8° to PtP$_2$ plane	14
(d)	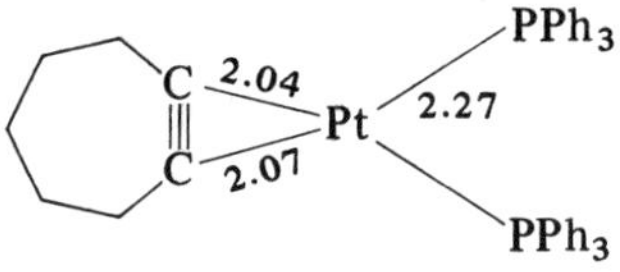 1.29	—	Free ligand has only transient existence	24
(e)	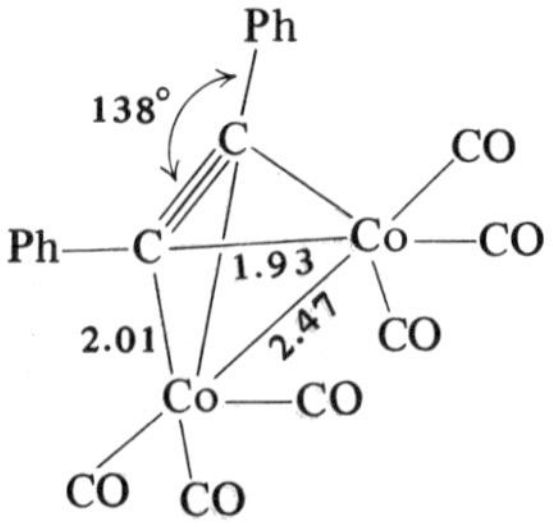 1.46	1.19	Bridging acetylene, cobalts roughly trigonal bipyramidal	25

susceptible to electrophilic attack, e.g., protonation. This property contrasts the attack by nucleophiles seen with Pt(II) and especially Pd(II) and Hg(II) complexes, and further emphasizes the increase in electron density on the ligand as a result of back-bonding. Electron-withdrawing substituents on the olefin or acetylene itself increase the "stability" of the complex by lowering the energy of the π^* orbital and thus increasing the importance of back-bonding in the complex.

With metals in higher oxidation states, such as Pt(II), Pd(II), Hg(II), Ni(II), Ag(I), and Cu(I), backbonding is less effective to all ligands, including olefins and acetylenes. Hence, coordination is favored by electron releasing substituents on the ligand, and upon coordination the ligand C—C bond lengths and stretching frequencies are perturbed relatively little. Olefins coordinated to these centers are relatively susceptible to *nucleophilic* attack. For example, Zeise's salt reacts with hydroxide to form acetaldehyde, and the Pd analog of Zeise's salt (or its dimer) is the catalyst in the Wacker process for oxidizing ethylene to acetaldehyde. (The reactions of coordinated olefins will be discussed further in Chapter XII.) In the solid, the C—C axis of the unsaturated ligand is perpendicular, or nearly so, to the coordination plane in Pt(II) and Pd(II) complexes, small distortions arising from crystal packing forces.

TABLE 11-3

Thermodynamic Parameters for Binding of Styrenes to Pd(II)[a]

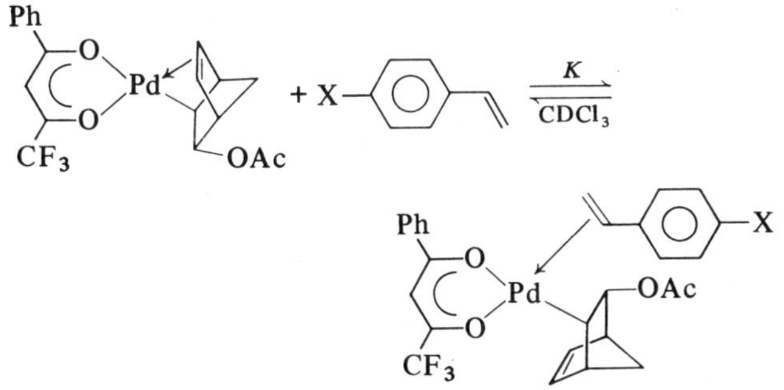

X	ΔG (kJ/mole)	ΔH (kJ/mole)	ΔS (J/mole deg)
$-NO_2$	3.3	-11.7	50
$-F$	1.5	-26.0	92
$-H$	1.0	-27.8	96
$-OMe$	-1.0	-33.0	105
$-NMe_2$	-4.5	-43.0	126

[a] Data from Ban *et al.*[27]

However, in solution, olefins coordinated to Pd(II), Pt(II), or in some cases Rh(I), rotate about the metal–olefin bond. The low energy barrier to rotation is thought to arise from the availability of several sets of metal orbitals which could possibly back-bond to the olefin, thereby making the energy of the complex less dependent on the orientation of the olefin. However, a more likely explanation is that the main contribution to the net metal–ligand bonding is the σ bond, which has axial symmetry.

Recent studies based on nmr measurement of equilibrium constants (Table 11-3)[27] clearly show that p-substituted styrenes are bound to Pd(II) more strongly when the para substituent, by means of resonance effects, increases the electron density of the styrene double bond. If back-bonding were important the opposite trend should have been observed.

CARBONYL AND NITROSYL COMPLEXES

Transition metal complexes of carbon monoxide have been known and studied for a very long time.[28] The first one described (in 1871) was the volatile platinum compound $Pt(CO)_2Cl_2$. Shortly thereafter binary iron and nickel carbonyls, $Fe(CO)_5$ and $Ni(CO)_4$, were prepared by direct union of CO and the metal at $100°–180°C$. Only for iron and nickel is this direct combination feasible, but this fact has important consequences: carbon monoxide under pressure and heat can readily corrode and weaken iron or nickel valves and storage tanks. A commercial process for obtaining nickel from nickel ore involves heating the ore with carbon monoxide, which reduces the metal and extracts it as volatile $Ni(CO)_4$.

In mononuclear carbonyl complexes the M–C–O unit is always linear, and in this case the bonding is described as a combination of σ donation and π back-bonding (cf. Fig. 5-3), similar to that in the isoelectronic end-on dinitrogen complexes (cf. Fig. 5-3 and Chapter X). In addition to this type of arrangement many examples of polynuclear carbonyl complexes are now known in which one CO may be simultaneously bonded to two or even three metals. When CO bridges two metals it is usually considered to involve localized M–C bonding, i.e., like an inorganic analog of a ketone. Bonding in the triply bridged case is more complex and requires molecular orbital treatment.

The nature of the metal–carbonyl interactions are reflected in the carbonyl stretching frequency. Nonbridging carbonyls generally absorb around 2000 cm^{-1}, compared to 2155 cm^{-1} for free CO and 1850 cm^{-1} for bridging carbonyls. However, these values are only nominal values, and there is considerable overlap of the ranges for the two types. The net charge (or effective charge) on the complex can also perturb the carbonyl stretching frequency by shifting the balance of the σ and π components of the net bonding. Greater electron density on the metal favors stronger back-bonding and lowers ν_{CO} (Table 11-6 and below).

$Mn(CO)_6^+$	$Cr(CO)_6$	$(CO)_6^-$
2096	2000	1859 cm^{-1}
$Ni(CO)_4$	$Co(CO)_4^-$	$Fe(CO)_4^{2-}$
2046	1883	1788 cm^{-1}

The above complexes are completely symmetrical and display only one carbonyl absorption; in less symmetrical complexes more than one carbonyl absorption is generally observed, depending on the number of CO ligands.

Carbon monoxide is a strong field ligand and causes spin pairing in its metal complexes. For metals which have an odd number of electrons, spin pairing is effected by dimerization, either through a metal–metal bond, as in $(CO)_5Mn–Mn(CO)_5$, or via bridging carbonyls as in $Co_2(CO)_8$ (Fig. 11-2).

Mononuclear
  Tetrahedral: $Ni(CO)_4$, $Ni(CO)_n(R_3P)_{4-n}$, $Co(CO)_4^-$, $Fe(CO)_4^{2-}$
  Trigonal bipyramidal: $Fe(CO)_5$, $Mn(CO)_5^-$, $Ru(CO)_5$
  Octahedral: $V(CO)_6^-$, $Cr(CO)_6$, $Mn(CO)_6^+$
        $Mo(CO)_6$, $W(CO)_6$, $Fe(CO)_4I_2$
Polynuclear

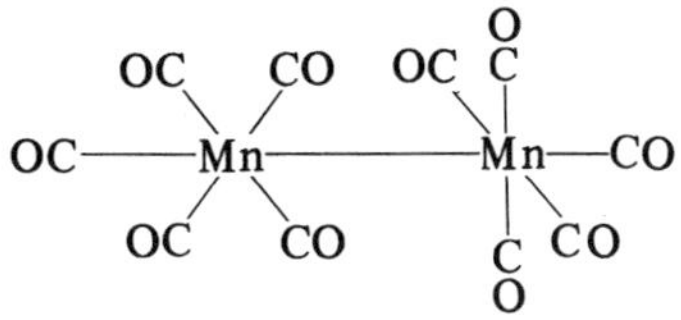

Carbonyls staggered

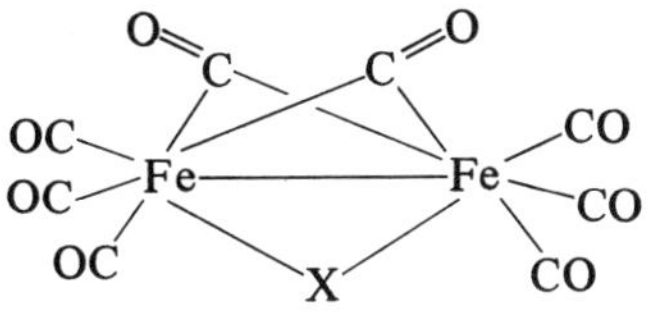

$Fe_2(CO)_9$, X = CO
$Fe_3(CO)_{12}$, X = Fe(CO)_4

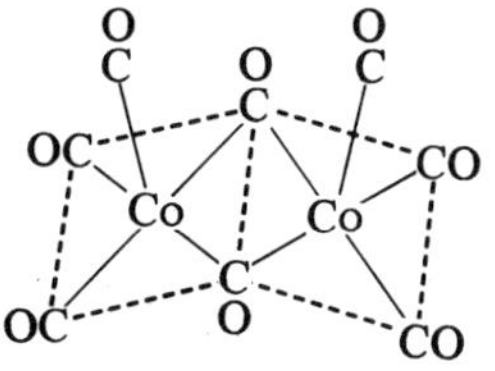

In solid and pentane solution

$(CO)_4Co–Co(CO)_4$

Only in solution

Fig. 11-2. Structures of representative metal carbonyls.

These dimers are readily reduced to form the corresponding stable anions $Mn(CO)_5^-$ and $Co(CO)_4^-$. Paramagnetic $V(CO)_6$ is anomalously monomeric, but it too forms a stable monoanion $V(CO)_6^-$.

Of the various chemical properties of coordinated CO, perhaps the two most significant are its susceptibility to nucleophilic attack (in many but not all complexes) and its ease of replacement (in many but not all complexes) by other ligands. Several examples of these processes are given in Scheme 11-8, and other examples may be found throughout this chapter and the next.

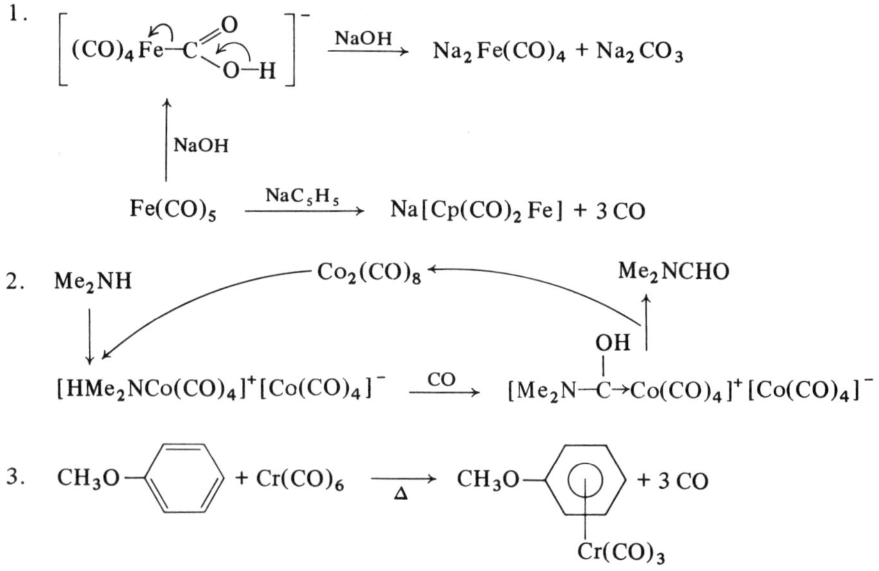

1.

$$\left[(CO)_4 Fe\!-\!C \!\!\! \diagup\!\!\!\!\!^{O}_{O\!-\!H} \right]^{-} \xrightarrow{\text{NaOH}} Na_2 Fe(CO)_4 + Na_2 CO_3$$

$$\uparrow \text{NaOH}$$

$$Fe(CO)_5 \xrightarrow{\text{NaC}_5\text{H}_5} Na[Cp(CO)_2 Fe] + 3\,CO$$

2. $Me_2NH \qquad\qquad Co_2(CO)_8 \qquad\qquad Me_2NCHO$

$$\downarrow \qquad\qquad\qquad\qquad \uparrow \text{OH}$$

$$[HMe_2NCo(CO)_4]^+[Co(CO)_4]^- \xrightarrow{CO} [Me_2N\!-\!C\!\rightarrow\!Co(CO)_4]^+[Co(CO)_4]^-$$

3. $CH_3O\!-\!\bigcirc \; + Cr(CO)_6 \xrightarrow{\;\;\Delta\;\;} CH_3O\!-\!\bigcirc \; + 3\,CO$

$$\Big| \atop \underset{\displaystyle Cr(CO)_3}{}$$

Scheme 11-8

The nitrosyl cation, NO^+, is isoelectronic with CO and bonds to transition metals in a similar fashion, i.e., linear M–N–O. However, for NO complexes an alternate mode of binding is also known, similar to organic nitroso compounds and corresponding formally to coordinated NO^-, i.e., bent M–N–O. The neutral species nitric oxide can thus either accept one electron from a metal to form a bent nitrosyl complex, or it can formally donate *three* electrons (one to form NO^+, then a pair from nitrogen to form a σ bond) and form a linear nitrosyl complex. Back-bonding from the metal reduces the net amount of charge transferred in both linear and bent forms. The N–O stretching frequency is somewhat lower for the bent nitrosyls than for the linear nitrosyls ($\geqslant 1700$ vs $\geqslant 1600\ cm^{-1}$, respectively), but the ranges overlap and ν_{NO} is affected by other variables such as electron density on the metal, just as is ν_{CO}.

One of the more interesting features of some nitrosyl complexes is the ready interconversion of linear and bent coordination geometries. This has the effect of changing the effective atomic number of the central metal by two electrons (reaction 11-10), a process which has important consequences for the catalytic

$$[M-N\equiv O \longleftrightarrow M=N=O] \rightleftharpoons M-N\overset{O}{\underset{\cdot\cdot}{\diagup}} \qquad (11\text{-}10)$$
$$(18) \qquad\qquad (18) \qquad\qquad (16)$$

activity of the complex. An interesting dinitrosyl complex in which both linear and bent modes of coordination are observed is shown in Fig. 11-3.[29] The two

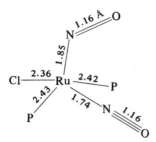

Fig. 11-3. Structure of the dinitrosyl complex ion, $Ru(NO)_2Cl(Ph_3P)_2{}^+$.[29]

nitrosyl groups interconvert easily as shown by the rapid scrambling of label when this complex is prepared from $Ru(^{15}NO)Cl(Ph_3P)_2$ and $^{14}NO^+PF_6{}^-$ in ethanol.[30] Linear-bent interconversion of nitrosyl groups has also been observed in solution (reaction 11-11).[31]

$$(11\text{-}11)$$

The chemical properties of coordinated NO parallel in many ways those of coordinated CO. Thus, linearly coordinated nitrosyls with $\nu_{NO} \geqslant$ ca. 1850 cm^{-1} undergo nucleophilic attack, while bent nitrosyls undergo electrophilic attack (reactions 11-12–11-14).[32] However contrast this to reaction 11-15.

$$[Ru(NO)(bipy)_2Cl]^{2+} + 2\,HO^- \longrightarrow Ru(NO_2)(bipy)_2Cl + H_2O \qquad (11\text{-}12)$$

$$[Ru(NO)(diars)_2Cl]^{2+} + N_2H_4 \longrightarrow Ru(N_3)(diars)_2Cl + H_2O + 2\,H^+ \quad (11\text{-}13)$$

$$OsCl(CO)(NO)(Ph_3P)_2 + HCl \longrightarrow OsCl_2(HNO)(CO)(Ph_3P)_2 \qquad (11\text{-}14)$$

$$Me_3N\cdot BMe_3 + Co(NO)_3 \longrightarrow Me_3N + Me_3B\cdot Co(NO)_3 \qquad (11\text{-}15)$$

MULTIELECTRON LIGANDS

A ligand molecule having two or more separated donor sites is said to be a multidentate or chelating ligand. The phenomenon of chelation per se has been more thoroughly studied among coordination complexes (*viz.*, Chapter VI) than among organometallic compounds. In the latter area chelating ligands most commonly involve "soft" π and/or σ donors such as carbon, phosphorus(III), or arsenic(III). Some of the more common chelating agents in organometallic chemistry include phosphorus and arsenic analogs of ethylenediamine, i.e., "diphos" and "diars," and 1,4- and 1,5-dienes such as 1,5-cyclooctadiene (COD), norbornadiene (NBD), and hexamethyldewarbenzene (HMDB).

diphos 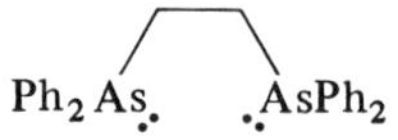   diars 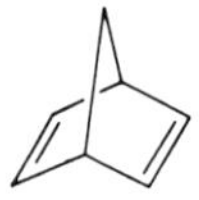   NBD

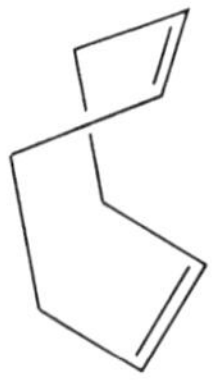

COD 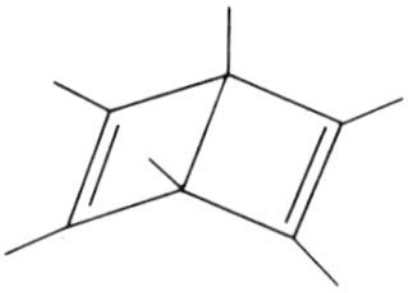   HMDB

In contrast to multidentate ligands, *multielectron ligands*[33,33a] have a conjugated π-electron system, all or part of which may overlap with appropriate orbitals on a metal center. Of course, metal to ligand π^* back-bonding is also involved, and the net bonding is difficult if not impossible to describe with anything less than molecular orbital treatment. Unsaturated multielectron ligands may contain heteroatoms such as nitrogen, but usually they consist of linear or cyclic arrays of sp^2-carbons. Those with an even number of carbons are the familiar unsaturated hydrocarbons ethylene, 1,3-butadiene, 1,3,5-hexatriene, benzene, etc. Neutral systems with an odd number of carbons, such as allyl, pentadienyl, and their cyclic congeners, are free radicals and hence are more commonly encountered as carbonium ions or carbanions in organic chemistry. When only a portion of a chelating or conjugated unsaturated ligand is involved in bonding to the metal, this can be specified by use of the term "hapto" (Greek, *haptein*, to fix or connect) as shown by the following examples.

Inasmuch as multielectron carbon ligands form a homologous series starting with σ-alkyl and π-olefin complexes, it is not surprising that many complexes of these ligands are interconvertible through proton or hydride ion addition to, or abstraction from, coordinated carbon ligand systems (reactions

Pentahaptocyclopentadienyl
$\eta^5-C_5H_5$
$\pi-C_5H_5$
$\pi-Cp$

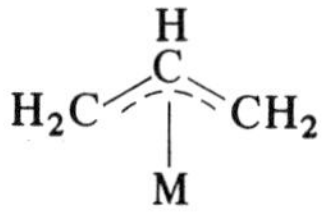

Trihaptoallyl
$\eta^3-C_3H_5$
$\pi-C_3H_5$
$\pi-$allyl

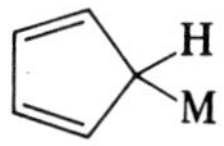

Monohaptocyclopentadienyl
$\eta^1-C_5H_5$
$\sigma-C_5H_5$
$\sigma-Cp$

Monohaptoallyl
η^1-allyl
$\sigma-C_3H_5$
$\sigma-$allyl

11-16–11-19).[33a] Another general route of synthesis involves replacement of labile ligands by diene or triene ligands (reactions 11-20 and 11-21). Reaction of metal (or carbon) halides with carbon (or metal) anions and oxidative addition of allylic halides to metal complexes are also important synthetic routes to complexes of multielectron ligands. In some cases the metal reagent serves to *generate* the ligand as well as to complex it (e.g., cyclobutadiene and trimethylene methane) (reactions 11-22–11-26).

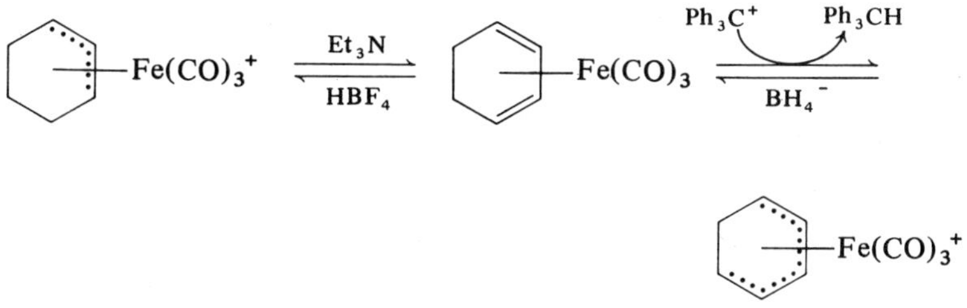

$$(CO)_5MnCH_2CH_3 \underset{BH_4^-}{\overset{Ph_3C^+}{\rightleftharpoons}} (CO)_5Mn^+ \!\!-\!\!\overset{CH_2}{\underset{CH_2}{\|}} \qquad (11\text{-}17)$$

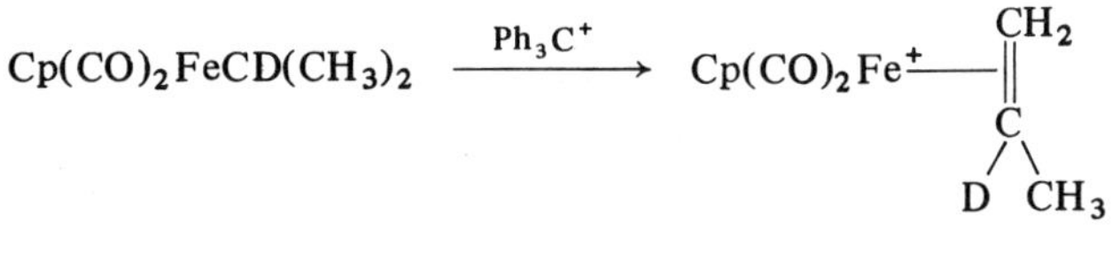

$$Cp(CO)_2FeCD(CH_3)_2 \xrightarrow{Ph_3C^+} Cp(CO)_2Fe^+\!\!-\!\!\overset{CH_2}{\underset{\overset{C}{\underset{D\ \ CH_3}{/\ \backslash}}}{\|}} \qquad (11\text{-}18)$$

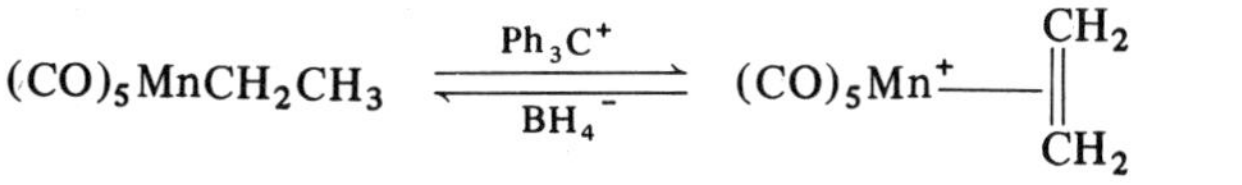

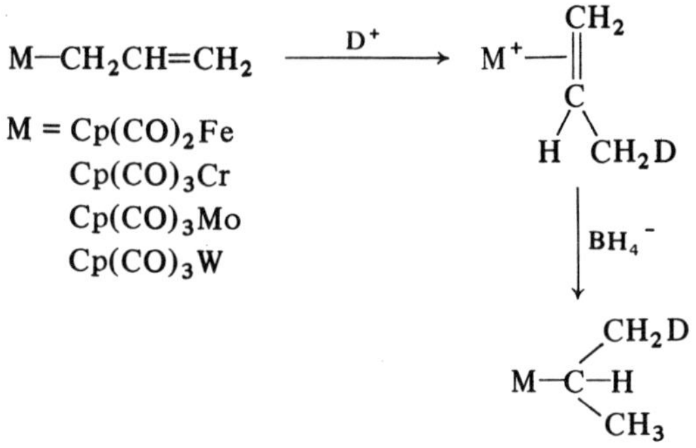

$$M-CH_2CH=CH_2 \xrightarrow{D^+} M^+ - \overset{\displaystyle CH_2}{\underset{\displaystyle \overset{|}{C}}{\|}}$$

$$M = Cp(CO)_2Fe$$
$$Cp(CO)_3Cr$$
$$Cp(CO)_3Mo$$
$$Cp(CO)_3W$$

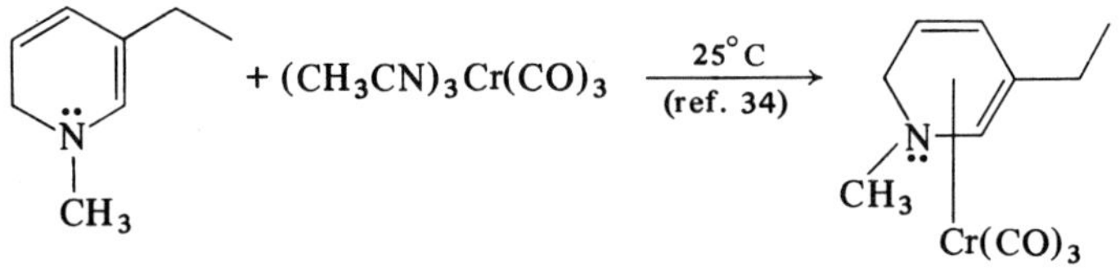

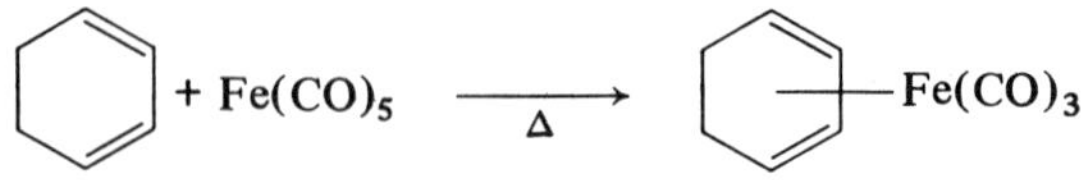

$$\downarrow BH_4^-$$

$$M-\overset{\displaystyle CH_2D}{\underset{\displaystyle CH_3}{\overset{|}{C}-H}} \tag{11-19}$$

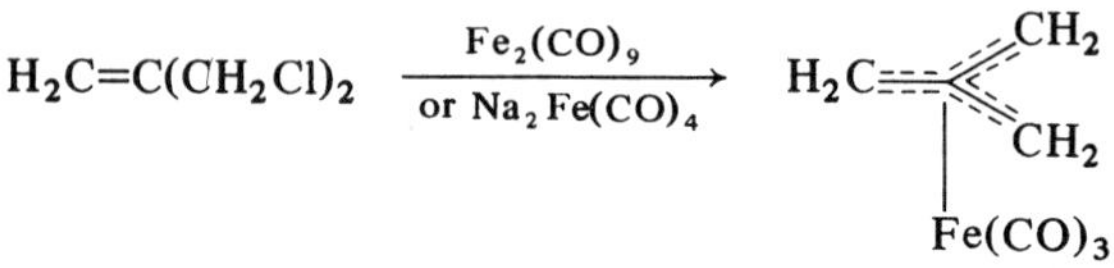

$$+ (CH_3CN)_3Cr(CO)_3 \xrightarrow[\text{(ref. 34)}]{25°C} \tag{11-20}$$

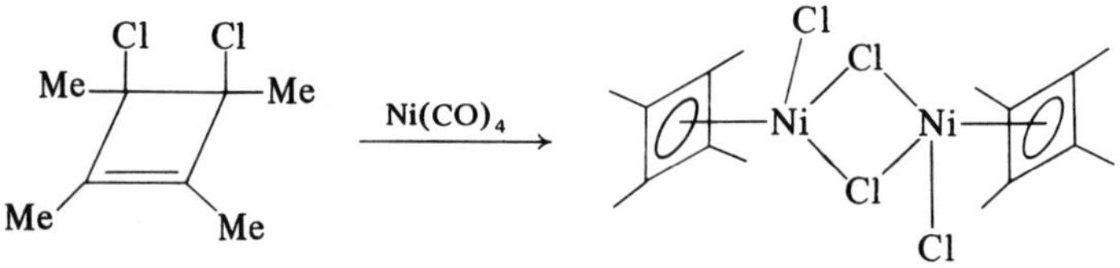

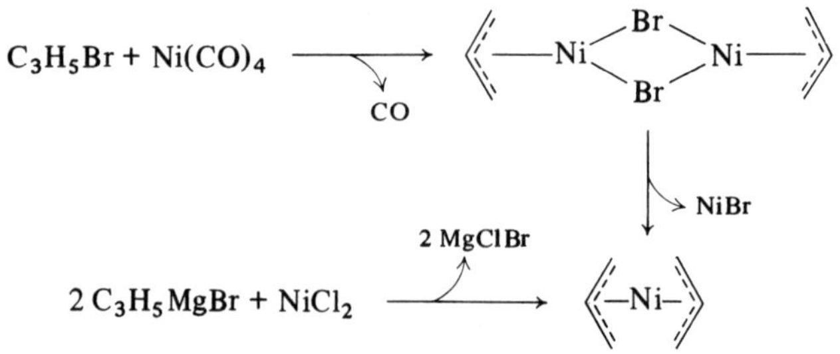

$$+ Fe(CO)_5 \xrightarrow{\Delta} \tag{11-21}$$

$$H_2C=C(CH_2Cl)_2 \xrightarrow[\text{or } Na_2Fe(CO)_4]{Fe_2(CO)_9} H_2C=\!=\!=\overset{\displaystyle CH_2}{\underset{\displaystyle \underset{\textstyle Fe(CO)_3}{CH_2}}{\diagup}} \tag{11-22}$$

$$\xrightarrow{Ni(CO)_4} \tag{11-23}$$

$$C_3H_5Br + Ni(CO)_4 \xrightarrow{} \tag{11-24}$$

CO

NiBr

$$2\,C_3H_5MgBr + NiCl_2 \xrightarrow{2\,MgClBr} \tag{11-25}$$

$$CpMgBr + Mo(CO)_6 \xrightarrow{} [CpMo(CO)_3]MgBr + 3\,CO \tag{11-26}$$

The structures of representative complexes of multielectron ligands are given in Figs. 11-4–11-6.[35–40] It is instructive to compare the various C–C and C–M bond distances in the various types of complexes, particularly in light of the effects metal groupings have in modifying the chemical reactivity of the organic ligand. For example, in the π-allyl complex (Fig. 11-4a) the C–C bonds

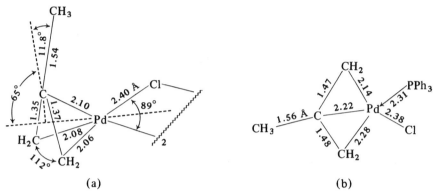

(a)   (b)

Fig. 11-4. Structures of two representative allyl complexes: (a) (π-2-methallyl PdCl)$_2$[35] and (b) π-allyl(triphenylphosphine)PdCl at −150°C.[36] a + 2 Ph$_3$P → 2b.

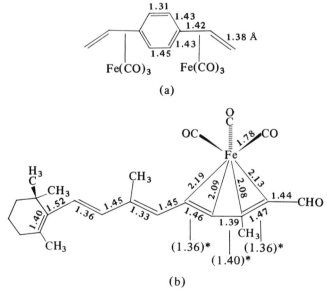

Fig. 11-5. Structures of (π-diene)iron tricarbonyl complexes: (a) Davis and Pettit;[37] (b) asterisk indicates values expected for free ligand, see text and Mason and Robertson.[38]

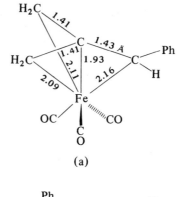

(a)

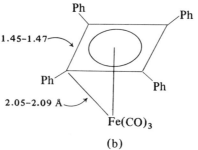

(b)

Fig. 11-6. Structures of tricarbonyliron complexes of unstable ligands. (a) Phenyl-trimethyl-enemethane, Ph plane forms 59° dihedral to plane of C_4 ligand, carbonyls and methylenes staggered.[39] (b) Tetraphenylcyclobutadiene, Ph planes twisted 30°–60° to square C_4 plane.[40]

of the allyl unit are equivalent and intermediate in length between C–C single and C–C double bonds, and the C–Pd bond lengths are nearly identical. However, in the related complex (Fig. 11-4b) the allyl ligand is bound in a highly unsymmetrical form, suggestive of independent metal-carbon σ and π bonds, i.e.,

$$\begin{array}{c}
CH_3 \diagdown \quad \diagup CH_2 \\
C \\
\| \quad \longrightarrow Pd \diagup PPh_3 \\
CH_2 \qquad \diagdown Cl
\end{array}$$

The structure of the iron tricarbonyl derivative of vitamin A aldehyde (Fig. 11-5b) reveals the normal alternation of bond lengths expected for a conjugated polyene, but where the ligand is bound to the metal center the expected pattern is *reversed*. This is attributed to the importance of metal–ligand π^* back-bonding. The structure of the bis(tricarbonyliron) complex of *p*-

divinylbenzene (Fig. 11-5a) shows that such effects can even intrude upon the sanctity of an "aromatic" ring, resulting in a pronounced localization (i.e., dearomatization) of the π system. On the other hand, the benzene ring in $(\eta^6\text{-}C_6H_6)Cr(CO)_3$ is symmetrical, showing no tendency toward localization to a 1,3,5-cyclohexatriene system. Similarly, cyclobutadiene derivatives have a square planar geometry when coordinated, although a recent X-ray study of a highly substituted but uncomplexed cyclobutadiene revealed a distinctly *rectangular* planar ring.[41]

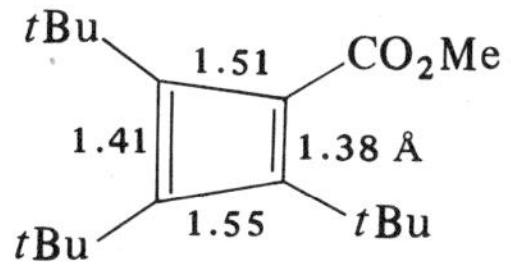

X-ray crystal studies fail to reveal a very important aspect of the structures of many complexes of multielectron ligands in solution, namely, the dynamic interconversion of various coordination isomers known as fluxional behavior.[42] For example, the complex $(\eta^1\text{-}C_5H_5)(\eta^5\text{-}C_5H_5)Fe(CO)_2$ has the structure shown in reaction (11-27) in the solid state. In solution at room temperature only two

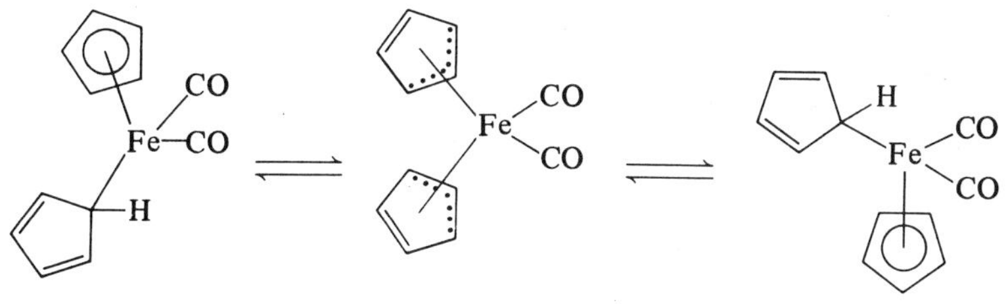

(11-27)

peaks are seen in the ^{1}H nmr spectrum, but upon cooling the sample to $-85°C$ one of the peaks shifts slightly to the normal chemical shift for a η^5-C_5H_5Fe unit, about 5.25, and the other collapses and splits into a pair of doublets and a singlet (5.6, 5.3, and 4δ, respectively) corresponding to the "frozen out" η^1-C_5H_5 substituent. Bis-π-allyl complexes of Ni, Pd, and Pt also show fluxional behavior which interconverts "cis" and "trans" isomers and scrambles "inner" and "outer" protons on the V-shaped allyl unit. Allyl Grignard reagents show similar behavior, but $(CO)_5Mn(\eta^3\text{-}C_3H_5)$ on the other hand is not fluxional at room temperature. Complexes of neutral, even-electron ligands may also exhibit fluxional behavior; for example, $(\eta^6\text{-}C_8H_8)(\eta^4\text{-}C_8H_8)Fe$, which gives a one-line ^{1}H nmr spectrum[43] that becomes progressively more complex below room temperature as various modes of fluxional interconversion are "frozen out."

TABLE 11-4

Elementary Organometallic Reactions[a]

Reaction	ΔNVE[b]	ΔOS[c]	ΔN[d]	Example	Reverse reaction	ΔNVE	ΔOS	ΔN
1. Lewis acid ligand dissociation	0	0	−1	$CpRh(C_2H_4)_2SO_2 \rightleftharpoons SO_2 + CpRh(C_2H_4)_2$	Lewis acid association	0	0	+1
or	0	−2	−1	$HCo(CO)_4 \rightleftharpoons H^+ + CoCO_4^-$	or	0	+2	+1
2. Lewis base ligand dissociation	−2	0	−1	$NiL_4 \rightleftharpoons NiL_3 + L$	Lewis base association	+2	0	+1
3. Reductive elimination	−2	−2	−2	$H_2IrCl(CO)L_2 \rightleftharpoons H_2 + IrCl(CO)L_2$	Oxidative addition	+2	+2	+2
4. Insertion	−2	0	−1	$MeMn(CO)_5 \rightleftharpoons MeCOMn(CO)_4$	Deinsertion	+2	0	+1
5. Oxidative coupling	−2	+2	0	$(C_2F_4)_2Fe(CO)_3 \rightleftharpoons$ 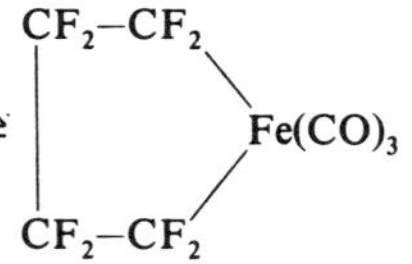 $\begin{array}{c} CF_2-CF_2 \\ \diagup \qquad \diagdown \\ \qquad\quad Fe(CO)_3 \\ \diagdown \qquad \diagup \\ CF_2-CF_2 \end{array}$	Reductive decoupling	+2	−2	0

[a] From Tolman.[44]

[b] Change in the number of metal valence electrons.

[c] Change in the formal oxidation state of the metal. The usual convention which regards hydrides, alkyls, π-allyls, and π-cyclopentadienyls as uninegative ions is used.

[d] Change in coordination number.

Elementary Organometallic Reactions

During the 1960's there was a great surge in interest in organometallic chemistry and the use of organometallic complexes as homogeneous catalysts for reactions which were or promised to be of considerable industrial significance, e.g., the Wacker process

$$2 C_2H_4 + O_2 \rightarrow 2 CH_3CHO$$

One result of this period of intense interest and investigation was the realization in the early 1970's that organometallic transformations and catalytic cycles could be analyzed as various sequences of a small number of elementary processes occurring at the metal center. Another important and unifying realization was that in virtually all cases the reactions involved diamagnetic complexes containing either 16 or 18 valence electrons. The "18-electron rule" had of course long been recognized as an upper limit electron configuration in transition metal chemistry, a consequence of the fact that only 9 stable orbitals can be obtained from s, d, and p atomic orbitals. The 16-electron case thus became referred to as "coordinative unsaturation," since in theory two more electrons or one Lewis base ligand could be accommodated at the metal center. Recently, Tolman has systematized the elementary organometallic reactions which occur at transition metal centers and integrated them with the concept of coordinative unsaturation in what he refers to as the 16- and 18-electron rule.[44] This rule proposes that

1. "Diamagnetic organometallic complexes of transition metals may exist in a significant concentration at moderate temperatures only if the metal's valence shell contains 16 or 18 electrons. A significant concentration is one that may be detected spectroscopically *or kinetically* and may be in the gaseous, liquid, or solid state.

2. "Organometallic reactions, including catalytic ones, proceed by elementary steps involving only intermediates with 16 or 18 metal valence electrons."

Tolman's classification of elementary organometallic reactions, listed in Table 11-4, is described below in Sections 1-5 in terms of the characteristics of each elementary process. The remainder of this chapter is then devoted to a discussion of some of the more important organometallic reagents and catalysts in terms of these elementary processes.

1. LEWIS ACID DISSOCIATION–ASSOCIATION

While most ligands are Lewis bases, electron-rich transition metal complexes with either 16 or 18 electrons can themselves be quite basic and hence form stable complexes with Lewis acids such as H^+, CH_3^+, BF_3, SO_2, or $HgCl_2$, and others. In general, low formal oxidation states ($\leqslant +1$) and electron donating ligands which are poorer back-bonders favor transition metal basicity. Basicity also increases as one descends a triad of the periodic table, which is opposite to the trend in the main group elements. The addition of a Lewis acid may or may not increase the *formal* oxidation state of the metal, but the coordination number always increases by one while the number of valence electrons remains constant. The addition of Lewis acids such as H^+, CH_3^+, or CH_3CO^+ is sometimes referred to as "oxidative addition," probably because in the adduct complex these ligands are usually regarded as H^-, CH_3^-, and CH_3CO^-, respectively. However, this should not be confused with the oxidative addition

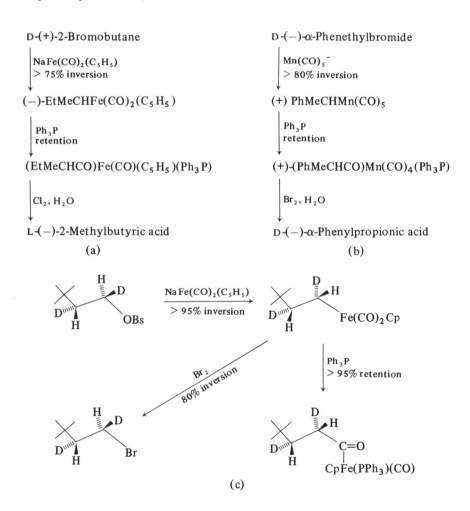

D-(+)-2-Bromobutane

$NaFe(CO)_2(C_5H_5)$
> 75% inversion

(−)-EtMeCHFe(CO)$_2$(C$_5$H$_5$)

Ph$_3$P
retention

(EtMeCHCO)Fe(CO)(C$_5$H$_5$)(Ph$_3$P)

Cl$_2$, H$_2$O

L-(−)-2-Methylbutyric acid

(a)

D-(−)-α-Phenethylbromide

$Mn(CO)_5^-$
> 80% inversion

(+) PhMeCHMn(CO)$_5$

Ph$_3$P
retention

(+)-(PhMeCHCO)Mn(CO)$_4$(Ph$_3$P)

Br$_2$, H$_2$O

D-(−)-α-Phenylpropionic acid

(b)

(c)

described in Section 3, in which the oxidation state, coordination number, and number of valence electrons each increase by two.

The proton occupies a special place as a Lewis acid in transition metal chemistry because the M–H bond is so important in isomerization,

TABLE 11-5

Acid Dissociation Constants for Some Metal Hydrides[a]

Complex	K_a
$HMn(CO)_5$	8.0×10^{-8}
$H_2Fe(CO)_4$	4×10^{-5} (K_1)
	4×10^{-14} (K_2)
$HCo(CO)_4$	~ 1
$HCo(CO)_3(PPh_3)$	$\sim 1 \times 10^{-7}$
$HV(CO)_6$	1
$HV(CO)_5(PPh_3)$	1.6×10^{-7}

[a] Data from Tolman,[1] Abel and Stone,[28] and Shriver.[45]

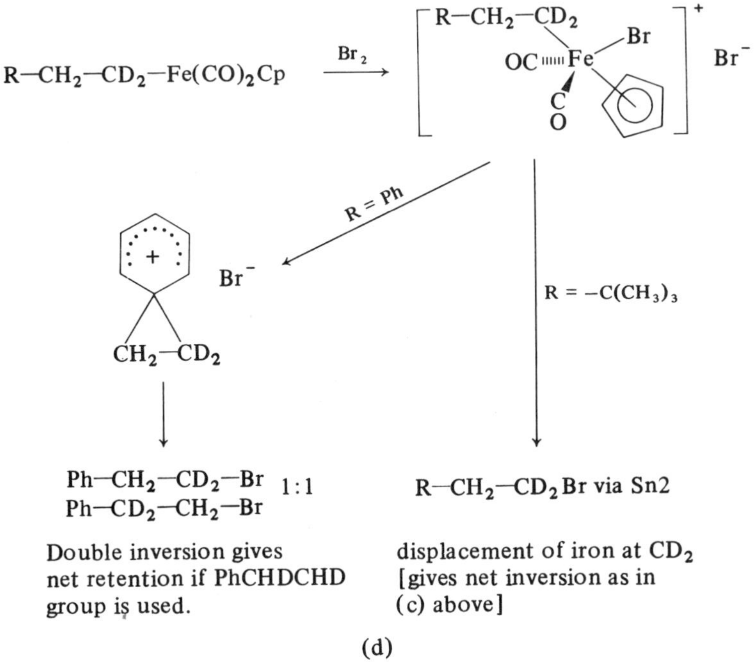
R—CH2—CD2—Fe(CO)2Cp
Br2
R—CH2—CD2
Br
OC
Fe
C
O
+
Br⁻
R = Ph
Br⁻
CH2—CD2
R = —C(CH3)3
Ph—CH2—CD2—Br
Ph—CD2—CH2—Br
1:1
R—CH2—CD2Br via Sn2
Double inversion gives net retention if PhCHDCHD group is used.
displacement of iron at CD2 [gives net inversion as in (c) above]
(d)

Fig. 11-7. Stereochemistry at carbon during formation and cleavage of some carbon-metal bonds: (a) and (b) from Johnson and Pearson,[46] (c) from Whitesides and Boschetto[47] and Bock *et al.*,[48] and (d) from Flood and Disanti.[49]

polymerization, and hydrogenation catalysis. The range of proton basicities of transition metal complexes is very large as can be seen from the data of Table 11-5.[1,28,45] Protonation of metal complexes leads to a major structural reorganization of the complex, as expected for a change in coordination number, and to the weakening of the bonding of other ligands to the metal. The latter effect can be seen in the increase in stretching frequencies for carbonyl groups from +75 to +175 cm^{-1} and in dramatic increases in the rate of exchange (via a dissociative mechanism) of coordinated carbonyls with ^{14}CO upon protonation of the metal.[1]

Basic metal complexes, especially metallate anions (so called "ate" complexes), are also nucleophilic and can be acylated or alkylated with acyl or alkyl halides. The usual reactivity pattern $X = I > Br > Cl$ and acyl > alkyl is

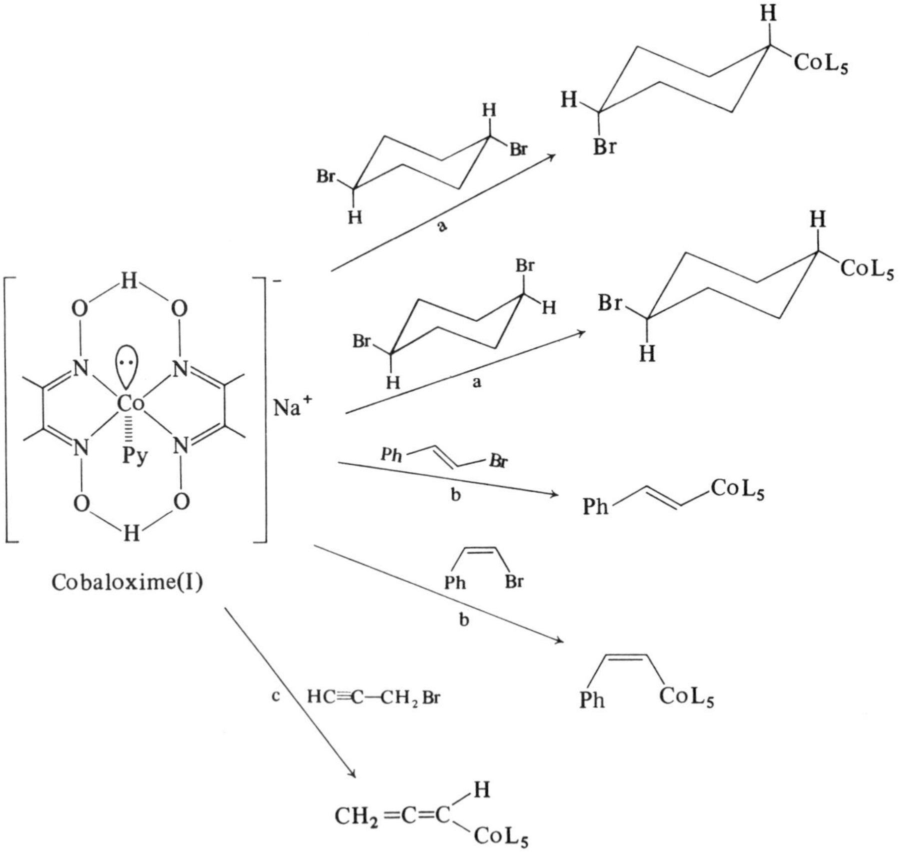

Fig. 11-8. Some nucleophilic reactions of cobaloxime(I). a, Jensen et al.[59]; b, Johnson and Meeks[8]; c, Johnson and Mayle.[51]

observed, and inversion of configuration at the alkylating carbon has been found in a number of cases (Fig. 11-7).[46-49] Metal complexes are made more nucleophilic by electron releasing ligands ($CN^- > Ph_3P > CO$), and complexes which can achieve a six-coordinate 18-electron-configuration are particularly nucleophilic. Two very important examples of the latter are the reduced form of vitamin B_{12} (B_{12}), and the cobaloxime(I) compound (Fig. 11-8)[8,50,51] developed by Schrauzer[52] as a chemical model for B_{12}. These complexes are among the strongest nucleophiles known.

Michael addition of nucleophilic metal complexes to unsaturated systems with electronegative substituents is also possible, as are addition–elimination sequences to give σ-bonded aryl, vinyl, allenyl, or acetylide complexes.

The chemistry of a number of other types of metal base-Lewis acid complexes has been reviewed by Shriver.[45] In many cases the adducts are thermodynamically very stable, yet the metal base can be readily displaced by a stronger Lewis base such as trimethylamine. Sulfur dioxide is an interesting Lewis acid because of its importance in both industrial processes and air pollution, and because it can also act as a Lewis base toward metal in higher oxidation states. For example, in the SO_2 adduct of Vaska's compound,[53] $IrCl(CO)(Ph_3P)_2(SO_2)$, the Ir–S bond makes an angle of $32°$ with the SO_2 plane, similar to the $22°$ angle in $Me_3N \cdot SO_2$, whereas in $[(NH_3)_4Ru(SO_2)Cl]Cl$ the $RuSO_2$ unit is planar,[54] presumably to allow for σ donation and π backbonding.

2. LEWIS BASE DISSOCIATION–ASSOCIATION

Ligand exchange reactions of organometallic complexes have been very thoroughly studied and reviewed. In general, 16-electron complexes react via an associative process involving an 18-electron intermediate. Conversely, an 18-electron complex generally reacts via a dissociative process and an unsaturated 16-electron intermediate; several examples are given in Chapter VI. Several "apparent exceptions" to this generalization have been interpreted by Tolman[44] in terms of nucleophilic attack on a coordinated ligand prior to a dissociation and deinsertion to give overall substitution, or in terms of an electronic rearrangement from an 18-electron complex to a 16-electron complex prior to nucleophilic attack (viz. nitrosyl complexes).

3. OXIDATIVE ADDITION—REDUCTIVE ELIMINATION

The term "oxidative addition" has been applied to a number of processes in which a metal center experiences an increase in oxidation state and coordination number. Since Lewis acid addition can also lead to these changes, the term "oxidative addition" is reserved for reactions in which there is also a net increase in the number of metal valence electrons. Hence, only complexes with 17 or fewer valence electrons are able to undergo oxidative addition processes similar to those shown in reactions (11-28) and (11-29).[55,56]

$$2[Co(CN)_5]^{3-} + X\text{—}Y \longrightarrow [Co(CN)_5 X]^{3-} + [Co(CN)_5 Y]^{3-}$$

$$XY = H_2, Br_2, HOOH, CH_3I, ICN, \text{etc.} \tag{11-28}$$

$$IrCl(CO)(Ph_3P)_2 + X\text{—}Y \longrightarrow IrCl(X)(Y)(CO)(Ph_3P)_2 \tag{11-29}$$

$$XY = H_2, Cl_2, HCl, CH_3I, RHgCl, R_3SiH, \text{etc.}$$

The ability of a coordinatively unsaturated metal complex to undergo oxidative addition depends on its ability to undergo oxidation per se, which in turn depends on the electron releasing tendency of the ligands (e.g., $C_5H_5^- >$ $R_3P > Ph_3P > CO$). Oxidative addition can be an *intramolecular* process involving C—H bonds, notably with arylphosphine, arylphosphite, arylamine, and azobenzene ligands, as shown in reactions (11-30) and (11-31)[57] and in Fig. 10-15. Oxidative addition may also relate formal canonical resonance forms

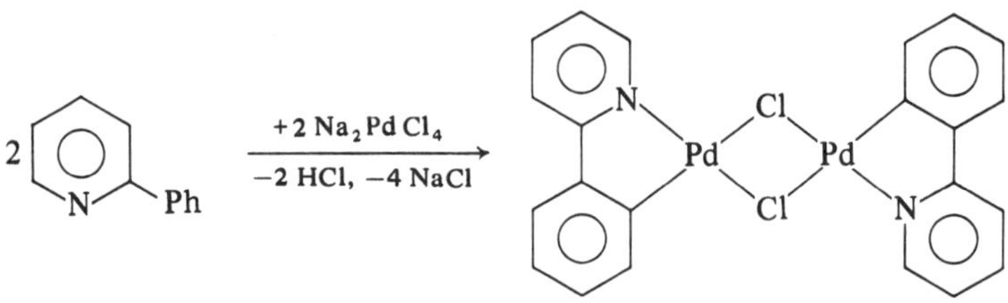

$$(11\text{-}30)$$

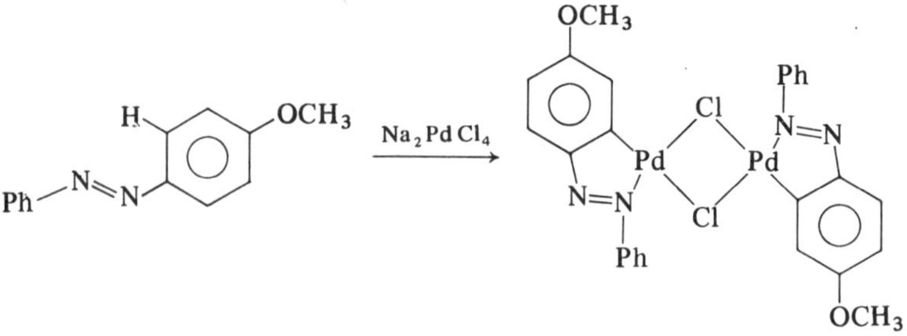

$$(11\text{-}31)$$

of a complex as in the butadiene iron tricarbonyl complex $(C_4H_6)Fe(CO)_3$ (11-32). The conversion of the $d^8Fe(0)$ complex to the $d^6Fe(II)$ complex is *formally* an oxidative addition, and with highly electronegative ligands like

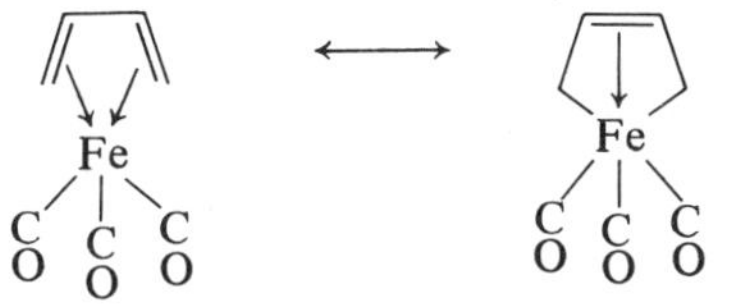

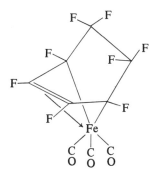

$$d^8(5) \qquad\qquad d^6(6)$$

(11-32)

perfluorobutadiene the $Fe(II)L^{2-}$ resonance form is more reasonable. An example of this type of bonding is seen below, in which the C_6 ring forms a 47° dihedral.[58] The notation $d^n(N)$ is a shorthand way of representing the number

of metal d electrons (n) and ligand donor pairs (N), and hence the total number of valence electrons (n + 2N), in a given transition metal complex.

The occurrence of an oxidative addition (or a Lewis acid addition) at a metal

TABLE 11-6

Variation of ν_{C-O} Upon Oxidative Addition and Lewis Acid Addition to *trans*-$IrCl(CO)(Ph_3P)_2$,[a]

$$IrCl(CO)(Ph_3P)_2 + XY \rightleftharpoons adduct$$

XY	ν_{C-O} (cm^{-1})	Relative oxidation state	Reversibility
—	1967	1.00, Ir(I)	—
O_2	2015	1.84	Easily
SO_2	2021	2.00	Easily
D_2	2034	2.24	Easily
C_2F_4	2052	2.57	Stable
TCNE	2057	2.67	Stable
BF_3	2067	2.85	Stable
I_2	2067	2.85	Irreversible
Cl_2	2075	3.00, Ir(III)	Irreversible

[a] Data from Vaska.[59]

center affects the interactions of the metal with its other ligands. An easily observable indicator of this is the C–O stretching frequency of a coordinated carbonyl group: As electron density on the metal decreases ν_{C-O} increases as a result of decreasing ability of the metal to back-bond. Table 11-6 shows some results for oxidative addition and Lewis acid addition to Vaska's complex. The arbitrary assignment of a "relative oxidation state" is based on interpolation of ν_{C-O} between that of the Ir(I) complex and that of the Cl_2 adduct, which is assumed to be fully oxidized to Ir(III). Note that "oxidative addition" of D_2 produces *less* of a change in ν_{C-O} than addition of the Lewis acid BF_3. As discussed in Chapter X, and as Table 11-6 shows, dioxygen is best regarded as a π Lewis acid, while TCNE and C_2F_4 add substantially more "oxidatively" to the Ir(I) complex.[59] Finally, for the oxidative addition of carboxylic acids to $Ir(X)(CO)L_2$ there is an approximately linear correlation[60] between log K and the pK_a of RCO_2H,

$$Ir(X)(CO)L_2 + RCO_2H \xrightleftharpoons{K} Ir(X)(RCO_2)(H)(CO)L_2$$

and the equilibrium constant decreases in the order $X = I > Br > Cl$ and $L = Me_3P > Me_2PhP > MePh_2P > Ph_3P$.

Oxidative additions to $d^7(5)$ and $d^8(4)$ complexes, e.g., reactions such as (11-28) and (11-29), have been studied in greater detail than any other type. Both the mechanisms and products of the reactions depend mainly on the nature of the X–Y addend and the electronic configuration of the metal center.

The $d^8(4) \rightarrow d^6(6)$ Case

In benzene solution the addition of H_2 to *trans*-$IrCl(CO)(Ph_3P)_2$, (**A** in

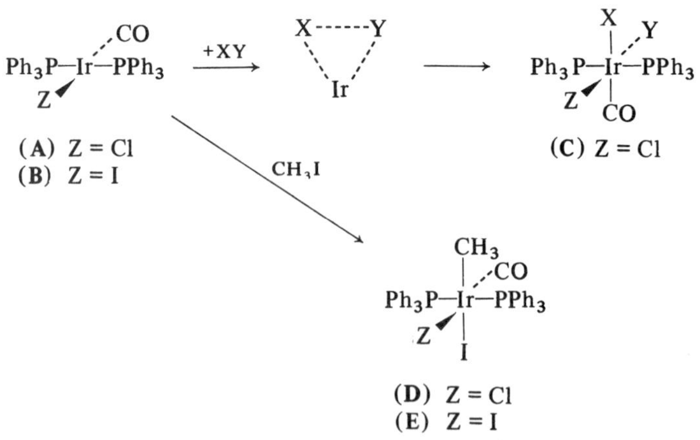

(A) Z = Cl
(B) Z = I

(C) Z = Cl

(D) Z = Cl
(E) Z = I

Scheme 11-9

Scheme 11-9), follows a simple bimolecular rate law first order in each reactant, and produces an octahedral complex having a *cis* H–Ir–H unit (**C** in Scheme 11-9). Addition of HCl and HBr to complex **A** is very fast in solution and also occurs in a *cis* fashion, as does the addition of gaseous HCl to *solid* **A**. Since ionic mechanisms would be unlikely to occur under these reaction conditions, these results have been interpreted in terms of a one-step addition via a three-center mechanism.

The addition of CH_3I to complex **A** also follows a bimolecular rate law in benzene, but in this case the product, **D**, contains a *trans*-CH_3–Ir–I unit. However, the formation of the same product from the reaction of *solid* **A** with gaseous CH_3I, and the lack of incorporation of $^{131}I^-$ into complex **E** during its formation in solution, argue against a two-step ionic mechanism involving Lewis acid (CH_3^+) addition followed by Lewis base (I^- or $^{131}I^-$) addition.[55]

According to Pearson, orbital symmetry considerations allow for either cis or trans addition to occur by one-step processes, as shown in Diagram 11-1.[61]

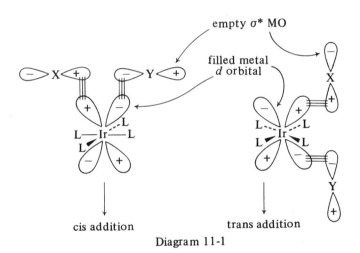

Diagram 11-1

This type of analysis suggests that oxidative addition should proceed with *retention* of configuration at carbon, in contrast to the inversion observed in the Sn2-like Lewis acid-metal nucleophile reactions (Fig. 11-7). Pearson and Muir[62] have shown that optically active $CH_3CHBrCO_2Et$ adds to *trans*-IrCl(CO)(Ph$_2$PMe)$_2$ to give an optically active adduct, and that treatment of the latter with Br$_2$ regenerates optically active bromoester with net retention of configuration. Thus this sequence involved either two inversions or two retentions of configuration. Since it has not been separately shown if cleavage of C–Ir(III) bonds occurs with retention or inversion of configuration, and since both types of cleavage are seen with other metals, the stereochemistry of oxidative addition at carbon is not completely settled. A free radical mechanism has been implicated for a related oxidative addition reaction[63] and ruled out in another.[64]

The $d^7(5) \rightarrow d^6(6)$ Case

Oxidative addition to low spin $d^7(5)$ ions usually is a two-step process involving homolysis of the X–Y bond since only one of the two electrons of the X–Y bond can add to the 17-electron metal center (reactions 11-33).[56] In the special

$$Co(CN)_5{}^{3-} + XY \xrightarrow{\ k\ } Co(CN)_5 X^{3-} + Y\cdot$$

$$Co(CN)_5{}^{3-} + Y\cdot \xrightarrow{\ \text{fast}\ } Co(CN)_5 Y^{3-}$$

$$\overline{2\,Co(CN)_5{}^{3-} + X\!-\!Y \longrightarrow Co(CN)_5 X^{3-} + Co(CN)_5 Y^{3-}} \tag{11-33}$$

$$\text{Rate} = k\,[Co(CN)_5{}^{3-}]\,[XY]$$

case XY = H_2, the H–H bond is too strong (104 kcal/mole) to be split homolytically by the Co species (Co–H = 58 kcal/mole), and the rate law shows the need for two cobalt species to participate in its cleavage (reaction 11-34).[56] High spin d^4 ions such as Cr(II) show reactivity patterns similar to

$$2\,Co(CN)_5{}^{3-} + H_2 \xrightarrow{\ k\ } 2\,CoH(CN)_5{}^{3-} \tag{11-34}$$

$$\text{Rate} = k\,[Co(CN)_5{}^{3-}]^2\,[H_2]$$

the low spin d^7 case. For example, chromous ion conveniently reduces 1,2-dibromides and epoxides to a mixture of cis and trans olefins.[65] In contrast, stereospecific reduction is possible with $NaCp(CO)_2Fe$ via nucleophilic mechanisms (Scheme 11-10).[66]

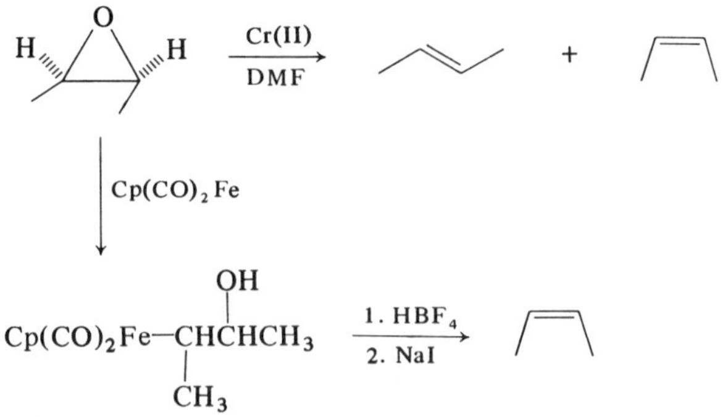

Scheme 11-10

The $d^{10}(3) \rightarrow d^8(5)$ Case

A recent study using the spin trap molecule $(CH_3)_3CNO$ has shown that a nonchain free radical reaction may be involved in the oxidative addition of alkyl halides RX to $Pt(Ph_2P)_3$. The proposed mechanism is shown in reactions (11-35–11-37).[67] In agreement with these reactions the oxidative addition of

$$(Ph_3P)_3Pt + RX \xrightarrow{\text{slow}} (Ph_3P)_3PtX + R\cdot \tag{11-35}$$

$$R\cdot + (Ph_3P)_3PtX \xrightarrow{\text{fast}} trans\text{-}(Ph_3P)_3RPtX \tag{11-36}$$

$$R\cdot + (CH_3)_3CNO \xrightarrow{\text{fast}} (CH_3)_3C\text{-}N\text{-}R \rightarrow ESR \tag{11-37}$$
$$\underset{\overset{|}{O\cdot}}{}$$

optically active 1-phenyl-2,2,2-trifluoroethyl chloride to $(Ph_3P)_3Pd$ produces a racemic adduct.[68]

4. INSERTION–DEINSERTION

Oxidative addition

$$X\text{–}Y + M \leftrightarrow X\text{–}M\text{–}Y$$

Insertion

$$Y + M\text{–}X \leftrightarrow M\text{–}Y\text{–}X$$

As shown by the above equations, the insertion reaction is in a sense complementary to the oxidative addition reaction. Collectively these two kinds of reactions are probably the most important group of elementary organometallic reactions, for their combination in sequences or catalytic cycles is opening exciting new horizons for synthetic organic chemistry. For example, a recent industrial process for acetic acid utilizing the unlikely raw materials CO and CH_3OH most likely involves conversion of methanol to methyl iodide with HI, oxidative addition of MeI to a Rh(I) species, insertion of CO into the Me–Rh

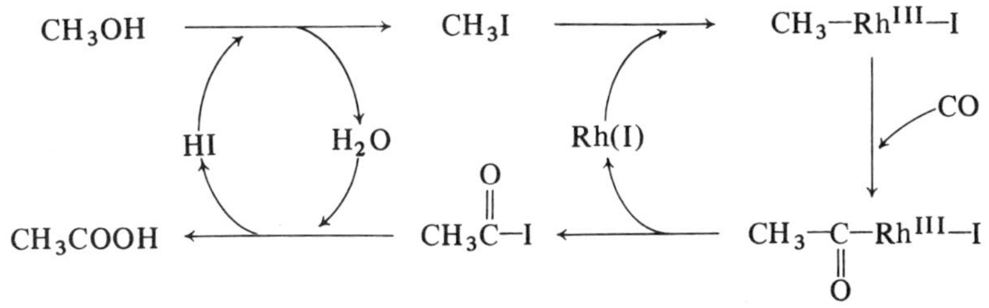

Scheme 11-11

bond, and reductive elimination of CH_3COI from a Rh(III) intermediate to form CH_3CO_2H and HI. The Rh(I) catalyst is formed from $RhCl_3 \cdot 3H_2O$ by reduction with methanol (Scheme 11-11).[69]

Other types of insertion reactions of considerable importance involve insertions of CO or C_2H_4 into metal hydride or metal alkyl bonds. These reactions are the basis of several important industrial processes which will be discussed later.

Mechanistically the insertion reaction can be regarded as a one-step direct insertion process, or it can be expressed as a two-step process involving initial coordination of the molecule being inserted, followed by an intramolecular migration giving the same overall result. For Lewis base molecules such as CO and C_2H_4 (and in *certain* cases SO_2) it appears that insertion involves the two-step process of coordination followed by migration. This of course necessitates that the complex in question be able to achieve a coordinatively unsaturated intermediate form. The insertion of CO into alkyl-metal bonds has been studied extensively and in all cases has been shown to proceed with *retention* of configuration at the migrating carbon (Fig. 11-7). The same has been shown for the reverse process of CO extrusion from optically active aldehydes catalyzed by $RhCl(Ph_3P)$,[70,71] and deuterated aldehydes RCDO give deuteroalkanes R—D (reaction 11-38). This reaction can be envisioned as a sequence of oxidative addition of RCO—H to $RhCl(Ph_3P)_3$, followed by loss of a phosphine to

$$R^*\!-\!\overset{\displaystyle O}{\overset{\|}{C}}\!-\!H \xrightarrow{\text{RhCl(Ph}_3\text{P)}_3} R^*\!-\!\overset{\displaystyle O}{\overset{\|}{C}}\!-\!Rh\!-\!H \longrightarrow R^*\!-\!H + CO \qquad (11\text{-}38)$$
$$(D) \qquad\qquad\qquad\qquad (D) \qquad\qquad\qquad (D)$$
$$\searrow$$
$$Rh(I)$$

give a $d^6(5)$ acylrhodium(III) hydride. Migration of the alkyl from CO to Rh followed by reductive elimination of alkyl and hydride gives optically active alkane R^*—H plus CO and regenerated catalyst.

A number of kinetic and mechanistic studies suggest that *both* CO insertion into alkyl-metal bonds and the reverse decarbonylation reaction involve two steps with a 16-electron intermediate. The carbonylation of $CH_3Mn(CO)_5$ follows a rate law consistent with a rate limiting alkyl migration to form $CH_3COMn(CO)_4$ which then reacts with CO or any other added ligand. The greater strength of the Ph—Mn bond prevents the occurrence of the migration and hence the carbonylation of $PhMn(CO)_5$, which is, in fact, produced by decarbonylation of $PhCOMn(CO)_5$ obtained from PhCOCl and $NaMn(CO)_5$. An elegant infrared study of the carbonylation of $CH_3Mn(CO)_5$ with ^{13}CO has shown that the acyl group is not formed from ^{13}CO, consistent with migration of CH_3 from Mn to CO prior to Lewis base addition (Scheme 11-12).[72] The same basic mechanism can be used to describe the carbonylation steps in Fig.

$$d^6 (6) \quad CH_3Mn(CO)_5 \quad \overset{\text{migration}}{\underset{k_{-1}}{\overset{k_1}{\rightleftharpoons}}} \quad CH_3CO-Mn(CO)_4 \quad d^6 (5)$$

$$k_2 \downarrow \quad \begin{matrix} L = {}^{13}CO, R_3P, (RO)_3P, R_3N \\ \text{addition} \end{matrix}$$

$$cis\text{-}CH_3COMn(L)(CO)_4$$

$$d[\text{product}]/dt = k_1 [CH_3Mn(CO)_5] \qquad \text{if } k_2[L] \gg k_{-1}$$

Scheme 11-12

11-7, as well as for the reverse direction. The importance of the 16-electron intermediate is shown by the fact that ease of decarbonylation of the 18-electron complexes shown in Scheme 11-13 parallels the ease with which they can dissociate a ligand to form a 16-electron species.[73]

$$CH_3CO-M \quad \longrightarrow \quad CH_3-M + CO$$

$$M = CO(CO)_4 > Fe(CH_3CO)(CO)_4 > Mn(CO)_5 > Fe(CO)_2(C_5H_5)$$

| fast at | heat | $h\nu$ |
| room temperature | required | required |

Scheme 11-13

Ethylene, olefins, and acetylenes can also be inserted into metal-hydride and metal-alkyl bonds. Again, the reaction is best viewed as a two-step sequence of addition to an obligate 16-electron intermediate followed by migration of hydride or alkyl to the coordinated substrate. The dimerization of ethylene to butenes by rhodium catalysts illustrates several characteristic features of olefin insertion reactions.[74] Rhodium hydrides are prepared at low temperatures by oxidative addition of HCl to Rh(I) complexes. These species add C_2H_4 and rearrange to give ethylrhodium(III) species in a facile reversible process leading to H/D exchange between deuterated ethylenes (Scheme 11-14). Alkyl groups

$$Rh-D + C_2H_4 \qquad\qquad\qquad Rh-H + CH_2=CHD$$
$$\updownarrow \qquad\qquad\qquad\qquad\qquad \updownarrow$$
$$Rh(D)(C_2H_4) \rightleftharpoons RhCH_2CH_2D \rightleftharpoons Rh(H)(CH_2=CHD)$$

Scheme 11-14

can also migrate from Rh(III) to coordinated ethylene (or other olefins or acetylenes) to give two-carbon chain elongation, but this irreversible migration is slower than hydride migration and considerably slower than exchange of coordinated olefins. Deinsertion of a new olefin after insertion of ethylene regenerates the rhodium hydride. This is a favored step and terminates the dimerization of ethylene before higher oligomers (e.g., hexenes) form because the equilibrium lies far to the right ($K \approx 10^3$) (reaction 11-39). Hence, the

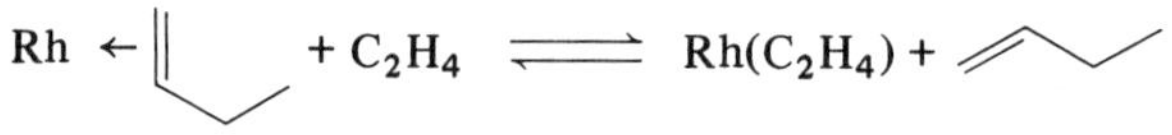

$$Rh \leftarrow \Vert \qquad + C_2H_4 \rightleftharpoons Rh(C_2H_4) + \diagup\diagdown \qquad (11\text{-}39)$$

overall reaction may be represented by Scheme 11-15. Metal hydrides and metal alkyls which are coordinatively saturated do not react with olefins without prior dissociation of a ligand. Sixteen-electron metal hydrides add

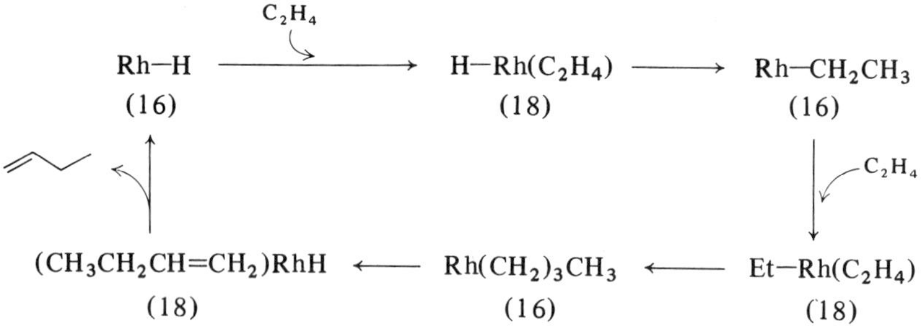

Scheme 11-15

olefins to give predominantly the metal alkyl with the least number of β-hydrogens since alkyls having a large number of β-hydrogens have the greatest opportunity to undergo olefin deinsertion, e.g., Scheme 11-16. However an

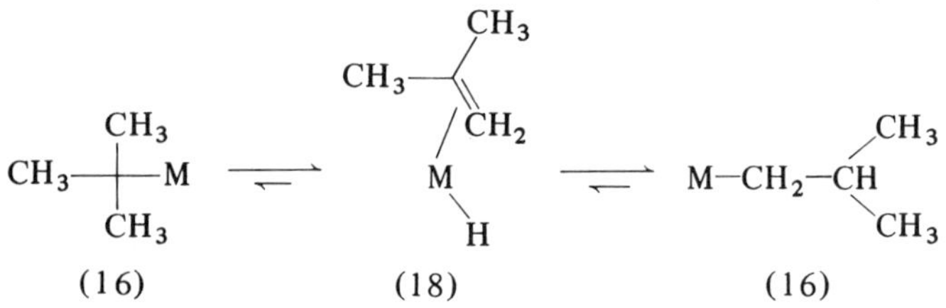

Scheme 11-16

18-electron iron complex containing a t-butyl ligand (A in Scheme 11-3), is stable since it cannot easily attain the 16-electron state required for an olefin deinsertion to occur.

The insertion of SO_2 in a metal-alkyl bond may differ from CO insertion in mechanistic detail. In at least one case involving an 18-electron complex it has been clearly shown that net *inversion* occurs at the carbon center involved (reaction 11-40).[47,48] This can be contrasted to the recently observed insertion of SO_2 with retention of configuration at carbon in a related 16-electron complex (reaction 11-41).[75] One explanation of this difference is that SO_2 inserts

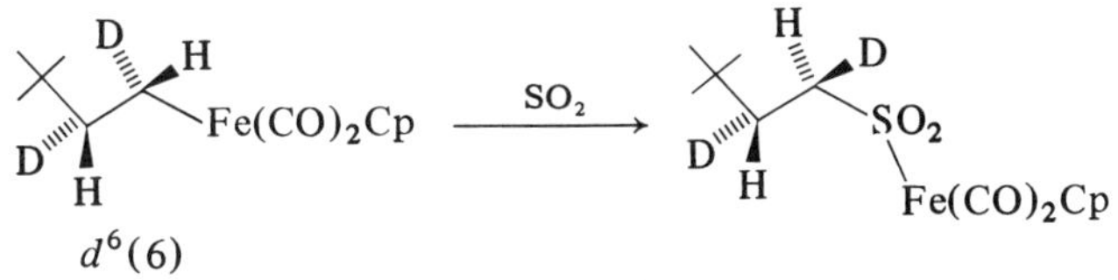

(11-40)

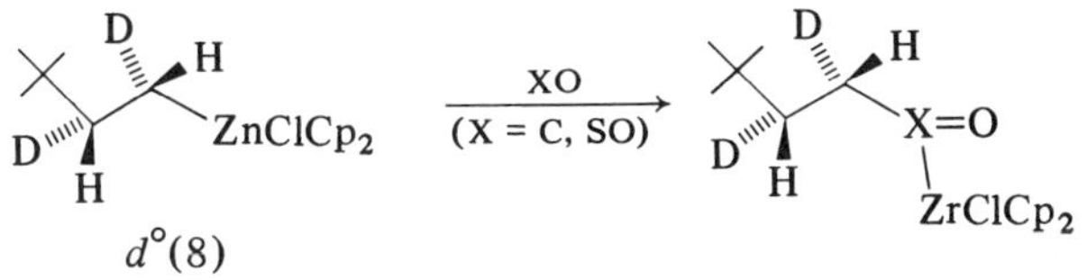

(11-41)

into the 16-electron Zr complex via the addition-migration sequence (CO also inserts with retention) while the 18-electron complex reacts via initial attack of SO_2 at carbon to give net inversion in the overall insertion process (reaction 11-40). However, it is also possible that a double migration occurs with net inversion in the iron complex (reaction 11-42).

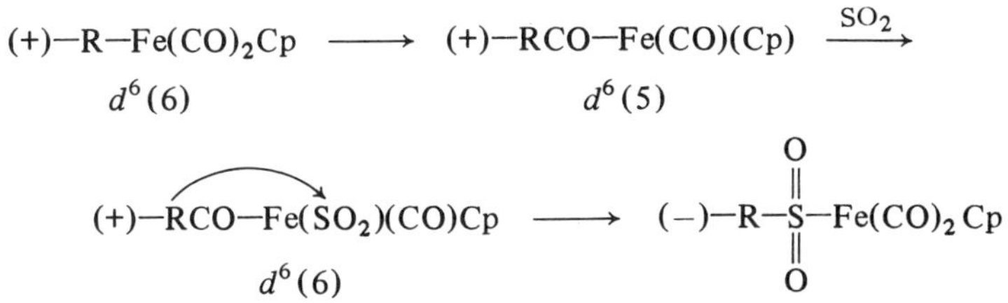

$$(+)-R-Fe(CO)_2Cp \longrightarrow (+)-RCO-Fe(CO)(Cp) \xrightarrow{SO_2}$$
$$d^6(6) \qquad\qquad d^6(5)$$

(11-42)

$$(+)-RCO-Fe(SO_2)(CO)Cp \longrightarrow (-)-R-\overset{\overset{O}{\|}}{\underset{\underset{O}{\|}}{S}}-Fe(CO)_2Cp$$
$$d^6(6)$$

5. OXIDATIVE COUPLING–REDUCTIVE DECOUPLING

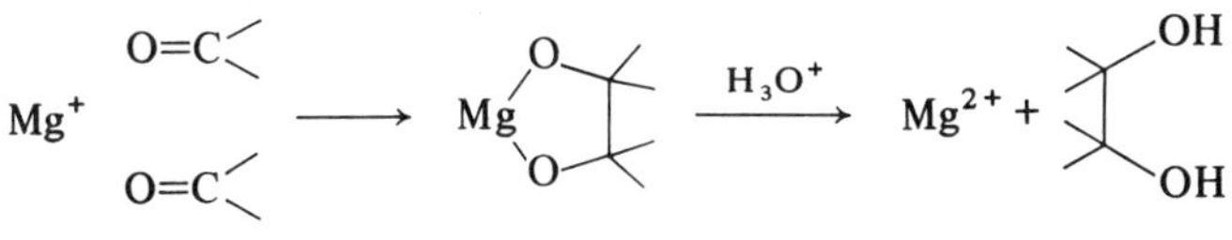

This process is really a special type of oxidative addition in which two ligands become one. A good "organic" analogy is the pinacol synthesis from magnesium metal and acetone, and another is the P(III) → P(V) oxidation with hexafluoroacetone (reactions 11-43 and 11-44). The reaction of $(MeO)_3P$ with biacetyl, however, is more strictly an oxidative addition. A number of cyclo-oligomerizations of olefins and acetylenes catalyzed by transition metal complexes of Cr, Co, Fe, Ir, Ni, Pd, and Rh almost certainly involve oxidative coupling and reductive elimination steps. Several examples are given in reactions (11-45)[76] and (11-46).[77]

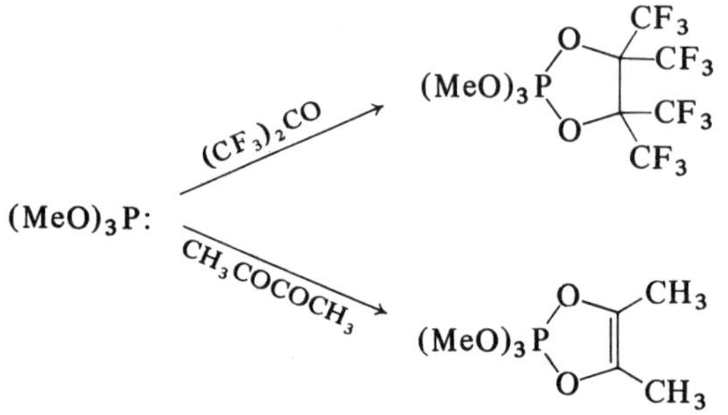

$$(11\text{-}43)$$

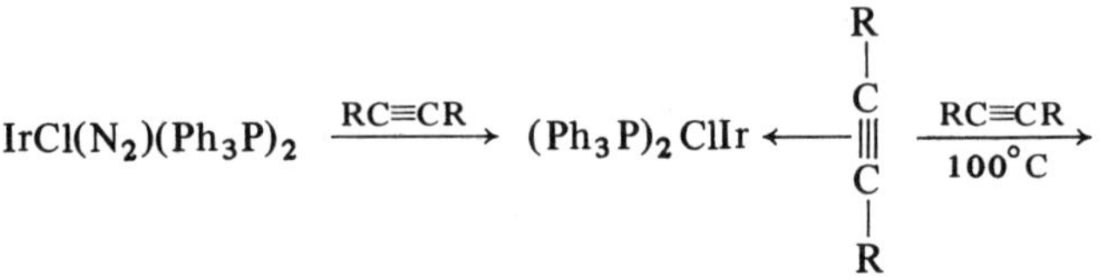

$$(11\text{-}44)$$

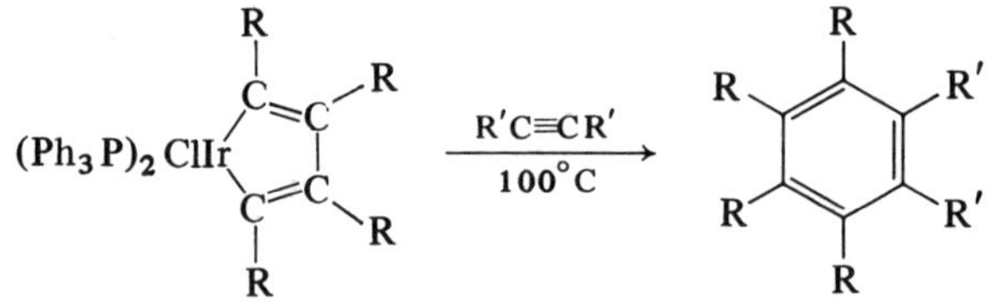

$$(11\text{-}45)$$

(isolable)

$$R = CO_2CH_3, \quad R' = CO_2CD_3$$

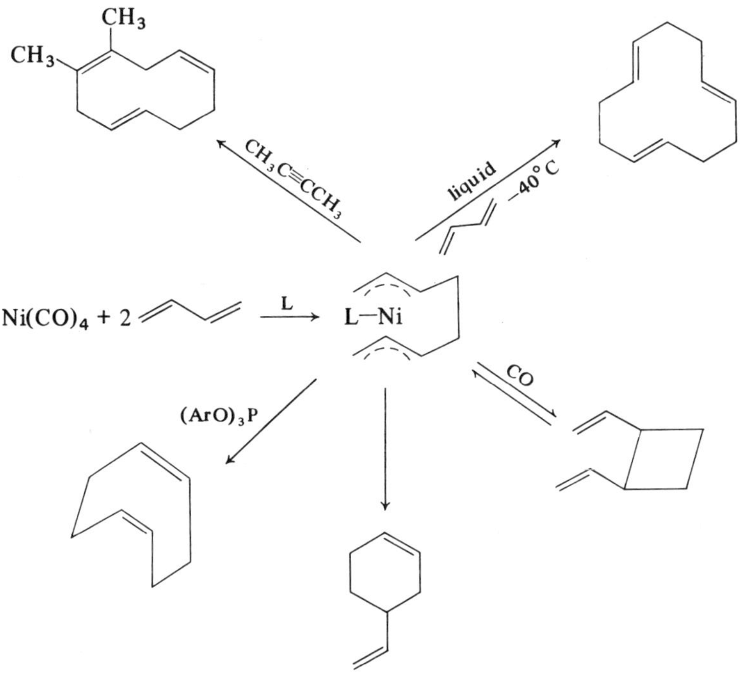

CH₃ ... (reaction scheme)

$$Ni(CO)_4 + 2 \xrightarrow{L} L-Ni \qquad (11\text{-}46)$$

The scheme shows: $CH_3C{\equiv}CCH_3$, liquid $-40°C$, $(ArO)_3P$, and CO as reaction conditions.

REFERENCES

1. C. A. Tolman, *in* "Transition Metal Hydrides" (E. L. Muetterties, ed.), Vol. I, pp. 271–331. Dekker, New York, 1971.
2. J. P. McCue, *Coord. Chem. Rev.* **10,** 265 (1973).
3. F. Basolo, J. Chatt, H. B. Gray, R. G. Pearson, and B. L. Shaw, *J. Chem. Soc. (London),* p. 2207 (1961).
4. G. Wilkinson, *Science* **185,** 109 (1974).
5. P. S. Braterman and R. J. Cross, *Chem. Soc. Rev.* **2,** 271 (1973).
6. M. Green, *MTP Int. Rev. Sci., Inorg. Chem., Ser. 1.* **6,** 171 (1972).
7. W. P. Giering and M. Rosenblum, *J. Organometal. Chem.* **25,** C71 (1970).
8. M. D. Johnson and B. S. Meeks, *J. Chem. Soc., B* p. 185 (1971); *Chem. Commun.* p. 1027 (1970).
9. D. J. Cardin, B. Cetinkaya, M. J. Doyle, and M. F. Lappert, *Chem. Soc. Rev.* **2,** 99 (1973).
10. R. Stewart and P. M. Treichel, *J. Amer. Chem. Soc.* **92,** 2710 (1970).
11. P. W. Jolly and R. Pettit, *J. Amer. Chem. Soc.* **88,** 5044 (1966).
12. O. S. Mills and A. D. Redhouse, *J. Chem. Soc., A* p. 642 (1968).
13. R. Mason, *Nature (London)* **217,** 543 (1968).
14. F. R. Hartley, *Angew. Chem., Int. Ed. Engl.* **11,** 596 (1972).
15. F. R. Hartley, *Chem. Rev.* **73,** 163 (1973).
16. J. A. Wunderich and D. P. Mellor, *Acta Crystallogr.* **8,** 57 (1955).
17. M. Black, R. H. B. Mais, and P. G. Owston, *Acta Crystallogr.* **B25,** 1753 (1969).

18. N. C. Baenziger, G. F. Richards, and J. R. Doyle, *Acta Crystallogr.* **18,** 924 (1965).
19. J. R. Holden and N. C. Baenziger, *J. Amer. Chem. Soc.* **77,** 4987 (1955).
20. C. Pedone and A. Sirigu, *Inorg. Chem.* **7,** 2614 (1968).
21. A. R. Luxmoore and M. R. Truter, *Acta Crystallogr.* **15,** 1117 (1962).
22. G. R. Davies, W. Hewertson, R. H. B. Mais, and P. G. Owston, *Chem. Commun.* p. 423 (1967).
23. J. O. Glanville, J. M. Stewart, and S. O. Grim, *J. Organometal. Chem.* **7,** P9 (1967).
24. M. A. Benett, G. B. Robertson, P. O. Whimp, and T. Yoshida, *J. Amer. Chem. Soc.* **93,** 3797 (1971).
25. W. G. Sly, *J. Amer. Chem. Soc.* **81,** 18 (1959).
26. R. Cramer, J. B. Cline, and J. D. Roberts, *J. Amer. Chem. Soc.* **91,** 2519 (1969).
27. E. M. Ban, R. P. Hughes, and J. Powell, *Chem. Commun.* p. 591 (1973).
28. E. W. Abel and F. G. A. Stone, *Quart. Rev. Chem. Soc.* **23,** 325 (1969); **24,** 498 (1970).
29. C. G. Pierpont and R. Eisenberg, *Inorg. Chem.* **11,** 1088 (1972).
30. J. P. Collman, P. Fornham, and G. Dolcetti, *J. Amer. Chem. Soc.* **93,** 1788 (1971).
31. C. P. Brock, J. P. Collman, G. Dolcetti, P. H. Farnham, J. A. Ibers, J. E. Lester, and C. A. Reed, *Inorg. Chem.* **12,** 1304 (1973).
32. R. Eisenberg and C. D. Meyer, *Accounts Chem. Res.* **8,** 26 (1975).
33. J. Powell, *MTP Int. Rev. Sci., Inorg. Chem., Ser. 1* **6,** 273 and 309 (1972).
33a. T. A. Stephenson, *MTP Int. Rev. Sci., Inorg. Chem., Ser. 1* **6,** 401 (1972).
34. C. A. Bear, W. R. Cullen, J. P. Kutney, V. E. Ridaura, J. Trotter, and A. Zanorotti, *J. Amer. Chem. Soc.* **95,** 3058 (1973).
35. R. Mason and A. G. Wheeler, *Nature (London)* **217,** 1254 (1968); *J. Chem. Soc., A* p. 2549 (1968).
36. A. E. Smith, *Acta Crystallogr., Sect. A* **25,** S3 and S161 (1969).
37. R. E. Davis and R. Pettit, *J. Amer. Chem. Soc.* **92,** 716 (1970).
38. R. Mason and G. B. Robertson, *J. Chem. Soc., A* p. 1229 (1970).
39. M. R. Churchill and K. Gold, *Chem. Commun.* p. 693 (1968).
40. R. P. Dodge and V. Schomaker, *Acta Crystallogr.* **18,** 614 (1965).
41. L. T. J. Delebaere, M. N. G. James, N. Nakamura, and S. Masamune, *J. Amer. Chem. Soc.* **97,** 1973 (1975).
42. F. A. Cotton, *Accounts Chem. Res.* **1,** 257 (1968).
43. A. Carbonaro, A. L. Segre, A. Greco, C. Tosi, and G. Dall'Asta, *J. Amer. Chem. Soc.* **90,** 4453 (1968).
44. C. A. Tolman, *Chem. Soc. Rev.* **1,** 337 (1972).
45. D. F. Shriver, *Accounts Chem. Res.* **3,** 231 (1970).
46. R. W. Johnson and R. G. Pearson, *Chem. Commun.* p. 986 (1970).
47. G. M. Whitesides and D. J. Boschetto, *J. Amer. Chem. Soc.* **91,** 4313 (1969).
48. P. C. Bock, D. J. Boschetto, J. R. Rasmussen, J. P. Demey, and G. M. Whitesides, *J. Amer. Chem. Soc.* **96,** 2814 (1974).
49. T. C. Flood and F. J. DiSanti, *Chem. Commun.* p. 18 (1975).
50. F. R. Jensen, V. Madan, and D. H. Buchanan, *J. Amer. Chem. Soc.* **92,** 1414 (1970).
51. M. D. Johnson and C. Mayle, *J. Chem. Soc., D* p. 192 (1970).
52. G. N. Schrauzer, E. D. Deutsch, and R. J. Windgassen, *J. Amer. Chem. Soc.* **90,** 2441 (1968).
53. S. J. LaPlaca and J. A. Ibers, *Inorg. Chem.* **5,** 405 (1966).
54. L. H. Vogt, J. L. Kutz, and S. E. Wiberley, *Inorg. Chem.* **4,** 1157 (1965).
55. A. J. Deeming, *MTP Int. Rev. Sci., Inorg. Chem., Ser. 1* **9,** 117 (1972).
56. J. Halpern, *Accounts Chem. Res.* **3,** 386 (1970).
57. G. W. Parshall, *Accounts Chem. Res.* **3,** 139 (1970).
58. M. R. Churchill and R. Mason, *Proc. Chem. Soc., London* p. 226 (1964).

59. L. Vaska, *Accounts Chem. Res.* **1**, 335 (1968).
60. A. J. Deeming and B. L. Shaw, *J. Chem. Soc., A* p. 1802 (1969).
61. R. G. Pearson, *Accounts Chem. Res.* **4**, 152 (1971).
62. R. G. Pearson and W. R. Muir, *J. Amer. Chem. Soc.* **92**, 5519 (1970).
63. J. S. Bradley, D. E. Connor, D. Dolphin, J. A. Labinger, and J. A. Osborn, *J. Amer. Chem. Soc.* **94**, 4043 (1972).
64. J. P. Collman, D. W. Murphy, and G. Dolcette, *J. Amer. Chem. Soc.* **95**, 2687 (1973).
65. J. K. Kochi and D. M. Singleton, *J. Amer. Chem. Soc.* **90**, 1582 (1968).
66. W. P. Giering, M. Rosenblum, and J. Tancrede, *J. Amer. Chem. Soc.* **94**, 7170 (1972).
67. M. F. Lappert and P. W. Lednor, *Chem. Commun.* p. 948 (1973).
68. J. K. Stille, L. F. Hines, R. W. Fries, P. K. Wong, D. E. James, and K. Lan, *Advan. Chem. Ser.* **132**, 90 (1974).
69. F. E. Paulik and J. F. Roth, *Chem. Commun.* p. 1578 (1968); *Chem. Eng. News* **49**, 19 (1971).
70. H. M. Walborsky and L. E. Allen, *J. Amer. Chem. Soc.* **93**, 5466 (1971); *Tetrahedron Lett.* 823 (1970).
71. J. Tsuji and K. Ohno, *Advan. Chem. Ser.* **70**, 155 (1968).
72. K. Noack and F. Calderazzo, *J. Organometal. Chem.* **10**, 101 (1967).
73. R. B. King, *Accounts Chem. Res.* **3**, 417 (1970).
74. R. Cramer, *Accounts Chem. Res.* **1**, 186 (1968).
75. J. A. Labinger, D. W. Hart, W. E. Seibert III, and J. Schwartz, *J. Amer. Chem. Soc.* **97**, 3851 (1975).
76. J. P. Collman, J. W. Kang, W. F. Little, and M. F. Sullivan, *Inorg. Chem.* **7**, 1298 (1968).
77. P. Heimbach, W. Jolly, and G. Wilke, *Advan. Organometal Chem.* **8**, 29 (1970).

XII

Reactions of Ligands in Organometallic Complexes

Factors Affecting Ligand Reactivity

The use of metal complexes as catalysts or reagents for organic synthesis clearly provides a major impetus for research in organometallic chemistry. Indeed, an enormous amount of such research has been carried out, described, and reviewed numerous times.[1–6b] The past 15 years has witnessed the development, application, and even commercialization of organometallic reagents and catalysts for processes which have no counterpart in organic chemistry. To attempt a comprehensive review of this area would require several volumes. Therefore we will follow the style set with Chapters VIII and IX, attempting to provide a broad spectrum of examples which illustrate the principles involved in governing the reactivity of *ligands* in organometallic complexes. As before, it is appropriate to begin here by mentioning briefly the factors which must be considered in discussing the modification of ligand behavior by metal complexation.

MULTIDENTATE VS MULTIHAPTO LIGANDS

With coordination complexes chelation is a major consideration in most reactions of coordinated ligands, the main exceptions being certain metal-assisted solvolysis reactions. With organometallic complexes chelation is the exception rather than the rule. More commonly one finds that *polyhapto* (multielectron) ligands are the focus of reactivity, and that the number and type (σ vs π) of metal–carbon interactions change during the reaction process.

LIGAND POLARIZATION

The enhanced susceptibility of ligands in coordination complexes to nucleophilic attack is attributed largely to the inductive effects of the metal cation. Ligand polarization effects also operate for organometallic complexes, but for several reasons the range of effects is much greater and includes both nucleophilic and electrophilic attack. The factors which govern this chemistry are largely (1) the oxidation state of the metal, (2) the availability of another stable oxidation state two units higher or lower, (3) the net charge on the reactant complex, and (4) the perturbation of the bonding on the ligand itself as a result of population of ligand antibonding orbitals through back-bonding from the metal. All of these effects are, of course, modulated by the nature of the coligands in the complex. In most cases the metal retains an 18- or 16-electron configuration, although 14-electron complexes are common where the energy of metal *d*-orbitals is relatively low, as with the coinage metals, cadmium, and mercury.

STEREOCHEMICAL EFFECTS

As with coordination complexes the chirality of the metal center, or more commonly the chirality of coligands, may be imposed on the reaction of another chiral or prochiral ligand. Recently, this has been pushed almost to the limit by a group at Monsanto who have obtained $\geqslant 95\%$ asymmetric reduction of a double bond using a chiral organometallic hydrogenation catalyst.

Another aspect of ligand stereochemistry unique to organometallic complexes is that for π-complexed ligands such as olefins and multielectron ligands both nucleophilic and electrophilic attack almost always occur from the side of the π-nodal plane opposite to the metal, i.e., exo or trans attack. However, in a few cases *protonation* may occur in an endo or cis fashion, probably as a result of protonation of the metal followed by transfer to the ligand. Thus, the principle of microscopic reversibility obtains, and the trans course of the reactions is dictated by the favorable energy of a transition state which maximizes orbital overlap between the metal, the ligand, and the entering or leaving group.

STABILIZATION OF REACTIVE INTERMEDIATES

Many organic species which are highly reactive and have at best a transient existence under ordinary conditions can be greatly stabilized through coordination to a suitable metal center. The stabilization achieved through complexation is often sufficient that chemical reactions can be carried out either on the ligand itself or by transfer of the ligand from the metal to form an adduct with another organic species. Examples of highly reactive ("nonexistent") ligands which have been isolated and manipulated as metal complexes include cyclobutadiene, cyclopentadienone, 2,4-cyclohexadienone (tautomer of phenol), dihydropyridines, small-ring acetylenes such as cyclohexyne, carbenes, and the trimethylenemethane diradical.

ELECTROPHILIC AND NUCLEOPHILIC ATTACK ON COORDINATED LIGANDS

Electrophilic and nucleophilic substitution are two of the most important reaction types in the realm of organic synthesis. Nevertheless, the scope of these reactions can be extended considerably by metal coordination of reactant species.[7] Reactions of coordinated CO will be taken up later in this chapter. For the most part we will consider the π-bonded multielectron ligands such as olefins, π-allyls, dienes, arenes, and homoaromatic ions such as tropillium $(C_7H_7^+)$. These ligands form a homologous series and their interconversion is possible through addition or elimination of a hydride ion or a proton. From the point of view of the metal center these processes are in a sense "vinylogous" Lewis acid or Lewis base associations or dissociations. The electrophilic and nucleophilic reactions of organometallic complexes were extensively reviewed and systematically described by White in 1968.[7] We will adopt a simpler organization and attempt to highlight potential applications of these reactions for organic synthesis.

HÜCKEL AROMATIC LIGANDS

Cyclic conjugated 6π electron systems such as $C_5H_5^-$, C_6H_6, and $C_7H_7^+$ are excellent ligands in organometallic chemistry, and a large number of their com-

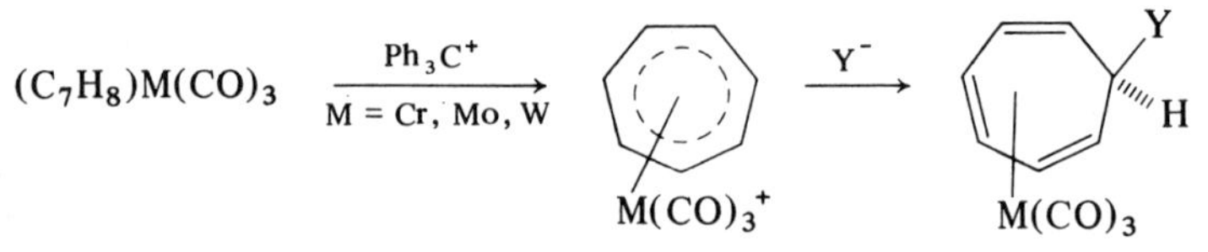

$$(12\text{-}1)$$

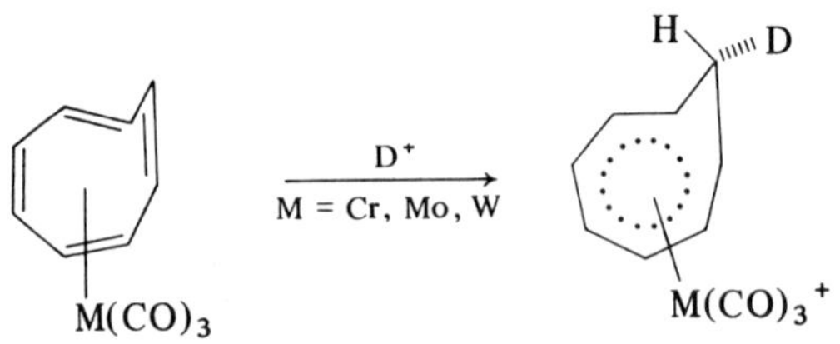

$$(12\text{-}2)$$

plexes are known. Characteristically, anionic complexes undergo electrophilic attack while cationic complexes undergo nucleophilic attack; neutral complexes may undergo both types of attack (reactions 12-1–12-6).

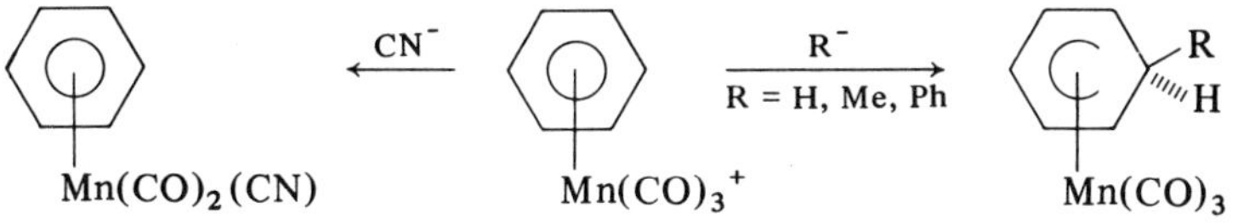

$$\underset{\substack{\text{Mn(CO)}_2\text{(CN)}}}{\text{[arene]}} \xleftarrow{\text{CN}^-} \underset{\substack{\text{Mn(CO)}_3{}^+}}{\text{[arene]}} \xrightarrow[R = H, Me, Ph]{R^-} \underset{\substack{\text{Mn(CO)}_3}}{\text{[arene-R,H]}} \qquad (12\text{-}3)$$

$$\underset{\substack{\text{Cr(CO)}_3}}{\text{[arene]}}\text{-Cl} \xrightarrow[65°C]{\text{MeO}^-} \underset{\substack{\text{Cr(CO)}_3}}{\text{[arene]}}\text{-OMe} \qquad (12\text{-}4)$$

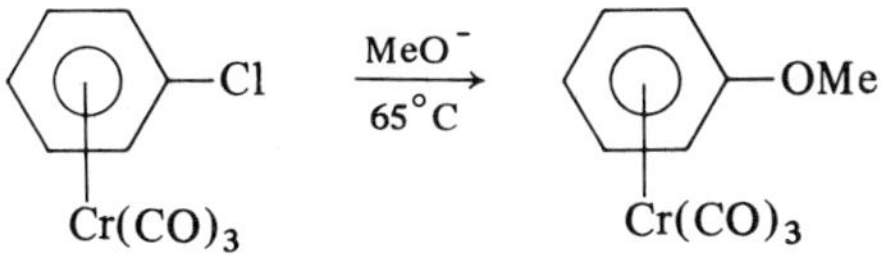

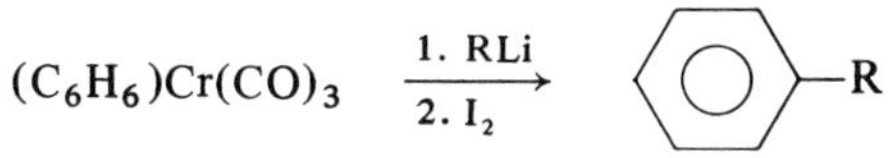

$$(\text{C}_6\text{H}_6)\text{Cr(CO)}_3 \xrightarrow[\text{2. I}_2]{\text{1. RLi}} \text{[arene]}\text{-R} \qquad \begin{array}{l}(12\text{-}5)\\ (\text{ref. 8})\end{array}$$

RLi (% R–Ph)[8]		RLi (% R–Ph)[8]	
$LiC(CH_3)_2CN$	(94)	$Li\text{–}C_6H_4CH_3$	(71)
$LiCH_2CN$	(68)	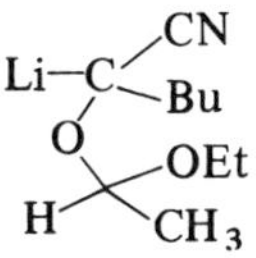 $\underset{H}{Li\text{–}C}\begin{smallmatrix}CN\\Bu\\OEt\\CH_3\end{smallmatrix}$	(90)
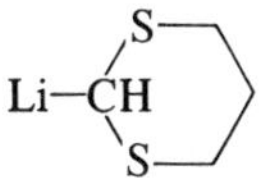 $Li\text{–}CH\langle\overset{S-}{\underset{S-}{}}\rangle$	(93)		
$Li\text{–}t\text{Bu}$	(97)		

$$\underset{\substack{\text{Fe}}}{\text{[ferrocene]}} \xrightarrow{X^+, Y^-} \underset{\substack{\text{Fe}}}{\text{[ferrocene]}}\text{-X} \;+\; \underset{\substack{\text{Fe}}}{\text{[ferrocene]}}\text{-X,X} \;+\; HY \qquad (12\text{-}6)$$

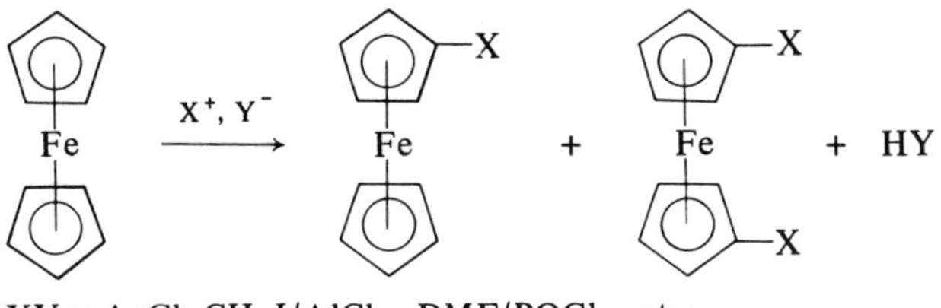

XY = AcCl, CH$_3$I/AlCl$_3$, DMF/POCl$_3$, etc.

Electrophilic substitution on ferrocene has been extensively studied and in principle it is analogous to benzene chemistry, except for the ready formation of ferricenium ions, Cp_2Fe^+ when oxidizing electrophiles such as Br^+ are used. Electrophilic reagents generally destroy (arene)Cr(CO)$_3$ complexes rather than lead to ring substitution. However, the electron-withdrawing effects of Cr(CO)$_3$ facilitate *nucleophilic* attack on coordinated arenes, thus opening many new possibilities for organic synthesis. Such chemistry is likely to be most useful, for example, in the synthesis of small amounts of labeled compounds when the route is dictated by availability of suitable precursors and the desire to introduce the label as late in the game as possible.

η^4-1,3-DIENE, η^3-ALLYL, AND η^5-PENTADIENYL LIGANDS

For the most part these ligands are found associated with the Fe(CO)$_3$ group. The diene complexes are readily available and easily converted to the others by nucleophilic or electrophilic reactions. These reactions often display pronounced stereospecificity and/or regiospecificity which can be used advantageously in synthetic and H/D exchange-labeling procedures (reactions 12-7–12-11).

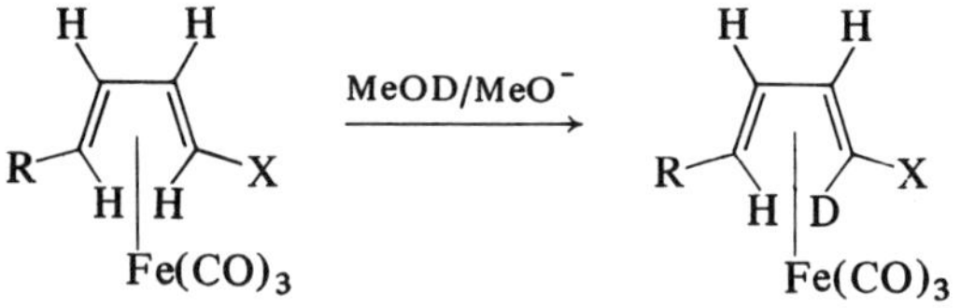

$$\text{X = CN, CHO, COPh, CO}_2\text{Me} \quad \text{(ref. 9)}$$

(12-7)

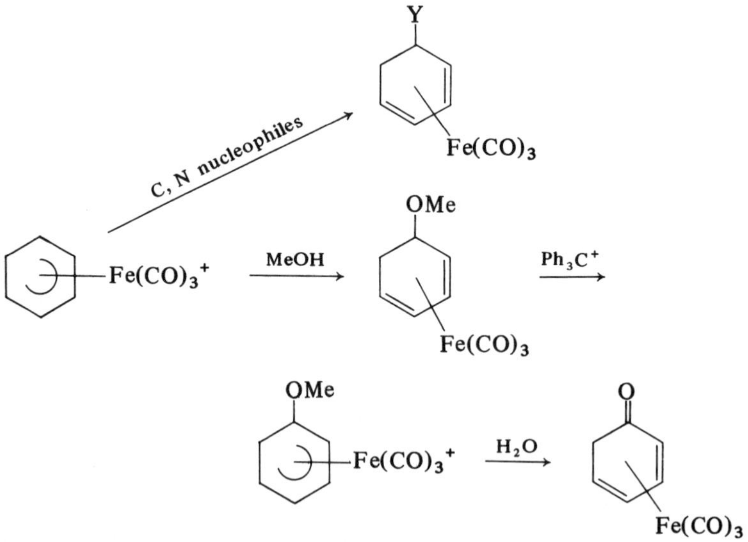

(12-8)
(ref. 10)

Nucleophilic attack usually occurs at the most substituted carbon. Attack of water on (η^5-pentadienyl)Fe(CO)$_3$ cations gives a (dienol)Fe(CO)$_3$ complex but (η^3-allyl)Fe(CO)$_3$ cations which should give coordinated primary or secondary allyl alcohols actually give aldehydes or ketones (cf. reactions 12-9 and 12-12).

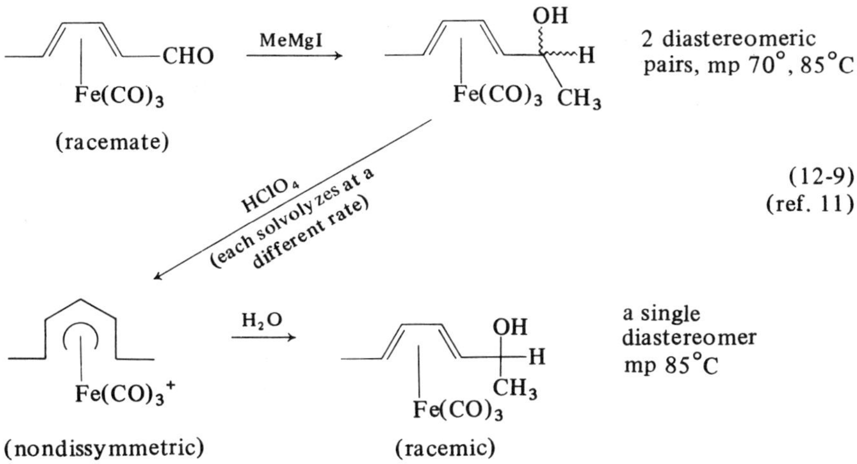

2 diastereomeric
pairs, mp 70°, 85°C

(12-9)
(ref. 11)

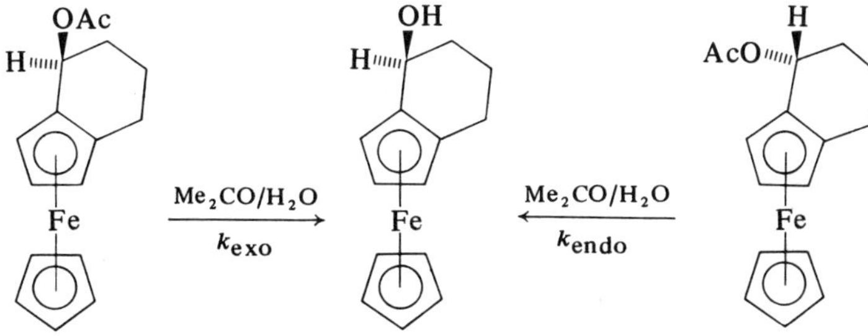

HClO$_4$
(each solvolyzes at a
different rate)

a single
diastereomer
mp 85°C

k_{exo}/k_{endo} = 2900, no endo alcohol formed

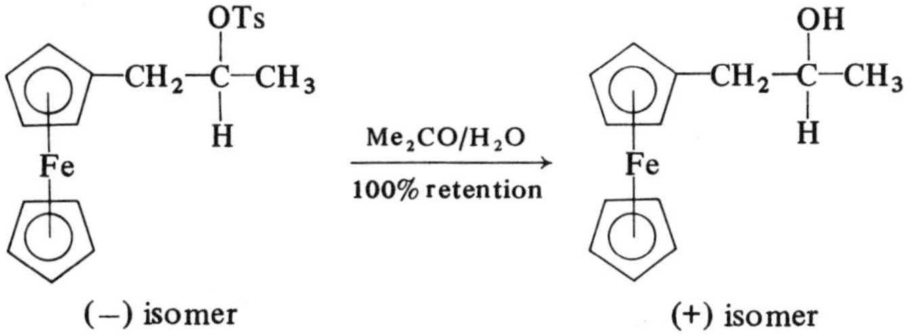

(12-10)
(ref. 12)

(12-11)
(ref. 13)

(−) isomer   (+) isomer

This probably occurs via oxidative addition of the intially formed allyl alcohols to give π-allyl metal hydride which undergoes reductive elimination to give the enolic isomer of the allyl alcohol (reaction 12-13). The dienol complexes are 18-electron complexes and cannot undergo the critical oxidative addition to form the π-allyl-metal hydride; steric factors can also block the reaction (reaction 12-13).

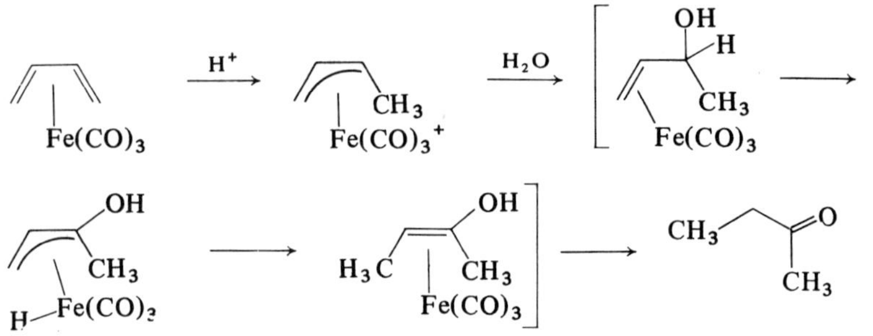

$$(12\text{-}12)$$

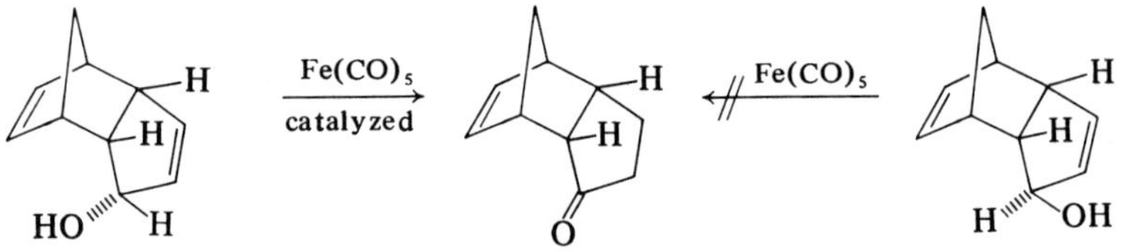

$$(12\text{-}13)$$
$$(\text{ref. } 14)$$

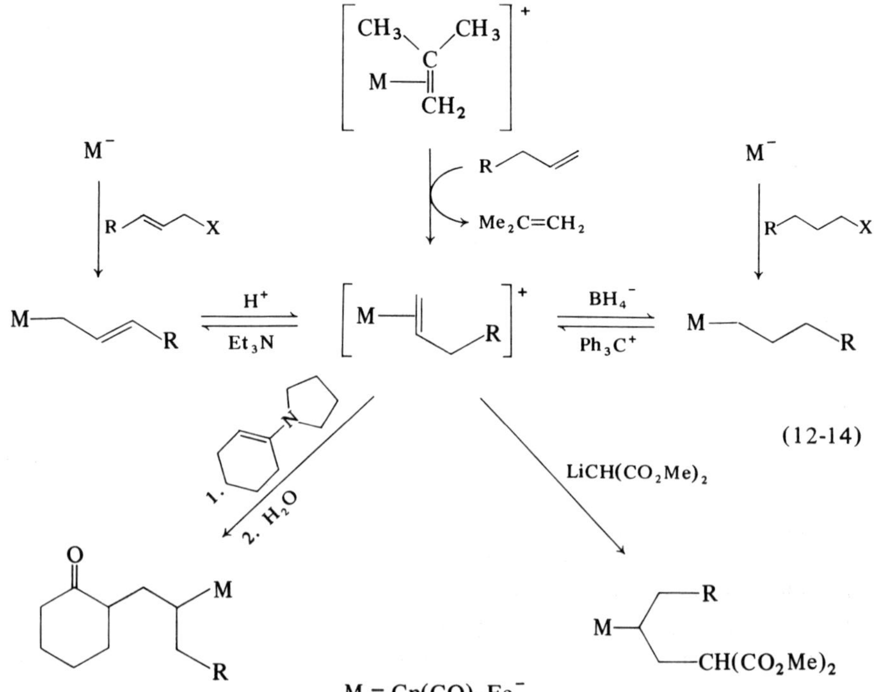

$$(12\text{-}14)$$

$$M = Cp(CO)_2Fe^-$$

π-OLEFIN, σ-ALKYL, AND η^3-ALLYL LIGANDS

Application of these types of complexes to organic synthesis holds considerable promise because of their ready availability, ease of interconversion via hydride or proton transfers, and differential susceptibility to nucleophilic and electrophilic attack depending on the metal center to which they are bound. For example, olefinic complexes of the type $Cp(CO)_2Fe(olefin)^+$ are susceptible to attack by a variety of O, N, and C nucleophiles including enolates and enamines (reaction 12-14).[15] Related reactions can be used to generate carbocyclic ring systems, and both hydride abstraction from cycloalkyl complexes and nucleophilic attack on cationic cycloolefin complexes have been shown to occur trans to the metal (reactions 12-15 and 12-16).

Of course, the synthetic utility of these reactions, most of which occur in quite good yield, depends on being able to remove the metal group in a useful

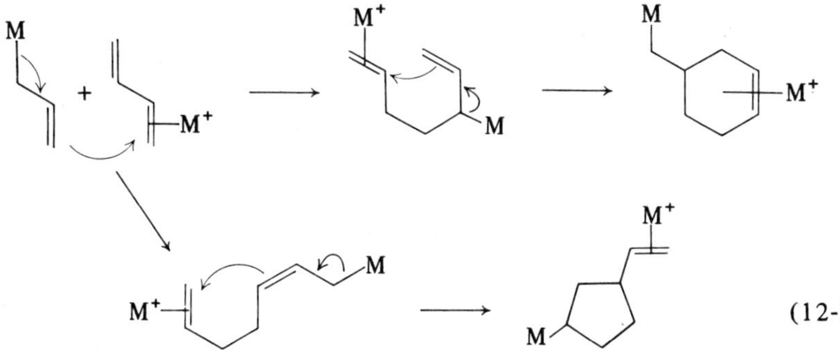

$$(12\text{-}15)$$

$$M = Cp(CO)_2Fe-$$

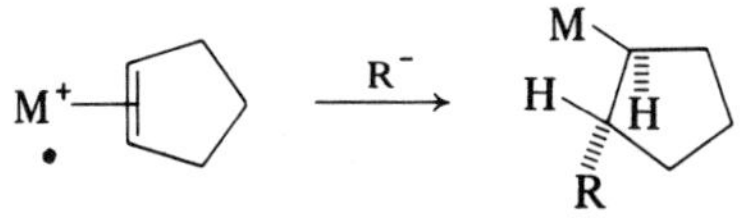

$$(12\text{-}16)$$

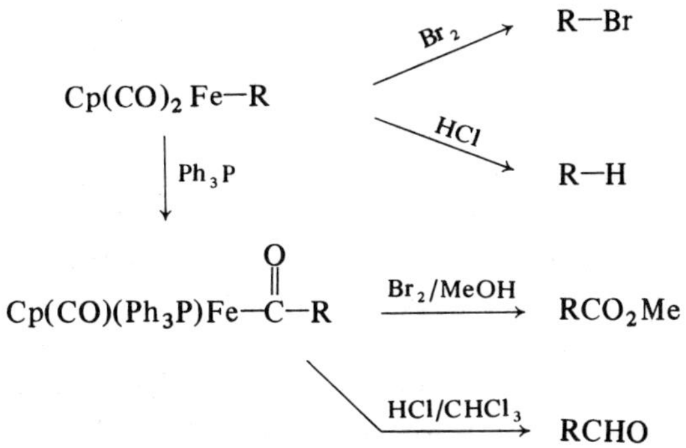

$$(12\text{-}17)$$

$$(12\text{-}18)$$

way. In this regard reactions 12-17–12-19 have all been applied with good
results (cf. Fig. 11-7).

$$Cp(CO)_2 Fe(olefin)^+ \xrightarrow[\text{acetone}]{\text{NaI/H}_2\text{O}} \text{olefin} + Cp(CO)_2 FeI \quad (12\text{-}19)$$

In contrast to the above nucleophilic reactions, neutral tricarbonyliron-olefin
and diene complexes undergo electrophilic attack leading to substitution under
conditions where the free ligand would be unstable or would give a mixture of
products. Removal of the $Fe(CO)_3$ group can be effected by oxidation of the
iron with aqueous ceric ion, or under milder conditions by oxidation with tertiary amine oxides[16] (reactions 12-20–12-22).

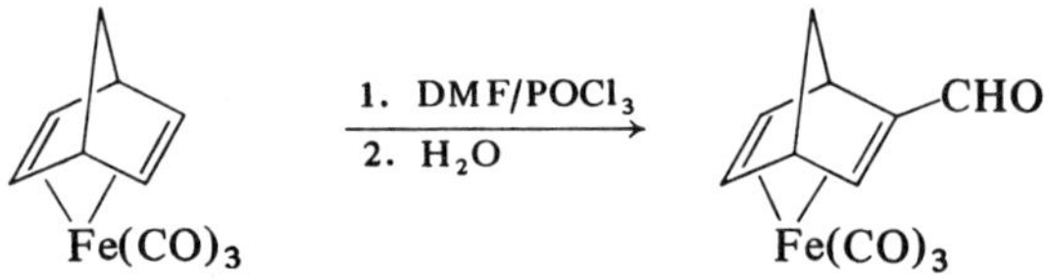

(12-20)
(ref. 17)

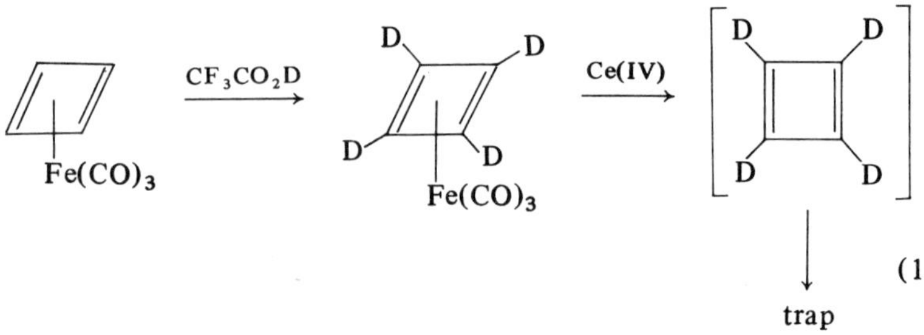

(12-21)

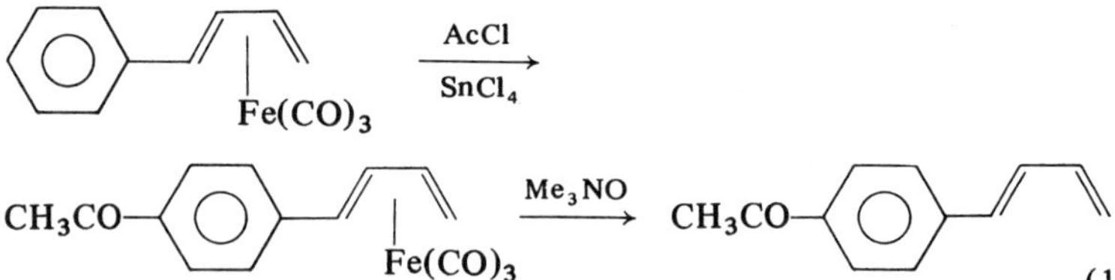

(12-22)

OXYMETALLATION REACTIONS

Nucleophilic attack on olefins coordinated to easily reduced metal ions such
as Ir(III), Rh(III), Pd(II), or Pt(II) is frequently followed by elimination of a
metal hydride which decomposes to H^+ and a reduced metal species. Thus the
above four ions stoichiometrically oxidize ethylene to acetaldehyde in the
presence of water or hydroxide. The Pd(II) reaction can be rendered catalytic
by linking the aerobic reoxidation of Pd(0) to a Cu(II)/Cu(I) couple, which is
the basis for the commercial Wacker process. The commonly accepted

mechanism for this reaction is one proposed by Henry,[18] as outlined in reactions (12-23). Deuterium studies have shown that the hydride transfer is intramolecular as shown, and the secondary isotope effect, $k_{(C_2H_4/C_2D_4)} = 1.07$, is

$$PdCl_4^{2-} + C_2H_4 \rightleftharpoons PdCl_3(C_2H_4)^- + Cl^-$$

$$PdCl_3(C_2H_4)^- + H_2O \rightleftharpoons PdCl_2(C_2H_4)(H_2O) + Cl^- \qquad (12\text{-}23)$$

$$PdCl_2(C_2H_4)(H_2O) + H_2O \rightleftharpoons PdCl_2(C_2H_4)(OH)^- + H_3O^+$$

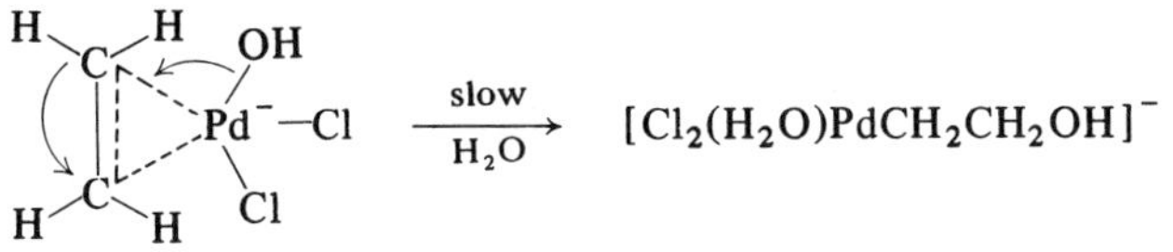

$$[Cl_2(H_2O)PdCH_2CH_2OH]^- \xrightarrow{\text{fast}} CH_3CHO + H_2O + Pd + 2\,Cl^-$$

consistent with a rate limiting conversion of a π complex to a σ complex with no breakage of C–H bonds. On the other hand, $k_{(H_2O/D_2O)} = 4.05$, and no deuterium is incorporated into the acetaldehyde.

Variations on this theme are the basis for several other important industrial processes. Many of these involve CO insertion and are covered in a later section. In reaction (12-25) the relative proportions of (A), (B), and (C) are

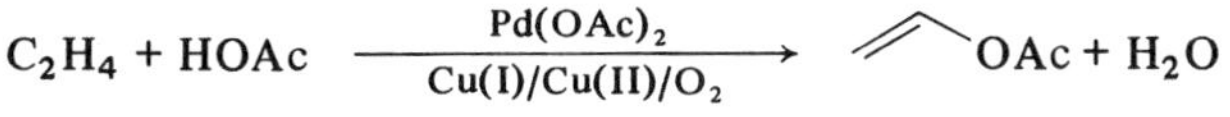

$$C_2H_4 + HOAc \xrightarrow[\text{Cu(I)/Cu(II)/O}_2]{\text{Pd(OAc)}_2} \qquad \diagdown\!\!\diagup OAc + H_2O \qquad (12\text{-}24)$$

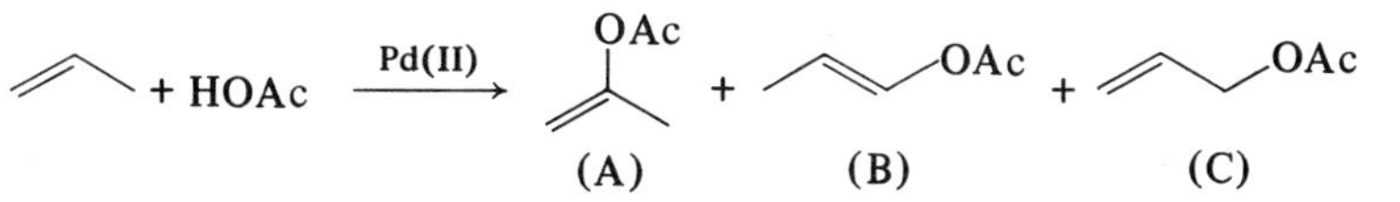

(12-25)

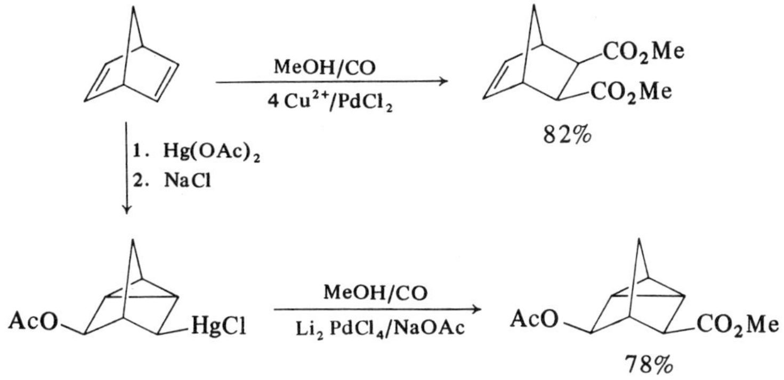

(12-26)

(ref. 20)

markedly sensitive to the relative concentrations of Pd(II), OAc⁻, and Cl⁻, as are, in fact, many other reactions catalyzed or promoted by Pd.[19] This sensitivity is attributed to the effects of these anionic ligands on equilibria involving monomeric and dimeric Pd complexes, the relative tendency of these to form olefin π complexes, and the relative rates of reaction of the different π complexes.

Diene chelates of Pd(II) undergo nucleophilic attack to form isolable σ complexes which may be carbonylated (q.v.) or decomposed by reductive elimina-

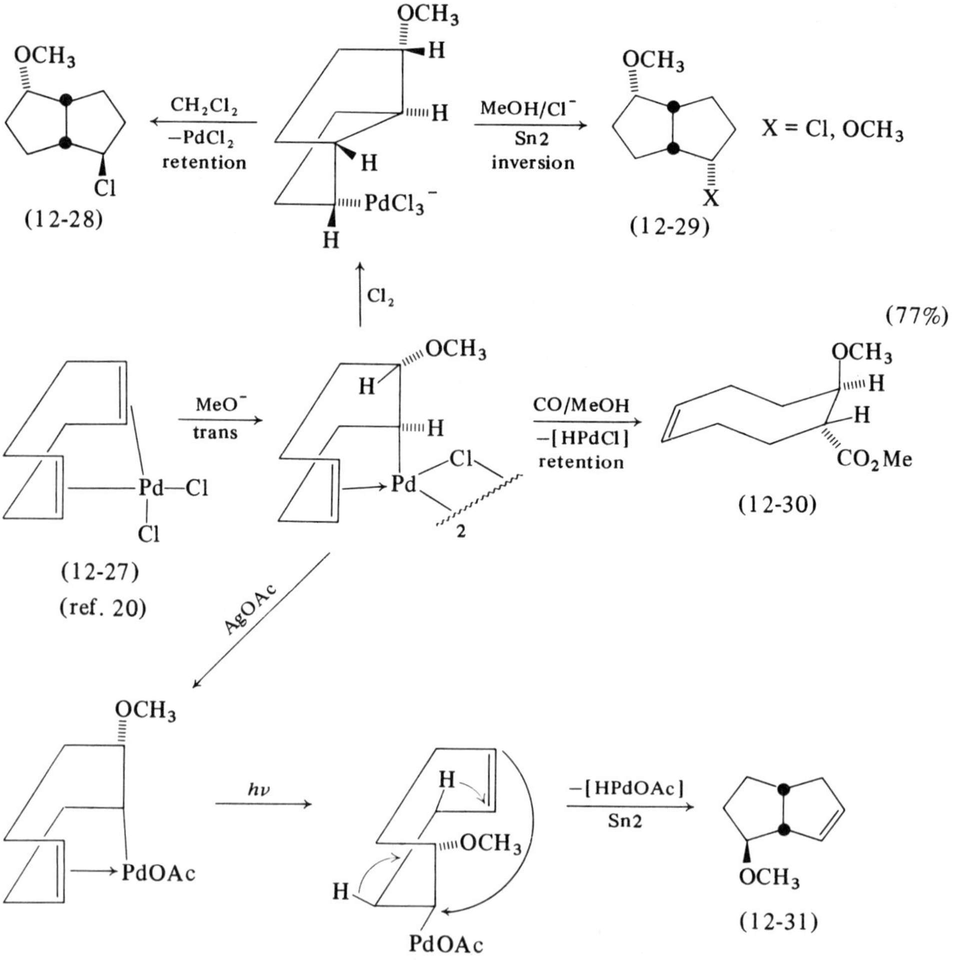

Scheme 12-1

tion of Pd(0) to give bicyclic compounds. Attack of solvated nucleophiles occurs trans to Pd while coordinated nucleophiles (e.g. $-H$ and $-CO_2CH_3$) give cis attack. The versatility of these reactions, particularly the carbonylations, is increased by utilizing easily obtainable alkylmercury precursors[20]; transmetallation occurs readily and with *retention* of configuration (reaction 12-26).[21]

A wide variety of bicyclo[3.3.0] systems can be obtained from (COD)PdCl$_2$ under various conditions. Following initial nucleophilic attack the cyclizations are effected by various combinations of oxidative addition, olefin insertion, and reductive elimination of Pd(0) or HPdX (Schemes 12-1 and 12-2).

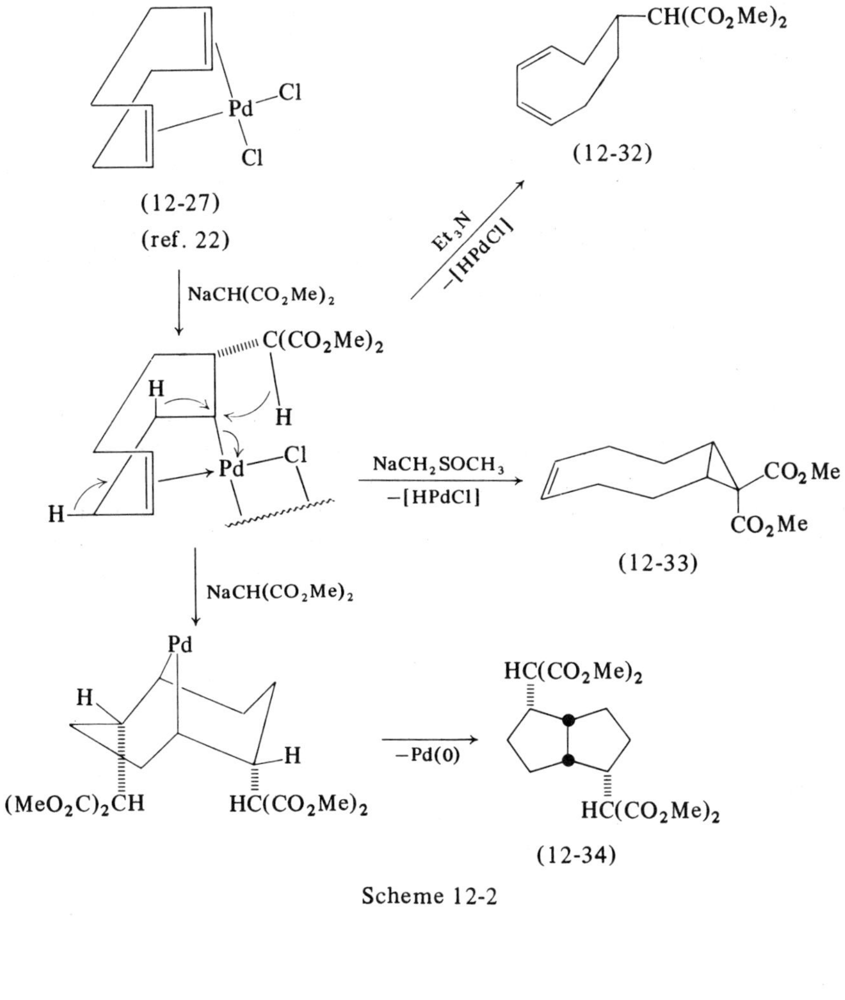

Scheme 12-2

Carbon–Carbon Coupling Reactions

A very large number of organometallic reactions lead to the formation of new carbon–carbon bonds. Many of these involve nucleophilic or electrophilic attack on coordinated ligands. In this section we will concentrate mainly on reactions which may be used to join two or more small molecules, principally by means of oxidative addition, insertion, and reductive elimination reactions. A number of these reactions are catalytic and some are the basis of large-scale industrial processes, e.g., Ziegler–Natta polymerization of ethylene, hydroformylation, and carbonylation of dimethylamine and methanol to form dimethylformamide (DMF) and acetic acid, respectively. Other reagents such as the π-allyl nickel complexes, the organocuprates, and $Na_2Fe(CO)_4$ have become important tools of the synthetic organic chemist. Undoubtedly the applications of organometallic complexes in both areas will continue to grow.

REDUCTIVE COUPLING OF ORGANIC HALIDES

Reductive elimination of two alkyl moieties from a metal center represents a way of forming C–C bonds which can be very convenient, especially when the dialkyl-metal complex can be prepared by the stepwise oxidative addition of, e.g., an organic halide to an alkyl- or allyl-metal complex. Such is the case for the "π-allyl halides"[23] and the "organocuprates,"[24,25] both of which have found widespread application in organic synthesis.

π-Allylnickel Reagents

π-Allylnickel reagents can be formed from allyl Grignard reagents and Ni(II) halides or by oxidative addition of allyl halides to Ni(0) complexes such as $(Ph_3P)_3Ni$ or $(cyclooctadiene)_2Ni$ (Scheme 12-3).[26] These are preferred to $Ni(CO)_4$ because side reactions involving CO insertion are avoided, although in other circumstances it is this reaction that is desired. In nonpolar solvents such as ether and benzene equilibration between σ- and π-allyl forms, and

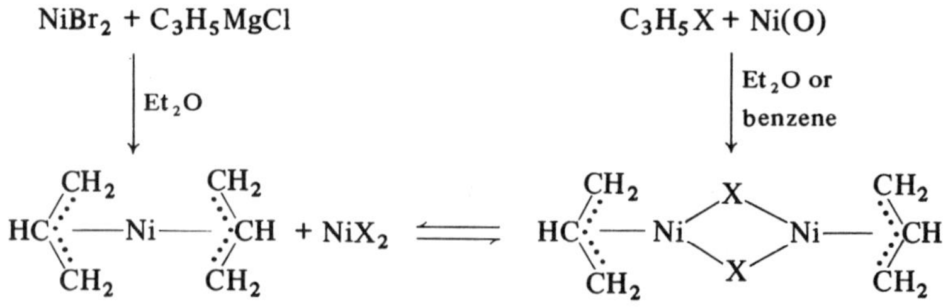

Scheme 12-3

between bis-π-allylnickel and π-allylnickel halide dimers, results in cis/trans isomerization of the allyl unit. The bis-π-allylnickel complexes can often be isolated by distillation or sublimation or by precipitation of $NiX_2 \cdot 6NH_3$ with ammonia gas. When the nonpolar solvent is replaced by a polar solvent such as DMF, the π-allylnickel halide complexes can undergo coupling reactions with organic halides, possibly by oxidative addition of the halide, followed rapidly by reductive elimination of NiX_2 to give coupling of the organic groups.[27] One of the organic groups is always allylic, but the second one may be from an aryl, vinyl, or alkyl as well as allyl halide (Scheme 12-4). It has

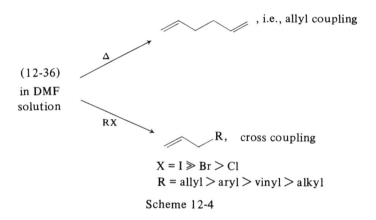

, i.e., allyl coupling

(12-36)

in DMF solution

R, cross coupling

$X = I \gg Br > Cl$

$R = allyl > aryl > vinyl > alkyl$

Scheme 12-4

recently been shown, however, that these couplings may involve a radical chain mechanism.[27a] Two disadvantages of π-allylnickel coupling reagents are that cis/trans isomerization and allyl–allyl exchange (Scheme 12-5) both occur

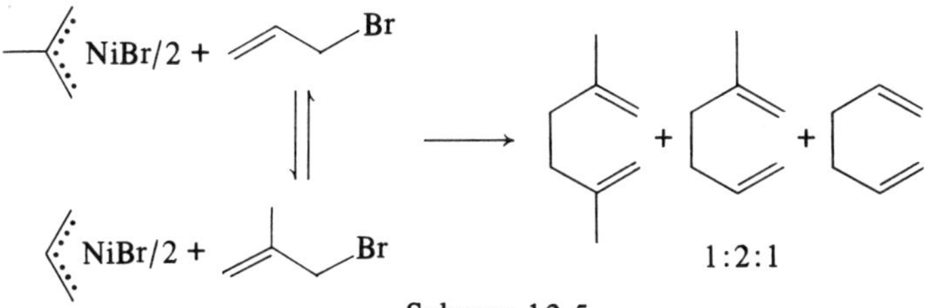

NiBr/2 + Br

NiBr/2 + Br

1:2:1

Scheme 12-5

faster than coupling. However, despite these limitations, π-allylnickel reagents are a powerful tool for organic synthesis. Some examples of compounds prepared with these reagents are listed in Table 12-1[27–30] in order to provide some perspective of the scope of this reaction.

TABLE 12-1

Syntheses Involving π-Allynickel Reagents

Reaction	Ref.

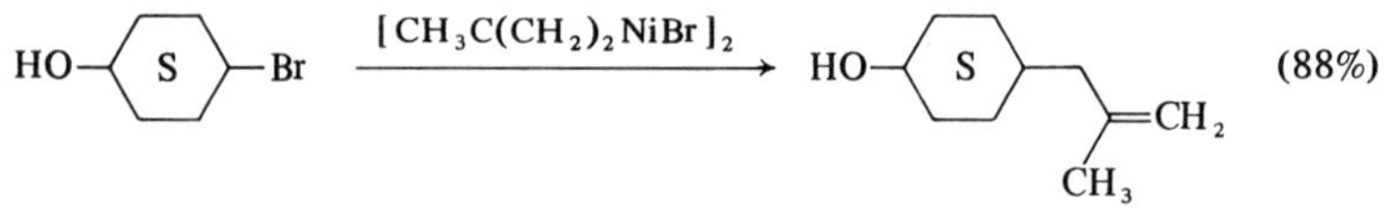

HO–⟨S⟩–Br $\xrightarrow{[CH_3C(CH_2)_2NiBr]_2}$ HO–⟨S⟩–CH$_2$ / C=CH$_2$ / CH$_3$ (88%) 27

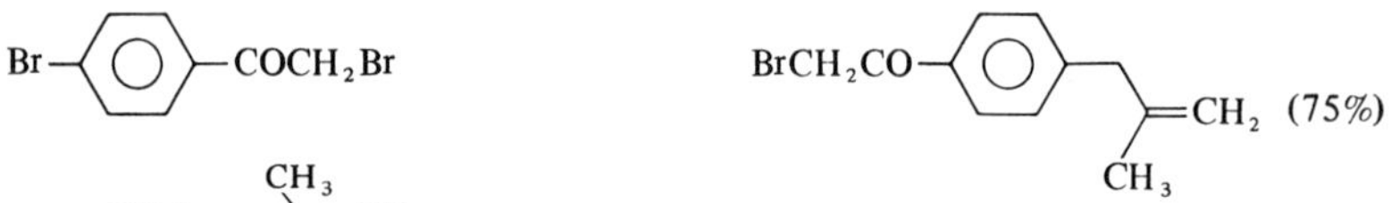

Br–⟨○⟩–COCH$_2$Br → BrCH$_2$CO–⟨○⟩–CH$_2$ / C=CH$_2$ (75%) / CH$_3$ 27

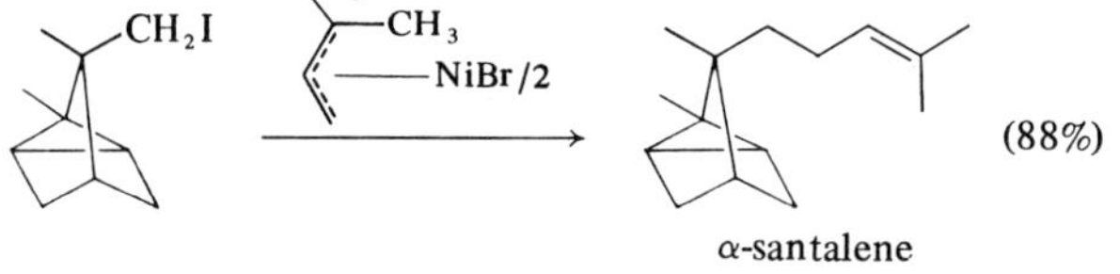

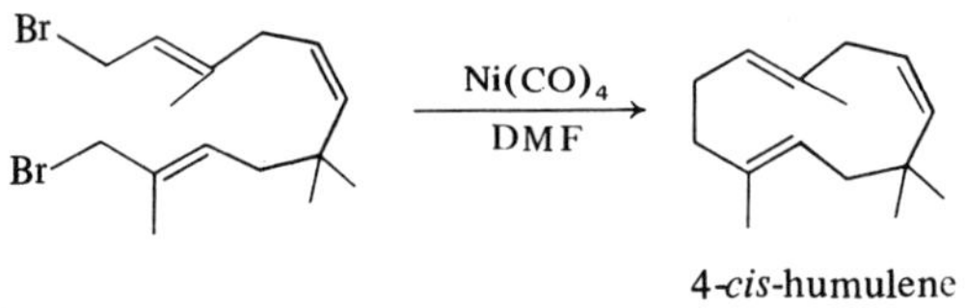

(α-santalene reaction) (88%) 27

α-santalene

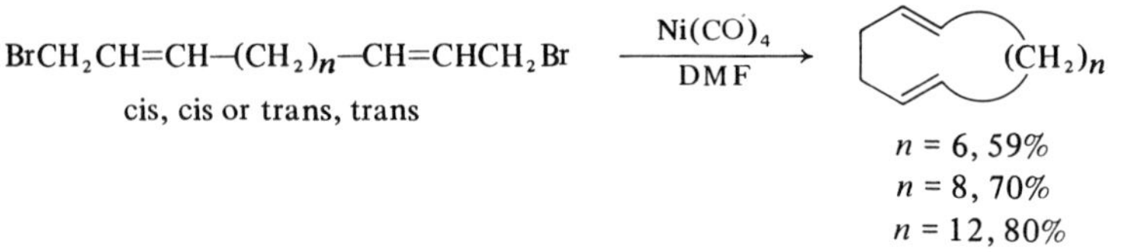

BrCH$_2$CH=CH–(CH$_2$)$_n$–CH=CHCH$_2$Br $\xrightarrow[DMF]{Ni(CO)_4}$ (cyclic diene (CH$_2$)$_n$) 28

cis, cis or trans, trans

$n = 6, 59\%$
$n = 8, 70\%$
$n = 12, 80\%$

(dibromide reaction) $\xrightarrow[DMF]{Ni(CO)_4}$ 4-cis-humulene 29

4-cis-humulene

(dibromo diester reaction) $\xrightarrow[DMF]{Ni(CO)_4}$ (macrocyclic diester) 30

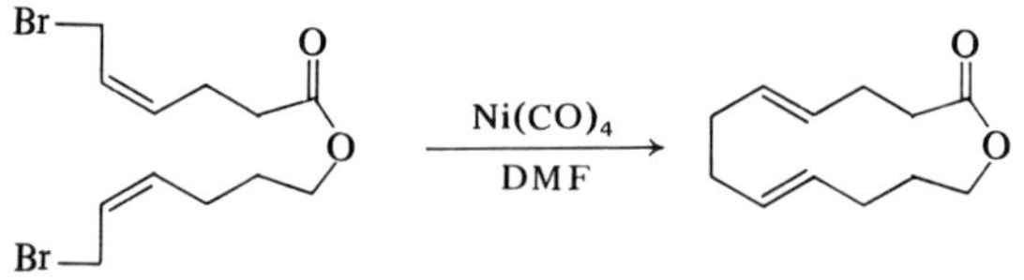

Organocopper Derivatives and Organocuprates

These reagents are an outgrowth of early observations on the effect of transition metal ions on the course of Grignard additions to α,β-unsaturated ketones.[24,25] Catalytic amounts of copper were known to favor 1,4 addition whereas with the pure Grignard reagent 1,2 addition predominates. Corey, House, and others later observed similar effects with organolithium agents and

Cu(I) salts. They also discovered and exploited the potential of the organocuprates [RCuR']Li as "nucleophilic" coupling reagents with organic halides.

Alkylcopper(I) reagents RCu are thermally rather unstable and tend to be polymeric and insoluble in ether. Both their thermal stability and solubility can be improved by the addition of soft nucleophilic ligands such as Bu_3P. Coordination also increases their reactivity toward other organic halides, although coupling in useful yields occurs only with acyl halides, aryl iodides, and α,β-unsaturated ketones. Thermal decomposition involves a chain reaction giving equal amounts of olefin and hydrocarbon but no R–R dimers, e.g.,

$$CH_3CH_2CuL \longrightarrow C_2H_4 + H\text{--}CuL$$

$$CH_3CH_2CuL + HCuL \longrightarrow 2L + 2Cu^\circ + C_2H_6$$

Alkenyl- and arylcopper(I) complexes are more stable thermally than the alkyl complexes, and ligands such as Bu_3P or quinoline further increase their stability. These complexes couple with acyl, allyl, aryl, and alkenyl halides and add 1,4 to unsaturated ketones, although for the most part they do not react with alkyl halides. They react with oxygen to give coupled biaryls or dienes, the latter being formed with retention of configuration of the double bond (reactions 12-37 and 12-38). Arylcopper(I) intermediates are probably also involved

$$\underset{}{\overset{Li}{\diagup\!\!\diagdown}} \quad \xrightarrow[\text{2. } O_2]{\text{1. Cu(I), } -30^\circ C} \quad \underset{\text{99\% cis, cis}}{\diagup\!\!\diagdown\!\!\diagup} \qquad (12\text{-}37)$$

$$PhLi \quad \xrightarrow[\text{2. } C_3H_5Br,\ -20^\circ C]{\text{1. CuBr, } -20^\circ C} \quad Ph\diagdown\!\!\diagup\!\!\diagdown \qquad (12\text{-}38)$$

in the Ullman reaction (12-39). Alkynylcopper(I) complexes are vastly more stable than any other type of organocopper(I) derivative, yet they readily react with acyl, allyl, vinyl, and aryl halides to give the coupled products. They are

$$2\,Ar\text{--}X \quad \xrightarrow[\text{quinoline, } \Delta]{Cu} \quad Ar\text{--}Ar + CuX_2 \qquad (12\text{-}39)$$

stable to water and even strong acids, but they undergo oxidation to give coupled diynes or other derivatives, in contrast to the purely thermal couplings of vinyl- and arylcopper compounds (reaction 12-40).

$$PhC\equiv CH \quad \xrightarrow{Cu(I)/NH_3/H_2O} \quad PhC\equiv C\text{--}Cu$$

$$PhC\equiv CI \quad \xleftarrow{\quad I_2 \quad} \qquad \Big\downarrow Cl_2,\ NBS,\ etc. \qquad (12\text{-}40)$$

$$PhC\equiv C\text{--}C\equiv CPh$$

Organocopper(I) derivatives are insoluble in ether and nonpolar solvents, but are both stabilized and solubilized by added ligands. If the added ligand is an anion, or even an excess of the carbanion used to form the organocopper complex, an organocuprate species of *stoichiometry* [RCuR′]Li is formed. The organocuprates are soluble, more stable, and more reactive toward organic halides than simple organocopper compounds. They give dimerized products (R–R′) upon oxidation and couple to organic halides, tosylates, epoxides, and even acetates (reactions 12-41–12-43). Vinyl halides and divinylcuprates react with retention of configuration[31] while alkyl derivatives undergo inversion during substitution, as in a normal Sn2 reaction.[32]

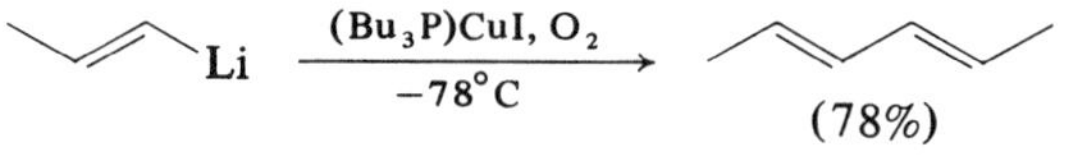

$$\text{(12-41)}$$

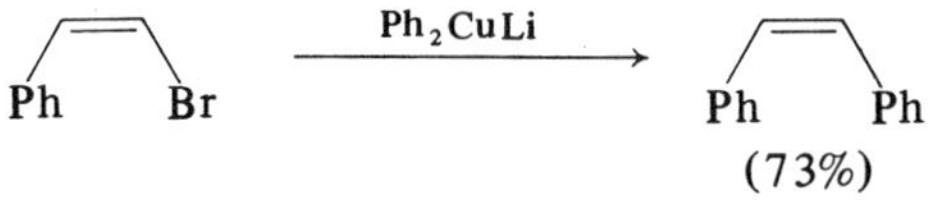

$$\text{(12-42)}$$

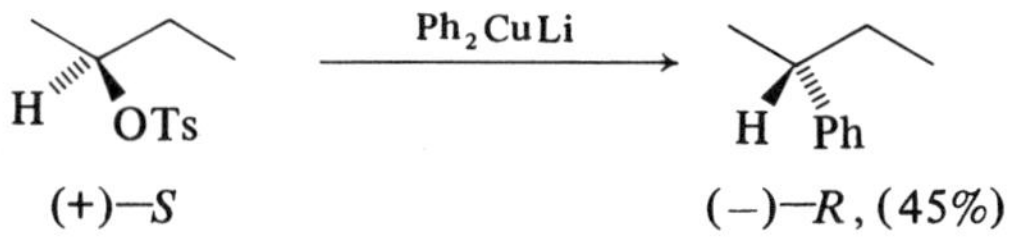

$$\text{(12-43)}$$

In polyfunctional molecules of the type shown below, organocuprates characteristically react at "soft" functional groups, in contrast to alkyllithium and Grignard reagents which react at "hard" functional groups,[33] although tosylates may be an exception (reaction 12-44).[32]

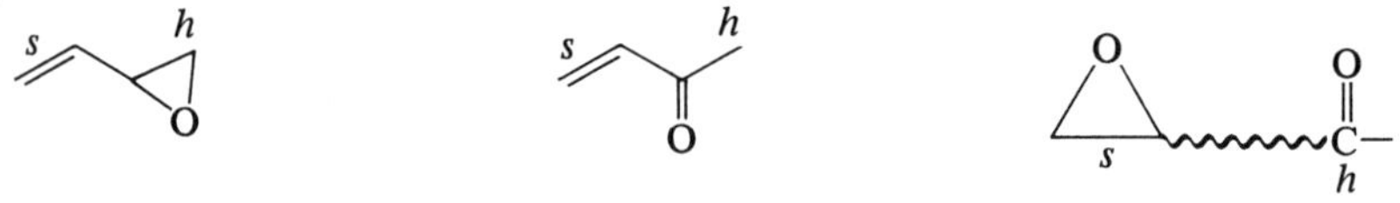

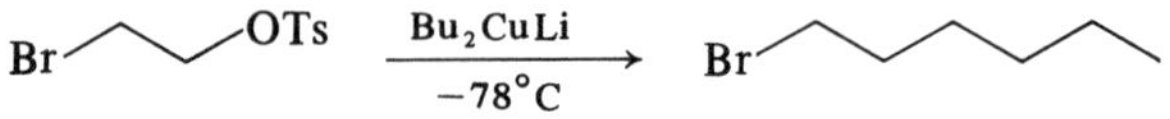

$$\text{(12-44)}$$

With the symmetrical organocuprates R_2CuLi, it is customary practice to use a fairly large excess of the reagent (e.g., 200–500%), and even then only one of the two R groups is available for transfer. Clearly this limits the utility of the method if the R group used to form the reagent derives from a multistep synthesis. This problem has recently been overcome through the use of heterocuprates such as [PhSCuR]Li. These reagents are typically used in only slight excess (e.g., 10–20%) and all of the R group is potentially available for transfer.[34]

Several mechanisms have been proposed to account for the reactions of organocopper compounds and organocuprates, and indeed more than one may be needed for the broad range of reactions encountered. For 1,4-addition reactions an electron-transfer–radical-recombination mechanism has been proposed (reaction 12-45a), based on the importance of the polarographic reduction

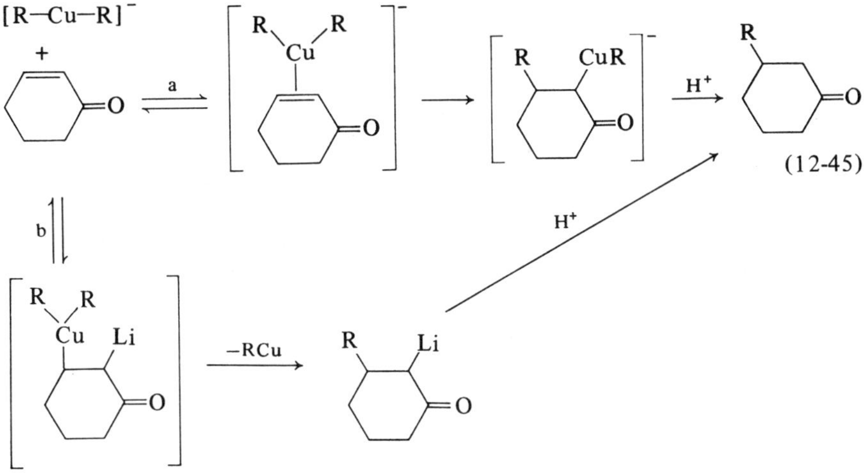

$$(12\text{-}45)$$

potential of the unsaturated ketone in determining whether the reagent would add to or reduce the substrate.[35,36] Since the $E_{1/2}$ value gives a measure of the stability of electrons in the π^*MO, and since Cu(I) is an electron-rich metal center, the correlation could also signify the importance of π complexation prior to transfer of the organic group as an insertion reaction. This is unlikely, however, since added excesses of other ligands which would compete with olefin for binding to copper do not interfere. Free radical reactions can be ruled out, based on the very clean stereospecificity of the coupling reactions (e.g., reactions 12-37 and 12-41–12-43). This leaves nucleophilic processes, e.g. reaction 12-45b, and oxidative addition to be considered, and the two are not incompatible since both retention and inversion at saturated carbon, and retention at sp²-carbon are known for oxidative additions and Lewis acid additions

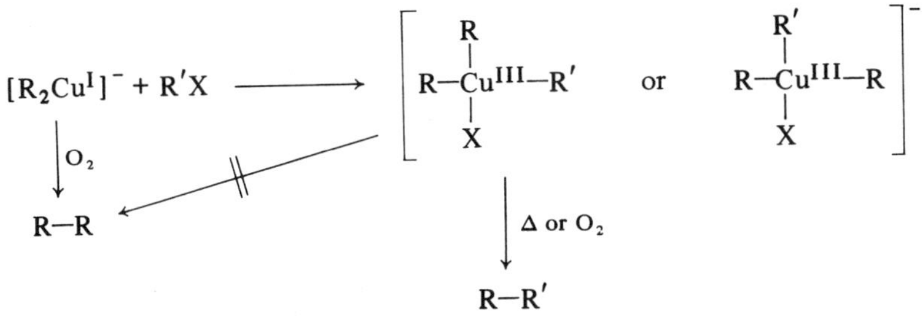

$$(12\text{-}46)$$

(see Chapter XI). The most unifying mechanism is that discussed by Johnson and Dutra,[32] who favor an oxidative addition process for the coupling reactions

$$\begin{array}{c} \text{PhLi} \\ + \\ \text{BuLi} \end{array} \xrightarrow[-78°C, O_2]{\text{ICuPBu}_3} \begin{array}{l} 33\% \text{ } n\text{-octane} + \\ 28\% \text{ biphenyl} + \\ 33\% \text{ butylbenzene} \end{array} \qquad (12\text{-}47)$$

$$\text{Bu}_2\text{CuLi} \xrightarrow{\text{C}_5\text{H}_{11}\text{Br}} \begin{array}{l} 99\% \text{ } n\text{-nonane} + \\ 0\% \text{ } n\text{-octane} \end{array} \qquad (12\text{-}48)$$

(reaction 12-46). They proposed that a transient Cu(III) intermediate could undergo reductive elimination to give the observed coupling products. However, its geometry [possibly square planar like isoelectric Ni(II)] and its mode of formation *and* decomposition must be specific to account for the failure to observe *symmetrical* coupling (12-48) of the two R groups originally present in the cuprate,[31] except when the complex is completely oxidized (12-47).[32] Obviously

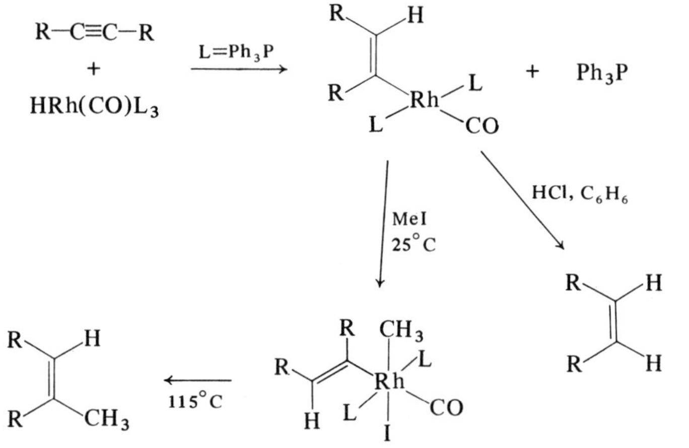

Acetylene	Halide	Product (yield)	
MeO₂C—C≡C—CO₂Me	MeI	MeO₂C⟍ ⁄R	(75%, 98% cis)
MeO₂C—C≡C—CO₂Me	EtI	MeO₂C⟍ ⁄H	(72%)
CH₃—C≡C—CO₂Me	MeI	CH₃ CO₂Me ⟍C=C⁄ CH₃	

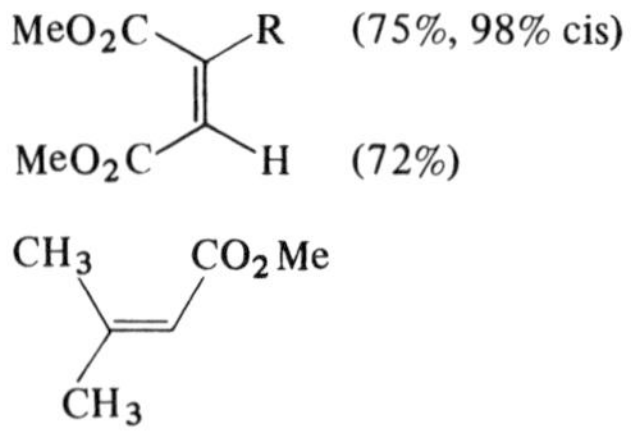

Fig. 12-1. Synthesis of trisubstituted olefins via alkylrhodium complexes.[37]

this mechanism is not yet a full explanation, but it remains attractive because it highlights the similarities of the copper reagents to the π-allylnickel halides discussed earlier, and to the chemistry of vinylrhodium complexes discussed below.

Rhodium(I) Reagents

The feasibility of achieving coupling of organic groups by a sequence of oxidative addition and reductive elimination at an appropriate metal center has been demonstrated by Schwartz using a hydridorhodium(I) complex as a starting point (Fig. 12-1).[37] The isolation and characterization of the intermediates leaves little doubt that this mechanism was involved, and the stereospecificity observed is consistent with other reactions of metal-vinyl derivatives.

Analogous alkyl- and arylrhodium(I) complexes undergo a similar addition–elimination sequence with acyl halides to give unsymmetrical ketones in good yields (Fig. 12-2).[38] Somewhat better yields are obtained with alkyllithium

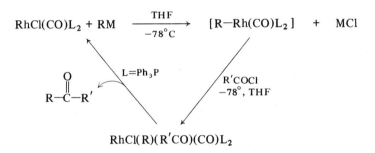

R'COCl	RM	RCOR' Product (%)
$CH_3(CH_2)_{10}COCl$	CH_3MgBr	58
	PhLi	96
$PhCH{=}CHCOCl$	nBuLi	68
PhCOCl	CH_3Li	82

Fig. 12-2. Synthesis of ketones via alkylrhodium complexes.[38]

reagents than with the analogous Grignard reagents. It is also claimed that vinylrhodium(I) complexes undergo coupling with acyl halides.

Collman's Reagent

Alkyl- and acyliron(0) carbonyls undergo oxidative attack leading to coupling with a wide variety of reagents, as shown in Fig. 12-3.[39-41] The commercially available reagent disodium tetracarbonyl-ferrate reacts with alkylating

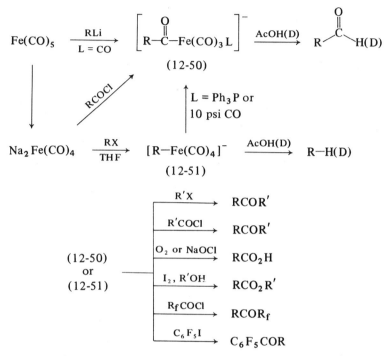

Fig. 12-3. Coupling reactions based on alkyl- and acyliron(0) complexes.[39-41]

agents to form alkyliron(0) complexes. This alkylation is a typical Sn2 reaction, with inversion at carbon and the normal ordering of substrate reactivity, i.e., $1° > 2° \gg 3°$ and ROTs > RI > RBr > RCl.[40,41] The alkyl complexes react with ligands via migratory insertion of CO to give acyliron(0) complexes. The latter can also be formed directly by reaction of $Na_2Fe(CO)_4$ with acyl halides.

Treatment of either the alkyl- or the acyliron(0) complexes with a wide variety of reagents leads to coupling reactions of the kind shown in Fig. 12-3. Since the Fe(0) complexes are coordinatively saturated, unlike the Cu(I), Rh(I), and Ni(II) reagents discussed previously, they cannot undergo oxidative addition per se. Nevertheless, it is reasonable to formulate the reaction as involving Lewis acid addition with concomitant oxidation, followed by reductive elimination of the coupled product, e.g., reaction (12-49). The method is particularly

$$[R-\underset{\underset{O}{\|}}{C}-Fe(CO)_4]^- \xrightarrow{AX} [(RCO)(A)Fe(CO)_4] \longrightarrow R-\underset{\underset{O}{\|}}{C}-A \qquad (12\text{-}49)$$

AX = R'OTs, HOAc, I_2, etc.

valuable for synthesis of perfluoracyl ketones, and a wide variety of other functional groups, such as $-CHO$, $-CO_2R$, $-OH$, $-Cl$, are stable and remain unaffected under the reaction conditions. The main problems encountered with the method are elimination of HX from the alkylating agent because of the basicity of $Fe(CO)_4^{2-}$ (comparable to HO^-), the resistance of benzyl ferrates to undergo carbonylation, and the decreased nucleophilicity of $RCOFe(CO)_4^-$ or $RCOFe(CO)_3(Ph_3P)^-$ relative to $Fe(CO)_4^{2-}$.

CYANATION

In a reaction related to organocuprate chemistry, CuCN will convert aryl halides to nitriles, but the reaction requires polar solvents, high temperatures, and a stoichiometric amount of CuCN. The same reaction can be run catalytically under mild conditions (30°–60°C) in methanol or acetone using $Ni(Ph_3P)_3$ as catalyst and NaCN or acetone cyanohydrin as the source of cyanide (reaction 12-52).[42] The reaction proceeds via oxidative addition of the

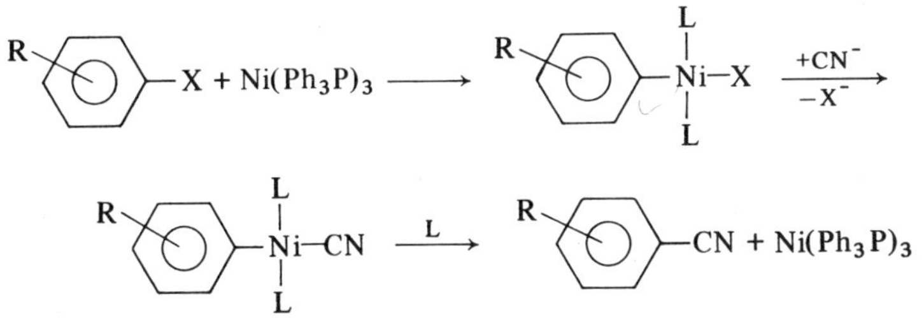

$$(12\text{-}52)$$

halide to give an arylnickel(II) halide complex, followed by exchange of CN^- for halide, and reductive elimination of the arylnitrile. The oxidative addition step is favored by electron-withdrawing substituents on the aryl halide ($\rho = 8.8$), while electron releasing substituents favor the elimination step. A number of functional groups (e.g., F, Cl, CHO, RCO, and CO_2Me) are unaffected by the reaction. Although it has not been reported, vinyl halides could also be expected to undergo catalytic cyanation.

INSERTION REACTIONS OF OLEFINS AND ACETYLENES

Insertion of olefins and acetylenes into metal-carbon bonds is an important way of homologating carbon chains, from dimerization of olefins to production of polyethylene. Often this chemistry begins with the insertion of an olefin

TABLE 12-2

Some Metal-Catalyzed Condensations of Olefins and Acetylenes[a]

Catalyst	Reaction
a. NiAcac$_2$ · Et$_3$Al or CrCl$_3$ · Et$_3$Al	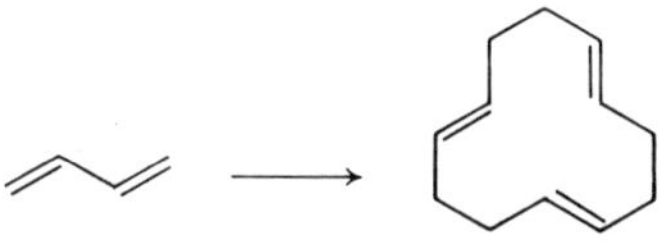
b. TiCl$_4$ · Et$_2$AlCl	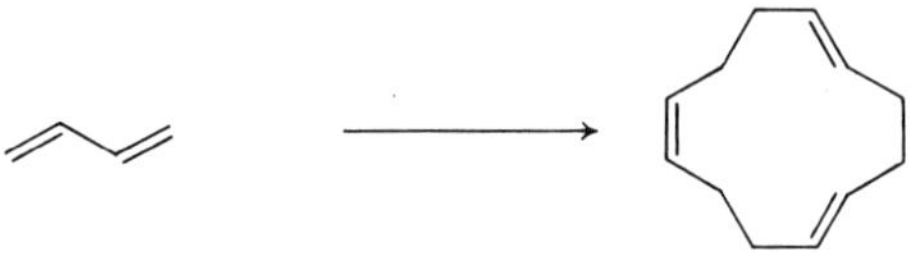
c. TiCl$_4$ · Et$_3$Al (Ziegler–Natta type, heterogeneous)	$C_2H_4 \longrightarrow$ polyethylene (MW 10^3 to 10^6) $C_3H_6 \longrightarrow$ isotactic polypropylene
d. (Ph$_3$P)$_2$Ni(CO)$_2$	$CH_2{=}C{=}CH_2 \longrightarrow$ 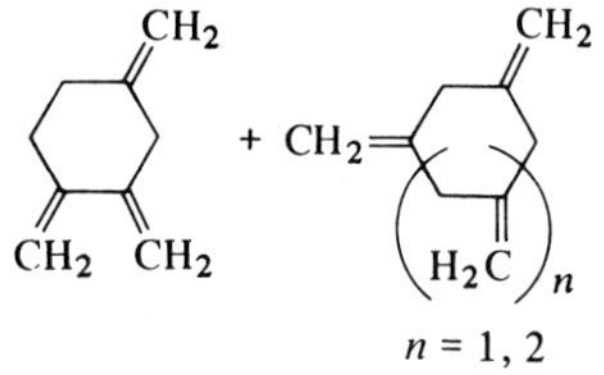
e. Ni(CN)$_2$ · THF	$HC{\equiv}CH \longrightarrow$; $PhC{\equiv}CH \longrightarrow$
f. AlCl$_3$/benzene	$CH_3C{\equiv}CCH_3 \longrightarrow$ 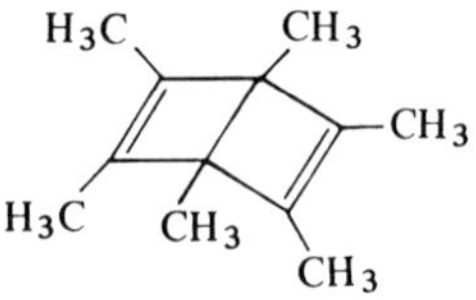
g. PdCl$_2$/HCl	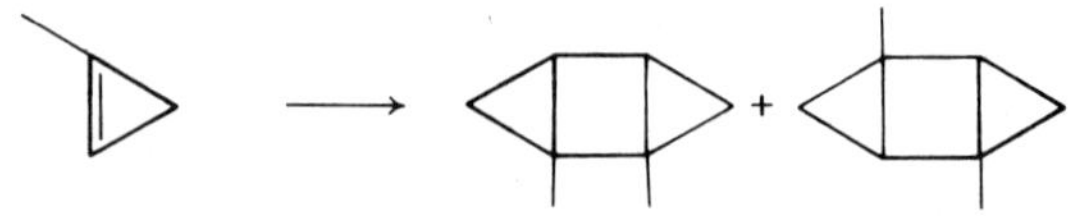

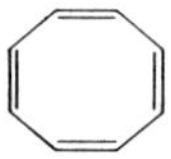

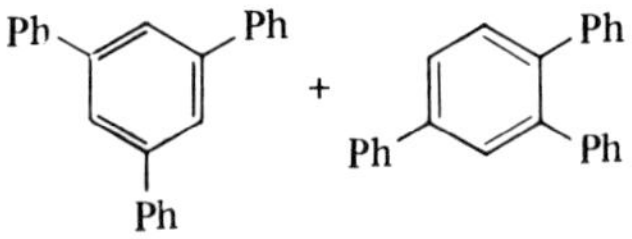

TABLE 12-2-continued

Catalyst	Reaction
g. PdCl$_2$/HCl	MeC≡CMe ⟶
h. C$_3$H$_5$Ni(Ph$_3$P)$^+$AlCl$_4^-$	C$_3$H$_6$ ⟶
i. NiCl$_2$·R$_3$P·Al$_2$Cl$_3$Et$_3$	C$_3$H$_6$ ⟶

^{a}Data from Davidson,[1] Ballard,[5] and Swan and Black.[6]

into a metal-hydrogen bond. Many important catalytic processes based on these types of reactions are related in the diagram shown in Fig. 12-4. A common feature to these reactions is the sequence: oxidative addition, migratory

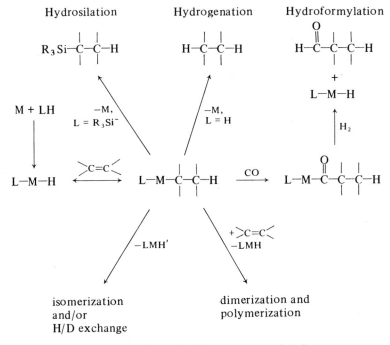

Fig. 12-4. Insertion and addition reactions of olefins.

insertion (sometimes followed by a second oxidative addition), and reductive elimination.

The dimerization of ethylene to 1-butene has been carefully investigated by Cramer (reaction 12-53).[43] In this reaction system the catalyst is a rhodium(I) species containing a halide ion, produced by reduction of $RhX_3 \cdot nH_2O$ with

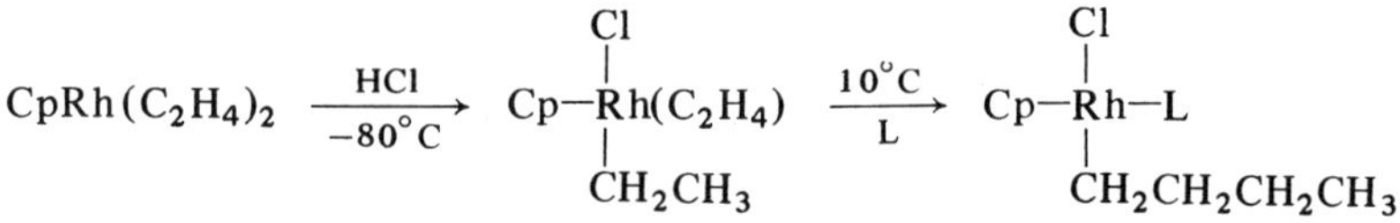

$$
CpRh(C_2H_4)_2 \xrightarrow[-80°C]{HCl}
\begin{array}{c} Cl \\ | \\ Cp-Rh(C_2H_4) \\ | \\ CH_2CH_3 \end{array}
\xrightarrow[L]{10°C}
\begin{array}{c} Cl \\ | \\ Cp-Rh-L \\ | \\ CH_2CH_2CH_2CH_3 \end{array}
\qquad (12\text{-}53)
$$

ethanol. In the presence of ligands other than olefin, it is possible to isolate rhodium hydrides, e.g. $(Ph_3P)_3RhCl_2H$. However, even at $-80°C$ hydride formed by oxidative addition of HA to Rh(I) reacts too fast to be observed prior to insertion of ethylene. At higher temperatures ethylene and other olefins can undergo insertion into the alkyl-Rh(III) bond. Reductive elimination of an Rh(III) hydride produces a new olefin and regenerates the catalyst.

The Rh(III) hydrides which dimerize olefins also catalyze their isomerization and H/D exchange reactions by means of insertion–elimination sequences. The isomerization of 1-butene to 2-butene actually occurs faster than ethylene dimerization, but its occurrence is minimized by keeping the $C_2:C_4$ ratio high, by the fact that Rh binds ethylene about 1000 times more strongly than 1-butene, and by the fact that Rh(III)-H adds to 1-butene mainly to give the *n*-butyl derivative which cannot lead to isomerization, although it can lead to H/D exchange (reaction 12-54).

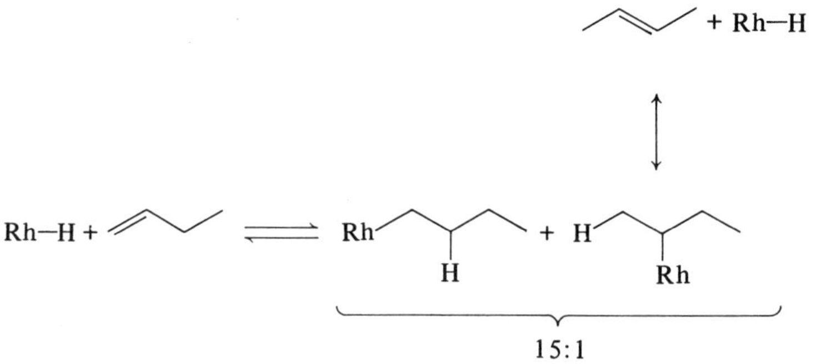

$$(12\text{-}54)$$

Numerous other transition metal catalysts will dimerize, oligomerize, or polymerize olefins. Acetylenes and dienes may also be inserted in a growing chain or cyclized into dimers or trimers. Several examples have been given in reactions 11-45 and 11-46, and other are given in Table 12-2.

CARBONYLATION REACTIONS

The carbon monoxide in many metal carbonyl complexes, especially if ν_{C-O} is greater than about 2000 cm^{-1}, is susceptible to nucleophilic attack. We have seen several examples of carbanion attack to give acyl complexes, e.g., in the synthesis of carbene complexes or in the iron-based ketone syntheses discussed

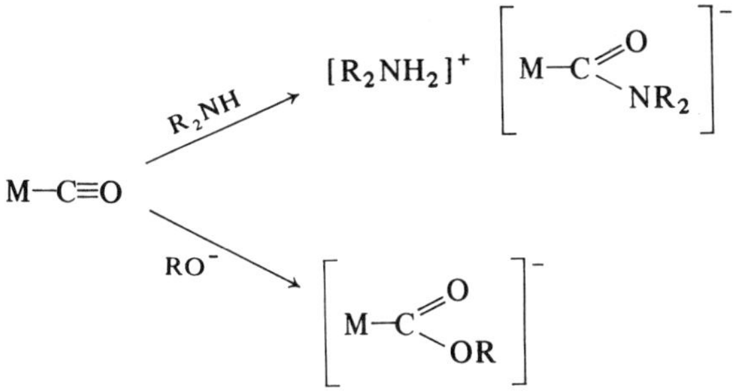

$$M-C\equiv O \xrightarrow{R_2NH} [R_2NH_2]^+ \left[M-C \begin{matrix} \diagup O \\ \diagdown NR_2 \end{matrix} \right]^- \qquad (12\text{-}55)$$

$$\xrightarrow{RO^-} \left[M-C \begin{matrix} \diagup O \\ \diagdown OR \end{matrix} \right]^- \qquad (12\text{-}56)$$

above. If the nucleophile is an amine or an alkoxide, carbamoyl or alkoxycarbonyl complexes, i.e., inorganic analogs of amides and esters, are produced (reactions 12-55 and 12-56). Such complexes are thought to be intermediates in a number of metal-catalyzed reactions of amines and CO to give formamides, ureas, and isocyanates, e.g., Scheme 11-8, 2 and reactions (12-57) and (12-58).[6a,44]

$$2\,RNH_2 + CO \xrightarrow{Mn_2(CO)_{10}} RNHCONHR + H_2 \qquad (12\text{-}57)$$

$$(RNH_2)_2PdCl_2 + 2\,CO \longrightarrow 2\,HCl + Pd^0 + 2\,RNCO \qquad (12\text{-}58)$$

Many of the chemical reactions of carbamoyl and alkoxycarbonyl complexes parallel those of their organic counterparts, i.e., interconversion, hydrolysis, and dehydration. Two reactions which are unique to the metal complexes involve cleavage of the metal-carbon bond by oxidizing agents (reactions 12-60 and 12-61). The mercuric chloride reaction may involve the intermediacy of a biscarbamoylmercury(II) species which reductively eliminates metallic mercury. It is known that mercuric ion in basic methanol absorbs CO to form a

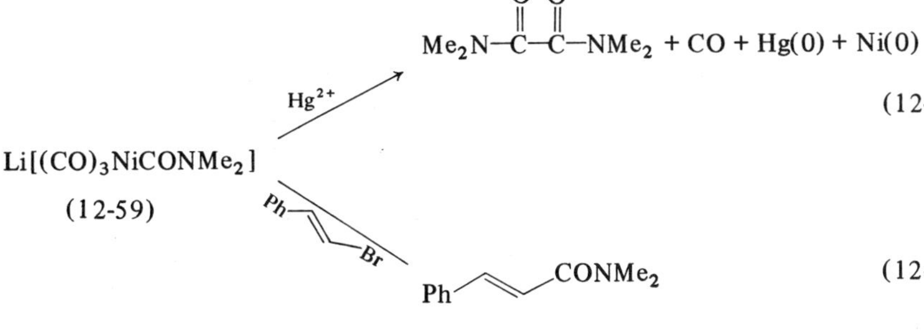

$$Li[(CO)_3NiCONMe_2] \quad (12\text{-}59)$$

$$\xrightarrow{Hg^{2+}} \underset{\substack{\| \quad \| \\ }}{Me_2N-C-C-NMe_2} + CO + Hg(0) + Ni(0) \qquad (12\text{-}60)$$

$$\xrightarrow{Ph \diagdown Br} Ph\diagup\diagdown CONMe_2 \qquad (12\text{-}61)$$

methoxycarbonyl complex.[45] This complex has been used to transfer the $-CO_2CH_3$ group to olefins and acetylenes with the aid of a reactive Pd(II) intermediate.[46] Unfortunately, stoichiometric amounts of $PdCl_2$ are needed. (See reactions 12-62 and 12-64.)

$$Hg(OAc)_2 \xrightarrow[\text{MeOH}]{\text{CO}} AcO-Hg-CO_2Me + HOAc \qquad (12\text{-}62)$$

$$AcOHgCO_2Me + PdCl_2 \longrightarrow [ClPdCO_2Me] + [ClHgOAc] \qquad (12\text{-}63)$$

(12-64)

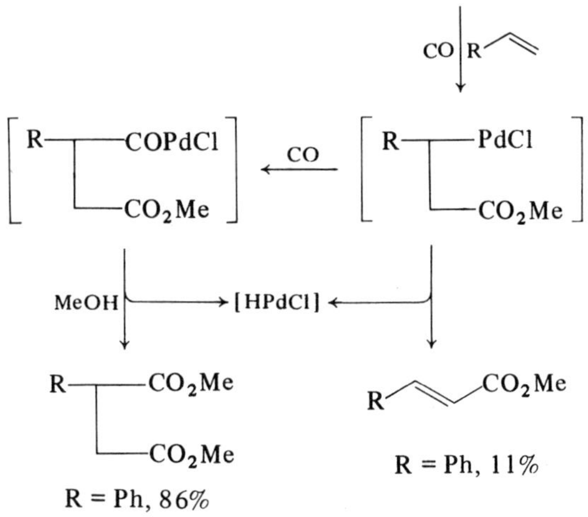

R = Ph, 86%

R = Ph, 11%

Phenylation of olefins can also be accomplished by the use of the unstable reagent [PhPdOAc] formed by transmetallation from the readily available mercury reagent.[47] The phenylation represents an electrophilic addition–elimination sequence, and even benzene can be converted to biphenyl by Pd(II) via [PhPdOAc] (Scheme 12-6).

Returning to the carbamoylnickel complex (12-59), another important reaction which it can undergo (after loss of one CO?) is the familiar oxidative addition–reductive elimination sequence to give coupling.[48] This results in the overall conversion of an organic halide to the homologous amide (reaction 12-16). Nickel tetracarbonyl can also be used with an alkoxide to convert vinyl or aryl halides to esters, and if the alcohol is wet, carboxylic acids are produced. Formally these reactions are analogous to the couplings of π-allylnickel halides with vinyl and aryl halides, but the formal oxidation states of nickel are different in the two processes. Alkoxycarbonylation with $Ni(CO)_4$ is a very powerful synthetic method because the aryl or vinyl halide groups are relatively inert to many other reagents and hence can be carried along a synthetic sequence where the related ester might not have survived. An elegant ex-

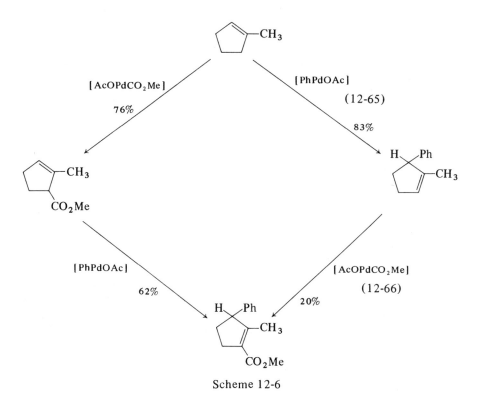

Scheme 12-6

ample of this is the stereospecific synthesis of α-santalol by Corey *et al.*, shown in reaction (12-67),[49] in which metal reagents played many key roles. Examples of other Ni(CO)$_4$ carbonylations are also given (reaction 12-68).

Reaction of carbanions and Ni(CO)$_4$ produces acylnickel(0) complexes RCONi(CO)$_3^-$. These undergo conjugate addition reactions to unsaturated ketones and esters, much like the organocuprates (reaction 12-69).[50]

(RBr)

1. LiCH$_2$C≡CSiMe$_3$
2. Ag$^+$/H$_2$O
3. KCN
4. RLi, 5. CH$_2$O

RCH$_2$C≡CCH$_2$OH

1. BuLi
2. iBu$_2$AlH
3. I$_2$

RCH$_2$ I

H   CH$_2$OH

1. MsCl; LiBr
2. Ni(CO)$_4$/MeO$^-$
3. NaBH$_4$/DMSO
4. AlH$_3$

CH$_2$OH

CH$_3$ (12-67)

α-santalol

$$R'-X \quad \xrightarrow[\text{ROH/RO}^-]{\text{Ni(CO)}_4} \quad R'-\overset{\overset{\displaystyle O}{\|}}{C}-OR \quad (12\text{-}68)$$

R'X	Product	(%)[a]
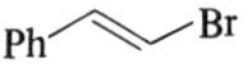 Ph⌒=⌒Br	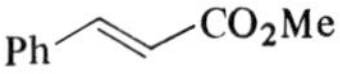 Ph⌒=⌒CO$_2$Me	95
	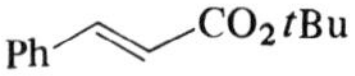 Ph⌒=⌒CO$_2$$t$Bu	60
PhI	PhCO$_2$Me	88
(cyclohexenyl)—Br	(cyclohexenyl)—CO$_2$Me	65
n-C$_7$H$_{15}$I	n-C$_7$H$_{15}$CO$_2$$t$Bu	66

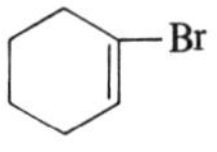

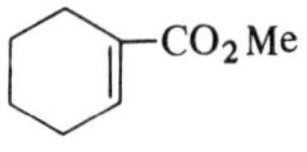

[a]Data from Corey and Hegedus.[48]

Another important group of carbon–carbon coupling reactions (12-70–12-74) involve the insertion of CO into a metal-carbon bond.[51,52] Most of these reactions involve metal-carbon bonds formed either by nucleophilic attack on coordinated olefins, or by addition of a metal hydride to an olefin or acetylene.

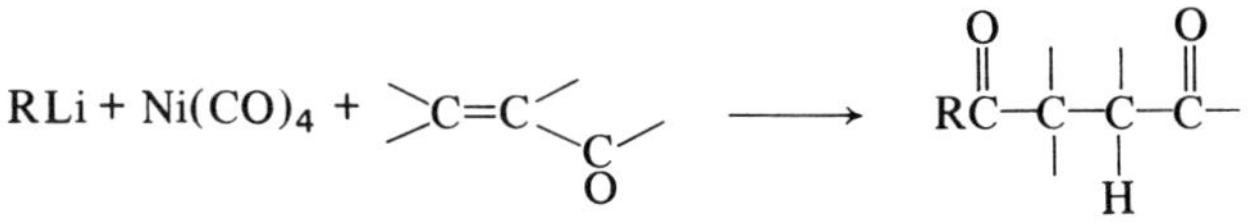

$$RLi + \text{Ni(CO)}_4 + \overset{}{\underset{\underset{\displaystyle O}{\overset{\displaystyle |}{C}}}{>}}C=C\overset{}{\underset{}{<}} \quad \longrightarrow \quad R\overset{\overset{\displaystyle O}{\|}}{C}-\overset{|}{\underset{|}{C}}-\overset{\overset{\displaystyle O}{\|}}{\underset{}{C}}- \quad (12\text{-}69)$$

R	Substrate	Product	(% isolated)[a]
Me	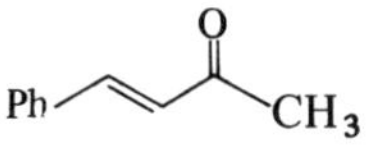 Ph⌒=⌒COCH$_3$	PhCH(COCH$_3$)CH$_2$COCH$_3$	82
	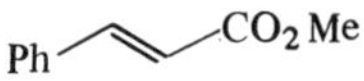 Ph⌒=⌒CO$_2$Me	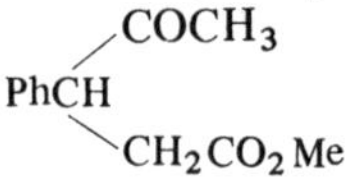 PhCH(COCH$_3$)CH$_2$CO$_2$Me	71
Bu	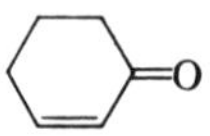 (cyclohexenone)	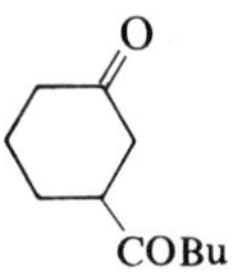 (cyclohexanone-COBu)	64
	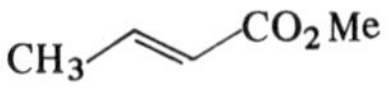 CH$_3$⌒=⌒CO$_2$Me	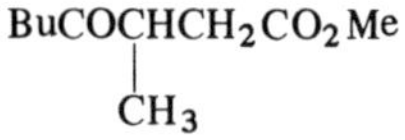 BuCOCH(CH$_3$)CH$_2$CO$_2$Me	76

[a]Data from Corey and Hegedus[50]

Many of these reactions are catalytic, and the stereospecificity is frequently quite good.

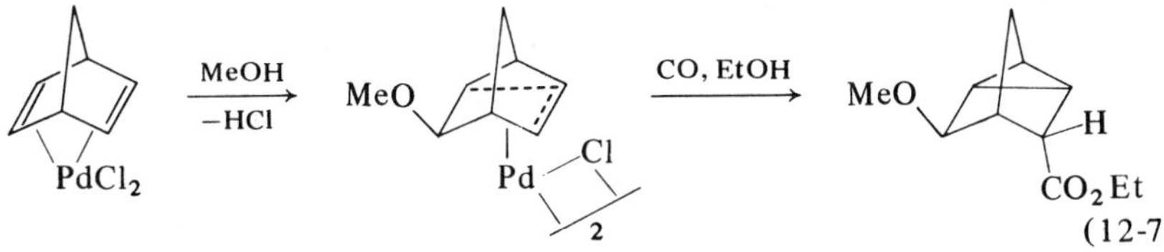

(12-70)

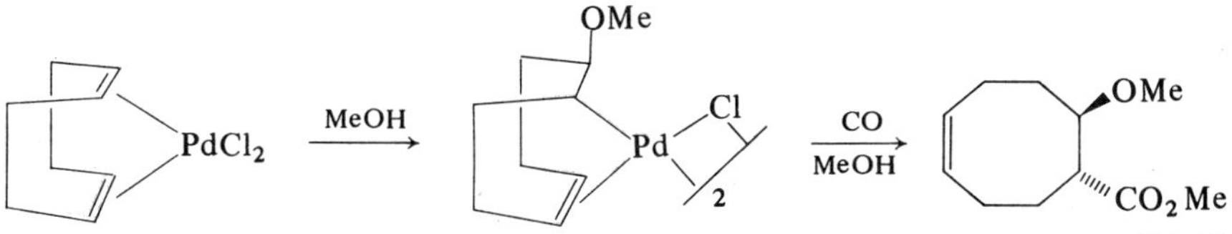

(12-71)

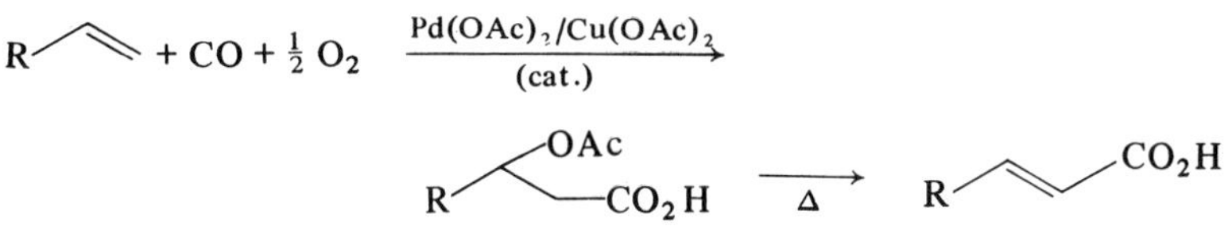

(12-72)

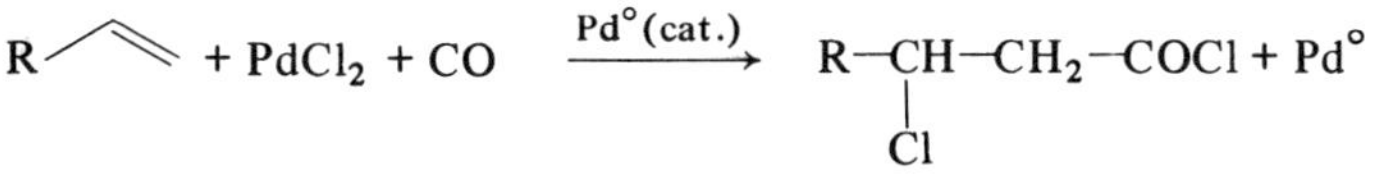

$$R \diagup\!\!\!\diagdown + PdCl_2 + CO \xrightarrow{Pd^\circ(cat.)} R-\underset{\underset{Cl}{|}}{C}H-CH_2-COCl + Pd^\circ$$

(12-73)

$$C_2H_4 + CO + EtOH \xrightarrow{Pd^\circ(cat.)} CH_3CH_2CO_2Et$$

(12-74)

π-Allyl complexes formed by oxidative addition to a metal center may be catalytically carbonylated because the final reductive elimination of the metal regenerates the catalyst; in practice it is common to start with a metal halide which is reduced *in situ* by carbon monoxide (reactions 12-75–12-78).[52]

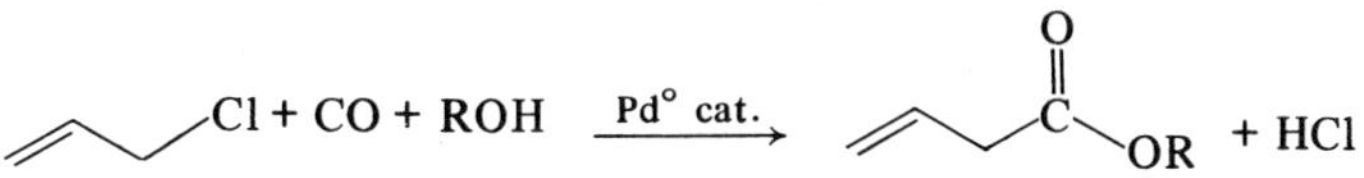

(12-75)

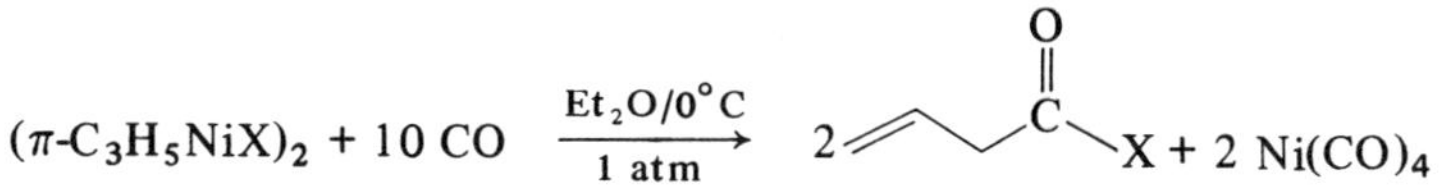

$$(\pi\text{-}C_3H_5NiX)_2 + 10\ CO \xrightarrow[\text{1 atm}]{Et_2O/0^\circ C} 2 \diagup\!\!\!\diagdown\overset{O}{\underset{\|}{C}}{-}X + 2\ Ni(CO)_4$$

(12-76)

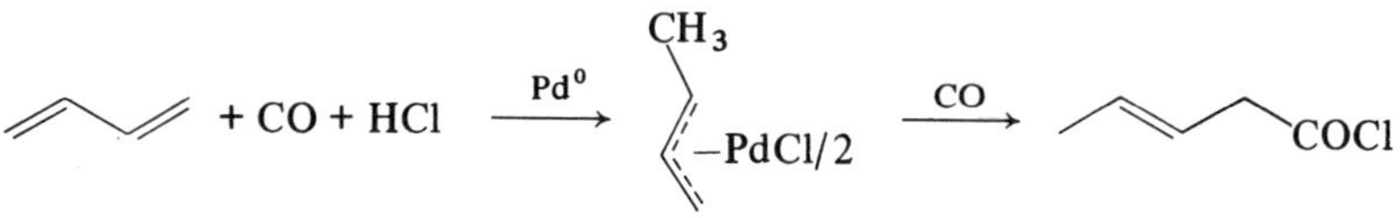

(12-77)

$$(\pi\text{-}C_3H_5NiBr)_2 + C_2H_2 + CO \xrightarrow{Pd^\circ} \diagup\!\!\!\diagdown\!\!\!\diagup\!\!\!\diagdown COBr$$

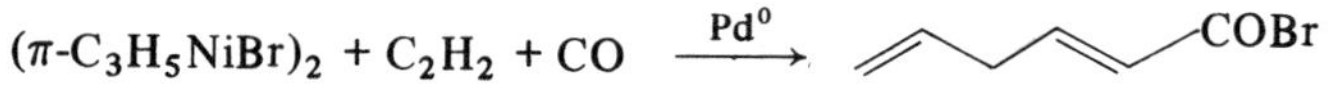

(12-78)

One of the oldest carbonylation reactions is the oxo- or hydroformylation process, which has been in use industrially for over 30 years (Fig. 12-5). The original oxo catalyst was $Co_2(CO)_8$, which was converted to $HCo(CO)_4$ with synthesis gas (H_2 + CO produced from red hot coke and steam). Olefins are converted to aldehydes at 120°–150°C and 200 atm total pressure, but at higher temperatures they are reduced to alcohols. The accepted catalytic cycle for hydroformylation, elucidated by Heck and Breslow,[53] is shown in Fig. 12-5. Recently, it was found that some rhodium analogs of the cobalt catalyst could be used under much milder conditions and gave much smaller amounts of branched products.[54,55] The selectivity may result from the larger metal hydride adding preferentially to give a primary alkyl complex. Since either CO or phosphine can dissociate from the rhodium complex, several catalytic cycles can be envisaged for these catalysts, but the overall patterns will be analogous to the cobalt reactions.

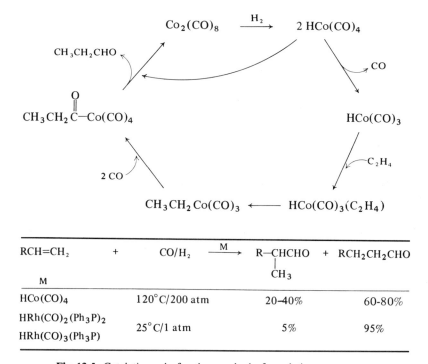

Fig. 12-5. Catalytic cycle for the oxo hydroformylation process.

A variation on hydro*formyl*ation is hydro*carboxyl*ation in which the acylcobalt bond is cleaved by alcoholysis, aminolysis, or hydrolysis rather than by hydrogenolysis.[56] Most of the examples of this reaction have involved

$Co_2(CO)_8$ as the catalyst, but there is no reason to suspect the rhodium catalysts would not offer advantages here similar to those in hydroformylation (reactions 12-79–12-81).

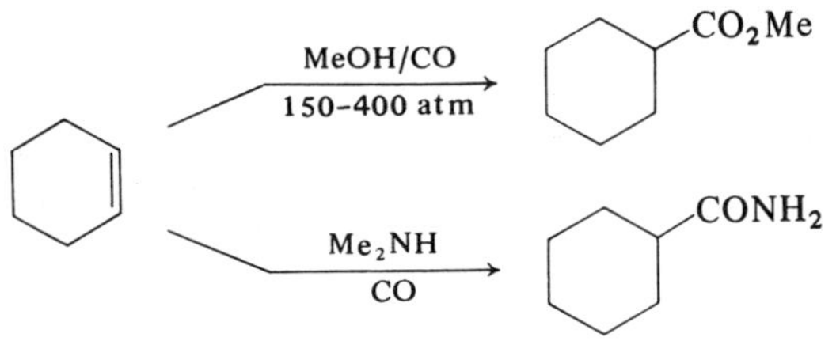

$$(12\text{-}79)$$

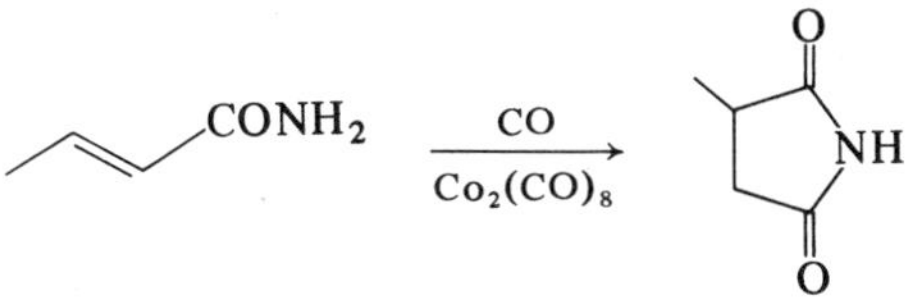

$$(12\text{-}80)$$

$$C_2H_4 + CO + H_2 \xrightarrow{\text{MeOH}} CH_3CH_2CO_2Me + CH_3CH_2COCH_2CH_3 \qquad (12\text{-}81)$$

Nickel carbonyl does not catalyze hydroformylation or hydrocarboxylation in reactions (12-79–12-81). However, under mild *acidic* conditions it is an effective hydrocarboxylating agent (Reppe reaction) (Fig. 12-6).[6b,57,58] The reaction most likely involves addition of a metal hydride to the substrate, followed by carbonyl insertion and hydrolytic regeneration of the Ni(0) catalyst, since the reaction results in the cis addition of $H–CO_2H$ to the multiple bond. Protonation of a coordinated olefin or acetylene followed by carbonylation should give trans addition, which is not observed. Acetylenes and strained olefins react much more readily than simple olefins. The direction of addition is again determined primarily by steric factors. Although additions to unsymmetrical stilbenes $XC_6H_4–CH=CH–C_6H_4Y$ give products which imply a M^+H^- polarization of the attacking metal hydride reagent, the effects are not large.[6b]

DECARBONYLATION REACTIONS

The insertion of CO into a metal-carbon bond is usually a reversible process. Although decarbonylations do not lead to the synthesis of new C–C bonds, they are included here because they are useful reactions difficult to accomplish by other means, and because they are related mechanistically to some of the carbonylation reactions discussed above. Both aldehydes and acyl halides can be decarbonylated[52] using d^8 metal complexes, chiefly of Rh(I), Ir(I), or Pd(II).

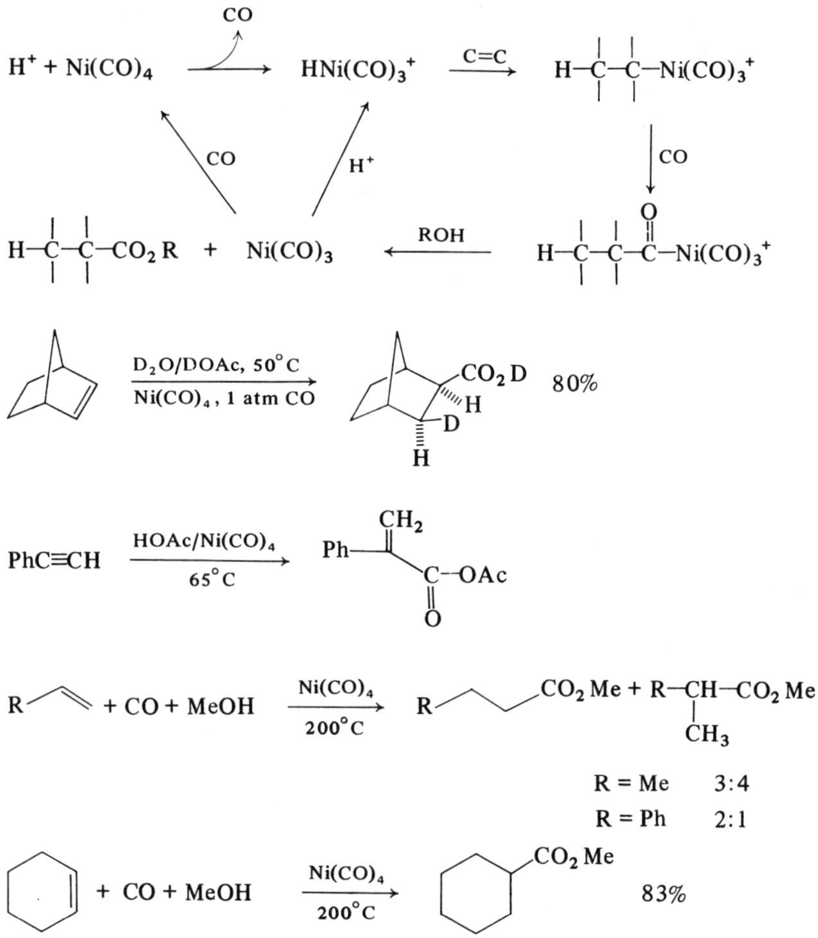

$$H^+ + Ni(CO)_4 \xrightarrow{CO} HNi(CO)_3^+ \xrightarrow{C=C} H-\overset{|}{\underset{|}{C}}-\overset{|}{\underset{|}{C}}-Ni(CO)_3^+$$

$$H-\overset{|}{\underset{|}{C}}-\overset{|}{\underset{|}{C}}-CO_2R \; + \; Ni(CO)_3 \xleftarrow{ROH} H-\overset{|}{\underset{|}{C}}-\overset{|}{\underset{|}{C}}-\overset{O}{\overset{||}{C}}-Ni(CO)_3^+$$

(with CO / H+ steps and CO step leading to the acyl intermediate)

Norbornene $\xrightarrow[\text{Ni(CO)}_4, \, 1 \text{ atm CO}]{\text{D}_2\text{O/DOAc, 50}^\circ\text{C}}$ norbornane-CO_2D, H, D, H 80%

$$PhC{\equiv}CH \xrightarrow[65^\circ C]{HOAc/Ni(CO)_4} Ph-\overset{CH_2}{\underset{\overset{|}{C-OAc}}{\parallel C}} $$
with $C{=}O$ (i.e. $Ph-C(=CH_2)-C(=O)-OAc$)

$$R\diagup\!\!\!\diagdown + CO + MeOH \xrightarrow[200^\circ C]{Ni(CO)_4} R\diagup\diagup CO_2Me + R-\overset{|}{\underset{CH_3}{CH}}-CO_2Me$$

R = Me   3:4
R = Ph   2:1

cyclohexene $+ CO + MeOH \xrightarrow[200^\circ C]{Ni(CO)_4}$ cyclohexane-CO_2Me 83%

Fig. 12-6. Catalytic cycle and examples of the Reppe hydrocarboxylation process.[6b,57,58]

The Rh(I) complexes are the best studied and appear to be the most useful, either as stoichiometric reagents at room temperature or as catalysts at elevated temperatures (Scheme 12-7).

Acyl halides are more reactive than aroyl halides, and olefins are usually formed by β elimination from the intermediate alkyl complex (12-83, A = halogen). Electron-withdrawing substituents on benzoyl halides increase both rate of decarbonylation and yield of aryl halide product (reactions 12-84–12-86),[59] which is typical for oxidative addition reactions of aroyl and aryl halides. Intermediates corresponding to (12-82) and (12-83) have been isolated and characterized; when heated they release aryl halides.

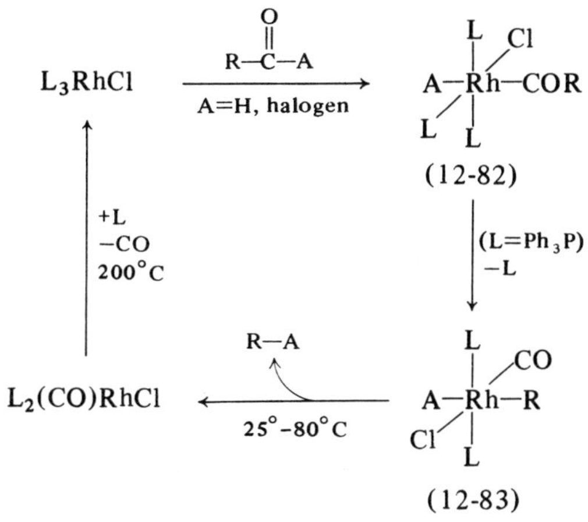

Scheme 12-7

Aldehydes are also smoothly decarbonylated to hydrocarbons (Fig. 12-7). Although the intermediates have not been isolated in this case, the observed retention of configuration[60] in the hydrocarbon product is consistent with a similar mechanism.

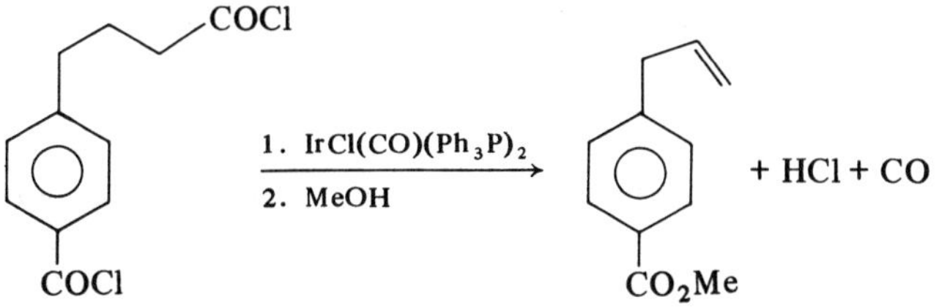

$$(12\text{-}84)$$

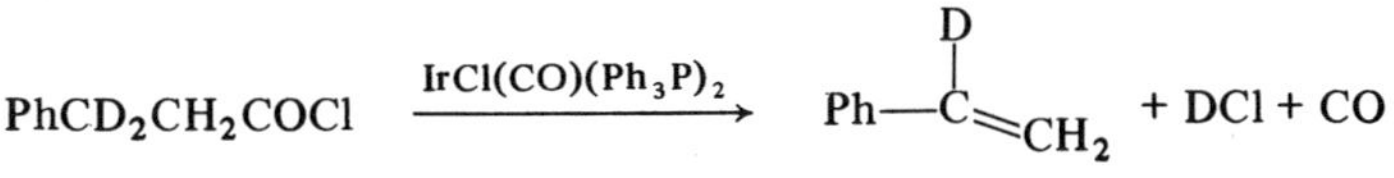

$$(12\text{-}85)$$

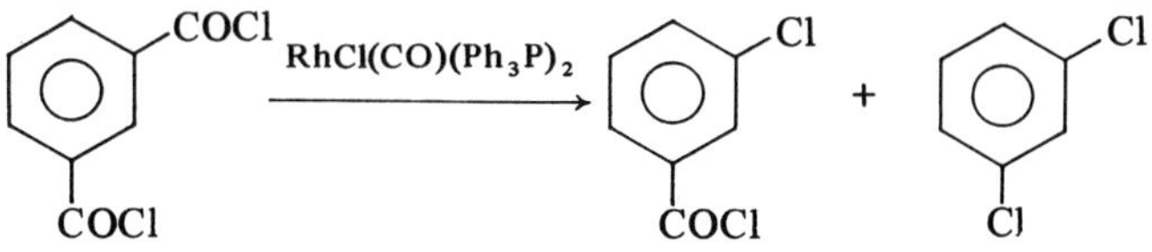

$$(12\text{-}86)$$

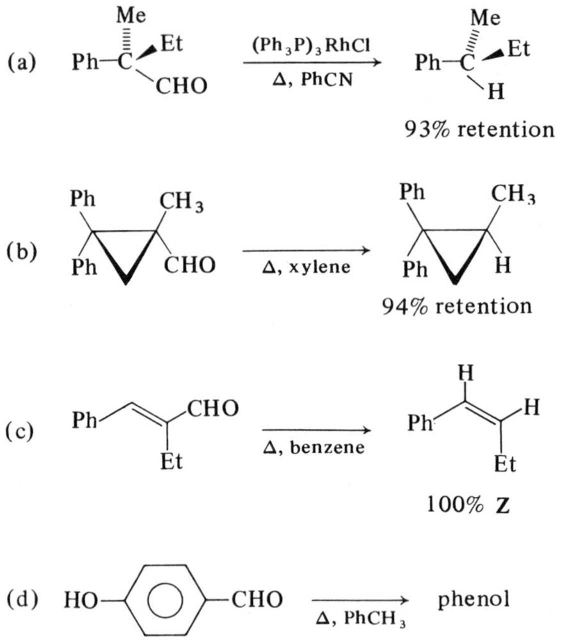

(a) to (d) reaction schemes

Fig. 12-7. Stereochemistry of aldehyde decarbonylation by $(Ph_3P)_3RhCl$.[60]

Addition Reactions of Olefins and Acetylenes

One of the most characteristic reactions of carbon–carbon multiple bonds is addition of small molecules HX. In many cases these additions involve ionic or free radical mechanisms. In the case of hydrogenation, however, the reaction requires a catalyst, for despite the favorable equilibrium the H–H bond is too strong to be broken easily by either ionic or free radical processes (cf. reaction 11-34). Traditionally, heterogeneous catalysts containing certain transition metals were employed. While these still play an important part in preparative chemistry, it has been most difficult to study the *mechanism* of the hydrogenation reaction with these catalysts. Since the introduction of soluble "homogeneous" hydrogenation catalysts, considerable progress in understanding their mechanism of action has been made. These catalysts also offer several distinct advantages over the heterogeneous types: reproducibility is usually better; they can be much more selective and specific for their substrates; H/D exchange is minimized; and if the catalyst contains chiral ligands, asymmetrical hydrogenation can be achieved.

HYDROGENATION

Hydrogenation is one of the most thoroughly studied areas of homogeneous catalysis, and the complex $ClRh(Ph_3P)_3$, known as Wilkinson's catalyst, is certainly the best studied as well as one of the most useful homogeneous hydrogenation catalysts.[61,62] It is sensitive to steric hindrance on the olefin, and generally only the following types of substrates are reduced:

$$R\diagup\!\!\!\diagdown\quad \sim \quad R-C\equiv C-R \quad > \quad \underset{R\quad\quad R}{\diagup\!=\!\diagdown} \quad > \quad R\diagup\!\!\!\diagdown\!\!\!\diagup^{R}$$

Reduction with H_2/D_2 mixtures gives only dihydro and dideuteroproducts, with no HD exchange in the gas phase, the substrate, or the products. The hydrogenation of both acetylenes and olefins involves cis addition, and dienes can be selectively monohydrogenated without isomerization of double bonds.

The catalyst $ClRh(Ph_3P)_3$ reacts in benzene with H_2 to form $H_2RhCl(Ph_3P)_2$, and with ethylene to form $(C_2H_4)RhCl(Ph_3P)_2$. Addition of olefins to the dihydride complex results in immediate reduction, with no detectable hydridorhodium-alkyl intermediate. Although such an intermediate is very likely involved, it must collapse to product much faster than rotation about a C–C single bond in order to explain the cis addition of D_2 to olefins such as maleic acid. The catalytic cycle in Fig. 12-8 has been proposed for

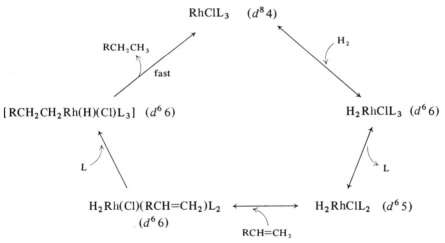

Fig. 12-8. The major catalytic cycle for olefin hydrogenation with Wilkinson's catalyst, $(Ph_3P)_3RhCl$.

Wilkinson's catalyst, although a similar cycle with the reverse order of H_2 and olefin addition may also be operative. Oxidative addition of H_2 is followed by Ph_3P dissociation and olefin uptake. Insertion of the olefin into a Rh–H bond, followed rapidly by reductive elimination of product, could be promoted by excess ligand, much like some carbonyl insertions into metal-alkyl bonds.

A fascinating and very practical modification of this basic chemistry is the use of chiral phosphine ligands (e.g., 12-87, 12-88, and 12-89) in order to

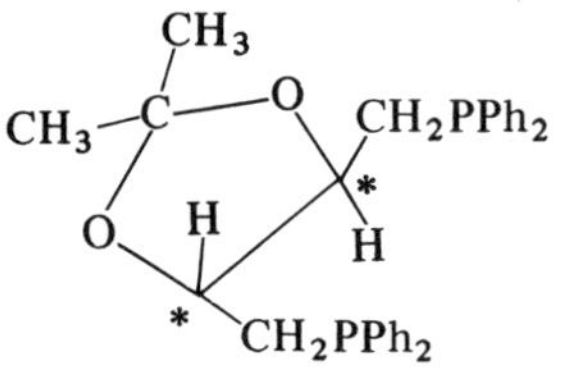

(12-87) 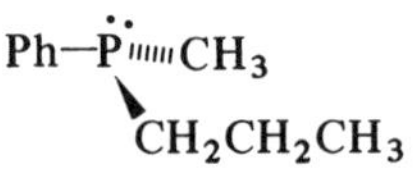   (12-88) 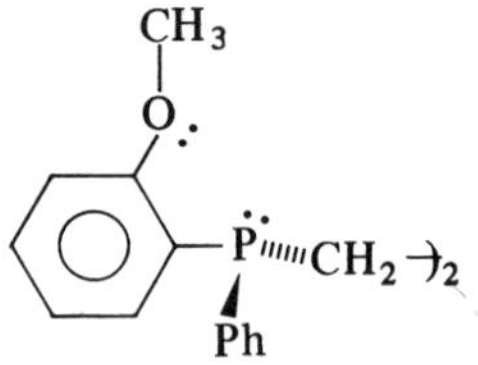   (12-89)

achieve asymmetric induction during hydrogenation. Several successful attempts have been made with various types of chiral phosphines and prochiral substrates.[63] In favorable cases with a phosphine ligand carefully designed to interact with the substrate, up to 95% enantiomeric excess may be achieved in the reduced product. A very good example of this approach is the Monsanto synthesis (reaction 12-90) of optically active (L) amino acids which are hard to obtain by fermentation, by asymmetric hydrogenation using a rhodium complex of (12-89).[64]

$$RCH=C-CO_2H \xrightarrow[\text{(12-89)}]{\text{cat.* } H_2} RCH_2\overset{*}{C}HCO_2H \qquad (12\text{-}90)$$
$$\quad\quad | \qquad\qquad\qquad\qquad\qquad\quad |$$
$$\quad NHAc \qquad\qquad\qquad\qquad\quad NHAc$$

$R = C_6H_5-,$

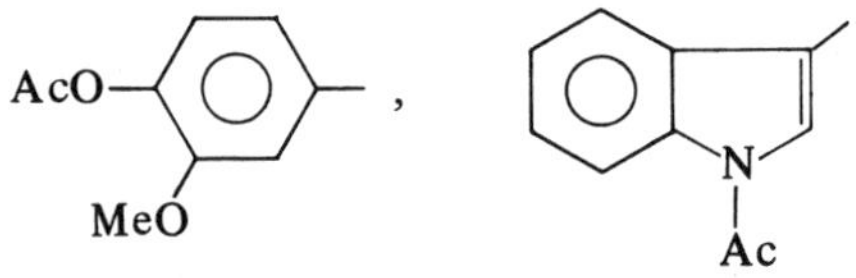

A large number of transition metal ions and complexes are known to "activate" molecular hydrogen and catalyze the reduction of many kinds of functional groups. Other mechanisms than the one discussed above are known, but since these have been extensively reviewed elsewhere,[62] they will not be discussed here.

HYDROZIRCONATION

The relatively new process of hydrozirconation, discovered by Schwartz and co-workers, embodies many of the organometallic reaction mechanisms which have been discussed in the previous sections.[65-67] The starting metal reagent is a coordinatively unsaturated metal hydride, $(\eta^5\text{-}C_5H_5)_2ZrHCl$ (12-91), which adds to olefins to give alkyl complexes. With terminal olefins the addition is regiospecific such that the metal is bonded to the least sterically hindered position. With internal olefins a process of addition–elimination–readdition rapidly effects the isomerization of the double bond such that the metal becomes bonded to the least sterically hindered carbon (Scheme 12-8).

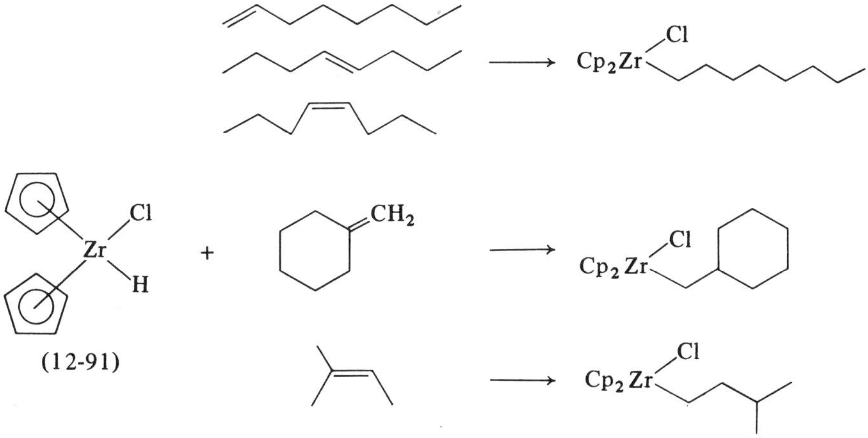

Scheme 12-8

In contrast to organoboron or -aluminum compounds in which the analogous migrations require high temperatures, the zirconium complexes rearrange rapidly at room temperature. Vinylic Zr(IV) complexes formed by addition of the hydride (12-91) to acetylenes do not undergo an analogous migration of the multiple bond, and addition of (12-91) to 1,3-dienes gives homoallylic derivatives as a result of 1,2 addition. Addition of (12-91) to unsymmetrical acetylenes gives two vinyl derivatives as *kinetic* products. A small excess of (12-91) equilibrates this mixture such that the major isomer, often ≥ 90%, is the one with the Zr bonded to the carbon with the smaller alkyl group. The large size of the ligands on the metal makes the position of equilibrium, i.e., the "regioselectivity," very sensitive to the size of the acetylene or olefin substituents, *viz.*, the results for addition to 2-pentyne in Fig. 12-9.

The real value of the alkyl- and alkenylzirconium(IV) complexes lies in the fact that the metal can be removed in a variety of ways to produce many kinds of functional groups. Studies with partially deuterated systems have shown that Zr–H adds cis to both olefins and acetylenes, and that cleavage of the Zr–C bond with a large variety of reagents proceeds with *retention* of configuration[68] (cf. Figs. 12-10 and 11-7 and reaction 11-41). Although hydrozirconation is a very new process, the comparative ease of handling of (12-91)

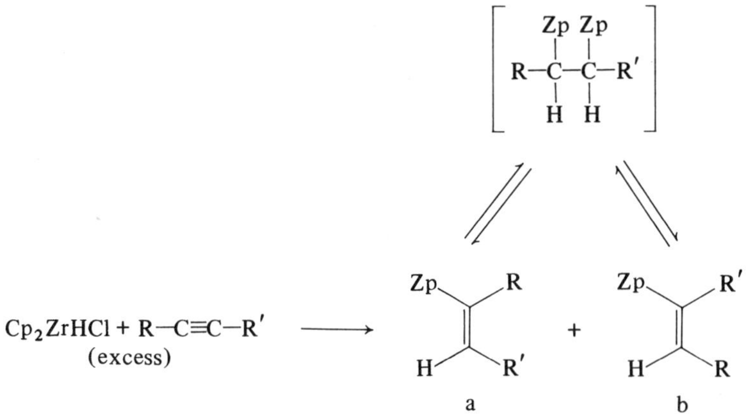

$$\left[\begin{array}{c} Zp\quad Zp \\ |\quad\quad | \\ R-C-C-R' \\ |\quad\quad | \\ H\quad H \end{array} \right]$$

$$Cp_2ZrHCl + R-C{\equiv}C-R' \longrightarrow$$
(excess)

a   +   b

$$Zp = Cp_2ZrCl-$$

		a/b ratio	
R	R'	Kinetic	Equilibrium
H	nBu	98:2	
CH$_3$	Et	55:45	89:11
CH$_3$	iBu	55:45	95:5
CH$_3$	iPr	84:16	98:2
CH$_3$	tBu	98:2	

Fig. 12-9. Regioselectivity for hydrozirconation of acetylenes.[67]

and its organic derivatives, together with the scope of functional groups which can be generated stereospecifically from the latter, may make hydrozirconation the method of choice for many organic synthetic applications. Several examples are given in Fig. 12-10.

Catalysis of "Symmetry-Forbidden" Reactions

Many transition metal ions or complexes are known to catalyze the migration of olefinic bonds and/or the valence bond reorganization of strained polycyclic hydrocarbon systems. Reactions such as shown in Scheme 12-9 have attracted considerable interest because in the absence of suitable metal catalysts, they are "thermally forbidden as concerted processes" according to the Woodward–Hoffmann rules.[69] These "rules" are deduced by considering the symmetry and occupancy of both reactant and product orbitals. If the symmetry and occupancy are conserved in the concerted process the reaction is

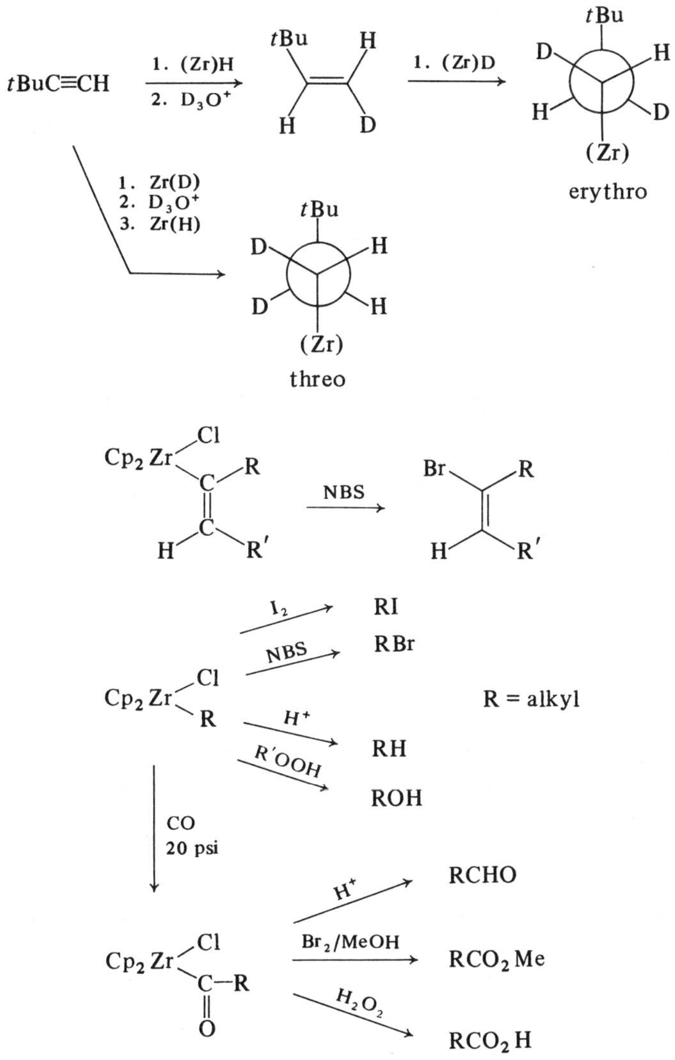

Fig. 12-10. Stereochemistry and synthetic versatility of the hydrozirconation reaction.[65-68]

"allowed," i.e., has a thermally accessible activation energy; if not, the process is "thermally forbidden" or "photochemically allowed." Considerable attention has been focused on the question of whether transition metal ions, as a result of bonding and back-bonding interactions with reactant molecules, might perturb the occupancy of various bonding or antibonding orbitals on the ligand such that the electron configuration of the *complexed* reactant might resemble an

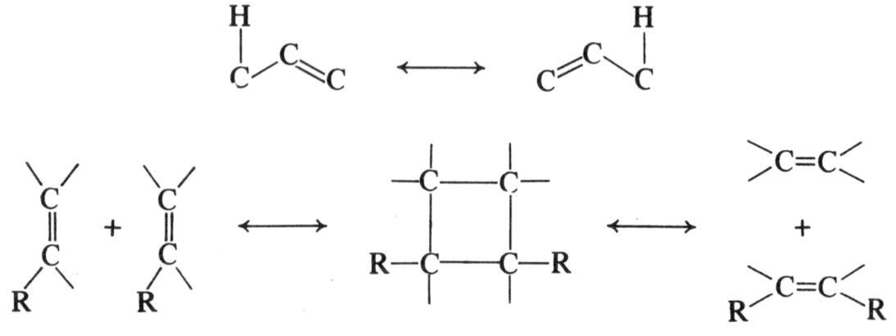

Scheme 12-9

excited state thus permitting it to undergo a thermally forbidden reaction.[70] Evidence that this is the case has not been forthcoming; rather, a number of *stepwise* mechanisms for "thermally forbidden concerted reactions" have been elucidated. In general, they fall into two groups which may involve substrate activation by oxidative addition processes or by Lewis acid-base processes, depending on the particular metal and whether or not it can undergo oxidative addition.

HYDROGEN MIGRATIONS

The migration of an olefinic bond by means of a net 1,3-hydride shift can be accomplished in various ways, with or without simultaneous exchange of hydrogens between the olefin and another reactant. Most isomerization catalysts are metal hydrides which either exist as such or are transient intermediates formed by reaction of a metal species with a "co-catalyst" or "promoter," usually an acid such as HCl, a hydride reducing agent such as $NaBH_4$, or hydrogen gas. These isomerization catalysts operate by addition–elimination sequences such as those already depicted for rhodium(I) hydride complexes (e.g., reaction 12-54) and for the hydrozirconation reaction discussed above. The isomerization and extensive H/D exchange sometimes

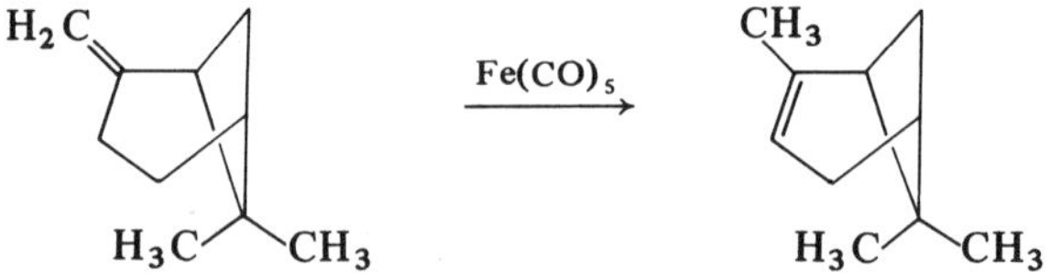

$$(12\text{-}92)$$

(−)-β-pinene

(−)-α-pinene
95% optically pure

$$CH_2\!=\!CHCD_2OH \xrightarrow{Fe(CO)_5} DCH_2CH_2\!-\!C{\displaystyle\mathop{\diagup}^{O}_{\diagdown D}} \qquad (12\text{-}93)$$

observed with olefins and heterogeneous Pt or Pd hydrogenation catalysts may involve similar mechanisms on the catalyst surface.

Another mechanism for 1,3-hydride transfer involves oxidative addition of the olefin to form a π-allyl-metal hydride intermediate.[17,71] In this case the transfer is stereospecific and intramolecular, and no H/D exchange is observed (cf. reactions 12-12, 12-13, 12-92,[72] and 12-93[73]).

VALENCE BOND ISOMERIZATIONS

The apparent stability of highly strained polycyclic hydrocarbons has been attributed to the constraints of the Woodward–Hoffman rules. In the reactions shown in Table 12-3,[74,75] and many others, certain transition metal ions cause a

TABLE 12-3

Examples of Metal-Catalyzed Valence Bond Isomerization Reactions

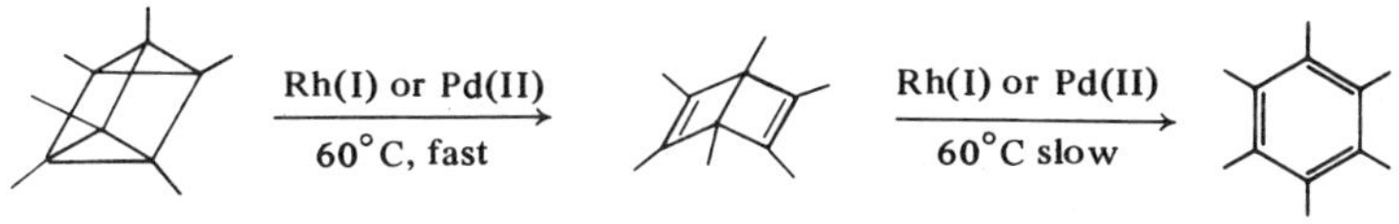

ΔE_{act} (kcal/mole)^a

Thermal 31.1
Rh(I) cat. 19.4

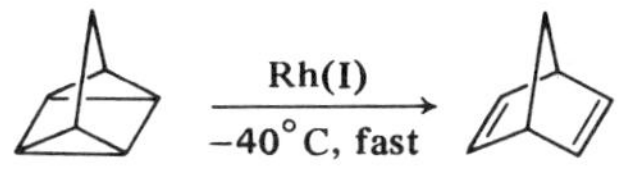

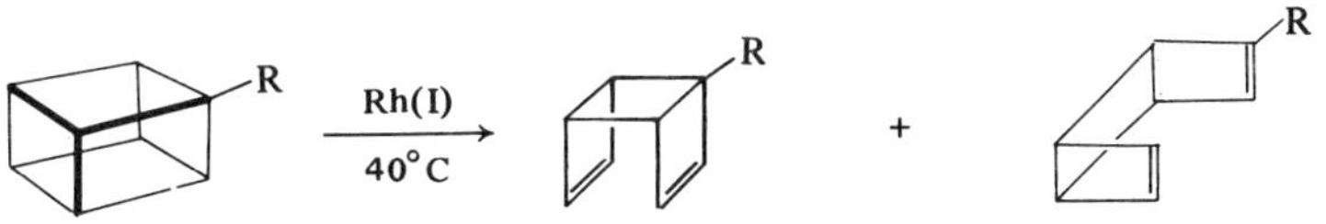

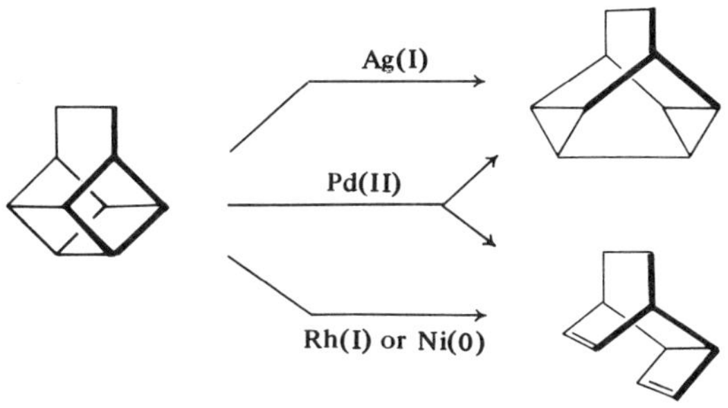

^{a}Data from Vogler and Hogeveen[74] and Vogler and Gaasbeek.[75]

remarkable lowering of the activation energies for the thermodynamically favored rearrangements. The two extremes of such reorganizations exemplified by reactions (12-94) and (12-95) (Table 12-3) also represent two extremes in terms of mechanism.

The first suggestion that the *retro* 2 + 2 cycloaddition reactions were stepwise processes came from the observation that while Rh(norbornadiene)Cl$_2$ catalyzed reaction (12-95), Rh(CO)$_2$Cl$_2$ reacted stoichiometrically with cubane to produce an isolable intermediate (12-96).[76] Treatment of this intermediate with Ph$_3$P produced the ketone (12-97) in 90% yield. Thus, it was postulated that Rh(I) catalysis of *retro* 2 + 2 cycloaddition reactions involved an oxidative addition–reductive decoupling sequence, in which the normally transient Rh(III) intermediate (12-98) could be captured by carbonylation (see Scheme 12-10). This mechanism has become generally accepted for reactions

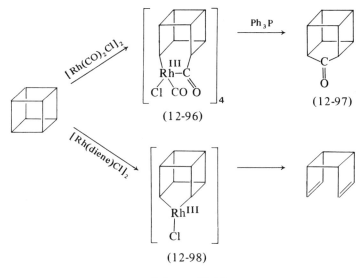

Scheme 12-10

of this type and is supported by extensive kinetic and product studies with substituted cubanes and related hydrocarbons. The steric effects of substituents generally outweigh electronic effects, indicating that little ionic character develops during the transition state.

In sharp contrast to the Rh(I) catalysis of reaction (12-94) is the Ag(I) catalysis of reaction (12-95) (see Table 12-3). In this case the electronic effects of substituents on the hydrocarbon skeleton are very dramatic, indicating that appreciable carbonium ion character develops in the transition state. Since Ag(I) is not very likely to undergo oxidative addition to an Ag(III) intermediate, it is proposed that it reacts as a soft Lewis acid with the strained hydrocarbon to form cationic alkylsilver(I) complexes such as (12-99). The latter undergo various cyclopropylcarbinyl rearrangements terminating in release of Ag+ and formation of rearranged hydrocarbon[77] (see Scheme 12-11). The intermediacy of carbonium ions is supported in some cases by the

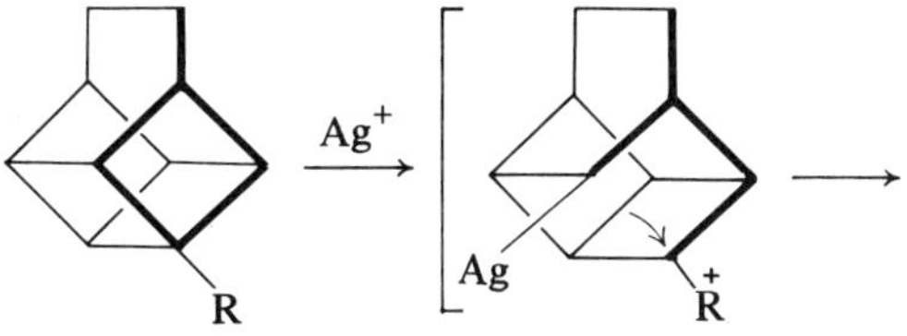

(12-99)

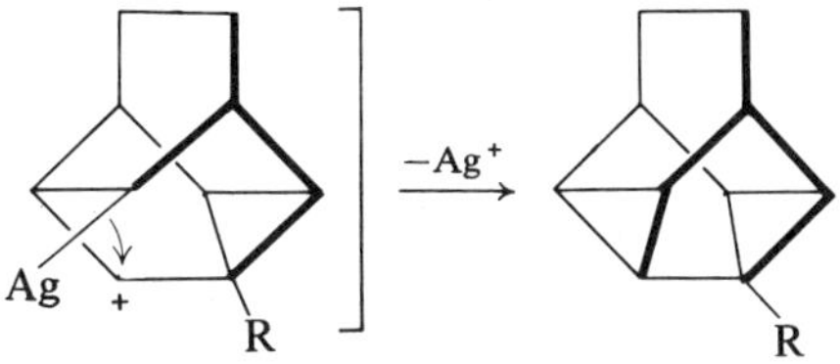

Scheme 12-11

isolation of the same nucleophile trapping products when several different catalysts are used.[78] In other cases products reminiscent of carbene insertion reactions are formed (reaction 12-100).

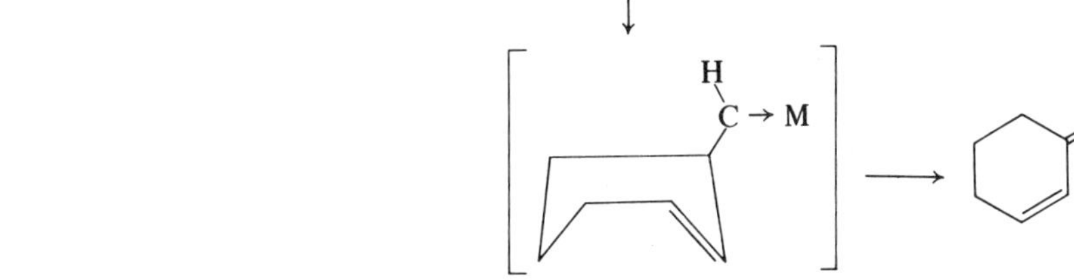

(12-100)

OLEFIN METATHESIS

A striking example of metal catalysis of a reaction which is thermally forbidden is the metathesis of olefins, as exemplified by the equilibration of propene with ethylene and 2-butene (cis and trans) (reaction 12-101).

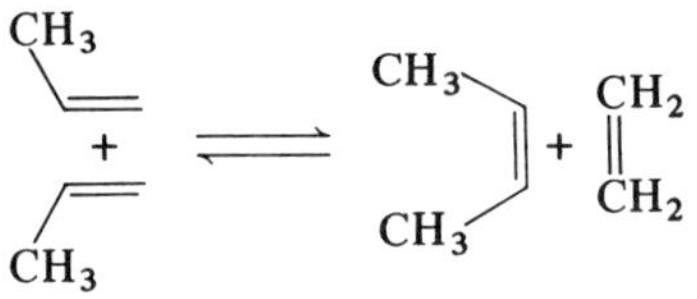

(12-101)

Catalyst	Conditions required[a]
None	$725°C$
$Mo(CO)_6$ on alumina support	$150°C$, 30 atm
$WCl_6 \cdot 4EtAlCl_2$ in ethanol	$25°C$, 1 atm

[a]Data from Haines and Leigh.[79]

Complexes or mixtures derived from a large number of different transition metals have been shown to exhibit some degree of catalysis of this reaction but by far the best, whether heterogeneous or homogeneous, are derived from metals having fewer than 6 d electrons. Metals in group VI are particularly

TABLE 12-4

Commercial Applications of Olefin Metathesis[a]

a.   Propene $\longrightarrow$ 2-butene + ethylene for polymerization

b.   [cyclopentene] $\longrightarrow$ trans-1,5-polypentamer, $-(CH_2CH=CHCH_2CH_2)_n$

c. $(CH_2)_n$ [with CH and CH] + [CH_2 and CH_2] $\longrightarrow$ $H_2C=CH-(CH_2)_n-CH=CH_2$, $n \geqslant 6$

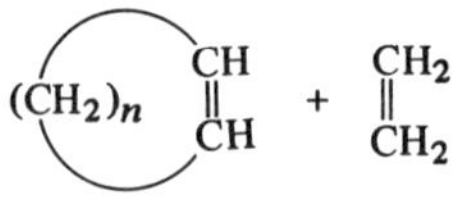

d.   Methyl stearate $\longrightarrow$ $MeO_2C-(CH_2)_7-CH=CH-(CH_2)_7-CO_2Me$
       $+$
       $CH_3-(CH_2)_7-CH=CH-(CH_2)_7-CH_3$

e. $CH_3CH=CH_2 + CH_2=CHCN$ $\longrightarrow$ $CH_3CH=CHCN + C_2H_4$

[a]Data from Haines and Leigh[79] and Hughes.[80]

effective. Various "promoters" and "co-catalysts," usually organometallic or hydridic reducing agents (e.g., BuLi, EtAlCl$_2$, or LiAlH$_4$) and/or Lewis acids (e.g., AlCl$_3$ or Me$_3$Al$_2$Cl$_3$), are often also required for catalysis, but their exact role is obscure. There is no sharp division between homogeneous and heterogeneous catalysts, and in many cases the catalytic species may be large aggregates of borderline solubility held together by bridging ligands.[79]

A variety of olefin types are susceptible to the metathesis reaction. In general, steric effects are most important (H$_2$C= > RHC= > R$_2$C=; R = phenyl, alkyl, CN) but heteroatoms which are good ligand donor sites poison the catalysis. The versatility and the rather large industrial potential of the metathesis reaction are indicated by the examples in Table 12-4[79,80] and by the fact that the first commercial metathesis operation went on stream only 3 years after the discovery of the first metathesis catalysts.

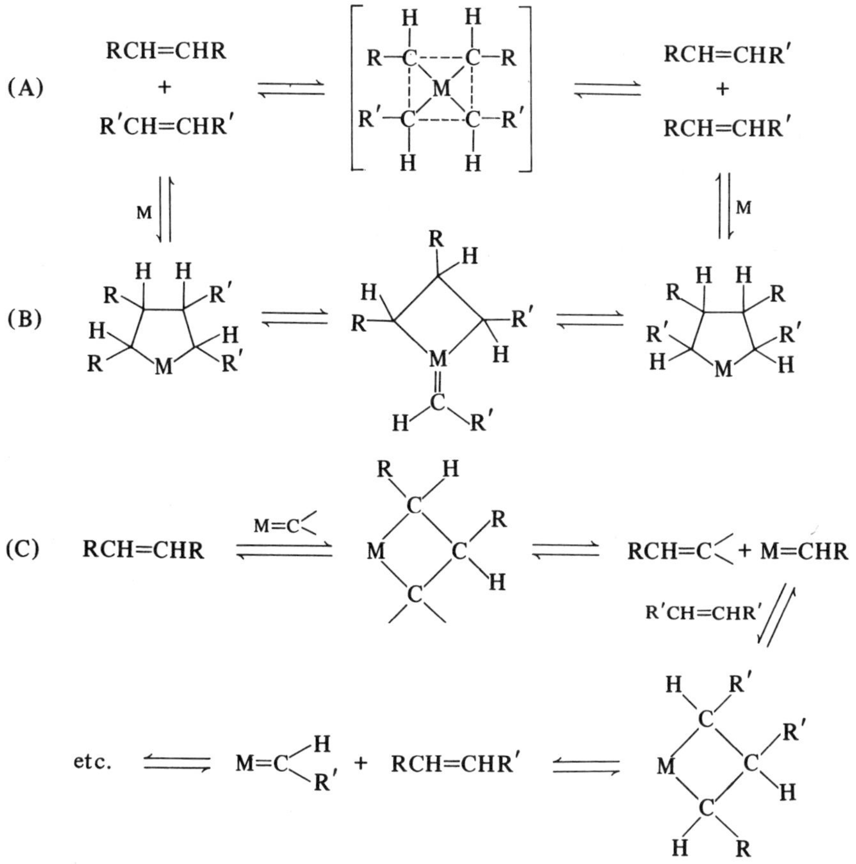

Fig. 12-11. Mechanisms proposed for the olefin metathesis reaction.

Studies of the mechanism of the metathesis reaction are still at an early stage. As might be expected from the large number of active catalysts and their apparent complexity in solution (not to mention the heterogeneous catalysts), more than one mechanism may be involved. The first mechanism to be proposed postulated a "quasi-cyclobutane" complex as a transition state (Fig. 12-11A), based largely on results with heterogeneous catalysts. That the reaction is actually an alkylidene exchange, and not an alkyl exchange, was shown by metathesis of 2-butene and 2-butene-d_8. Metathesis of mixtures of internal olefins produces the statistically expected mixture of olefins in their cis/trans equilibrium ratio, thus showing that the process is random. However, since cyclobutanes are rarely formed under metathesis conditions and cyclobutane itself is not very efficiently split to ethylene, the quasi-cyclobutane mechanism has not received strong acceptance.

Another mechanism proposed for the metathesis reaction is the "tetramethylene metallocycle" pathway (Fig. 12-11B), which has obvious analogy to the valence bond isomerizations discussed above. The main experimental support for this mechanism is the isolation of stable tetramethylene metallocycles such as $(Ph_3P)_2Pt(CH_2)_4$ and observations on the conversion of 1,4-dilithiobutane to ethylene by WCl_6 in benzene, presumably via a cyclic mechanism. Experiments with $LiCH_2CD_2CD_2CH_2Li$ and WCl_6 give deuterated ethylenes $C_2H_2D_2:C_2H_3D:C_2H_4$ in a $6:88:6$ ratio, showing that exchange occurs with metathesis or fragmentation.[81] However, this type of exchange is not characteristic of most metathesis systems, and only fluorinated olefins are known to undergo the initial oxidative cyclization to form the metallocycle.

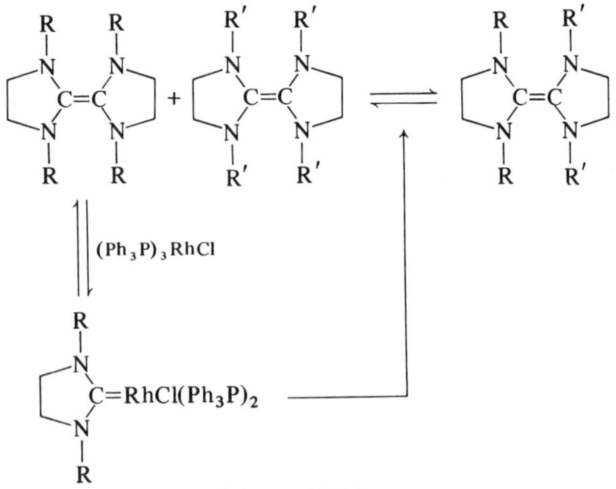

Scheme 12-12

The currently favored mechanism for olefin metathesis involves a carbene-transfer chain sequence, i.e., a true transalkylidenation process (Fig. 12-11C).[82,83] Several lines of evidence support this view. First, catalytically active Rh(I)–carbene complexes can be isolated from the reaction during the metathesis of electron-rich olefins by $(Ph_3P)_3RhCl$, although this need not necessarily hold for normal olefins (Scheme 12-12)[84]. Second, it has been shown that metal carbene complexes can react with olefins to form cyclopropanes (see next section), which are not infrequent by-products of metathesis reactions, and that alkylidene exchange with an olefin can also occur to give a more stable carbene complex. (Scheme 12-13).[85] Finally, statistical

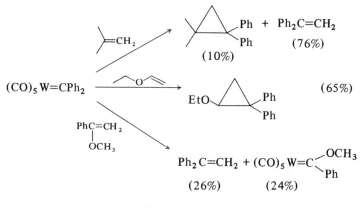

Scheme 12-13

analysis of deuterated ethylenes formed by metathesis of mixtures of 1,7-octadiene and 1,7-octadiene-1,1,8,8-d_4 are in agreement with predictions based on a carbene chain transfer mechanism (Scheme 12-14).[83] Furthermore, the

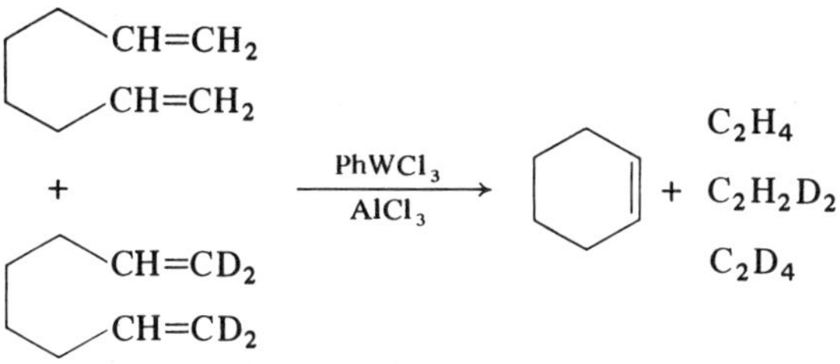

Scheme 12-14

formation of $C_2H_2D_2$ from the above olefins indicates nonpairwise exchange, which is not easily compatible with mechanisms A or B in Fig. 12-11.

Stabilization and Transfer of Reactive Ligands

Metal groups are very useful for converting highly reactive molecules into more stable forms, i.e., complexes, which can be more easily handled and studied, both by chemical as well as spectroscopic means. Although complexation does perturb the ligand, in many cases this is no more serious than that caused by heavily substituting the molecule with trifluoromethyl, phenyl, or *t*-butyl groups, which is the usual "organic" approach to taming reactive species (viz. methyl tri(*t*-butyl)cyclobutadienecarboxylate in Chapter 11). Several examples of "nonexistent" molecules which have been stabilized and isolated as metal complexes have already been encountered and are listed along with several others in Table 12-5.[10,84,86-88]

Another important application of metal groups is their use as blocking or protecting groups for sensitive *portions* of molecules, thus allowing organic transformations to be carried out selectively at other sites in the molecule. This

TABLE 12-5

Metal Complexes of Highly Reactive Ligands

Cyclobutadienes[a]:

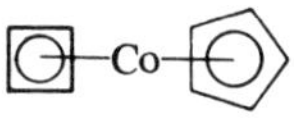

Trimethylenemethanes[a]:

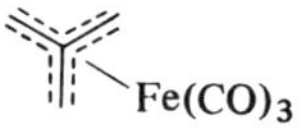

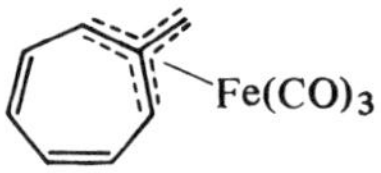

Carbenes[b]: 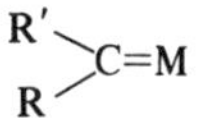   Oxepins[c]:

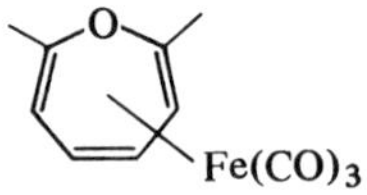

Nonaromatic tautomers[d]: 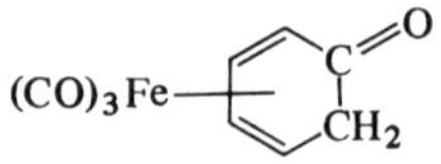   Small ring acetylenes[e]·

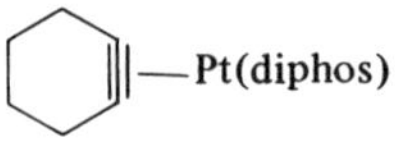

[a] Powell.[86]
[b] Cardin.[84]
[c] Fisher.[87]
[d] Birch.[10]
[e] Bennett and Yoshida.[88]

may be contrasted to situations discussed earlier where complexation was used to *increase* reactivity, as, for example, in electrophilic or nucleophilic substitution reactions (q.v.). Thus, 1,3-dienes are readily protected by complexation with a $Fe(CO)_3$ group, as shown by reactions (12-22), (12-102), and (12-103).

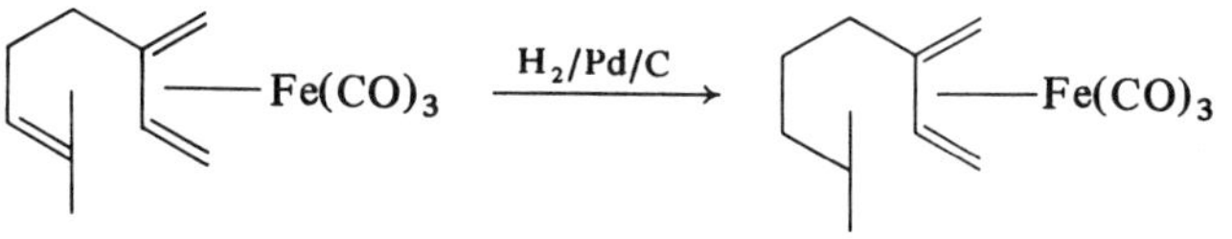

$$\text{(12-102)}$$

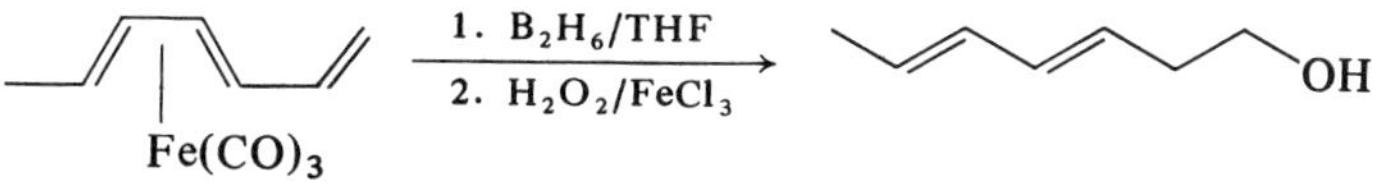

$$\text{(12-103)}$$

Isolated olefinic bonds are conveniently protected by complexation to a $Cp(CO)_2Fe^+$ group (Fp^+), as shown by Scheme 12-15.[89] In this case the "free" double bond is also slightly deactivated since it is not epoxidized by peracids.

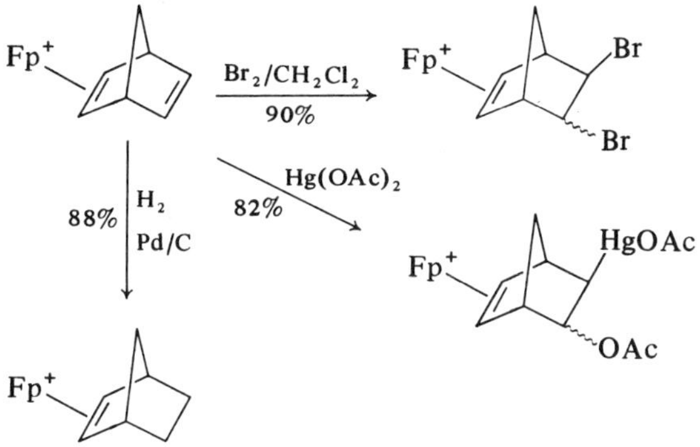

Scheme 12-15

The installation and removal of these iron-based protecting groups was discussed earlier in connection with their effects on electrophilic and nucleophilic attack on coordinated olefin and diene ligands (cf. reactions 12-14–12-19).

Another important feature of some complexes of reactive ligands is that they can be useful synthetically as sources for the ligand species. Often a ligand just becoming free of a metal complex has different chemical properties than when it is generated or introduced by other means. A case in point is carbene and its derivatives, the usual sources of which are the corresponding diazo compounds. Their conversion to carbenes can be accomplished photochemically, thermally, through catalysis by copper metal surfaces, or by homogeneous copper catalysts. As with hydrogenation, homogeneous

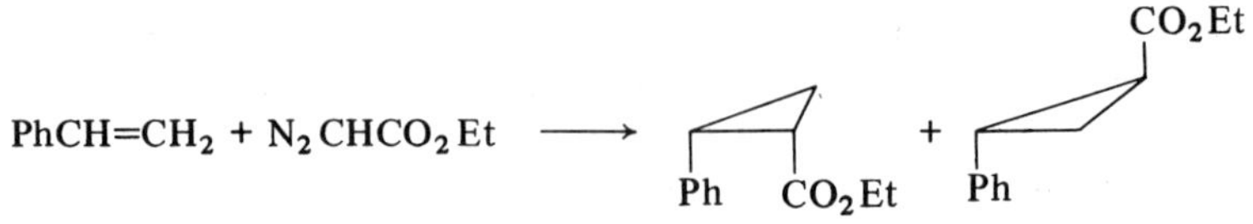

$$PhCH=CH_2 + N_2CHCO_2Et \longrightarrow$$

$$(12\text{-}104)$$

Catalyst	% Asymmetrical induction[90a] (%)
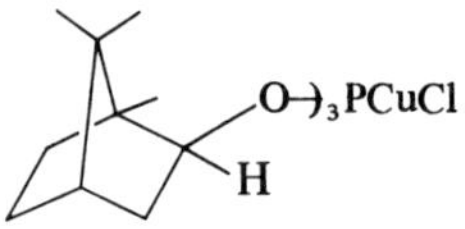	3
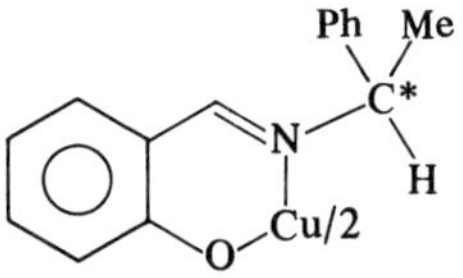	6

[a] Data from Moser.[90]

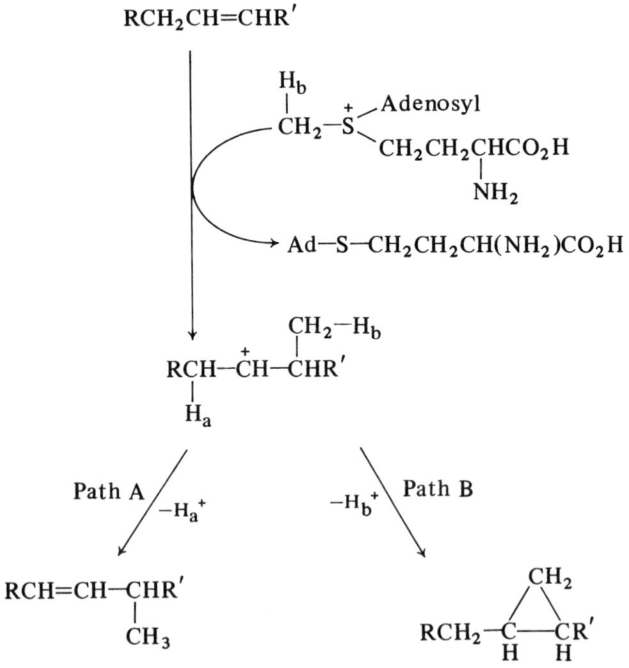

Scheme 12-16

catalysts offer greater selectivity and specificity than do heterogeneous catalysts. For example, in the addition of carbethoxycarbene to cyclohexene, catalyzed by (trialkylphosphite)CuCl, the endo/exo ratio varies with the size of the phosphite ligand, and optically active products may be obtained using copper complexes with chiral ligands (reaction 12-104).[90]

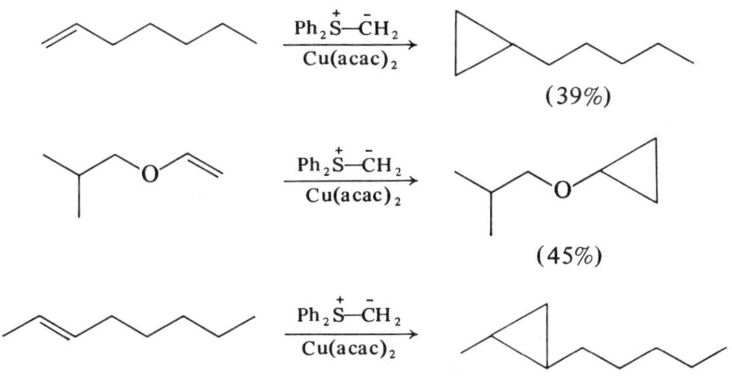

(39%)

(45%)   (12-105)

(41%)

Coordinated carbenes may also be important intermediates in the olefin metathesis reaction discussed above, and chemical model experiments indicate they could have a role in the enzymatic formation of cyclopropanoid fatty acids from unsaturated fatty acids.[91] Tracer experiments have shown that the source of the extra carbon is the methyl group of methionine via the coenzyme S-adenosylmethionine (SAM), which is also known to be a good biological cationic methylating agent and is important in forming branched fatty acids by mechanisms such as path A in Scheme 12-16. Path B invokes a rather unusual

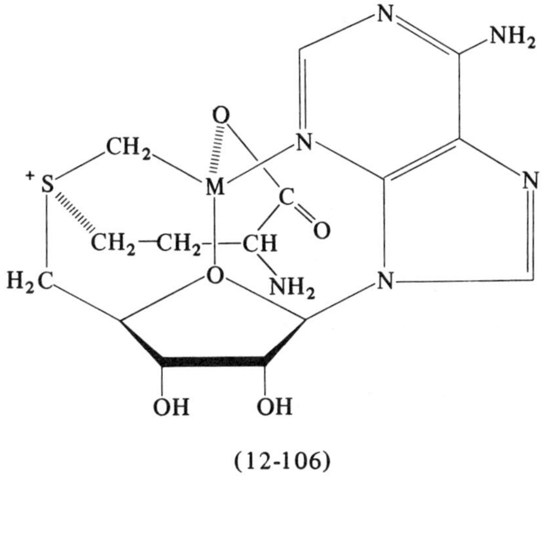

(12-106)

"electrophilic substitution at saturated carbon" to account for enzymatic cyclo-propane formation from SAM with no isomerization or migration of the double bond. Recently, it has been shown that sulfonium ylids can act as carbene donors in the presence of catalytic amounts of Cu(acac)$_2$[91] (reactions 12-105). Since the conversion of sulfonium salts to ylids can occur under conditions which could exist in the active site of an enzyme, it has been suggested that an ylid-chelate derived from SAM (12-106), or an ylid-metal-enzyme equivalent, could be involved in enzymatic cyclopropanation reactions.[91]

Other reactive species generated and perhaps transferred via metal com-plexes include 1,3-dipolar species related to allene oxides and cyclopropanones. These may be involved in 2 + 3 and 4 + 3 cycloaddition reactions of α,α'-dibromoketones with olefins and dienes as shown in Scheme 12-17. However,

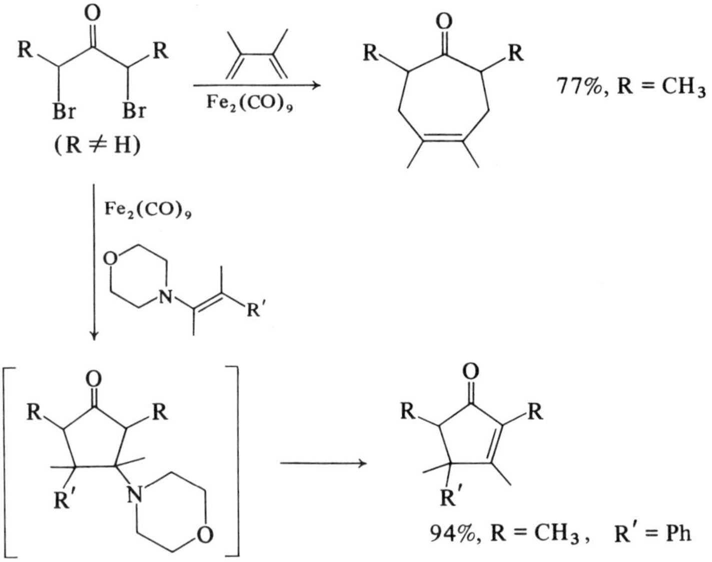

Scheme 12-17

cyclopropanones do not react in this system, and Zn·Cu will not replace Fe(CO)$_5$ or Fe$_2$(CO)$_9$ as a reducing agent. The intermediate which transfers the 1,3-dipole may be an oxa analog of a trimethylenemethane-iron complex.[92] Alternatively, since (diene)Fe(CO)$_3$ complexes themselves react with the dibromoketones, a series of oxidative addition and insertion steps may be in-volved. Nevertheless, these reactions provide synthetically useful routes to cyclopentanones and tropones.

REFERENCES

1. J. M. Davidson, *MTP Int. Rev. Sci. Inorg. Chem., Ser. 1* **6**, 347 (1972).
2. G. N. Schrauzer, ed., "Transition Metals in Homogeneous Catalysis." Dekker, New York, 1971.
3. C. A. Tolman and J. P. Jesson, *Science* **181**, 501 (1973).
4. M. M. Taqui Kahn and A. E. Martell, "Homogeneous Catalysis by Metal Complexes." Academic Press, New York, 1974.
5. D. G. H. Ballard, *Chem. Brit.* **10**, 20 (1974).
6. J. M. Swan and D. St. C. Black, "Organo-metallics in Organic Synthesis." Chapman & Hall, London, 1974.
6a. A. Rosenthal and I. Wender, *in* "Organic Syntheses via Metal Carbonyls" (I. Wender and P. Pino, eds.), pp. 405–467. Wiley (Interscience), New York, 1968.
6b. C. W. Bird, "Transition Metal Intermediates in Organic Synthesis," pp. 149–205. Academic Press, New York, 1967.
7. D. A. White, *Organometal. Chem. Rev., Sect. A* **3**, 497 (1968).
8. M. F. Semmelhack, H. T. Hall, M. Yoshifuji and G. Clark, *J. Amer. Chem. Soc.* **97**, 1247 (1975).
9. H. W. Whitlock, C. R. Reich, and R. L. Markezick, *J. Amer. Chem. Soc.,* **92**, 6665 (1970).
10. A. J. Birch, P. E. Cross, J. Lewis, D. A. White, and S. B. Wild, *J. Chem. Soc. (London), Sect. A* p. 332 (1968).
11. J. E. Mahler and R. Pettit, *J. Amer. Chem. Soc.* **88**, 3955 (1963).
12. J. H. Richards and E. A. Hill, *J. Amer. Chem. Soc.* **83**, 4216 (1961).
13. M. J. Nugent, R. E. Carter, and J. H. Richards, *J. Amer. Chem. Soc.* **91**, 6145 (1969).
14. F. G. Cowherd and J. L. von Rosenberg, *J. Amer. Chem. Soc.* **91**, 2157 (1969).
15. A. Rosan, M. Rosenblum, and J. Tancrede, *J. Amer. Chem. Soc.* **95**, 3062 (1973).
16. Y. Shvo and E. Hazam, *Chem. Commun.* p. 336 (1974).
17. F. G. Cowherd and J. L. von Rosenberg, *J. Amer. Chem. Soc.* **91**, 2157 (1969); *Chem. Commun.* p. 271 (1973).
18. P. M. Henry, *Advan. Chem. Ser.* **70**, 126 (1968).
19. P. M. Henry and R. N. Pandey, *Advan. Chem. Ser.* **132**, 33 (1974).
20. J. K. Stille, L. F. Hines, R. W. Fries, P. K. Wong, D. E. James and K. Lau, *Advan. Chem. Ser.* **132**, 90 (1974).
21. J. E. Backvall and B. Akermark, *Chem. Commun.* p. 82 (1975).
22. J. Tsuji, *Accounts Chem. Res.* **2**, 144 (1969).
23. M. F. Semmelhack, *Org. React.* **19**, 115 (1972).
24. G. H. Posner, *Org. React.* **19**, 1 (1972).
25. J. F. Normant, *Synthesis* p. 63 (1972).
26. U. Birkenstock, H. Bönnerman, B. Bogdanović, D. Walter, and G. Wilke, *Advan. Chem. Ser.* **70**, 250 (1968).
27. E. J. Corey and M. F. Semmelhack, *J. Amer. Chem. Soc.* **89**, 2755 (1967).
28. E. J. Corey and E. K. W. Wat, *J. Amer. Chem. Soc.* **89**, 2757 (1967).
29. E. J. Corey and E. Hamanaka, *J. Amer. Chem. Soc.* **89**, 2758 (1967).
30. E. J. Corey and H. A. Kirst, *J. Amer. Chem. Soc,* **94**, 667 (1972).
31. G. M. Whitesides, J. SanFillipo Jr., C. P. Casey, and E. J. Panek, *J. Amer. Chem. Soc.* **89**, 5302 (1967).
32. C. R. Johnson and G. A. Dutra, *J. Amer. Chem. Soc.* **95**, 7777 and 7783 (1973).
33. C. R. Johnson, R. W. Herr, and D. M. Wieland, *J. Org. Chem.* **38**, 4263 (1973).
34. G. H. Posner, C. E. Whitten, and J. J. Sterling, *J. Amer. Chem. Soc.* **95**, 7788 (1973).
35. H. O. House and W. J. Fisher, *J. Org. Chem.* **33**, 949 (1968).

36. H. O. House, L. E. Huber, and M. J. Umen, *J. Amer. Chem. Soc.* **94**, 8471 (1972).
37. J. Schwartz, D. W. Hart, and J. L. Holden, *J. Amer. Chem. Soc,* **94**, 9269 (1972).
38. L. S. Hegedus, S. M. Lo, and D. E. Bloss, *J. Amer. Chem. Soc.* **95**, 3040 (1973).
39. M. P. Cooke, *J. Amer. Chem. Soc.* **92**, 6080 (1970).
40. J. P. Collman, S. R. Winter, and D. R. Clark, *J. Amer. Chem. Soc.* **94**, 1788 (1972); W. O. Siegl and J. P. Collman, *J. Amer. Chem. Soc.* **94**, 2516 (1972).
41. J. P. Collman, S. R. Winter, and R. G. Komoto, *J. Amer. Chem. Soc.* **95**, 249 (1973); J. P. Collman and N. W. Hoffman, *J. Amer. Chem. Soc.* **95**, 2689 (1973).
42. L. S. Cassar, S. Ferrara, and M. Foa, *Advan. Chem. Ser.* **132**, 252 (1974).
43. R. Cramer, *Accounts Chem. Res.* **1**, 186 (1968).
44. R. J. Angelici, *Accounts Chem. Res.* **5**, 335 (1972).
45. J. Halpern and S. F. A. Kettle, *Chem. Ind. (London)* p. 668 (1961).
46. R. F. Heck, *J. Amer. Chem. Soc.* **94**, 2712 (1972).
47. R. F. Heck, *J. Amer. Chem. Soc.* **93**, 6896 (1971).
48. E. J. Corey and L. S. Hegedus, *J. Amer. Chem. Soc.* **91**, 1233 (1969).
49. E. J. Corey, H. A. Kirst, and J. A. Katzenellenbogen, *J. Amer. Chem. Soc.* **92**, 6314 (1970).
50. E. J. Corey and L. S. Hegedus, *J. Amer. Chem. Soc.* **91**, 4926 (1969).
51. L. F. Hines and J. K. Stille, *J. Amer. Chem. Soc.* **94**, 485 (1972).
52. J. Tsuji and K. Ohno, *Advan. Chem. Ser.* **70**, 155 (1968).
53. R. F. Heck and D. S. Breslow, *J. Amer. Chem. Soc.* **83**, 4023 (1961).
54. C. K. Brown and G. Wilkinson, *J. Chem. Soc., A* p. 2753 (1970).
55. D. Evans, J. A. Osborn, and G. Wilkinson, *J. Chem. Soc., A* p. 3133 (1968).
56. J. Falbe, *Angew. Chem., Int. Ed. Engl.* **5**, 435 (1966).
57. M. M. Taqui Khan and A. E. Martell, "Homogeneous Catalysis by Metal Complexes," pp. 293–351. Academic Press, New York, 1974.
58. G. N. Schrauzer, ed., "Transition Metals in Homogeneous Catalysis," pp. 147–215. Dekker, New York, 1971.
59. J. Blum and S. Kraus, *J. Organometal. Chem.* **33**, 227 (1971).
60. H. M. Walborsky and L. E. Allen, *J. Amer. Chem. Soc.* **93**, 5466 (1971); *Tetrahedron Lett.* p. 823 (1970).
61. J. A. Osborn, F. H. Jardine, J. F. Young, and G. Wilkinson, *J. Chem. Soc., A* p. 1711 (1966).
62. M. M. Taqui Kahn and A. E. Martell, "Homogeneous Catalysis by Metal Complexes," pp. 1–72. Academic Press, New York, 1974; J. M. Davidson *MTP Int. Rev. Sci., Inorg. Chem., Ser. 1* **6**, 355–360 (1972); G. N. Schrauzer, ed., "Transition Metals in Homogeneous Catalysis," pp. 13–50. Dekker, New York, 1971.
63. J. W. Scott and D. Valentine, *Science* **184**, 943 (1974).
64. W. S. Knowles, M. J. Sabacky, B. D. Vineyard, and D. J. Weinkauff, *J. Amer. Chem. Soc.* **97**, 2567 (1975); W. S. Knowles, M. J. Sabacky, and B. P. Vineyard, *Advan. Chem. Ser.* **132**, 274 (1974).
65. D. W. Hart and J. Schwarz, *J. Amer. Chem. Soc.* **96**, 8116 (1974).
66. C. A. Bertello and J. Schwartz, *J. Amer. Chem. Soc.* **97**, 228 (1975).
67. D. W. Hart, T. F. Blackburn, and J. Schwartz, *J. Amer. Chem. Soc.* **97**, 678 (1975).
68. J. A. Labinger, D. W. Hart, W. E. Seibert, and J. Schwartz, *J. Amer. Chem. Soc.* **97**, 3851 (1975).
69. R. B. Woodward and R. Hoffmann, "The Conservation of Orbital Symmetry." Verlag Chemie, Weinheim, 1970.
70. F. Mango, *Advan. Catal.* **20**, 291 (1969); G. N. Schrauzer, ed., "Transition Metals in Homogeneous Catalysis," pp. 223–270. Dekker, New York, 1971.
71. C. P. Casey and C. R. Cyr, *J. Amer. Chem. Soc.* **95**, 2248 (1973).
72. P. A. Spanninger and J. L. von Rosenberg, *J. Org. Chem.* **34**, 3658 (1969).

73. W. T. Hendrix, F. G. Cowherd, and J. L. von Rosenberg, *Chem. Commun.* p. 97 (1968).
74. H. C. Vogler and H. Hogeveen, *Rec. Trav. Chim. Pays-Bas* **86**, 830 (1967).
76. L. Cassar, P. E. Eaton, and J. Halpern, *J. Amer. Chem. Soc.* **92**, 3515 (1970).
77. L. A. Paquette, R. S. Beckley, W. B. Farnham, J. S. Ward, and R. A. Boggs, *J. Amer. Chem. Soc.* **97**, 1084, 1089, 1101, 1112, and 1118 (1975).
78. M. Sakai, H. H. Westbery, H. Yamaguchi, and S. Masamune, *J. Amer. Chem. Soc.* **93**, 4611 (1971).
79. R. J. Haines and G. J. Leigh, *Chem. Soc. Rev.* **4**, 155 (1975).
80. W. B. Hughes, *Advan. Chem. Ser.* **132**, 192 (1974).
81. R. H. Grubbs and T. K. Brunk, *J. Amer. Chem. Soc.* **94**, 2538 (1972).
82. T. J. Katz and J. McGinnis, *J. Amer. Chem. Soc.* **97**, 1592 (1975).
83. R. H. Grubbs, P. L. Burk, and D. D. Carr, *J. Amer. Chem. Soc.* **97**, 3265 (1975).
84. D. J. Cardin, B. Cetinkaya, M. J. Doyle, and M. F. Lappert, *Chem. Soc. Rev.* **3**, 99 (1973).
85. C. P. Casey and T. J. Burkhardt, *J. Amer. Chem. Soc.* **96**, 7808 (1975).
86. J. Powell, *MTP Int. Rev. Sci., Inorg. Chem., Ser. 1* **6**, 309 (1972).
87. E. O. Fisher, C. G. Kreiter, H. Rühie and K. E. Schwarzhans, *Chem. Ber.* **100**, 1905 (1967).
88. M. A. Bennett and T. Yoshida, *J. Amer. Chem. Soc.* **93**, 3798 (1971); **95**, 3030 (1973).
89. K. M. Nicholas, *J. Amer. Chem. Soc.* **97**, 3254 (1975).
90. W. R. Moser, *J. Amer. Chem. Soc.* **91**, 1135 and 1141 (1969).
91. T. Cohen, G. Herman, T. M. Chapman, and D. Kuhn, *J. Amer. Chem. Soc.* **96**, 5627 (1974).
92. R. Noyori, S. Makino, and H. Takaya, *J. Amer. Chem. Soc.* **93**, 1272 (1971); R. Noyori, K. Yokoyama, S. Makino, and Y. Hayakawa, *J. Amer. Chem. Soc.* **94**, 1772 (1972).

Subject Index

A

Absorption spectra, 85, 91, 100–102, 111, 264, 265, *see also* Spectrochemical series

Acetylene complexes, 302–308

Acetylenes, reduction of, 288, *see also* Hydrogenation

σ-Acetylide complexes, 299

π-Acid ligands, 263, 305

Acidity of coordinated ligands
  alcohols, 244
  amines, 247
  ammonia, 114–115
  water, 208–212

Aconitase, 203
  mechanism of action, 226–229

Action potential, 57

Acyl metal complexes, 299–301

Ag/AgCl electrode, 135–137

Alcohol(s)
  coordinated
    isomerization of, 344, 378
    solvolysis of, 343
  reaction with Mg halides, 50
  as reducing agents, 145, 297, 330

Alcohol dehydrogenase, 145

Aldol condensations, 196–199

Aldolase enzymes, 199

Alkene complexes, *see* Olefin complexes

σ-Alkenyl complexes, 299

Alkyne complexes, *see* Acetylene complexes

Allosteric effects, 55, 273

π-Allylnickel reagents, 350–352

Anchimeric assistance, 113–115 *see also* Sn1cb mechanism

π-Arene complexes
  reactions of, 340–343
  structures of, 314–316

Ascorbic acid, 147
  oxidation of, 173

Asymmetric induction, 339, *see also* Enantiomer discrimination
  examples of, 200–202
  in mixed-ligand complexes, 131
  use of chiral ligands, 48

Atomic orbitals, 66–71

Autoxidation, 139, 167

Axial bases, 257–260, 272–276, 322

Aziridine, coordinated, 247

Azotobacter, 287

B

Back-bonding
  chemical consequences of, 99–103
  examples of, 85, 86, 99–103, 195, 263, 280, 301, 305–311, 315–317
  spectrochemical series, 100

Bailar-Twist mechanism, 228

Bilayer lipid membranes, 57–62

395

A 6
B 7
C 8
D 9
E 0
F 1
G 2
H 3
I 4
J 5